WITHDRAWN
from stock

CONVERSION FACTORS FROM U.S. CONVENTIONAL UNITS TO METRIC (SI) UNITS

Quantity	To convert from USCS units			Multiply by	To obtain SI unit	
	Symbol	Name			Symbol	Name
Length	in	inch		*2.54	mm	millimeter
	ft	foot		*0.3048	m	meter
Area	in²	square inch		*6.4516×10^{-4}	m²	
Volume	in³	cubic inch		1.639×10^{-5}	m³	
	gallon	US gallon		3.785×10^{-3}	m³	
Time	min	minute		*60	s	second
Velocity	fpm	ft/min		*5.08×10^{-3}	m/s	
Mass	lb	pound		0.4536	kg	kilogram
Acceleration (gravitational)	ft/sec² (32 ft/sec²)			*0.3048 *(9.80665)	m/s² m/s²	
Force	lbf (or lb)	pound force		4.448	N	newton
	tonf	ton force (2000 lb)		8.9	kN	
Stress (pressure)	lbf/in²	psi		6.895×10^3	Pa	pascal (=N/m²)
	kips (or kpsi)	1000 psi		6.895	MPa	(or N/mm²)
Torque (work)	lbf·ft	foot-pound		1.356	N·m	newton-meter
Energy (work)	Btu	British thermal unit		1055	J	joule (=N·m)
	cal	gram calorie		*4.1868	J	
Power	hp	550 ft·lb/sec		745.7	W	watt (=J/s)
Viscosity	P	poise (dyn·s/cm²)		0.1	Pa·s	(or N·s/m²)
Temperature interval	F	Fahrenheit		0.5555	C or K	Celsius degree or kelvin
Temperature	t_F			$(t_F - 32)\, 5/9$	t_C	degree Celsius
	t_C			$t_C + 273.15$	t_K or T	absolute degrees

Notes: Exact conversion factors are recorded with an asterisk.

The Celsius degree is often written °C to avoid confusion with C (coulomb)

Most frequently used multipliers:

	Prefix	Symbol
10^6	mega	M
10^3	kilo	k
10^{-3}	milli	m
10^{-6}	micro	μ

The International Committee of Weights and Measures (CIPM) modernized the metric system in 1960. The resulting SI units are now used worldwide in the literature; all industrialized nations have already committed themselves to conversion to the International System (SI).

For a detailed discussion see, for example, *ASME Orientation and Guide for Use of Metric Units*, 3d ed., American Society of Mechanical Engineers, New York, or *The International System of Units*, National Bureau of Standards SP330 (SD cat. no. C13.10:330/2), Government Printing Office, Washington, D.C.

INTRODUCTION TO MANUFACTURING PROCESSES

Second Edition

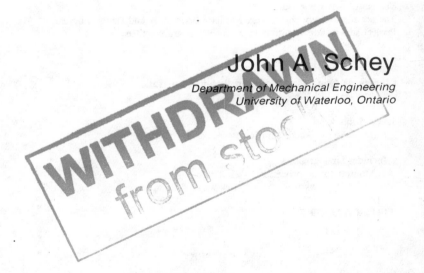

John A. Schey

Department of Mechanical Engineering
University of Waterloo, Ontario

McGraw-Hill Book Company

New York   St. Louis   San Francisco   Auckland   Bogotá   Hamburg
London   Madrid   Mexico   Milan   Montreal   New Delhi
Panama   Paris   São Paulo   Singapore   Sydney   Tokyo   Toronto

TO GITTA
INTERNATIONAL EDITION

This book was set in Times Roman by Science Typographers, Inc.
The editor was Anne Duffy.
The production supervisors were Marietta Breitwieser and Denise Puryear.
Project supervision was done by Science Typographers, Inc.

Library of Congress Catalog in Publication Data

Schey, John A.
 Introduction to Manufacturing Processes

 Includes bibliographies.
 1. Manufacturing processes.   I. Title

TS 183.S33   1987   670   86–10500
ISBN 0–07–055279–7

When ordering this title use ISBN 0–07–100311–8

Printed and Bound in Singapore by Fong & Sons Printers Pte Ltd

CONTENTS

ABOUT THE AUTHOR

Dr. John A. Schey was educated in his native Hungary. He received his Dipl. Met. Ing. degree from the Jozsef Nador Technical University, Sopron, in 1946, and was awarded the Candidate Techn. Sci. (Ph.D.) degree by the Academy of Sciences, Budapest, in 1953. He subsequently held positions at the Metal Works in Csepel, Budapest (1947–1951), the Technical University of Miskolc, Hungary (1951–1956), the Research Laboratories of the British Aluminium Co. Ltd., England (1956–1962), IIT Research Institute, Chicago (1962–1968), and was Professor of Metallurgical Engineering at the University of Illinois at Chicago (1968–1974).

He is currently a Professor in the Department of Mechanical Engineering at the University of Waterloo in Ontario, Canada. His work at the university involves developing courses and conducting research in the areas of manufacturing technology, deformation processes, and tribology. His interests include the social consequences of changing technologies.

He is a Member of the National Academy of Engineering, Washington, D.C., Fellow of the American Society for Metals, Fellow of the Society of Manufacturing Engineers, member of the Canadian Institute of Mining and Metallurgy, The North American Manufacturing Research Institute of SME, and Society of the Sigma Xi. He is also Curriculum Director of the Institute for Forging Die Design of the Forging Industry Association. In 1966 he was awarded the W. H. A. Robertson Award and Medal by the Institute of Metals, London, England, in 1974 the Gold Medal Award by the Society of Manufacturing Engineers, and in 1984 the Dofasco Award by the Canadian Institute of Mining and Metallurgy.

He is a certified manufacturing engineer and a professional engineer registered in Ontario.

He is the author of a number of publications and books on hot deformation, strip rolling, tube manufacture, cold forming, sheet metalworking, forging, lubrication and friction in plastic deformation, and on manufacturing processes, and he has eight patents.

PREFACE TO THE SECOND EDITION

Since the appearance of the first edition, significant changes occurred in the world. The microelectric revolution moved into high gear and, for a while, it was believed by many that industrialized countries will move in the direction of an information society in which manufacturing will play little role. These visions of the future have now given way to a more realistic view in which manufacturing, in a more sophisticated form and with more information content, remains a critical vehicle of wealth generation. Much of this manufacturing will be controlled by computer-based systems and—if we accept that one can control only that which one understands—the people who will program and operate these systems will need to know what the processes are really about. It has also become abundantly clear that good designers will have to know about the processes by which their designs are transformed into competitive products.

The present edition aims to help in educating engineers and technologists for this future, while also providing the professional with a volume that answers many questions arising in daily practice. The emphasis is still on principles; for the benefit of those who lack some of the background or need a little refreshing of their knowledge, principles are expanded upon in greater detail. New material is also added, particularly on polymers, high-technology ceramics, and the fabrication of microelectronic devices. Opportunities for automatic control are indicated.

The subject can be comfortably covered in a two-semester course. In many institutions, where only a one-semester course is given, the instructor may tailor the course for the needs of the students, taking into account the level of their preparation. Suggestions for this will be found in the Instructor's Manual which also contains worked problems, examination questions, derivations of equations, and a list of available audiovisual aids.

As in the first edition, I enjoyed the support of many colleagues who generously sacrificed their time to read specific sections of the manuscript. I am indebted to K. G. Adams, S. Chamberlain, S. El-Gizawy, D. French, S. George-Cosh, F. Ismail, J. Jachna, H. W. Kerr, R. Komanduri, K. M. Kulkarni, P. Niessen, K. F. O'Driscoll, A. Plumtree, J. F. Schey, R. Sowerby, D. Weckman, and I. Yellowley. I am grateful to Anne Duffy and Anne Brown of McGraw-Hill for their contributions and to Cheryl Kranz of Science Typographers for her editing. For any remaining errors only I am responsible; I should greatly appreciate to have the views and remarks of users.

My wife Gitta has worked with me for long hours in preparing the manuscript. To her I dedicate this volume.

McGraw-Hill and the author would like to thank the following reviewers who evaluated the second edition: Amit Bagchi, Ohio State University; David L. Bourell, University of Texas, Austin; Marvin F. DeVries, University of Wisconsin, Madison; Warren R. DeVries, Rensselaer Polytechnic Institute; David Dornfeld, University of California, Berkeley; O. R. Fauvel, University of Calgary; and K. J. Weinmann, Michigan Technological University.

John A. Schey

PREFACE TO THE FIRST EDITION

Manufacturing is the lifeline of all industrialized societies. Without it, few nations could afford many of the amenities that improve the quality of life for their citizens yet all too often are taken for granted.

Despite its obvious importance to society in general and engineering in particular, manufacturing has been neglected in most engineering schools, particularly in North America. The hopeful assumption of the educational theory prevailing in the 1960s was that the young engineer, well versed in the fundamentals, would readily apply technical knowledge to the real problems he encountered in his professional career. Consequently, the more applied—and often purely descriptive—courses, including those on manufacturing, were eliminated from most curricula. It soon became clear, however, that the transition from fundamentals to practice was by no means as natural and easy as had been hoped for, and the practicing engineer often came to question the relevance of the tidy, well-defined solutions characteristic of engineering science courses to the rather messy, open-ended problems of real life. Hence, the integration of basics into courses of applied orientation is again becoming a required part of engineering education.

The task is not easy in any specific field, but it becomes especially demanding in manufacturing. Many processes—developed over the centuries through the perseverance of gifted "natural" engineers—defy exact solutions, and when a solution is found, it may be of little help in dealing with practical problems. This does not mean, however, that fundamentals cannot be applied; on the contrary, they can aid not only in understanding existing processes but also in developing new ones.

Manufacturing is really nothing else but the art and science of transforming materials into usable—and salable—end products. Starting from the premise that this transformation is best achieved with the cooperation, rather than against the objection, of the material, the treatment in this book builds on the interactions between material properties and process conditions. Processes and equipment are described in only as much detail as is essential for an understanding of the more fundamental arguments that finally evolve into a judgment of the advantages and limitations of various processes. Manufacturing technology is inextricably interwoven with concerns for cost and productivity; therefore, an attempt is made to develop also an understanding of the competitive nature of processes, and thus to help in arriving at a reasoned choice of a technologically sound and economically attractive manufacturing process.

There are a great number of books available that discuss in detail the "how" of manufacturing; this book aims at the "why" and "under what conditions." It builds, therefore, on advances made in recent years in various scientific and engineering disciplines, but always with an eye on applicability and relevance to manufacturing processes. Many of the principles discussed are too complex to be expressed mathematically. Others are amenable to quantitative analysis but, within the confines of the present treatment, only the most useful and relevant methods of calculation can be given, and then without proof.

The material is kept to the minimum and can be fitted into an intensive one-term course if some of the more descriptive passages are treated as reading assignments. At the same time, expansion to a full year needs only the added emphasis of a field close to the instructor's interests. It is assumed that the student will have had introductory courses in engineering materials and in mechanics, although most concepts are defined, however briefly, to aid recapitulation and self-study.

Most readers will perhaps agree that the inch system is dead, but one must also admit that it is still a rather vigorous corpse, particularly in North America. To make the transition less painful, conventional USCS and SI units are used side-by-side, except in those tables that would become too awkward to handle.

I would hope that practicing engineers too will find some food for thought in this book, if not in their own specialty, in the ever-broadening fields of alternative and competitive processes and materials.

At this point I would like to express my gratitude to numerous colleagues, among them W. Rostoker (University of Illinois at Chicago Circle), K. J. Schneider (California State Polytechnic University), and Z. Eliezer (University of Texas at Austin) who read the entire text, K. G. Adams, I. Bernhardt, H. W. Kerr, H. R. Martin, P. Niessen, K. F. O'Driscoll, A. Plumtree, D. M. R. Taplin, B. M. E. van der Hoff (University of Waterloo), H. W. Antes (Drexel University), G. F. Bolling (Ford Motor Company), and M. Field (Metcut Research Associates) who read various sections and offered helpful criticism. I am also indebted to G. E. Roberts for his assistance, and to many students, most notably S. M. Woodall, D. L. Agarwalla and J. V. Reid, for their help with problem solutions. My wife, Gitta, has worked with me for many months, and without her help this book would have remained but an unrealized plan.

John A. Schey

LIST OF RECURRENT SYMBOLS

A	contact area, cross-sectional area (instantaneous)	M	atomic weight
A_0	original (starting) cross-sectional area	M_s	temperature at which martensite transformation begins
A_1	cross-sectional area after deformation	N	rotational frequency (rpm)
C	strength coefficient in hot working	P	applied force
C	Taylor constant (cutting speed in fpm for 1 min tool life)	P_a	force in axial upsetting of cylinder
		P_b	bending force
C_E	composition of a eutectic	P_c	cutting force
C_L	composition of liquid	P_d	deep-drawing force
C_0	alloy composition	P_{dr}	drawing force
C_S	composition of solid crystals	P_e	extrusion force
D_p	punch diameter	P_i	indentation force
E	Young's modulus	P_n	normal force on cutting-tool face
E_c	specific cutting energy	P_r	rolling force
E_1	specific cutting energy for 1-mm undeformed chip thickness	P_s	shearing force (maximum)
		P_t	thrust force on cutting tool
F	shear force	R	universal gas constant
F	frictional force	R_e	extrusion ratio
F_s	shear force in shear plane	Q	activation energy
G	shear modulus	Q	pressure-multiplying factor
I	current	Q_a	Q for axial upsetting of a cylinder
K	strength coefficient in cold working	Q_c	Q for impression and closed-die forging
L	length of contact zone between tool and workpiece	Q_{dr}	Q for drawing of wire
		Q_e	Q for extrusion (ram pressure)
M	mass	Q_{fe}	multiplying factor for energy requirement in forging

Q_i	Q for inhomogeneous deformation	h_1	height after deformation
Q_p	Q for plane-strain upsetting	i	ISO tolerance unit
R	resistance	j	current density
R	radius of tool	k	constant; heat conductivity
R_a	average surface roughness	l	instantaneous length
R_b	radius of bending die	l_f	final length at fracture
R_d	draw-die radius	l_0	original length
R_e	extrusion ratio, reduction ratio	m	strain-rate sensitivity exponent
R_f	final radius of bent part	m^*	frictional shear factor
R_{min}	minimum bending radius	n	strain-hardening exponent
R_p	punch radius	n	Taylor exponent
R_q	root-mean-square surface roughness	p	interface pressure
R_t	maximum surface roughness	p_a	average interface pressure in axial upsetting
S	solubility		
T	temperature	p_c	specific cutting stress
T_E	eutectic temperature	p_e	average extrusion (ram) pressure
T_L	liquidus temperature	p_i	indentation pressure (on punch)
T_S	solidus temperature	p_p	average interface pressure in plane-strain upsetting
T_g	glass-transition temperature		
T_m	melting point (K)	q	reduction of area in tension test
T_1	intermediate temperature	r	r value (a measure of anisotropy)
V	volume	r_0	r value in rolling direction
V_t	rate of material removal	r_{45}	r value at 45° to rolling direction
W	weight	r_{90}	r value transverse to rolling direction
W_b	opening of bending die		
W_e	weight per unit area removed by an electric current	r_c	average roll pressure; radial pressure in powder pressing; crack radius
W_f	final weight	r_m	mean r value
W_0	original weight	$\bar{r}$	mean r value
Z_f	multiplying factor for feed	t_r	reference tool life
Z_v	multiplying factor for cutting speed	t_s	solidification time
a	side dimension of HCP prism	v	velocity of deforming tool
c	height of HCP prism	v_s	standard (reference) velocity of cutting
c	specific heat		
d	average grain size	w	instantaneous width
d	instantaneous diameter of workpiece	α	solid solution species
d_0	original diameter of workpiece	α	die half-angle in extrusion or wire drawing
e_c	compressive engineering strain		
e_f	tensile engineering strain at fracture	α	cutting tool rake angle
e_t	tensile engineering strain	α_b	angle of bending
e_u	engineering tensile strain at necking	α_{max}	angle of acceptance in rolling
		β	solid solution species
f	feed (in cutting)	γ	shear strain
g	gravitational acceleration	γ	interfacial energy
h	instantaneous height, undeformed	γ_{SL}	interfacial energy between solid and liquid
h_c	chip thickness		
h_m	mean or average height	γ_{SV}	interfacial energy between solid and vapor
h_0	original height or thickness		

$\dot{\gamma}$	shear strain rate	ρ	density; electrical resistivity
ϵ	natural (logarithmic) strain	σ	normal stress; true stress
ϵ_m	mean (average) strain	σ_{eng}	engineering stress $= P/A_0$
ϵ_u	uniform strain in tension	σ_f	flow stress
$\dot{\epsilon}$	strain rate (instantaneous)	σ_{fm}	mean flow stress
$\dot{\epsilon}_m$	mean (average) strain rate	$\sigma_{0.2}$	yield stress for 0.2% plastic strain
η	dynamic viscosity	τ	shear stress
η	efficiency	τ_f	flow strength in shear
θ	cutting tool relief angle	τ_i	interface shear strength
θ	wetting angle	ϕ	shear angle in cutting
μ	coefficient of friction	ϕ	inhomogeneity factor in drawing
ν	Poisson's ratio; kinematic viscosity	ψ	friction angle in cutting

CHAPTER 1

INTRODUCTION TO MANUFACTURING

Manufacturing is a human activity that pervades all phases of our life. Derived from the Latin (*manus* = hand, *factus* = made), the word is used to describe "the making of goods and articles by hand or, especially by machinery, often on a large scale and with division of labor." An understanding of the role of manufacturing in human development is essential for everyone involved in its study and practice.

1-1 HISTORICAL DEVELOPMENTS

The history of manufacturing is marked by gradual developments; however, some developments were of such substantial social consequences that they can rightly be regarded as revolutionary.

1-1-1 Early Developments

Manufacturing has been practiced for several thousand years, beginning with the production of stone, ceramic, and metallic articles. The Romans already had factories for the mass production of glassware, and many activities, including mining, metallurgy, and the textile industry have long employed the principle of division of labor. Nevertheless, much of manufacturing remained for centuries an essentially individual activity, practiced by artisans and their apprentices. The ingenuity of successive generations of artisans led to the development of many processes and to a great variety of products (Table 1-1), but the scale of production was necessarily limited by the available power. Water power supplemented muscle power in the Middle Ages, and then only to the extent allowed by

TABLE 1-1 HISTORICAL DEVELOPMENT OF MANUFACTURING UNIT PROCESSES

Year	Casting	Deformation	Joining	Machining	Ceramics	Plastics	Machine Control
4000 B.C.	Stone, clay molds	Bending, forging (Au, Ag, Cu)	Riveting	Stone, emery, corundum, garnet, flint	Earthenware	Wood, natural fibers	Wedge Manual
2500	Lost wax (bronze)	Shearing, sheet forming	Soldering, brazing	Drilling, sawing	Glass beads, potter's wheel		Wheel
1000		Hot forging (iron), wire drawing (?)	Forge welding, gluing	Iron saws	Glass pressing, glazing		Lever, pulley
A.D. 0		Coining (brass), forging (steel)		Turning (wood), filing	Glass blowing		Screw press, crank
1000		Wire drawing			Stoneware, porcelain (China)	Protein glues	Waterwheel
1400	Sand casting, cast iron	Water hammer		Sandpaper	Majolica, crystal glass		Connecting rod, flywheel
1600	Permanent mold	Tinplate can, rolling (Pb)		Wheel lathe (wood)			
1800	Flasks	Deep drawing, rolling (steel), extrusion (Pb)		Boring, turning, screw cutting	Plate glass; porcelain (Germany)		Steam engine
1850	Centrifugal, molding machine	Steam hammer, tinplate rolling		Shaping, milling, copying lathe	Window glass from slit cylinder	Vulcanization	Mechanization
1875		Rail rolling, continuous rolling		Turret lathe, universal mill, vitrified wheel		Celluloid, rubber extrusion, molding	

Date	Metal casting	Metal forming	Joining	Machining	Glasses and ceramics	Polymers	Automation and control
1900		Tube rolling, extrusion (Cu)	Oxyacetylene, arc welding, electrical resistance welding	Geared lathe, automatic screw machine, hobbing, high-speed steel, synthetic SiC, Al_2O_3			Electric motor
1920	Die casting	W wire (from powder)	Coated electrode			Bakelite, vinyl acetate, casting, cold molding, injection molding	Hard automation (electrical)
1940	Lost wax for engineering parts, resin-bonded sand	Extrusion (steel)	Submerged arc, structural adhesives		Automatic bottle making	PVC, Acrylics, PMMA, PE, polystyrene nylon, synthetic rubber, polyesters, transfer molding, foaming	
1950	Ceramic mold, nodular iron, semiconductors	Cold extrusion (steel)	TIG welding, MIG welding, electroslag	EDM	Glass ceramics	ABS, silicones, fluorocarbons, polyurethane	Numerical control (NC)
1960	Rapid solidification		Plasma arc, electron beam	Manufactured diamond	Float glass	Acetals, polycarbonate, polypropylene, cyanoacrylate	Computer-NC
1970	Isothermal forging		Laser	CBN		Polyimide, aramids polybutylene	Adaptive control, programmable controller

the availability of swift water; this limited the location of industries and the rate of growth of industrial production.

1-1-2   The First Industrial Revolution

At the end of the 18th century, the development of the steam engine made power available in large quantities and at many locations. This spurred advances in manufacturing processes (Table 1-1) and facilitated the growth of production, providing an abundance of goods and, with the mechanization of agriculture, of agricultural products. As a result, society was also transformed, and later these developments came to be recognized as the industrial revolution. It was characterized by mechanical power supplementing the physical power of the worker, with many machines driven by belts from a common drive shaft.

Toward the middle of the 19th century, some functions of the worker were taken over by machines in which mechanical components such as cams and levers were ingeniously arranged to perform relatively simple and repetitive tasks. Such mechanization or "hard automation" eliminated some jobs, but the workers thus displaced—together with those made redundant in agriculture—usually found jobs in the expanding manufacturing and service sectors of the economy. Around the turn of the 20th century, development was further aided by the introduction of electric power: Machines could now be individually driven and controls based on electric circuits allowed a fair degree of sophistication.

1-1-3   The Second Industrial Revolution

Beginning with the second half of the 20th century, further developments have taken place. Computers have begun to offer hitherto undreamt of computational power, and solid-state electronics—growing out of the transistor—permitted the fabrication of devices of great versatility at ever decreasing costs. In the early 1970s the availability of the microchip, with thousands of electronic components crammed onto a tiny silicon wafer, made it possible to perform computational, control, planning, and management tasks at high speeds, very often in real time (i.e., while the process to be controlled takes place) and at low cost. The consequences have been far-reaching in every facet of our lives and are now beginning to be felt in manufacturing; the limits of developments are still only dimly perceived. It has been recognized, nevertheless, that the social consequences of these changes will be as fundamental as those wrought by the 19th-century industrial revolution, and most observers now agree that we are in the midst of the second industrial revolution.

A characteristic of the second industrial revolution is that, in addition to the possibility of replacing most—or potentially all—physical labor, it is now also feasible to enhance and sometimes even replace mental effort. In manufacturing this means the introduction of true automation, with appropriate sensors providing the feedback which then allows the control device to take "intelligent" action. Such closed-loop control is not really new; the mechanical governor on the steam engine fulfilled the same function. The difference is that microelectronics allows the control of a large number of variables at a substantial level of sophistication.

Some consequences of these developments are already noticeable: Many dangerous, physically demanding, or boring jobs are performed by machines or robots equipped with programmable controllers; product variety is increasing; quality is improving; productivity—as expressed by output per unit labor—is rising; demand on natural resources is decreasing. There are also signs of possible undesirable consequences, in particular, the decline in the number of people employed in traditional fields. It is not clear whether the new technologies will create an adequate number of new jobs. We are facing an uncertain economic future, in which the role of manufacturing is often hotly debated.

1-2   THE ECONOMIC ROLE OF MANUFACTURING

Manufacturing has often been cast as the villain on the stage of human development. Indeed, the first industrial revolution began with little concern for the very people who made this revolution possible. Yet the factory was the alternative willingly chosen by the masses seeking to escape their rural existence: The idyllic, pastoral qualities of rural life extolled by poets and writers were mostly imaginary, while reality was burdened with famine and disease. Modern demographic studies show that the misery of rural life prompted people to crowd into cities even before the first industrial revolution. Since then, the excesses of the early industrial revolution have been moderated and the growth of manufacturing has led to undeniable advances, not only in providing an abundance of material possessions, but also in creating the economic basis for genuine improvements in the quality of life.

There are no universal measures to express well-being but, in the absence of better measures, the gross national product (GNP; the sum of the value of all goods and services produced in a national economy) can be taken as a measure of material well-being. Even for this, it is an imperfect measure because it excludes the value of all work performed in the home, by voluntary organizations, etc. Thus it presents a distorted picture in favor of industrially developed nations.

If one analyzes the components of the GNP, it is evident that material wealth comes from only two substantial, basic sources: material resources, and the knowledge and energy that people apply in utilizing these resources. Agriculture and mining are of prime importance, yet they represent only 5–8% of the GNP of industrially developed nations. Manufacturing had claimed the largest single share, at least until the 1950s. Indeed, one could make the argument that the mark of an industrially developed nation is the proportionately large contribution of manufacturing to national wealth. A review of typical data for industrialized and developing nations is instructive. The data plotted in Fig. 1-1 should not be taken at face value, since they are distorted by the differences in the purchasing power of local currencies and by the exclusion of all unpaid services which make up such a large portion of wealth generated, especially in the industrially less developed countries. Nevertheless, the overall conclusion is clear: For nations

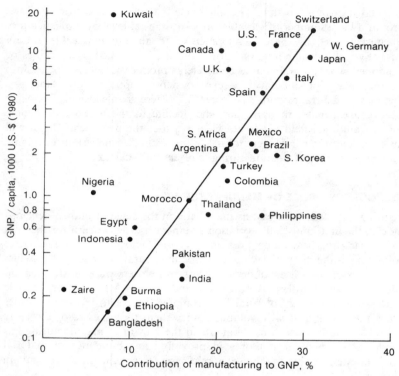

FIGURE 1-1
In general, nations more intensively engaged in manufacturing enjoy a higher standard of living, at least as it is expressed by the per capita output of the economy. (*Data compiled from World Development Report, World Bank / Oxford University Press, New York, 1982.*)

otherwise similarly endowed with natural resources and human talent, there are large differences in the material standard of living, and these differences can be, very approximately, related to the contribution of manufacturing to overall economic activity.

With the advent of the second industrial revolution, some observers argue that the importance of manufacturing is or will be diminishing. They note the rapid increase of the component of the GNP attributable to information processing and to services in general. They speak of a post-industrial society and argue that "low technology" production (such as textile, shoe, agricultural machinery, and automobile production) should be allowed to move to low-wage countries while post-industrial societies should concentrate on selected "high technology" industries such as aerospace and biotechnology and—more and more—on information technology and, in particular, on information processing.

This book starts from the premise that, while information technology is undoubtedly gaining in importance, it alone cannot create the wealth required to maintain high living standards, nor can it pay for the importation of all manufactured products. Manufacturing has lost none of its importance: While employment in manufacturing may well decline in relative terms, the contribution of manufacturing to the GNP must be maintained. For this, it is essential that manufacturing should be *competitive*, not only locally but—with the shrinking of our world—on a global basis. Indeed, one measure of economic development is the proportion of manufactured goods and information services in the export trade of a nation.

Competitiveness can be achieved only by attaining a high level of productivity. *Manufacturing productivity* is a key issue of economic development, and nations falling behind in this respect find their living standards gradually eroding. Exceptional natural resources may, for a short time, boost living standards but, judging from experience to date, only manufacturing can create a permanent basis of economic well-being. Manufacturing includes, of course, the production of nondurables and semidurables. In the narrower sense adopted here, we will limit ourselves to the manufacture of "hardware," articles of production and consumption, both durable and semidurable.

1-3 MANUFACTURING AS A TECHNICAL ACTIVITY

Without manufacturing, there would be little need for engineers and technologists or, indeed, many of the people who are engaged in supporting activities. Manufacturing is a central function of most technically educated people, although in more recent times they have often failed to recognize this themselves. The reason is to be found in the rapid expansion of knowledge which has, inevitably, led to a fragmentation of engineering and technology into many disciplines and subdisciplines. Creative people active in many of the basic engineering disciplines have often forgotten that the ultimate purpose of all engineering activity is to produce something tangible and salable, hopefully for the benefit of humanity. Today, in a climate of intense international competition, we cannot afford to ignore this truth. It is, therefore, essential to recognize that every technical decision carries a manufacturing implication, and it has a marked effect on costs. To quote but a few examples:

1 The jet engine is a machine designed on the basis of thermal and fluid engineering principles. The engine consists of essentially three sections: air is compressed in the compressor section; fuel is introduced and burned with the compressed air in the combustor section; hot gases generated in the combustor section drive the turbine which turns the compressor and provides the thrust (Fig. 1-2). Thermodynamic efficiency increases with increasing turbine entry temperature, and fuel consumption drops (Fig. 1-3a). Adding a large fan to the intake end allows air, ducted around the outside of the engine, to mix with the hot gases coming from the turbine. Such fan engines are more efficient and also less noisy.

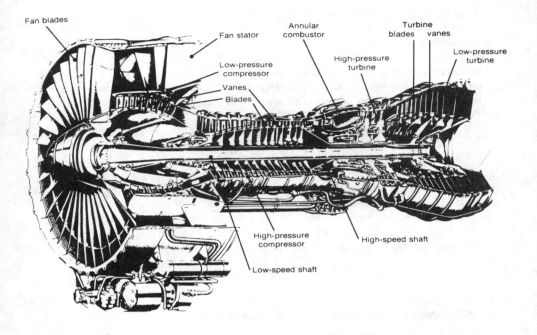

FIGURE 1-2
Today's jet engine is a highly sophisticated manufactured product; many of its parts must operate at high temperatures and stress levels, demanding advanced methods of manufacturing. (PW2037 turbofan, *courtesy United Technologies, Pratt & Whitney.*)

Compression raises the temperature of air and, in recent engines, the final-stage compressor blades are required to run red hot (Fig. 1-3b). This means that the material and technologies used for making compressor blades had to change over the years.

An even more demanding application is the turbine where entry temperatures were limited by the temperature capability of the blade and vane materials (Fig. 1-4a) until cooling of the airfoils was introduced by passing cooler uncombusted compressor air through the turbine airfoils (Fig. 1-4b). Advanced cooling techniques involve the use of small cooling holes; thus, in addition to developing new manufacturing approaches for processing increasingly difficult-to-manufacture alloys, techniques for making very fine, deep holes in very hard materials had to be developed too.

2 Designers of passenger automobiles have been required to introduce many changes to satisfy new demands regarding safety, pollution level, gasoline consumption, durability, and quality of the product. These changes have affected the choice of materials and manufacturing techniques. For example, early automobile

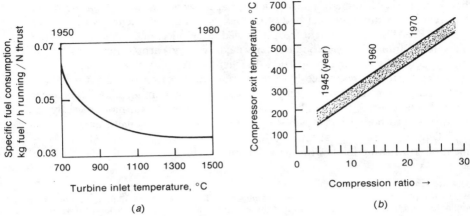

FIGURE 1-3
Higher operating temperatures result in higher efficiency and lower fuel consumption. Hence, (a) specific fuel consumption decreases with increasing turbine entry temperatures and (b) higher compression ratios lead to higher temperatures even in the compressor section. Both stages require new materials and manufacturing techniques. [(a) *Reprinted with permission from M. F. Ashby and D. R. H. Jones, Engineering Materials, Pergamon Press, 1980; (b) based on data by G. W. Meetham, The Metallurgist and Materials Technologist, 8(11):589–593, 1976.*]

FIGURE 1-4
Turbine inlet temperatures (T.I.T.) can be increased by cooling the blades (a); for this, new techniques of making holes in hard materials had to be developed. D. S.: directional solidification. (*After M. F. Ashby and D. R. H. Jones, as Fig. 1.3a.*) In the example shown (b) the holes are molded into a precision-cast turbine blade.

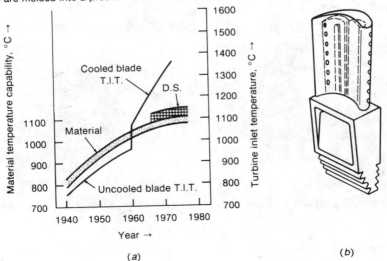

bodies had a steel frame to which wooden panels were attached. Soon the wood was replaced with an all-steel body that was, however, still secured to a heavier frame. The desire for weight reduction led to monocoque construction; the all-welded, frameless steel bodies were made of low-carbon steel which had highly desirable forming properties. Further efforts at weight reduction and increased corrosion resistance have led to the introduction of galvanized steel, high-strength low-alloy steels and, to some extent, aluminum alloys. Polymers have also been used, first as fiberglass-reinforced epoxies and, more recently, as mass-produced body parts attached to a precision-machined steel frame.

3 One of the most frequently performed orthopedic surgeries is the replacement of arthritic hip joints with surgical implants (Fig. 1-5a). Materials had to be

FIGURE 1-5
Hip joint replacements are advanced manufactured products. (a) The titanium-alloy stem is implanted into the femur (thigh bone) and the cup into the acetabulum (socket of the hip bone), with a wear-resistant polymer lining providing the sliding surface. (*Zimmer Inc., Warsaw, Ind.*). (b) Pure titanium wire of 0.25-mm diameter is diffusion bonded into a porous mass to (c) provide room for the in-growth of fresh bone tissue which fixes the implant in the bone of an experimental animal. (*Courtesy Dr. W. Rostoker, University of Illinois at Chicago.*)

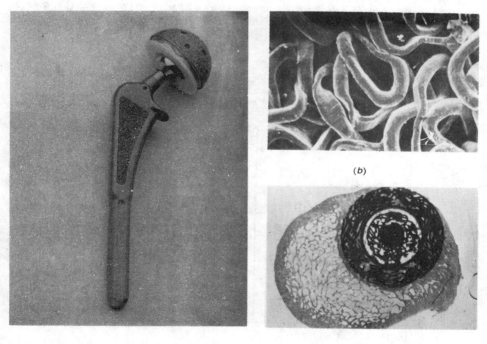

(a)

(b)

(c)

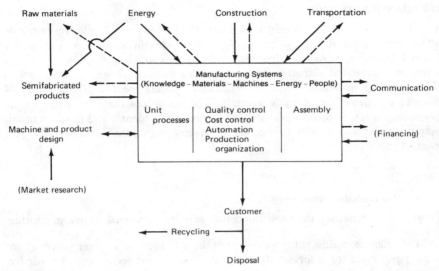

FIGURE 1-6
Manufacturing is an integral and indispensable part of the economy. It draws upon many other activities, while providing all the hardware necessary for those activities to take place.

found that could be implanted into the body without adverse reactions and that could withstand the substantial dynamic loading (of millions of cycles per year) imposed by ever younger and more active patients. Techniques of making the parts had to be found as well as techniques for fastening the replacement in the bone. Grouted implants frequently fail after some years of service. A more recent and still experimental approach uses the regenerative capacity of the bone to establish the bond: Intricate channels are provided (Fig. 1-5b) into which new tissue can grow (Fig. 1-5c). This too required the development of new manufacturing processes.

4 Microelectronics, which is at the heart of the second industrial revolution, had its origins in physical phenomena that could be exploited only by adapting old and creating new manufacturing technologies for new materials; these technologies will be discussed in Chap. 10.

Beyond its technological significance, manufacturing is also a cornerstone of any modern, industrialized society. Figure 1-6 is a simplified sketch to show some of the interactions between various economic activities. It will be noted that many activities provide essential inputs to manufacturing; at the same time, manufacturing creates all the machines that are needed for the conversion of energy and raw materials, and for construction, transportation, and communication activities. Thus, these industries and businesses, together with the individual consumer, dictate the range of products that manufacturing has to provide.

1-4 MANUFACTURING

The definition of manufacturing as the making of goods and articles reveals little about the complexity of the problem. A more specific definition is given by CAM-I (Computer Aided Manufacturing International, Arlington, TX): "A series of interrelated activities and operations involving design, materials selection, planning, production, quality assurance, management, and marketing of discrete consumer and durable goods." This recognizes that, from the simple beginnings when an artisan provided all the necessary mental and physical input, manufacturing has grown to become a *system* with many components that interact in a dynamic manner.

1-4-1 The Manufacturing System

Figure 1-7 summarizes the most important activities involved in manufacturing.

1 A manufacturing entity (a company or a branch of a larger corporation) usually possesses some special strengths, such as specific technology, knowledge, or equipment. Exploitation of these strengths requires that the appropriate *markets* be identified, their magnitudes estimated, and the existing and potentially emerging competition appraised. After the market and its future development are projected, products are identified oi developed. The sales organization then secures orders, sometimes with the aid of samples or prototypes, at other times on the basis of specifications backed by an established record of performance. Either way, the creative act of design is usually completed—at least in a general sense—at this stage.

2 The product, whether it be a machine tool, household machine, building product, computer, automobile, aircraft, chemical processing plant, power station, oil drilling rig, cookware, or soft-drink container, is *designed* to fulfill its intended function; that is, to operate satisfactorily over its expected life, subject to a number of constraints:

a *Industrial designers* strive to create a visually appealing, functional product that the customer will be willing to buy.

b *Mechanical and electrical designers and analysts* ensure that the product will properly function. This requires the choice of appropriate materials, often in cooperation with materials specialists.

c The product must serve the customer, with due regard to the physical capabilities and limitations of operators or consumers. This way it will optimize performance and ensure well-being in the workplace. These aspects are the subject of the relatively new science of *ergonomics* which takes a general, holistic approach to the relationship between people and machines.

d The product must fulfill its function at a *reasonable cost*. Therefore, neither excess performance nor excess life are needed, and performance must be optimized. In general, it is found that a product satisfying minimum requirements

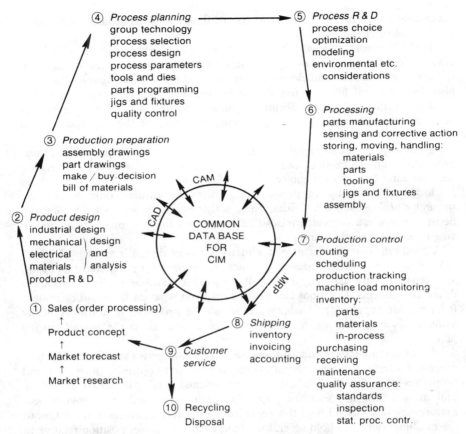

FIGURE 1-7
Manufacturing entails a large variety of activities, many of which have become specialities on their own. More properly, manufacturing is regarded as a system with interdependent activities. Interaction can be strengthened by the use of the computer, leading to computer-integrated manufacturing (CIM).

can be produced at some minimum cost. Performance can often be increased—and thereby the selling price substantially raised—with relatively little increase in the cost of manufacturing. Further improvements may lead to much higher manufacturing cost and only marginally increased customer appeal; thus, the selling price cannot be raised proportionately. Consequently, there is always a point beyond which performance cannot be economically improved. This point is determined through the cooperation of marketing, design, and manufacturing teams.

e The product must be easy to *maintain* over its intended life. It must be readily *disposed* of or *recycled* at the end of its life, in a safe and ecologically acceptable way. This often calls for ease of disassembly or separation by some mechanical or chemical means.

f Most importantly, all the above criteria must be satisfied while also ensuring *ease of manufacture*. This requires not only close cooperation between industrial, mechanical, and electrical designers and manufacturing specialists, but also demands that all designers should be aware of the manufacturing consequences of their decisions. Seemingly minor changes may often present (or remove) enormous manufacturing problems, thus affecting the cost, quality, and reliability of the product.

g An important consideration is the *number of units* to be produced, in one production run (batch size) and over the projected life of the product, because this will enter into process choice.

h The performance of products can often be ensured only by *product research and development* activities. These activities are essential when a company desires to secure a competitive position by introducing new products or improving established product lines.

A comprehensive treatment of the many facets of design is given by Dieter.*

3 Once a product is designed, production *drawings* (or computer data bases) are prepared of the assembly and of all parts other than standardized, mass-produced components. Decisions can then be made on what parts should be bought from outside suppliers and what parts should be produced in-house. A *bill of materials* is prepared which, in many ways, is central to the manufacturing process.

4 For components produced in-house, *process design* is carried out: The best process is selected and process parameters are chosen to optimize the quality and properties of the finished product. Dies are designed, tooling is chosen, and, if the tool must follow a prescribed path, this path is selected and programmed. Fixtures are designed to hold the workpiece in the correct position in relation to the machine tool or to hold several workpieces in the correct position relative to each other. Jigs perform a similar function but also incorporate guides for the tool. Activities in this group are often described in the narrower sense as manufacturing engineering. For components produced by a vendor, these functions are usually performed by the vendor, ideally in cooperation with the purchaser. Again, process design is not an isolated activity. For example, it must facilitate inspection for quality control.

5 The choice of the appropriate manufacturing technique and its optimization are important functions. A strong competitive position also requires that new processes be developed and old ones improved through *process research and development*. New processes often make it possible to develop new products, thus further increasing competitiveness. Process development on the production scale can be very expensive. Therefore, the fundamentals of processes are often explored in the laboratory. Models of a process can be used to explore the

---

Engineering Design: A Materials and Processing Approach, McGraw-Hill, New York, 1983.

influence of process parameters. Two approaches are possible:

a In *physical modeling* the process is conducted on a reduced scale or simulating materials are used that are easier and cheaper to work with than the real materials.

b In *mathematical modeling* equations are set up that express the response of the process to changes in process parameters. Such models usually require lengthy computations which are made off-line (in the laboratory).

Whichever modeling approach is used, a sound understanding of the physical realities is essential for success.

In choosing and developing processes, their impact on the environment (air and water pollution, noise, vibration, etc.) and on the safety and health of operators and other people must be considered. Manufacturing often involves high temperatures, molten metal, highly stressed tooling, flammable or toxic liquids, and involves activities that generate noise, smoke, fumes, gases, or dust. It is imperative that appropriate precautions and remedial measures are taken. Beyond the social responsibilities of the engineer and technologist, there are also legal requirements, such as the regulations of the Environmental Protection Agency (EPA) and the Occupational Health and Safety Act (OSHA) in the USA and corresponding measures in other countries.

6 The actual process of *production* takes place on the workshop floor, which is arranged according to some plant layout. In the course of production, critical characteristics of processes are observed; the dimensions, quality, etc., of parts are systematically checked and, when needed, corrective actions are taken. The most important auxiliary function is the timely movement of raw materials, partly finished parts, tooling, and jigs and fixtures. Finally, manufactured and purchased parts are assembled into products which, after checking, are ready for shipping. All these functions are still in the domain of manufacturing specialists, many of whom are formally trained as industrial engineers.

7 The complex sequences of production require a strong *manufacturing organization*. Raw materials, parts, and tools must be routed to their destination and scheduled to arrive when required. The status of production must be known. Formal methods of quality assurance must be established, together with a plan for preventive maintenance of equipment. An up-to-date inventory of parts in process, combined with inventories of purchased materials and parts must be maintained to ensure that no shortages develop that could delay production and assembly. For a running analysis of performance, machine loading (utilization) and machine and labor performance are monitored. Many of these activities are in the domain of industrial engineering, while others are regarded primarily as management tasks. Superior organization, even of existing technology, can lead to substantial competitive advantage.

8 The completed products are shipped; *inventory control* feeds information back to the production process on the basis of sales performance.

9 *Field service* ensures the continuing performance of products delivered to the customer. It is here that the information loop is finally closed: Feedback is valuable in sharpening production practices and, if necessary, changing design.

10 At the end of its useful life, the product is disposed of in an ecologically and economically acceptable manner, or it is recycled to reclaim the materials of construction. This aspect has assumed great significance since the tightening of energy supplies. For most materials, recycling of scrap requires much less energy than production from raw materials.

1-4-2 Impediments to Systems Approach

Even this very brief and in some ways incomplete discussion of the manufacturing system indicates the complexity of the problem. Efficient and competitive manufacturing requires close cooperation between the various activities, so that they truly become parts of an *interacting, dynamic system*. Unfortunately, this ideal has seldom been reached for a number of reasons.

1 Companies are often organized into departments clearly separated according to functions. Such organization hinders interaction between the design, manufacturing, and production-control departments and supporting departments such as maintenance, quality assurance, purchasing, etc.

2 Excessive specialization has prevented people from appreciating the manufacturing consequences of their decisions. Once a problem is solved from the narrow point of view of one specialist, it is passed on to the next specialist to solve both the underlying problem and problems added by the "solutions" introduced at earlier stages.

3 The study of manufacturing has been neglected in universities and colleges, especially in the English-speaking world; hence, many people involved in manufacturing have lacked the background that would have allowed them to make well-informed decisions.

4 The complexity of the problem increases with every "solution." Documentation requires a vast flow of paperwork, with diminishing probability that all parts of the system will be consistent. The number of potential errors increases: Changes made at one point in the system fail to diffuse through the organization, and contradictory measures surface too late, when their reconciliation involves vast effort and cost.

The traditional approach views manufacturing essentially as a sequential activity. The shortcomings of this view became painfully evident when, in response to competitive pressures, many companies attempted to rationalize their operation. Fortunately, a powerful tool for improvement has become available in the form of the computer.

1-4-3 Computers in Manufacturing

Computers have been used in manufacturing, as in other businesses, since the 1960s for bookkeeping, accounting, purchasing, and inventory control functions. Gradually, with the rapidly increasing speed of operation, larger memories, and decreasing costs, the use of computers has spread to provide a number of other functions:

1 All phases of product design can take place on a computer, with the design displayed on a VDT (video-display terminal), most often a CRT (cathode-ray tube). With the aid of *geometric modeling and analysis*, the designer can explore a number of options and analyze them with the aid of software packages (including those for *finite-element analysis*, FEM). Design can be optimized in a much shorter time, rapid design changes become possible, and changing consumer demands can be satisfied. Both assemblies and parts can be designed with assurance that they will properly fit together. A library of standard components can be built up. The bill of material is also generated and, if required, drawings can be prepared by driving a pen under the commands of the computer. Thus, activities indicated in blocks 2 and 3 of Fig. 1-7 are performed in what has become known as CAD (*computer-aided design*).

2 Computers can be used for programming the machine-tool motions necessary for generating the geometric shape of the part. Beginning with the 1950s, information contained in drawings could be transformed into digital form for the *numerical control* (NC) of machines. When the part geometry is created by CAD, the data base already exists and can be directly employed in NC. Exchange of the graphic data base between various systems is still a problem but standard formats are being developed.* Beginning with the 1970s, the computer has been used also for process optimization and control, materials management, material movement (including transfer lines, robots, etc.), scheduling, monitoring, etc. This entire field is now generally called CAM (*computer-aided manufacturing*); it encompasses blocks 4, 5, 6, and 7 in Fig. 1-7.

3 The benefits of CAD and CAM can be fully realized only if an effective interface is established between them, creating what is usually referred to as CAD/CAM. Information flow in both directions ensures that parts and assemblies will be designed with the capabilities and limitations of materials and manufacturing processes in mind. Superior products can be created and tremendous competitive advantages attained. A company that uses integrated CAD/CAM can develop new products more rapidly and at a lower cost, thus increasing its market share at a time when product life cycles are declining and efforts required to develop new products are increasing; it can be more responsive to customer requests both in providing quotations and in accommodating special needs; it can reduce the effort spent on design and process changes by ensuring that such changes are entered in the common data base and are thus immediately recognized at all stages of design and production. An important benefit is that the introduction of CAD/CAM forces a review and improvement of existing design and manufacturing practices and production planning.

4 The critical tasks of inventory management are also facilitated by the computer. *Material requirements planning* (MRP) is based on the master produc-

---

*For example, IGES, "Initial Graphics Exchange Specification," American National Standards Institute ANSI Y14.26M.

tion schedule, the bill of materials, and inventory records relating to raw materials, purchased components, parts to be delivered, in-process materials and parts, finished products, and tools and maintenance supplies. With this information as the input, MRP manages the inventories with due regard to the timing of material requirements. In detail, it provides notices for releasing orders, scheduling and rescheduling, cancellation, inventory status, performance reports, deviations from schedules, etc. A good MRP system results in reduced lead times, minimum inventory, faster response to customer requests, and increased productivity. A further development is *manufacturing resource planning* (also called MRP-II or closed-loop MRP) which integrates a complete manufacturing control system. Production planning, master scheduling, capacity requirements planning, and functions necessary for executing the production plan (including vendor schedules and dispatch lists) are incorporated, and provisions are made for continuous updating.

5 The management aspects of manufacturing lend themselves to computerization even without CAM, and are often regarded as elements of a *management information system* (MIS). More recently, MIS is taken to stand for *manufacturing information system*.

6 Most recently, manufacturing has come to be treated as a single system comprising hardware and software (as in Fig. 1-7) in which the complex interactions are followed with the aid of the computer. In such *computer-integrated manufacturing* (CIM) all actions take place with reference to a common data base. *Data base management* is a complex but not insurmountable task. Drawings serve only to visualize the geometry of parts; no changes are allowed to be made on them. If changes in design, process, scheduling, bill of materials, quality standard, etc. are to be made, they are made in the data base; thus they reflect throughout the organization. The data base is continuously updated by most recent information on production, sales, etc. For many industries, total CIM is still in the future, but beginnings have been made.

One of the major remaining tasks is modeling of the manufacturing process. Many of the most successful processes are largely based on experience and intuition, and formalization of knowledge in a form useful for computer control will take considerable time. A significant development is the modeling of the entire manufacturing process by the IDEF methodology. It was developed by SofTech, Inc. (Waltham, MA) under the sponsorship of the U.S. Air Force in the Integrated Computer Aided Manufacturing (ICAM) program. IDEF stands for ICAM Definition System, and incorporates a model of the manufacturing system in a hierarchical form, using a graphical representation in which the manufacturing functions are shown (together with the inputs, mechanisms, controls, and outputs) by the gradual expansion of diagrams into lower level diagrams. An exposition of the approach is given by Harrington in *Understanding the Manufacturing Process*.

In the forseeable future, much of manufacturing will still be conducted under human control, and it is important to recognize that *linking a process to a computer or some microelectronic device does not automatically guarantee advantages*. A process that is ill-conceived, inapplicable, or outdated will not become competitive by such computerization. Only a truly systems view can be productive. For this, *the physical basis of processes must be understood*, the variables of importance identified, and a control strategy established. If necessary, better, more competitive processes must be introduced. Only then can the mechanical, electrical, and electronic devices and software be linked into a functioning system. The magnitude of the undertaking must not be underestimated, but a complete review of the manufacturing system often proves profitable because problem areas that can be resolved even without the aid of the computer are revealed.

The purpose of this book is to provide the physical understanding; in doing so, opportunities for and difficulties in the application of computers and microelectronics will also be shown. For readers who have only a passing familiarity with computers, necessary terms are introduced in Appendix A.

1-4-4 Control of Manufacturing Processes

We will discuss this topic again in Chap. 11, after gaining a closer acquaintance of processes. However, we have to clarify a few terms at this point so that the potential of process control can be pointed out throughout the discussion of individual processes.

Control Strategies The different approaches to control can be best explained by reference to a simple example, that of turning a cylindrical component on a lathe. The principles apply to any process.

1 To understand what task a control system is expected to perform, we must first examine some of the actions of a skilled lathe operator. Let us assume that the part to be machined has been mounted in the chuck and the cutting speed and the feed (the axial movement of the tool during each revolution of the workpiece) have been set. The task to be followed is that of maintaining the diameter of the finished part between specified minimum and maximum values and to ensure that the surface finish satisfies specifications.

The trained and experienced operator possesses knowledge that allows a decision on the depth of cut (the thickness of layer removed in one cut). Operator knowledge is often supplemented or even replaced by instructions provided on the basis of past experience or published data. The important point is that information is stored in some form.

Next, the operator checks the setting of the cross slide by reading the micrometer dial (Fig. 1-8a). In other words, the current status of the machine is sensed.

The operator then determines what changes are needed and makes logical decisions.

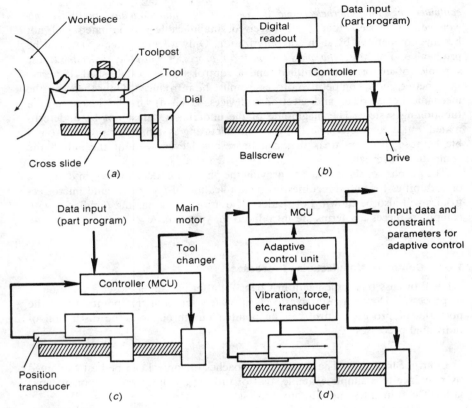

FIGURE 1-8
All manufacturing processes must be controlled. The example is for cutting on a lathe, under (a) manual, (b) open-loop, (c) closed-loop, and (d) adaptive control. Adaptive control takes actions in a manner a highly skilled machinist would.

The operator communicates these decisions to the system by actuating the screw to set the position of the cross slide. At this point, a correct part will be made, assuming that the dial is correctly calibrated and machine deflections are negligible.

A skilled operator will go further and stop the machine after the beginning of the cut, check the part diameter, and make necessary adjustments.

A highly skilled operator will observe the surface produced, listen to the sound of the machine, and, generally, sense changes that are often difficult to describe accurately. For example, under given conditions, vibration (chatter) may develop which causes the surface finish to vary in a periodic manner, resulting in an objectionable surface finish. The operator will then change cutting conditions (speed, feed, support of the part or tool) until the undesirable condition disappears. The operator will also compensate for tool wear, change the tool when needed, and will make sure that the machine tool is not overloaded.

A control system will take over several or all of the functions of the operator.

2 In *open-loop control* an action is taken without verifying the results of this action. Actuators may be mechanical (cam, lever, linkage), electromechanical (dc or ac motor, stepping motor), or hydraulic or pneumatic (motor or cylinder). For example, the cross slide of the lathe may be moved by a cam, stepping motor, or hydraulic cylinder to a predetermined position. The setting will be repeated for each part, but it still takes an operator or setup person to confirm that the part is within tolerance and, if not, to reset the cam, mechanical stop, microswitch, or change the program instruction (Fig. 1-8b).

3 In *closed-loop control*, sensors provide *feedback* to the system. In the simplest case, a high-resolution position transducer is added to confirm that the intended position of the cross slide has indeed been reached (Fig. 1-8c). The signal from the transducer is processed by a comparator that compares it with the control signal and then issues an error signal to correct the position. In other applications, the control would maintain a speed or other parameter at a set level (as mentioned, the oldest example of closed-loop control is a purely mechanical device, the centrifugal governor invented by Watt in 1788 for maintaining a preset speed on a steam engine, irrespective of the load imposed on it). A simple closed-loop control system is ignorant of possible secondary inputs to the system and will go on producing parts even with a worn or broken tool or under conditions of chatter.

4 *Adaptive control* is the highest level of control which, in its fullest development, can replace the operator entirely. Sensors are used to provide *feedback of secondary inputs* (in the case of the lathe, in-process measuring devices check the diameter of the part, load cells measure forces, vibration transducers give signals characteristic of the existing cutting conditions, etc.). The feedback signal is then processed so that the control unit can take appropriate corrective action (Fig. 1-8d). Obviously, the corrective action will accomplish its intended purpose only if the effect of process variables on the finished part are known. Interrelations between process variables can be extremely complex, and full adaptive control can be successful only if a sufficiently quantitative model of the process can be formulated. Even if a simple model is used, the constraints of the process or system (maximum force, speed, etc.) must be obeyed (*adaptive control with constraints*, ACC). A more complex model allows optimization (*adaptive control with optimization*, ACO), for example, for maximum production rate.

5 The power of the computer can be used to endow the control with some measure of *artificial intelligence* (AI). As the name implies, the control program is designed to solve a problem the way humans solve it; it is capable of some reasoning, can learn from experience, and, ultimately, can do some self-programming. Alternatively or additionally, elements of expert programs may be incorporated in the control system.

Automation The word automatic is derived from the Greek and means self-moving or self-thinking. The word *automation* was coined to indicate a form of manufacturing in which production, movement, and inspection are performed or controlled by self-operating machines without human intervention. In general, one may distinguish between several levels of automation. Here we will make a distinction between:

1 *Mechanization.* This means that something is done or operated by machinery and not by hand. Feedback is not provided; thus, one deals with open-loop control. An example would be the use of a cam to move the cross slide in Fig. 1-8*a*.

2 *Automation.* This will imply closed-loop control and, in its advanced form, adaptive control. Automation utilizes programmable devices, the flexibility of which can be quite different:

a *Hard automation* refers to methods of control that require considerable effort to reprogram for different parts or operations. For example, a limit switch could be manually reset if a cylinder of different diameter had to be machined.

b *Soft* or *flexible automation* implies ease of reprogramming, usually simply by changing the software.

An important aspect of automation in manufacturing is automation of material movement. We shall come back to this topic in Sec. 11-1; however, we have to give here some brief definitions: *Manipulators* are mechanical devices for the movement of materials, tools, and parts, and *robots* are programmable manipulators.

Numerical Control Electrical controls may be analog, as when the voltage generated by a transducer is compared to the control voltage. However, the greatest advances in manufacturing control have been made by the introduction of *numerical control* (NC). In the most general sense, NC is the use of symbolically coded instructions for the automatic control of a process or machinery. Various forms of NC have been developed:

1 The hardware for basic NC includes the *machine control unit* (MCU, Fig. 1-8), which contains the logic required to translate information into appropriate action; *actuators*; and, if control is closed-loop, *feedback devices* and associated circuits. The plan of action is provided to the MCU in the form of a program on a punched tape or magnetic tape or disk. Programs are usually prepared by a programmer or the machine-tool operator, and are read into the MCU, incrementally, by a tape reader. The MCU is hard-wired to perform various functions. For example, the machine tool or other mechanical device may be expected to move from one point to another. This may be accomplished in several ways:

a If the machine tool is equipped with two actuators arranged in *xy* coordinates, the simplest MCU moves first the *x* then the *y* actuator by the prescribed

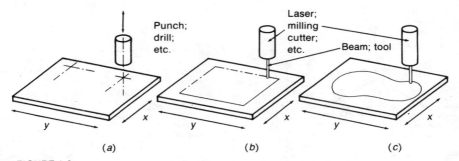

FIGURE 1-9
Control methods may provide (*a*) simple position or point-to-point control or (*b*) control over tool or workpiece movement in a straight cut or (*c*) along a contour.

distances, without controlling the motion itself (*point-to-point* or *positioning system*, Fig. 1-9*a*); when the programmed position is reached, the operation is performed (say, a hole is drilled). A slightly more complex system also moves first in one and then the other direction, but this time with full control of the rate of movement (*straight-cut system*, Fig. 1-9*b*) while an operation such as cutting, milling, or welding takes place.

b NC is particularly valuable when a complex contour is to be followed (Fig. 1-9*c*). In *contouring systems* the MCU is programmed to break up the contour into shorter segments and to interpolate between the endpoints of segments. Linear interpolation approximates the curved profile in small straight lengths; better approximation is obtained with circular paths.

Information is read in blocks, and a buffer memory (buffer register) prevents discontinuity of operation which, in the case of machining, welding, etc., would result in visible stop marks on the surface.

2 In *computer numerical control* (CNC) the functions of the MCU are partly or fully taken over by a dedicated computer (a mini- or microcomputer assigned to the machine tool, Fig. 1-10*a*). The entire program is read into memory. Since computers can be readily reprogrammed, much greater flexibility of operation is obtained. For example, it is possible to trace a complex curve without any breaks in continuity, and thus attain the closest approximation to the desired contour. Furthermore, programs can be added that provide technological functions, perform adaptive control, and incorporate some elements of a process model. The microprocessors used in place of the hard-wired NC circuits are more reliable and can have self-diagnostic features. In general, the part or process program is still received on tape or disk, although many CNC systems allow direct programming. The computer has sufficient memory to serve not only as a buffer but also to store the programs necessary for extended operation.

Both NC and CNC raise productivity and reproducibility, thus raising accuracy, quality, and reliability of the end product. CNC minimizes the errors

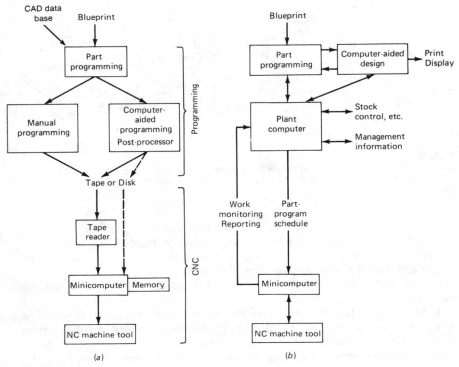

FIGURE 1-10
Many NC machines, with structures similar to those shown in Fig. 1-8, are now controlled by (a) a dedicated microcomputer (CNC) or (b) by a hierarchy of computers (DNC).

introduced by the tape reader since the tape is read only once; it also reduces overhead relative to NC.

3 In *direct numerical control* (DNC) several machine tools are connected to one larger, central computer which stores all programs and issues the NC commands to all machines (Fig. 1-10b). No machine tool must ever wait for an instruction; therefore, there are often satellite control computers interposed between the central computer and smaller groups of NC units, and the central computer is used only to store, download, edit, and monitor programs, and to provide supervisory and management functions. Such hierarchical control breaks down even very complex tasks into manageable elements. The task of real-time computation and sensory processing is allocated to the first-level computers. The NC units may be of the conventional hard-wired type, with the tape reader replaced by a direct communication line to the central computer (*behind-the-tape-reader systems*), or specialized units which, like CNC units, use a minicomputer as the MCU. Obviously, the latter allow much greater flexibility.

It should be noted that, to exploit all the benefits of computer control, it is usually necessary to improve the mechanical performance of the system (see Sec. 8-7-3). The integration of mechanical and electronic aspects is sometimes termed *mechatronics*.

NC Programming   Programming of the machine tool has been greatly simplified over the years and has spread from machining to other processes. Programming starts by defining the optimum sequence of operations and the process conditions for each operation. The geometric features of the part are then used to calculate the tool path. The resulting program can be quite general and must be converted, with the aid of a program called the postprocessor, into a form acceptable for the particular machine tool control. The output is a punched tape or other storage medium. An important step is tape verification which reveals programming errors and ensures the production of correct parts. Basically there are four approaches:

1 *Manual programming*: All elements of the program are calculated by a skilled parts programmer who puts them into standardized statements. Programming is laborious and is now largely limited to simple point-to-point programs.

2 *Computer-assisted programming*: The programmer communicates with a software system in a special-purpose language that uses English-like words. The most comprehensive of these languages, APT (automatically programmed tools) was developed in the 1950s at the Massachusetts Institute of Technology under U.S. Air Force sponsorship, and was expanded in the 1960s, under sponsorship of a consortium of users, at IIT Research Institute, and then at CAM-I. Many simplified languages and languages designed for specific processes have since been developed. Programming languages translate the input into a form understandable to the computer so that it can perform the necessary computations, including compensation for tool dimensions (cutter offset in machining). Tape verification must be done on the machine tool or a drafting machine.

3 *CAD/CAM*: When parts are designed by CAD, the numerical data base can be used to generate the program on the graphics terminal, either by a programmer or by the designer of the part with the aid of the CAD/CAM software. The program can be immediately verified by viewing on a VDT the path of the tool relative to the part. Programming is fast and relatively inexpensive, justifying its use even for single parts or, as it is often called, for *one-off production*.

4 *Manual data input*: Many CNC machine tools are equipped with a CRT display and a powerful software that prepares the part program. In response to queries, the operator enters data to define the part geometry, material, and tooling. Standard English words are used, and the software does the rest. The technique is particularly suitable for CNC lathes, and it is highly economical when it allows programming while another job is running.

With the spread of CNC and manual data input, the trend is to entrust more programming to the machine-tool operator; however, conventional computer-assisted and CAD/CAM programming is still performed in programming departments.

1-5 SCOPE AND PURPOSE OF BOOK

With the proliferation of computer applications in manufacturing, it is tempting to conclude that knowledge of the physical principles is losing significance and that information processing—taken in the narrower sense of data processing—is becoming the central activity in manufacturing. Nothing could be further from the truth. Information processing is very important indeed, but it is still only a tool; a tool that in itself cannot ensure competitiveness unless applied to a physically sound process. Even the ultimate computerization of a process will be of no avail if a more original mind meanwhile develops a new process that wipes out the competitive advantages of the old one. Computer control of a given process can be effective only if the role of process variables is understood and if at least an elementary yet physically sound model of the process can be formulated. For all these reasons, this book is devoted to developing an understanding of the physical background of the various unit processes that are used for the manufacture of parts. One must, of course, recognize that parts will have to be assembled into finished products, but these assembly operations will be touched upon only briefly.

In choosing the particular approach adopted in this volume, the guiding principle was that fundamental, general principles are more powerful than details. The number of manufacturing processes in existence defies enumeration, let alone description, in a single volume. There are already a number of encyclopedic books available in which details on individual processes can be found.

Thus, the purpose is not to give detailed information, but to impart a knowledge of principles which can then be used to improve existing processes, create new ones, and interpret the information presented in books and, increasingly, computer data bases. To quote Sherlock Holmes: "a man should keep his little brain attic stocked with all the furniture that he is likely to use, and the rest he can put away in the lumber room of his library, where he can get it if he wants it."*

In dealing with principles, it is recognized that many engineers and technologists are needed to make up a manufacturing team. Some team members may be specialists in manufacturing, but others are experts in materials, mechanical, industrial, or systems engineering and technology. Some students may embark upon their exploration of manufacturing after taking a course in the properties of materials and in the strength of materials; others may have no more preparation than high school physics and chemistry courses. The book is constructed so as to cater to both groups. Those who have the appropriate preparation may simply

---

*A. Conan Doyle, *Adventure of the Five Orange Pips*, Crown Publishers, New York, 1976.

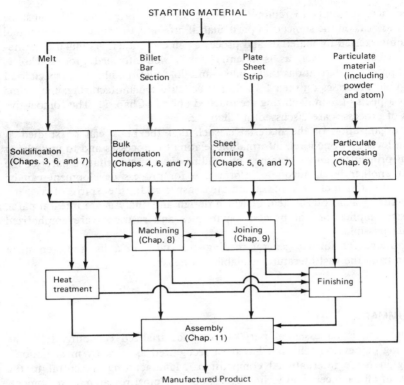

FIGURE 1-11
This text emphasizes the physical principles of unit manufacturing processes. These principles, properly applied, ensure that a finished product, satisfying service requirements, will be
produced competitively.

skip the background material or read it as a refresher; others may use the
background material given here as a jumping-off point for further studies. The
aim is that, at the minimum, a student should acquire a knowledge of process
principles to the level where useful interaction with specialists is possible. At a
higher level, the foundations for specialization will be laid.

Some of the background material is given in Chap. 2 with reference to the
service properties of manufactured products. Other background is included in
Chaps. 3–10 which are devoted to major classes of discrete-part (unit) processes,
as shown in Fig. 1-11. The reader interested in further details or in the theories of
processes will find ample material in the readings suggested at the end of each
chapter. These readings have been chosen to alleviate the dilemma of depth
versus breadth of treatment. A fully quantitative treatment would require excessive length and could easily obscure the larger issues; a fully qualitative treatment
would give little guidance to intelligent process selection. The compromise

adopted here attempts to retain the sound scientific principles that must be brought to bear on the subject; in particular, it attempts to emphasize the mutual constraints exerted by materials and processes on each other. As much quantitative information is given as is essential for a well-informed process choice, without proofs or derivations that could—and do—fill up a library of specialized books. Unit processes on their own do not constitute manufacturing; thus some broader aspects of manufacturing are touched upon in Chap. 11. The competitive aspects of processes are discussed in Chap. 12.

The organization of the material is such that the book can be studied at different levels. At one level, information relating to processes and to underlying physical principles can be obtained by reading the nonquantitative parts of the text. At another level, adequate information for process development can be obtained by the inclusion of quantitative aspects. In the spirit of viewing manufacturing as a total system of which design and analysis are integral parts, the design implications of process capabilities and restraints are emphasized whenever possible.

Suggestions for further general reading are given, in a list representing a selection from the vast literature available in English.

1-6 SUMMARY

Manufacturing is an essential part of any industrialized economy. It is the mainspring of development and has been recognized as such by most nations, resulting in fierce international competition. Manufacturing is central to the activities of all engineers and technologists, because most research, development, design, and management activity finally results in some manufactured product. If an industrial unit (company) or nation is to be successful in the worldwide competition (and, indeed, if humankind is to be best served by plentiful, high-quality, high-value manufactured products), it is essential to recognize some very general features of manufacturing:

1 Manufacturing involves many steps from market research through the development, design, analysis, and control of products and processes to the delivery, service, and finally, disposal of the manufactured products. Gradually, the many activities associated with these stages have become specialized, compartmentalized, and disjointed, resulting in great inefficiencies.

2 More recently, manufacturing has come to be viewed as a system, with all parts of the system interacting in an organic manner.

3 The complex interactions within the system are facilitated by a common computer data base, essential for the development of CIM.

4 Subsystems such as CAD, CAM, MRP, and MIS have been based on the computer for some time now, with many benefits in improved productivity, quality, equipment utilization, reduced inventory, and faster delivery. The full benefits require integration of these actions.

5 Computers and other microelectronic devices, such as programmable controllers, have been used extensively for the control of production processes and machinery with the aid of NC, CNC, and DNC. A better understanding of processes and the development of appropriate transducers allow control in the adaptive mode, responding to changes in process conditions the same way or better than a highly skilled operator could.

6 The application of the computer to an outdated or basically defective process cannot solve the basic problems. Therefore, if anything, it has become even more important to acquire a sound understanding of the physical principles upon which process control can be based. A knowledge of these principles is also essential if an interface between mechanical equipment and electronic devices is to be built.

This book addresses itself to the physical principles, laying the foundations for further in-depth studies. Opportunities for the application of computer techniques are shown wherever appropriate, without going into detail.

FURTHER READING

A History

Aitchison, L.: *A History of Metals*, Macdonald and Evans, London, 1960.

Derry, T. K., and T. I. Williams: *A Short History of Technology*, Oxford University Press, London, 1961.

Kingery, W. D. (ed.): *Ancient Technology to Modern Science*, American Ceramic Society, Columbus, Ohio, 1985.

Simpson, B. L.: *History of the Metalcasting Industry*, 2d ed., American Foundrymen's Society, Des Plaines, Ill., 1969.

Woodbury, R. S.: *History of the Lathe*, MIT Press, Cambridge, Mass., 1961.

B General Manufacturing Textbooks

Alting, L.: *Manufacturing Engineering Processes*, Dekker, New York, 1982.

Amstead, B. H., P. F. Ostwald, and M. L. Begeman: *Manufacturing Processes*, 7th ed., Wiley, New York, 1977.

Bolz, R. W.: *Production Processes*, 5th ed., Industrial Press, New York, 1981.

DeGarmo, E. P., J. T. Black, and R. A. Kohser: *Materials and Processes in Manufacturing*, 6th ed., Macmillan, New York, 1984.

Doyle, L. E., C. A. Keyser, J. L. Leach, G. F. Schrader, and M. S. Singer: *Manufacturing Processes and Materials for Engineers*, 3d ed., Prentice-Hall, Englewood Cliffs, N.J., 1985.

Kalpakjian, S.: *Manufacturing Processes for Engineering Materials*, Addison-Wesley, Reading, Mass., 1984.

Kenlay, G., and K. W. Harris: *Manufacturing Technology*, Arnold, London, 1979.

Lindberg, R. A.: *Processes and Materials of Manufacture*, 3d ed., Allyn and Bacon, Boston, 1983.

Pollack, H. W.: *Manufacturing and Machine Tool Operations*, 2d ed., Prentice-Hall, Englewood Cliffs, N.J., 1979.

Radford, J. D., and D. B. Richardson: *Production Engineering Technology*, 3d ed., Macmillan, London, 1980.

Yankee, H. W.: *Manufacturing Processes*, Prentice-Hall, Englewood Cliffs, N.J., 1979.

C General Coverage with Details of Processes

Baumeister, T. (ed.): *Marks' Standard Handbook for Mechanical Engineers*, 8th ed., McGraw-Hill, New York, 1984.

Lankford, W. T. Jr., et al. (eds.): *The Making, Shaping and Treating of Steel*, 10th ed., U.S. Steel Corp./Association of Iron and Steel Engineers, Pittsburgh, 1985.

Ryffel, H. H. (ed.): *Machinery's Handbook*, 22d ed., Industrial Press, New York, 1984.

Standen, A. (ed.): *Kirk-Othmer Encyclopedia of Chemical Technology*, 3d ed., Wiley-Interscience, New York, from 1978 on.

D Conference Proceedings and Reviews (Up-To-Date Coverage)

Annual Review of Materials Science (includes up-to-date reviews of manufacturing technologies), Annual Reviews, Inc., Palo Alto, Calif., since 1971.

Blake, P. L. (ed.): *Advanced Manufacturing Technology*, North-Holland, Amsterdam, 1980.

Bruggeman, G., and V. Weiss (eds.): *Innovations in Materials Processing*, Plenum, New York, 1985.

Colwell, L. V., et al. (eds.): *International Conference: Manufacturing Technology*, American Society of Tool and Manufacturing Engineers, Dearborn, Mich., 1967.

Gardner, L. B. (ed.): *Automated Manufacturing* (Proc. 1st Int. Symp. on Automated Manufacturing), STP 862, American Society for Testing and Materials, Philadelphia, Penn., 1985.

Hollier, R. H., and J. M. Moore (eds.): *The Production System: An Efficient Integration of Resources*, Taylor and Francis, London, 1977.

Kops, L. (ed.): *Manufacturing Solutions Based on Engineering Sciences*, American Society of Mechanical Engineers, New York, 1981.

Lane, K. A. (ed.): *Proceedings of the First International Machine Tool Conference*, IFS (Publications) Ltd./North Holland, Amsterdam, 1984.

Proceedings of the Fourth International Conference on Production Engineering, Japan Society for Precision Engineering, Tokyo, 1980.

Proceedings of the North American Metalworking (since 1983: *Manufacturing*) *Research Conference*, proceedings of annual conferences: McMaster University, 1973; University of Wisconsin-Madison, 1974; subsequent volumes published by Society of Manufacturing Engineers, Dearborn, Mich.

Shaw, M. C., et al. (eds): *International Research in Production Engineering*, American Society of Mechanical Engineers, New York, 1963.

Tobias, S. A., F. Koenigsberger, J. M. Alexander, and B. J. Davies (eds.): *Advances in Machine Tool Design and Research*, proceedings of annual conferences, since 1960, Pergamon, Oxford; from 1972, Macmillan, London.

Treatise on Materials Science and Technology (also includes articles on manufacturing), Academic Press, New York, since 1972.

E Selected Journals with General Coverage

Advanced Manufacturing Technology
Advanced Materials and Processes
American Machinist
Artificial Intelligence
Canadian Machinery and Metalworking
Computer-Aided Engineering
Computer and Industrial Engineering
Computers in Mechanical Engineering
International Journal of Machine Tool Design and Research
International Journal of Production Research
Journal of Engineering for Industry (Trans. ASME)
Journal of Engineering Materials and Technology (Trans. ASME)
Journal of Manufacturing Systems
Machine and Tool Blue Book
Machinery
Machinery and Production Engineering
Manufacturing Engineering
Manufacturing Engineering Transactions
Metal Progress
Metals Forum
Microtechnic
Precision Engineering
Production
Production Engineer (Contains abstracts of papers in all fields.)
Robotics and Computer-Integrated Manufacturing
SAMPE Quarterly
SME Transactions

F Abstract Journals

Applied Mechanics Reviews
Applied Science and Technology Index
Engineering Index
Metals Abstracts

G Computers in Manufacturing (See also Chap. 11):

Aleksander, I.: *Designing Intelligent Systems*, Unipub, New York, 1984.
Begg, V.: *Developing Expert CAD Systems*, Unipub, New York, 1984.
Besant, C. B.: *Computer Aided Design and Manufacture*, 3d ed., Wiley, New York, 1985.
Childs, J. J.: *Principles of Numerical Control*, 3d ed., Industrial Press, New York, 1982.
Glossary of Computer Aided Manufacturing Terms, 2d ed., Computer Aided Manufacturing-International, Arlington, Tex., 1978.
Groover, M. P., and E. W. Zimmers, Jr.: *CAD/CAM: Computer-Aided Design and Manufacturing*, Prentice-Hall, Englewood Cliffs, N.J., 1984.

Gunn, T. G.: *Computer Applications in Manufacturing*, Industrial Press, New York, 1981.

Halevi, G.: *The Role of Computers in Manufacturing Processes*, Wiley, New York, 1980.

Hall, D. W.: *Computer Numerical Control for Machine Tools*, Macmillan, New York, 1984.

Harrington, J. Jr.: *Computer Integrated Manufacturing*, Industrial Press, New York, 1973/Krieger, Malabar, Fla., 1979.

———: *Understanding the Manufacturing Process*, Dekker, New York, 1984.

Hatvany, J. (ed.): *World Survey of CAM*, Butterworths, London, 1983.

Hitomi, K.: *Manufacturing Systems Engineering*, Taylor and Francis, London, 1979.

Kochhar, A. K.: *Development of Computer-Based Manufacturing Systems*, Arnold, London, 1979.

Kochhar, A. K., and N. D. Burns: *Microprocessors and their Manufacturing Applications*, Arnold, London, 1983.

Koren, Y.: *Computer Control of Manufacturing Systems*, McGraw-Hill, New York, 1983.

Krause, J. K.: *What Every Engineer Should Know About Computer-Aided Design and Computer-Aided Manufacturing*, Dekker, New York, 1982.

Landau, I. D.: *Adaptive Control*, Academic Press, New York, 1979.

Miller, R. K.: *Artificial Intelligence Applications for Manufacturing*, IFS (Publications) Ltd., Bedford, England, 1985.

Nicks, J. E.: *Basic Programming Solutions for Manufacturing*, Society of Manufacturing Engineers, Dearborn, Mich., 1981.

Pao, Y. C.: *Elements of Computer-Aided Design and Manufacturing, CAD/CAM*, Wiley, New York, 1984.

Pressmann, R. S., and J. E. Williams: *Numerical Control and Computer-Aided Manufacturing*, Wiley, New York, 1977.

Pusztai, J., and M. Sava: *Computer Numerical Control*, Reston (Prentice-Hall), Reston, Va., 1983.

Rembold, U., M. K. Seth, and J. S. Weinstein: *Computers in Manufacturing*, Dekker, New York, 1977.

Roberts, A. D. and R. C. Prentice: *Programming for Numerical Control Machines*, 2d ed., McGraw-Hill, New York, 1978.

Simon, W.: *The Numerical Control of Machine Tools*, Arnold, London, 1973.

Simons, G. L.: *Computers in Engineering and Manufacture*, National Computing Centre, Manchester, 1982.

Smolik, D. P.: *Material Requirements of Manufacturing*, Van Nostrand Reinhold, New York, 1983.

Subczak, T. (ed.): *Glossary of Terms for Computer-Integrated Manufacturing*, Society of Manufacturing Engineers, Dearborn, Mich., 1984.

Taraman, K. (ed.): *CAD/CAM Integration and Innovation*, Society of Manufacturing Engineers, Dearborn, Mich., 1984.

———: *CAD/CAM: Meeting Today's Productivity Challenge*, Society of Manufacturing Engineers, Dearborn, Mich., 1984.

Teicholz, E. (ed.): *CAD/CAM Handbook*, McGraw-Hill, New York, 1984.

H Sources of Equipment and Services

Productivity Equipment Series (collections of manufacturers catalogue pages of products, both hardware and software, in several volumes), Society of Manufacturing Engineers, Dearborn, Mich.

I Trade, Business, and Commercial Organizations

Abrasive Engineering Society (AES), 1700 Painters Run Road, Pittsburgh, Penn. 15243.

The Aluminum Association (AA), 818 Connecticut Avenue N.W., Washington, D.C. 20006.

Aluminum Extruders Council (AEC), 4300-L Lincoln Avenue, Rolling Meadows, Ill. 60008.

American Bureau of Metal Statistics (ABMS), 400 Plaza Drive, Secaucus, N.J. 07094.

American Die Casting Institute (ADCI), 2340 Des Plaines Avenue, Des Plaines, Ill. 60018.

American Gear Manufacturers Association (AGMA), 1901 N. Ft. Meyer Drive, Arlington, Va. 22209.

American Iron and Steel Institute (AISI), 1000 16th Street N.W., Washington, D.C. 20036.

American Metal Stamping Association (AMSTA), 27027 Chardon Road, Richmond Heights, Ohio 44143.

American National Standards Institute (ANSI), 1430 Broadway, New York, N.Y. 10018.

American Powder Metallurgy Institute (APMI), 105 College Road E., Princeton, N.J. 08540.

Computer Aided Manufacturing International (CAM-I), 611 Ryan Plaza Drive, Arlington, Tex. 76011.

Copper Development Association (CDA), 405 Lexington Avenue, New York, N.Y. 10174.

Ductile Iron Society (DIS), 615 Sherwood Parkway, Mountainside, N.J. 07092.

Electronic Industries Association (EIA), 2001 Eye Street N.W., Washington, D.C. 20006.

Forging Industry Association (FIA), 55 Public Square, Cleveland, Ohio 44113.

Grinding Wheel Institute (GWI), 712 Lakewood Center N., Cleveland, Ohio 44107.

International Copper Research Association (INCRA), 708 Third Avenue, New York, N.Y. 10017.

International Lead Zinc Research Organization (ILZRO), 292 Madison Avenue, New York, N.Y. 10017.

International Organization for Standardization (ISO), rue de Varembe 1, CH-1211 Geneve 20, Switzerland.

Investment Casting Institute (ICI), 8521 Clover Meadow, Dallas, Tex. 75243.

Iron Casting Research Institute (ICRI), 870 W. Third Avenue, Columbus, Ohio 43212.

Iron Castings Society (ICS), 455 State Street, Des Plaines, Ill. 60016.

Metal Powder Industries Federation (MPIF), 105 College Road E., Princeton, N.J. 08540.

National Machine Tool Builders Association (NMTBA), 7901 Westpark Drive, McLean, Va. 22102.

Robotic Industries Association, 20501 Ford Road, Dearborn, Mich. 48121.

Society of Die Casting Engineers, 2000 N. 5th Avenue, River Grove, Ill. 60171.

Steel Founders' Society of America (SFSA), 455 State Street, Des Plaines, Ill. 60016.

Wire Machinery Builders Association (WMBA), 7297 Lee Highway, Falls Church, Va. 22042.

J Professional and Technical Societies

American Ceramic Society (ACerS), 65 Ceramic Dr., Columbus, Ohio 43214.

American Foundrymen's Society (AFS), Golf and Wolf Roads, Des Plaines, Ill. 60016.

American Institute of Mining, Metallurgical and Petroleum Engineers (AIME), 345 E. 47th Street, New York, N.Y. 10017.

American Society for Metals (ASM), Metals Park, Ohio 44073.

American Society for Nondestructive Testing (ASNT), 4135 Arlingate Plaza, Columbus, Ohio 43228.

American Society for Quality Control (ASQC), 230 W. Wells Street, Milwaukee, Wis. 53203.

American Society for Testing and Materials (ASTM), 1916 Race Street, Philadelphia, Penn. 19103.

American Society of Lubrication Engineers (ASLE), 838 Busse Highway, Park Ridge, Ill. 60068.

American Society of Mechanical Engineers (ASME), 345 E. 47th Street, New York, N.Y. 10017.

American Welding Society (AWS), 550 N.W. LeJeune Road, Miami, Fla. 33126.

Association for Integrated Manufacturing Technology (AIMT) (formerly: Numerical Control Society), 111 E. Wacker Drive, Chicago, Ill. 60601.

Institute of Electrical and Electronic Engineers (IEEE), 345 E. 47th Street, New York, N.Y. 10017.

Institute of Industrial Engineers (IIE), 25 Technology Park/Atlanta, Norcross, Ga. 30092.

International Institution for Production Engineering Research (CIRP), 10, rue Mansart, 75009 Paris, France.

Iron and Steel Society of AIME (ISS), 410 Commonwealth Drive, Warrendale, Penn. 15086.

The Metallurgical Society of AIME (TMS), 420 Commonwealth Drive, Warrendale, Penn. 15086.

Robotics International (RI/SME), 1 SME Drive, Dearborn, Mich. 48128.

Society for the Advancement of Material and Process Engineering (SAMPE), Box 613, Azusa, Calif. 91702.

Society of Automotive Engineers (SAE), 400 Commonwealth Drive, Warrendale, Penn. 15086.

Society of Manufacturing Engineers (SME), 1 SME Drive, Dearborn, Mich. 48128.

Society of Plastics Engineers (SPE), 14 Fairfield Drive, Brookfield Center, Conn. 06805.

Welding Institute (WI), North American Office, P.O. Box 5268, Hilton Head Island, S.C. 29928.

CHAPTER 2

ATTRIBUTES OF MANUFACTURED PRODUCTS

The purpose of manufacturing is the production of usable and salable end articles. The properties which make these products valuable are called service attributes, the knowledge of which is important from a manufacturing point of view because:

1 Service properties often dictate the choice of materials or, at least, narrow the choice of alternative materials that can be considered. Since *optimum processes are different for different materials*, the choice of manufacturing processes is also affected.

2 Properties of materials are changed by processing, and the sequence of *manufacturing processes must be chosen for any given material so that the desired end properties will be reached*.

Beyond service properties, there are other attributes that can be satisfied only by the choice of the appropriate manufacturing technique. Foremost are geometrical attributes such as shape, dimension, dimensional tolerances, and surface roughness.

Each manufacturing process has specific capabilities and limitations, and the aim of process selection and control is to produce parts *satisfying all service requirements at minimum cost*. The acceptability of the finished product is judged on the basis of tests in which conformance to specifications is checked. Therefore, properties and attributes are routinely checked as part of the quality-assurance program. In this chapter, we shall review service properties and dimensional attributes as well as methods for their measurement. The treatment will be general; the most comprehensive single-volume source of data is *Metals Handbook*

Desk Edition* (referred to in the following as MHDE). Even though the tests are described here primarily with reference to service properties, some of the same tests are extensively used for manufacturing control, often in combination with technological tests that simulate the conditions imposed on the material during manufacture.

2-1 MECHANICAL PROPERTIES

A most obvious property of manufactured products is that they are capable of supporting loads. *Loads* (*forces*) may be of many kinds; accordingly, there are many test methods that are designed to reproduce loading in service. In many applications the load is static, i.e., constant and stationary, and several tests are conducted at such low speeds that the application of force can be regarded as static.

2-1-1 Tension

Most frequently, properties are tested in the *tension test* (also called *tensile test*), which is subject to ASTM (American Society for Testing and Materials) Standard E8-82. (From time to time, these standards are revised and then the last two digits change to show the year.)

Test Setup The test specimen is machined with larger heads at its ends to ensure secure gripping. On imposing a load, the weaker part of uniform cross section (the *gage length*) deforms (Fig. 2-1). There are standard specimen geometries for round and flat (sheet) specimens. The gage length is usually accurately marked by scribing the surface.

The specimen is held in self-aligning heads which ensure that only pure tensile loads (and no bending) will be imposed. The *test machine* is essentially a press in which a moving crosshead is displaced in a controlled manner (such as a preset speed) by an actuator. In Fig. 2-1 the actuator is a hydraulic cylinder, but it could be a screw and nut or other mechanism too. The movement of the crosshead develops a force P which is balanced by the reaction force P. The magnitude of P is measured with an instrument called a *dynamometer*. This could be a steel beam, the deflection of which is measured with a dial gage, but most machines are equipped with a load cell that gives an electric signal proportional to the applied load. All load cells are calibrated against another load cell of known accuracy or directly by the application of weights.

Extension of the specimen is measured by attaching an *extensometer* to the gage length. A dial gage needs frequent reading; therefore, transducers giving an output proportional to the elongation Δl are normally used.

*American Society for Metals, Metals Park, Ohio, 1985.

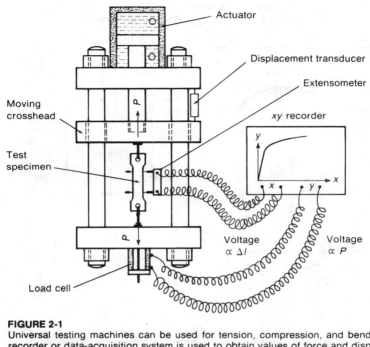

FIGURE 2-1
Universal testing machines can be used for tension, compression, and bending tests. A recorder or data-acquisition system is used to obtain values of force and displacement; the latter may be obtained from an extensometer attached to the specimen or a displacement transducer attached to the moving crosshead.

In the course of testing, both load and extension change continuously. Most conveniently, transducer outputs are used to drive an xy recorder so that a force (dependent variable) versus extension (independent variable) recording is obtained (Fig. 2-2). Outputs can be directly digitized with a data-acquisition system linked to a computer; thus the analysis of results is speeded up. Even so, there is merit in a visual recording which often reveals features that may be obscured by numerical processing.

Stress-Strain Curve The force-displacement diagram shown in Fig. 2-2*a* is typical of materials such as copper tested at room temperature. If specimens of different diameters were tested, different curves would be obtained, simply because it takes a greater force to deform a larger specimen. Therefore, results can be normalized by dividing the force P by the area A over which the force acts. In general, stress is defined as the internal force per unit area in an object subjected to external forces. A *normal stress* acts perpendicular to the cut surface and is

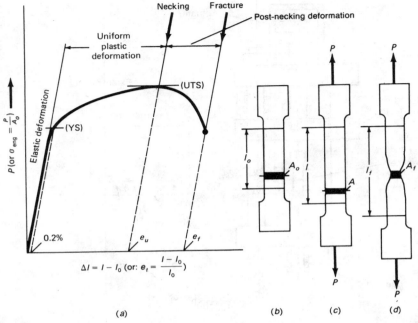

FIGURE 2-2
(a) The force-displacement (or engineering stress-strain) curve obtained on testing a ductile material reflects the sequence of events: (b) a specimen of A_0 initial cross section first suffers elastic deformation, then (c) deforms plastically—more or less uniformly within the gage length—and (d) subsequently it necks and finally fractures.

denoted by σ. Its value is

$$\sigma = \frac{P}{A} \left(\frac{N}{m^2} \equiv Pa \quad \text{or} \quad \frac{lbf}{in} \equiv psi \right) \tag{2-1}$$

It should be noted that the SI unit N/m^2 (also called pascal, Pa) represents a very small stress; therefore, MN/m^2 or MPa is often used. The unit MPa is equal to N/mm^2, which is more convenient for many calculations. Similarly, psi is a small unit, and a thousandfold value (often denoted as ksi but, more logically, written as kpsi in this book) is more customary. In the old metric system, the unit was kg/mm^2, which roughly equals $10 \ N/mm^2$ (the kg stands for kg force).

In the course of the tension test the specimen is forcefully elongated. To a first approximation, most engineering materials are incompressible; thus, their volume V remains virtually constant during plastic deformation:

$$V = A_0 l_0 = A_1 l_1 = Al \tag{2-2}$$

where A and l are instantaneous cross-sectional area and length, respectively. The subscript 0 refers to the starting dimensions, the subscript 1 to final dimensions. Because of *constancy of volume*, there is a reduction in cross-sectional area corresponding to the increase in length (compare Fig. 2-2c with Fig. 2-2b). However, the exact value of the cross-sectional area is not immediately known. For convenience, the convention has been adopted that the force P is divided by the original cross-sectional area A_0. By definition, stress is force acting on unit area; since here we divide force by a nonexistent area, the result is distinguished from a true stress by calling it *nominal, conventional,* or *engineering stress* (σ_{eng} or S)

$$\sigma_{eng} = \frac{P}{A_0} \qquad (2\text{-}3)$$

Elongation can be normalized too by taking the change in length and dividing it by the original length; this is usually termed *engineering tensile strain* e_t

$$e_t = \frac{l - l_0}{l_0} \qquad (2\text{-}4a)$$

where $l - l_0 = \Delta l$, the change in length. For convenience, a percentage value is often quoted

$$e_t(\%) = \frac{l - l_0}{l_0} 100 \qquad (2\text{-}4b)$$

Inspection of the engineering stress-strain curve shows a number of critical points which can be used to characterize a material.

Strength Properties

1 Elastic modulus. At the beginning of the test, the force increases rapidly and proportionately to strain: The stress-strain curve obeys *Hooke's law*

$$\sigma = Ee_t \qquad (2\text{-}5)$$

The proportionality constant (the slope of the curve) is called the *elastic modulus* or *Young's modulus* E

$$E = \frac{\sigma}{e_t} \text{ (MPa or psi)} \qquad (2\text{-}6)$$

If the specimen is unloaded in this range, it will return to its original length, i.e., all deformation is elastic. Most structures are designed so that they should never suffer permanent deformation and E then determines the change in the length of a component for a given load. The elastic modulus reflects the basic structure and bond strength of materials; therefore, data given in Table 2-1 are typical also of alloys of the pure metals listed.

TABLE 2-1
SELECTED PRODUCTION DATA AND PROPERTIES FOR MATERIALS
OF MANUFACTURING

Material	World production,[*] 10^6 tonnes		Melting point, °C	Density, kg/m^3	Elastic modulus, 10^3 MPa	Resistivity at 20 °C, 10^{-8} Ω·m
	1972	1982				
Iron (steel)	634	654	1536	7900	210	9.7
Aluminum	11.0	15.1	660	2700	70	2.7
Copper	7.0	8.1	1083	8900	122	1.7
Zinc	5.2	6.0	419	7100	90	5.9
Lead	3.6	5.2	327	11300	16	21.0
Nickel	0.6	0.6	1455	8900	210	6.8
Magnesium	0.26	0.26	649	1700	44	4.0
Tin	0.2	0.2	232	5800	42	11.0
Titanium	0.06	0.07	1670	4500	106	6.8
Plastics	43.3[†]	60[†]		900–2200	3–10	(10^{12})

[*]Compiled from *Metal Statistics 1974 and 1984*, American Metal Market, Fairchild Publications, Inc., New York, 1974 and 1984.
[†]1973 data, from *Encyclopaedia Britannica Book of the Year, 1975*, Chicago, 1975; 1982 data, estimates.

2 Yield strength. At some higher stress the slope of the curve changes, and this stress is termed the proportional limit. However, its determination is quite difficult; therefore, it is customary to choose a point at which a metallic specimen deforms permanently. The corresponding engineering stress is called the *yield strength* YS or S_y. For most metallic materials, 0.2% permanent deformation is taken as the threshold because it is relatively easily measured, and then the yield strength is denoted as $\sigma_{0.2}$ (or $S_{0.2}$)

$$\sigma_{0.2} = \frac{P_{0.002}}{A_0} \qquad (2\text{-}7)$$

Note that, by definition, the strain of 0.002 is all *plastic* (permanent) strain; therefore, the corresponding force $P_{0.2}$ is found by drawing a line from $e_t = 0.002$ *parallel to the elastic line*. If the specimen were unloaded at this point, all elastic deformation would be recovered, at a slope equal to the initial slope of the force-displacement curve. By drawing the parallel line, the contribution of elastic deformation to total strain is eliminated (Fig. 2-2a).

Yield strength is an important design quantity. To prevent even the slightest plastic deformation of an engineering structure, the design stress is often kept to some fraction of $\sigma_{0.2}$ by the use of a safety factor, or design is made to some lower value such as $\sigma_{0.02}$.

3 Tensile strength. On further loading and elongation, the gage section of the specimen elongates (and its cross section reduces) uniformly along its entire length (Fig. 2-2c) yet the force gradually increases. For reasons to be explained in Sec. 4-1-2, the material becomes stronger with deformation (it strain hardens). At some critical deformation level typical of the material and its processing history, strain hardening cannot counterbalance the loss of strength resulting from the ever-decreasing cross-sectional area, and a neck forms at the weakest point. Since the cross section is now locally reduced, the force sustained by this weakened section is less, and the force P declines while deformation is concentrated in the already necked zone. Finally, fracture occurs.

The engineering or conventional stress at the maximum load is called the *tensile strength* (TS or S_u) or often also *ultimate tensile strength* (UTS)

$$TS = \frac{P_{max}}{A_0} \tag{2-8}$$

The TS is not a true stress (because force is divided by a nonexistent area), but it has great practical value for quality-control purposes. It is also a measure of the maximum force a component can sustain before catastrophic failure.

Example 2-1

An aircraft component, made of 7075-T6 aluminum alloy, can be represented as a bar of diameter 25 mm and length 400 mm, loaded in pure tension. Calculate (a) the extension of the bar under an imposed load of 80 kN, (b) the load at which the bar suffers permanent deformation, and (c) the maximum load the bar can take without fracture.

From Table 2-1, $E = 70$ GPa; from MHDE (p. 6.46), YS = 496 MPa; TS = 558 MPa.

(a) The cross-sectional area of the bar is $A_0 = 20^2 \pi/4 = 314$ mm^2. The imposed tensile stress is $80\,000/314 = 255$ N/mm^2 ($= 255$ MPa) and is thus less than the YS: deformation will be purely elastic. From Eq. (2-5), $e_t = \sigma/E = 255/70\,000 = 0.0036$ or 0.36%.

(b) YS $= \sigma_{0.2} = 496$ N/mm^2. From Eq. (2-7), $P_{0.002} = (\sigma_{0.2})(A_0) = (496)(314) = 156$ kN.

(c) From Eq. (2-8), $P_{max} = (TS)(A_0) = (558)(314) = 175$ kN.

Measures of Ductility   The engineering stress-strain curve also provides information on the ductility of the material, i.e., its ability to deform without damage.

1 Uniform elongation. Since prior to necking the cross section reduces roughly uniformly along the gage length, the engineering strain sustained at the point of maximum load is called *uniform elongation*, denoted by e_u

$$e_u = \frac{l_u - l_0}{l_0} \tag{2-9a}$$

where l_u is the gage length at the point of necking.

2 Elongation. More frequently, the total *elongation to fracture* (also called *total elongation* or simply, and somewhat misleadingly, *elongation*) is measured, most often by placing the broken parts of the specimen together and measuring the distance l_f between the gage marks

$$e_f = \frac{l_f - l_0}{l_0} \tag{2-9b}$$

Note that, if l_u or l_f (or any length during deformation) is measured from a recording, the elastic contribution to elongation must be taken out by drawing a line parallel to the elastic loading line (Fig. 2-2a).

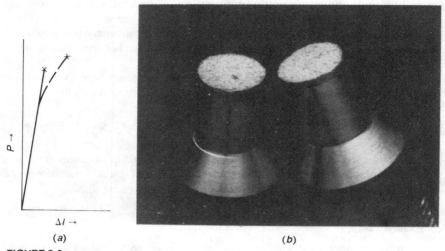

$P \uparrow$

$\Delta l \rightarrow$

(a)   (b)

FIGURE 2-3
A brittle material shows (a) little or no evidence of plastic deformation in the tension test and (b) fracture often occurs along grain boundaries or other weakening features. (*The example shown is a Zn-12 Al alloy. Courtesy Dr. P. Niessen, University of Waterloo.*)

As visible from Fig. 2-2a and d, e_f is the sum of uniform elongation and elongation in the neck. Thus, it is sensitive to gage length; A shorter gage length will make the same material appear to have a larger elongation. For this reason, the gage length must always be stated; otherwise, total elongation—a readily measurable quality-control indicator—would lose its meaning.

3 Reduction in area. The most sensitive measure of the ductility of materials is the *reduction in area* measured at fracture. Basically, materials can be brittle or ductile.

a A completely *brittle material* deforms only elastically; at some critical stress, separation (fracture) occurs suddenly (Fig. 2-3a), usually in a plane perpendicular to the axis of load application (Fig. 2-3b). Fracture often originates from some minute crack that locally raises the stress (see Sec. 2-1-5).

b A *ductile material* is capable of plastic deformation; thus, imposed stresses are redistributed and, even if there is some local imperfection, deformation continues beyond necking (Fig. 2-4a). Necking causes tensile stresses to develop in all directions (the stress state becomes triaxial; see Sec. 4-3-3). Triaxial tensile stresses literally tear the material apart: they open up cavities, voids in the center of the neck cross section. On further straining, voids interlink, roughly on a plane perpendicular to the axis. Once the remaining annular cross section is insufficient to carry the load, it fails in shear to give the lip of the characteristic cup-and-cone fracture (Fig. 2-4b).

Cavities form earlier if there are points of weakness, such as inclusions, in the material. As seen in Fig. 2-4a, strength is not necessarily affected, but post-necking deformation (which determines reduction in area) is greater for the cleaner

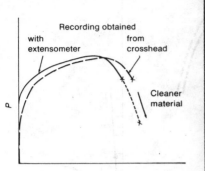

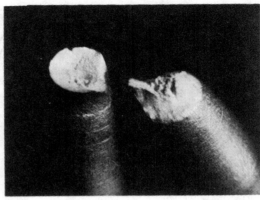

(a)

(b)

FIGURE 2-4
A ductile material (a) undergoes plastic deformation before and beyond necking and (b) the
fractured surface shows the formation and interlinking of voids and a characteristic
cup-and-cone configuration. (The example shown is a 70/30 brass.)

material; hence, *cleanliness* is an important aim of manufacturing control. Very
clean ductile materials may reduce to a point before separation, which is then
sometimes called rupture.

The minimum cross-sectional area of the fractured test specimen, A_f can be
measured and reduction in area q (or R.A.) can be calculated as

$$q = \frac{A_0 - A_f}{A_0} \tag{2-10}$$

4 Toughness. The area under the stress-strain curve has the dimension of force
times distance, i.e., work. Thus it can be regarded as a measure of *toughness*, i.e.,
the energy absorbed by the material prior to fracture.

It should be noted that data-acquisition systems can be added to all test
equipment, and all processing of the force-displacement data can be made by a
computer.

Example 2-2

A tension-test specimen of 6.35-mm thickness and 6.38-mm width was machined from an annealed
80Cu–20Ni alloy plate. The gage length of $l_0 = 25.0$ mm was lightly marked with a scriber. The
test was performed on a 10000-kgf (98-kN) capacity testing machine, with an extensometer
attached to the gage length. The curve shown as a full line was recorded. To obtain a better

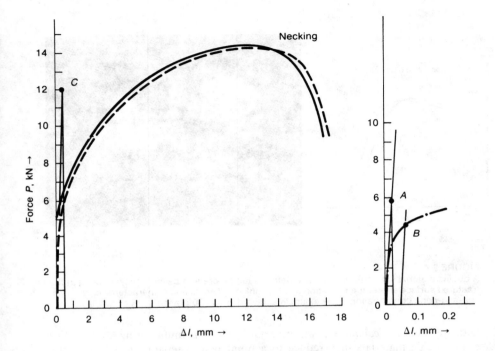

resolution at low strains, the test was repeated with a 20 times higher gain on the extension axis (dash-dot line). The fractured specimen halves were placed together and the distance between the scribe marks was $l_f = 42.2$ mm. The fractured cross section was 2.85 mm $\times$ 3.50 mm. Calculate (a) Young's modulus; (b) $\sigma_{0.2}$; (c) TS; (d) elongation; and (e) reduction in area.

(a) Select a convenient point on the dash-dot line. In the example chosen (point A), $P = 5.7$ kN and $A_0 = (6.35)(6.38) = 40.5$ mm$^2 = 40.5(10^{-6})$ m^2. Extension is $\Delta l = 0.025$ mm; hence, $e_t = 0.025/25.0 = 0.001$. From Eq. (2-6),

$$E = \frac{5700}{40.5(10^{-6})(0.001)} = 141 \text{ GPa}$$

(b) For a strain of 0.2%

$$e_t = 0.002 = \frac{\Delta l}{l_0} = \frac{\Delta l}{25}$$

thus, $\Delta l = 0.002(25.0) = 0.05$ mm. For the dash-dot curve, draw a line from this point, parallel to the elastic line. It intersects the recording at point B, where $P = 4.4$ kN. Thus, from Eq. (2-7)

$$\sigma_{0.2} = \frac{4400}{40.5} = 109 \ \frac{\text{N}}{\text{mm}^2} \ (\equiv \text{MPa})$$

(c) The maximum load is $P_{max} = 14.2$ kN. From Eq. (2-8)

$$TS = \frac{14\,200}{40.5} = 351 \ \frac{N}{mm^2}$$

(d) Elongation, from Eq. (2-9b), using the l_f measured on the specimen,

$$e_f = \frac{42.2 - 25.0}{25.0} = 0.688 = 68.8\%$$

From the recording (full line), $l_f - l_0 = 17.0$ mm, which slightly less than the measured l_f because perfect fitting of the broken halves is difficult.

(e) Fracture area is $A_f = (2.85)(3.5) = 9.98$ mm^2. From Eq. (2-10)

$$q = \frac{40.5 - 9.98}{40.5} = 0.75 \text{ or } 75\%$$

It is always advisable to check results against published data. In this case, agreement with MHDE (p. 7.4) is acceptable for a soft material, except that the measured elongation is 69% versus 40% quoted in MHDE. Note, however, that elongation is given in MHDE for a gage length of 2 in, whereas the gage length was only 25 mm in the present test.

Process / Equipment Interactions For quality-control purposes it is usually sufficient to determine $\sigma_{0.2}$, TS, e_f, and q. No extensometer is then needed, and the force-displacement curve can be simply recorded, either by driving a chart at a constant speed while recording P, or by driving the chart from the moving crosshead. The recording obtained by the latter method (Fig. 2-4a, broken line), is similar to the one obtained with an extensometer but with one significant difference: The initial elastic slope of the curve is now much lower. Inspection of Fig. 2-1 will show that when the force P is applied to the specimen, not only will the gage length (and the rest of the specimen) deform, but also the machine: The stationary crossheads are bent and the columns are compressed. Even though the machine is much sturdier than the specimen, the length over which deformation occurs (the *elastic loading path*) is much longer: The machine behaves as a very long spring, attached in series with the short spring representing the specimen. In consequence, the deformation of the machine is added to the elastic deformation of the specimen, and the initial slope of the force-displacement curve represents the sum of the two.

This is an important observation because, in most manufacturing processes, elastic deformation of the machine is large enough to affect dimensional control of the parts produced.

Example 2-3

The tension test of Example 2-2 is repeated but this time the recording is made from the crosshead movement (broken line in the recording). Calculate the spring constant of the system.

Taking a convenient point at, say, $\Delta l = 0.4$ mm (point C), the force is $P = 12.0$ kN. Thus, the overall spring constant $K = P/\Delta l = 30$ kN/mm. The total elastic deformation is the sum of deformations in the specimen and the machine: The contribution of each can be calculated if the spring constant of the machine is known or the spring constant of the specimen is calculated (see Prob. 2-4).

2-1-2 Compression

For reasons to be explained later, some materials, such as gray cast iron and concrete, are weak in tension but strong in compression. When design of the structure ensures that only compressive loads will be imposed, *compression testing* (ASTM E9) is most relevant.

The test equipment is again a press (or a universal testing machine), this time arranged so that the specimen is compressed between two well-lubricated, flat, parallel, hardened platens (Fig. 2-5a). Because of constancy of volume (Eq. (2-2)),

FIGURE 2-5
In (a) compression testing, the cross-sectional area continually increases, therefore, (b) the recorded force increases even if the material does not harden with deformation. (c) The derived stress-strain curve in this instance shows strain hardening. (d) Brittle materials fracture after initial elastic compression although some plastic deformation is sometimes observed (broken line).

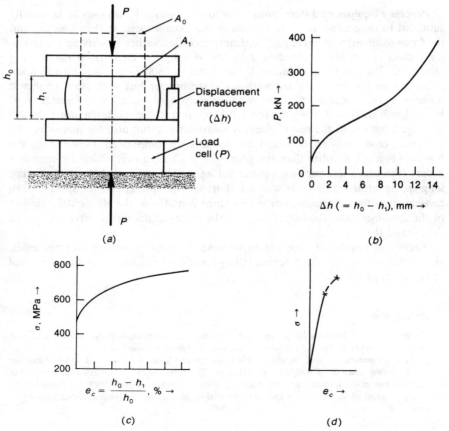

the cross-sectional area of the specimen must increase in proportion to the decrease in height and the recorded force rises not only because of strain hardening (if present) but also because of the increasing area (Fig. 2-5b). The instantaneous cross-sectional area A can be calculated from the instantaneous height h obtained from the output of the displacement transducer ($h = h_0 - \Delta h$)

$$A = \frac{A_0 h_0}{h} = \frac{V}{h}$$ (2-11)

At any point of the press stroke, the *die pressure* is force divided by area; we shall see in Sec. 4-4-1 that, if friction effects are negligible, this equals the (true) *compressive strength*

$$\sigma = \frac{P}{A}$$ (2-12)

The engineering *compressive strain* is

$$e_c = \frac{h_0 - h}{h_0} = \frac{A - A_0}{A}$$ (2-13)

from which a stress-strain curve can be plotted (Fig. 2-5c).

Engineering components or structures are seldom allowed to deform substantially; therefore, the compressive stress corresponding to some small strain (say, 0.2% or 0.5%) is usually taken as the basis of design. Brittle materials fail suddenly on reaching a critical stress (Fig. 2-5d); fracture often occurs on a 45° diagonal.

Example 2-4

The recording shown in Fig. 2-5b was made while compressing, at room temperature, a steel cylinder of diameter 15.00 mm and height 22.5 mm, made of hot-rolled AISI 1020 steel. A graphited grease was used to reduce friction. Force P readings at six points are given below together with the instantaneous heights h. Calculate the true stress σ and compressive strain e_c.

The volume of the specimen is $(15.0^2 \pi/4)(22.5) = 3976$ mm³. The instantaneous area A is calculated from Eq. (2-11); σ from Eq. (2-12); and e_c from Eq. (2-13).

Point no.	h, mm	P, kN	A, mm²	σ, MPa	e_c	ϵ
0	22.5	. . .	177	. . .	. . .	. . .
1	20.5	115	194	593	8.9	0.09
2	17.5	158	227	695	22.2	0.25
3	14.5	200	274	729	35.6	0.44
4	12.5	235	318	739	44.4	0.59
5	10.5	290	379	766	53.3	0.76
6	8.5	370	468	791	62.2	0.97

The resulting plot is given in Fig. 2-5c.

2-1-3 Bending

Brittle materials are also used in applications where tensile stresses are imposed, either in pure tension or in bending. Testing in tension is difficult because the slightest misalignment in the jaws imposes bending which increases stresses in an unknown manner. Therefore, it is preferable to test in pure bending (Fig. 2-6).

The specimen is supported at two points (ASTM F417). In the *three-point test* (Fig. 2-6a) a force P is applied at the center. The specimen bends, and the outer (lower) half is put into tension, whereas the inner half is put into compression. Tensile stresses reach their maximum at the outer surface, midway between the supports. Failure (fracture) occurs when the maximum tensile stress reaches a critical value, often called the *rupture strength* (or *modulus of rupture*). For a rectangular beam

$$\sigma_B = \frac{3}{2}\frac{Pl}{bh^2} \tag{2-14a}$$

For a round specimen

$$\sigma_B = \frac{2.546Pl}{d^3} \tag{2-14b}$$

Alternatively, the deflection for a given load (or load for a specified deflection) is given, especially for plastics.

FIGURE 2-6
Less-ductile materials are often subjected to (a) three-point or (b) four-point bending tests. Tensile stresses peak at the center in the three-point test but are distributed uniformly between the two loading points in the four-point test.

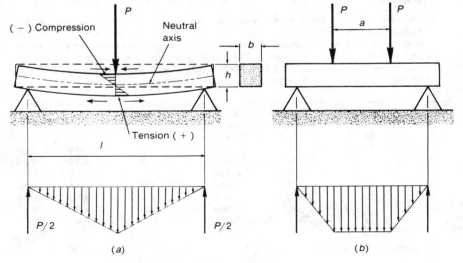

The *four-point test* (Fig. 2-6b) generates uniform tensile stresses between the loading points. If $a = l/3$, the modulus of rupture for a rectangular specimen is

$$\sigma_B = \frac{Pl}{bh^2} \qquad (2\text{-}14c)$$

Less-ductile materials may have minute defects, cracks in the surface or body of the specimen (Sec. 2-1-5). The modulus of rupture is then a function of testing method and is highest—and shows the greatest scatter—in three-point bending because there is a low probability of a defect residing at the point of maximum stress. The uniform stress distribution in the four-point test makes it more likely that a defect will be found; hence, the modulus of rupture is lower but more consistent.

Example 2-5

A high-technology ceramic (hot-pressed silicon nitride, Si_3N_4) was tested by bending 3.2-mm-thick, 6.4-mm-wide specimens loaded over a 38-mm span. Fracture occurred at a load of 1070 N in three-point bending and at 1250 N in four-point bending. Calculate the maximum stresses in each case.

From Eq. (2-14a), in the three-point test $\sigma_B = 3(1070)(38)/(2)(6.4)(3.2)^2 = 930$ MPa. From Eq. (2-14c), in the four-point test $\sigma_B = (1250)(38)/(6.4)(3.2)^2 = 725$ MPa. We will see in Sec. 6-3-2 that ceramics may have minute imperfections, cracks in the surface: The probability of finding such a crack increases with an increasing length over which a high stress is developed; hence, the measured strength is lower in four-point bending than in three-point bending.*

2-1-4 Hardness

The resistance of a material to deformation is most conveniently tested by indentation (Fig. 2-7). The specimen must be large enough to keep deformation highly localized, so that the material displaced by the indentor is pushed up around the indentation but does not deform the entire thickness of the specimen. A great advantage is that a relatively small local indentation may be permissible even on a full-size part; thus, there is no need to destruct the part to obtain a reading. Tests are standardized, including the geometry and dimensions of the indentor, the magnitude of the applied load, and the rate of load application.

1 In the *Brinell hardness test* (ASTM E10) the indentor is a steel (or, for harder materials, tungsten carbide) ball (Fig. 2-7a). After the load is applied, the mean diameter of the impression is measured. Force divided by the *surface area* of the indentation gives the *Brinell hardness number* (HB or BHN), which is still quoted in the old metric units of kg/mm^2. Since surface area is not a linear function of impression diameter, tables are available to simplify the calculation.

---

*Data for this example were taken from D. W. Richerson, *Modern Ceramic Engineering*, Dekker, New York, 1982.

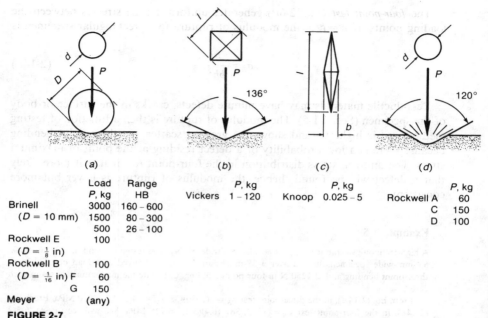

	Load P, kg	Range HB		P, kg		P, kg		P, kg
Brinell	3000	160 – 600	Vickers	1 – 120	Knoop	0.025 – 5	Rockwell A	60
(D = 10 mm)	1500	80 – 300					C	150
	500	26 – 100					D	100
Rockwell E	100							
($D = \frac{1}{8}$ in)								
Rockwell B	100							
($D = \frac{1}{16}$ in) F	60							
G	150							
Meyer	(any)							

FIGURE 2-7
Hardness tests have the advantage that information on the compressive strength can be obtained by localized deformation, without destroying the workpiece.

Very deep indentations must be avoided; hence, the load is reduced for softer materials to keep the indentation diameter between 2.50 and 4.75 mm.

2 Impressions made by a pyramid (Fig. 2-7b) remain geometrically similar independent of load; therefore, they can be used for a wide range of hardnesses. In the *Vickers hardness test* (ASTM E92-82) the hardness number (HV or VHN, in kg/mm^2) is again obtained on dividing the force by the surface area, calculated from the diagonal of the impression.

Fundamentally more correct is *Meyer's hardness test* (Fig. 2-7a) in which the load on the ball is divided by the *projected area* of the indentation; unfortunately, the test has not become popular.

3 *Microhardness tests* (ASTM E384) are used to explore localized variations in hardness within a body and close to the edges. Loads have to be very small; hence, the surface must be prepared by polishing, taking care not to cause local deformation which would increase the hardness. In the *Knoop test* (Fig. 2-7c) hardness is calculated from the long diagonal of the indentation. Because of elastic recovery at low loads, the hardness (HK, in units of kg/mm^2) is not a linear function of the diagonal.

4 For quality-control purposes, the *Rockwell hardness test* (ASTM E18) is most widespread because of the convenience of using the test apparatus. The indentor is a ball or a diamond cone (Fig. 2-7a and d). After preloading to minimize surface roughness effects, the main load is applied. The apparatus automatically measures the depth of indentation and gives a readout on arbitrary scales of which the A, B, and C scales are used most frequently (reported as HRA, HRB, HRC, etc.). Conversion to other units is possible, particularly for materials of low strain hardening such as heat-treated steels. Conversion tables are given in ASTM E140 and in MHDE; a nomograph is given in App. B.

5 The hardness of large parts can be measured with a *scleroscope* (ASTM E448), a portable instrument which relates hardness to the rebound of a small weight (hammer) dropped from a standard height; a diamond indentor is attached to the hammer.

6 The hardness of brittle materials is measured in a comparative *scratch test* and is reported on the Mohs scale which is based on the scratch resistance of selected minerals (see App. B).

For reasons to be explained in Sec. 4-4-2, the hardness of materials is approximately three times their TS (*but only if both are expressed in consistent units*). The relationship works best for materials of low strain hardening (such as heat-treated steels) and for HV; less well for HB.

Example 2-6

A cold-drawn steel bar has a Brinell hardness of HB = 190. What TS is to be expected?

$$TS = 190/3 = 63.3 \ kg/mm^2 = 620 \ N/mm^2$$

Converting to conventional units, we get 620 N/mm² = 90 kpsi = 90 000 psi. Hence the statement that TS(psi) = 500 BHN.

Example 2-7

A large rolling-mill roll was supposed to be heat treated to a hardness of HRC 55. How could one check if this has indeed been done?

The roll is too large to be placed in a Rockwell hardness tester, and no specimen can be cut from it. Therefore, a Model C Shore scleroscope is used. It gives a reading of 78. From a conversion table (e.g., MHDE, p. 1.60) or App. B, this corresponds to HRC 58, thus the heat treatment has indeed been carried out.

2-1-5 Impact and Fracture Toughness

In Sec. 2-1-1 we mentioned that the energy per unit volume is sometimes used as a measure of toughness. It is found, however, that some normally ductile, tough materials suffer brittle fracture when they are in the form of a notched specimen or component and are exposed to sudden loading (impact force), especially below

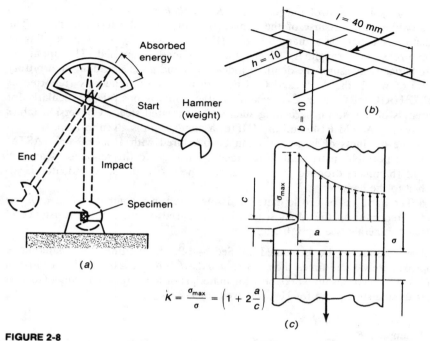

FIGURE 2-8
(a) The Charpy impact test gives a measure of the fracture toughness of the material. (b) The specimen is notched; (c) the notch serves to cause stress concentration.

some critical *ductile-to-brittle transition temperature*. This can be a problem, for example, in arctic service of welded structures such as ships, drilling rigs, and pipelines which may contain planar welding defects and also residual stresses.

Several standard *impact tests* (ASTM E23) exist, each using a different test geometry and loading method. In impact tests a load is suddenly applied, for example, by a swinging pendulum (Fig. 2-8a). The impact energy absorbed by the specimen (the energy lost from the pendulum) is reported (in units of joule). At the transition temperature the energy absorbed drops more or less suddenly. Also, the appearance of the fracture surface changes.

It will be noted from Fig. 2-8b that the impact specimen is notched. The notch causes a *stress concentration*, i.e., a local increase in stress to σ_{max}. The *stress concentration factor* K is the ratio of σ_{max} to the stress σ that would prevail in a smooth body (Fig. 2-8c) and can reach very high values when the notch radius is small. When the maximum stress or strain reaches some critical value, a crack propagates at high speed through the part. Thus, the presence of cracks on the surface or inside the body may severely reduce the tensile stress that a material can withstand without fracture. This *fracture stress* σ_{fr} can be shown to depend on a crack radius r_c and crack depth (crack length) a as

$$\sigma_{fr} = \left(C \frac{r_c}{a} \right)^{1/2} \tag{2-15a}$$

where C is a material constant. For truly brittle materials, such as glass, r_c is on the order of atomic radii and then Eq. (2-15a) reduces to the *Griffith criterion*

$$\sigma_{fr} = \left(C \frac{1}{a} \right)^{1/2} \tag{2-15b}$$

Because of the great sensitivity of the impact test results to specimen geometry and preparation, the impact energy quoted is only a comparative value between materials tested under identical conditions. It is a very useful quality control indicator but cannot be used for design calculation purposes.

In a given material system, the highest strength can usually be attained only at the expense of ductility and, thus, increased sensitivity to brittle fracture. This is true, for example, of high-strength aluminum alloys used in aircraft construction and of the highest-strength steels. The need to design with this danger in mind has led to the development of a *linear elastic fracture mechanics* approach. Special tests are used to determine the *plane-strain fracture toughness* K_{Ic}

$$K_{Ic} = \alpha\sigma\sqrt{\pi a_c} \tag{2-16}$$

where α is a factor depending on specimen and crack geometry, σ is stress or a function of the stress field, and a_c is the critical crack length below which fracture will not occur. Thus a structure can be designed to the allowable stress if the likely crack length is known, or the maximum allowable crack length may be specified for a given design stress.

Because cracks or notches can be a problem in all but the most ductile materials, one of the aims of manufacturing processes is to *prevent the formation of cracks* or, if this is not possible, *keep cracks in compression* during the service of the part, either by allowing only compressive loading or by inducing compressive residual stresses (Sec. 2-1-8).

Example 2-8

During the Second World War, a large number of transport ships (the liberty ships) were constructed. The traditional riveted structure was abandoned in favor of welding, thus greatly speeding up the rate of production. Of the over 4000 ships built, about 24 had serious cracking and about 12 broke into two in the cold waters of the North Atlantic. Fractures were of the brittle type, even though the steel was ductile in room-temperature tension and impact tests. Research into the causes of the problem did much to shed light on ductile-to-brittle transition and has led to the specification and manufacture of steels with guaranteed impact energies at low temperatures.

2-1-6 Fatigue

In many instances, materials are subjected to repeated load applications. Even though each individual loading event is insufficient to cause permanent deformation and, even less, fracture, the repeated application of stress can lead to *fatigue* failure. Fatigue is the result of cumulative damage, caused by stresses much smaller than the tensile strength. Fatigue failure begins with the generation of small cracks, invisible to the naked eye, which then propagate on repeated loading until brittle fracture occurs or the remaining cross-sectional area is too small to carry the load. Fractured surfaces bear evidence of this sequence of events (Fig. 2-9a).

The suitability of a material can be judged from experiments (ASTM E206) in which a specimen is exposed to a preset level of stress S until fracture occurs after N cycles (Fig. 2-9b and c). The results are reported in fatigue diagrams or *SN* diagrams (Fig. 2-10a). In some environments, some materials such as steel may sustain some minimum stress level indefinitely; this is called the *fatigue limit* or *endurance limit*. There is also a limiting stress for nonferrous materials. It is better, however, to specify the stress which can be sustained for a given number of cycles (say, 2 million cycles).

Because fatigue involves the propagation of cracks under an imposed tensile stress, the number of cycles to failure (or the stress sustained for a given number

FIGURE 2-9
Repeated application of even relatively small stresses can result in fatigue; (*a*) the fractured surface shows evidence of crack initiation and propagation. Materials are tested by subjecting specimens to (*b*) cyclic tension, tension and compression, or (*c*) bending in rotation. (*The example shown is of a high-strength steel sealing ring subjected to fluctuating internal pressure. Courtesy Dr. D. J. Burns, University of Waterloo.*)

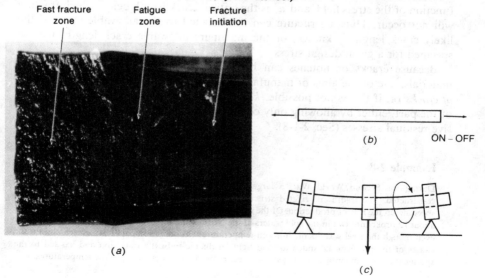

| Fast fracture zone | Fatigue zone | Fracture initiation |

(*a*)

(*b*)   ON – OFF

(*c*)

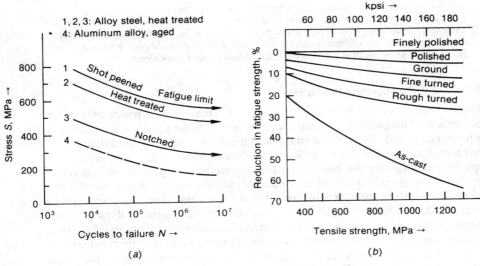

FIGURE 2-10
With an increasing number of loading cycles, (a) the stress at which fracture occurs drops, although some materials show an indefinite life at some stress level, the so-called fatigue limit or endurance limit. Fatigue strength is greatly impaired by the presence of surface cracks or notches and (b) even by a rough surface. [(a) *From various sources; (b) from E. S. Burdon, SCRATA Proceedings 1968 Annual Conference, Steel Castings Research and Trade Association, London, 1968, Paper No. 3, with permission.*]

of cycles) is greatly reduced if there are preexisting cracks (Fig. 2-10a), internal defects, or inclusions of a brittle nature. The surface roughness produced in some processes acts similarly; therefore, fatigue strength is reduced if the surface is rough, especially in high-strength materials that are less ductile (Fig. 2-10b).

Repeated loading at elevated temperatures—such as is caused by differential expansion and contraction of the surface of a part—may lead to *thermal fatigue*. It is particularly troublesome in forging tools and casting dies because cracking (crazing) of the surface is reproduced on the surface of the part.

A special form of failure occurs when certain materials are exposed to a chemically aggressive (corrosive) environment. Surface cracks form and, in combination with the applied stress, lead to *stress-corrosion cracking*. If there are residual tensile stresses on the surface of the part (as in Fig. 2-14), cracks develop in a part even in the absence of external loads.

Example 2-9

The first commercial jet-powered plane was the British Comet. Several planes disintegrated in flight with the loss of all lives. Fatigue failure due to cyclic hoop stresses, generated by repeated pressurization of the cabin, was suspected. Therefore, a complete airframe which had been through 1230 flights, was submerged in water in a test tank and was subjected to pressure cycles. After 1830 cycles the cabin failed by fatigue cracks that grew at the corners of cabin windows. The lesson was well learned; since then, great advances have been made in the science of designing for fast fracture in materials of limited fracture toughness as well as in manufacturing methods for improved fracture toughness. Nevertheless, planes are still regularly inspected for evidence of cracks; they are designed so that cracks too small to be detected in one inspection will not grow so fast that they would result in catastrophic failure before the next inspection.*

2-1-7 High-Temperature Properties

Many components are expected to operate at some elevated temperature. Our perception of temperature is conditioned by our response to it; similarly, there is a temperature scale for each material that is much more relevant for it than any of our temperature scales. Using a Greek derivation (*homos* = the same, *legein* = to speak) this is termed the *homologous temperature* scale, signifying that it corresponds to the specific points of relevance for each material. Not surprisingly, one of the endpoints is absolute zero, the other is the melting point T_m (expressed in degrees Kelvin). Below roughly $0.5T_m$, most metals and polymers show a "cold" behavior (Fig. 2-11): strength is high, ductility is relatively low. Above $0.5T_m$, they exhibit typically "hot" properties: strength is lower, ductility higher; substantial deformation may occur after necking, with the neck spreading over the entire length of the specimen (Fig. 2-12). The structural reasons for this behavior will become evident in Secs. 4-1-4 and 7-2-4. It should be noted here that $0.5T_m$ is a very rough dividing line and alloying can push the onset of hot behavior to higher temperatures.

At this point it is important to note that deformation in the hot temperature range involves substantial rearrangement of atoms or molecules, processes which take time. Therefore, properties are also a function of the rate of load application or, more correctly, of the imposed *strain rate* $\dot{\epsilon}$ which in the tension test is simply

$$\dot{\epsilon} = \frac{v}{l} \tag{2-17}$$

where v is the crosshead velocity (see Fig. 2-1) and l is the *instantaneous deforming length* (the gage length prior to necking, but *the length of the necked portion after necking*). High strain rates allow less time for atomic or molecular rearrangement; hence, stresses are higher and ductility is lower.

This also means that, if even a small load is applied over a long period of time at elevated temperatures, some deformation may occur. We say that the material suffers *creep*. This is very important when deformation is unacceptable (e.g., the growth of a turbine blade would cause damage in a jet engine, or a plastic hot-water pipe would sag or even burst in a home).

In the typical *creep test* (ASTM E139), a tension-test specimen is exposed to a preset (constant) load at a constant temperature. There is rapid initial extension (*primary creep*), followed by slower deformation at a constant rate (*secondary creep*) and, finally, when structural damage occurs, creep accelerates and the part

---

*J. K. Williams, in *Fatigue Design Procedures*, E. Gassner and W. Schutz (eds.), Pergamon, Oxford, 1969.

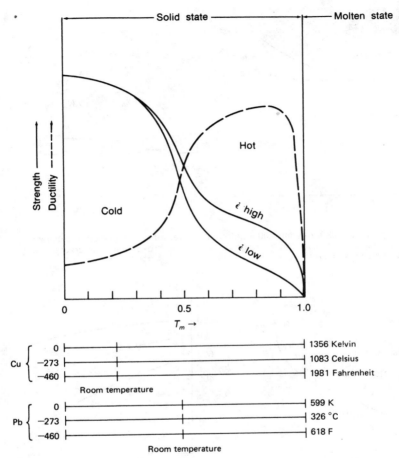

FIGURE 2-11
Materials such as metals have their own, built-in, homologous temperature scale. In the "cold" regime they are strong but less ductile, whereas in the "hot" regime they are less strong but more ductile. In the hot regime their strength is greater at higher rates of loading.

fails (*tertiary creep*, Fig. 2-13*a*). For parts expected to give long service, design is based on the stress that produces a linear creep rate of 1% per 10 000 h (e.g., for jet engine components) or per 100 000 h (e.g., for steam turbine components). Alternatively, the minimum creep rate is plotted against stress (or vice versa) on log-log paper.

To accelerate testing and to obtain design data for components that may be allowed to creep but must not fracture, tests are conducted at higher stresses to total failure. In *stress-rupture tests* the time required for rupture is determined at various stress levels (Fig. 2-13*b*).

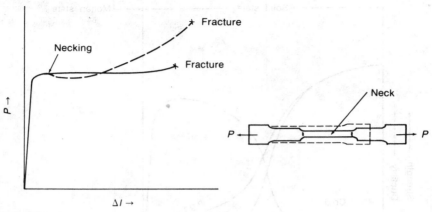

FIGURE 2-12
In the hot regime, a neck forms after little deformation, yet total deformation is substantial because the neck spreads over the entire gage length. The broken line is characteristic of tough polymeric materials (plastics).

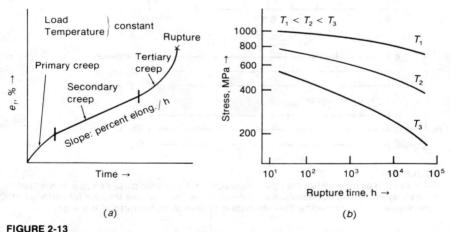

FIGURE 2-13
(a) In the hot regime, materials suffer deformation even under low stresses: they creep. Ultimately, fracture occurs. (b) Fracture (rupture) sets in faster at higher stress levels and temperatures.

Example 2-10

Temperatures in the turbine stage of jet engines are limited by creep deformation of the turbine blade (Fig. 1-4). From 1940 to 1960, gradual improvements in superalloys permitted gradually increasing temperatures; after 1960, a jump increase in temperature became possible with the introduction of internally cooled blades. First, cooler air was ducted from the compressor stage through holes provided in the length of the blade; later, cool air was passed over the surface of the blade to give a cooling boundary layer (Fig. 1-4b). The creep resistance of the blade material was improved too by novel manufacturing techniques such as directional solidification (Sec. 3-8-2). Yet higher temperatures can be obtained with ceramics (Sec. 6-3).

2-1-8 Residual Stresses

Stresses imposed externally on a component are not necessarily the only active stresses. As a result of manufacturing operations, there may also be stresses, called *internal stresses* or *residual stresses*, locked into the part or structure.

To understand how internal stresses arise, consider a cylindrical component. Assume that it had been made by joining a shorter tube and a longer, closely fitting core (Fig. 2-14a). Also assume that while joining was performed, the core was compressed to the length of the tube (Fig. 2-14b). Upon completion of the joint the core was released, whereupon the cylinder assumed a new length: The core wanted to expand to its original length, while the tube also wished to retain its original length. The mutually exerted forces must reach a balance. Since core and tube are of the same material and were chosen to have the same cross-sectional area, the cylinder will take up a length halfway between the original lengths of the tube and core (Fig. 2-14c). The tube will be expanded relative to its original length and will thus be subjected to (residual or internal) tensile stresses, while the core will be compressed and subjected to compressive stresses. Even though the cylinder is solid and sound, its surface is in tension. When the component is then put in tension, the applied stress is added to the surface residual stress. This would be dangerous for a material of limited ductility,

FIGURE 2-14
If a bar (*a*), longer than a tube into which it fits, is (*b*) joined to the tube while compressed to the same length as the tube, the assembly—released from compression—will (*c*) occupy an intermediate length, and surface tensile stresses will be generated.

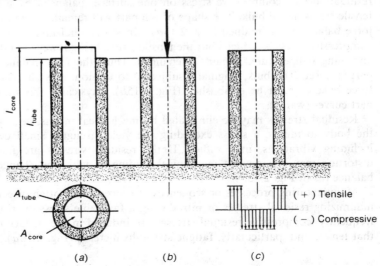

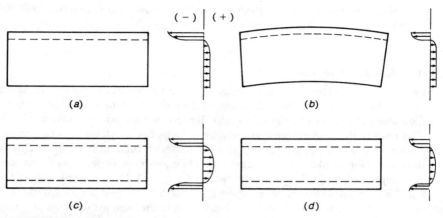

FIGURE 2-15
A rectangular part, (*a*) produced with a residual compressive stress on one of the surfaces, (*b*) will distort when subjected to stress-relief anneal. In contrast, (*c*) a workpiece with equal residual stresses on both surfaces (*d*) will retain its shape.

because any surface defects present would propagate much earlier, and the cylinder would fail in tension or fatigue at lower loads than a cylinder free of internal stresses. Stress corrosion may also occur.

Internal stresses can be reduced by heating to some higher temperature (*stress-relief anneal*). As visible in Fig. 2-11, the strength of materials drops at higher temperatures; therefore, internal stresses are reduced to the YS prevailing at the stress-relief anneal temperature. This may have undesirable consequences. Take, for example, a manufactured part of rectangular shape which has a high residual surface compressive stress on one surface, balanced by a much lower tensile stress in the bulk. The shape of the part will remain stable as long as the force balance is maintained (Fig. 2-15*a*). On stress-relief annealing, the surface compressive stress is reduced but the tensile stress, being lower than the YS at the annealing temperature, remains unchanged. Thus, the force balance within the part is upset: The bulk, originally subjected to tension, now shrinks, and a new force balance must be established (Fig. 2-15*b*). Physically, this means that the part curves (warps).

Residual stresses may be eliminated by mechanical means, i.e., by deforming the body to induce a stress exceeding the yield strength. Small deformations, including vibrations, may suffice. If the residual stresses are nonsymmetric, distortion may occur when mechanical loading or vibration changes the force balance (as in Fig. 2-15*b*).

Manufacturing processes or sequences of processes are often directed at either minimizing residual stresses or introducing a favorable stress distribution. Most frequently, compressive residual stresses are induced in the surface of a part so that tensile and, particularly, fatigue strengths increase (Fig. 2-10*a*). If compressive stresses are equal on both surfaces (Fig. 2-15c), the shape remains unchanged even if stresses are partially relieved (Fig. 2-15d).

Residual stresses can be determined by drilling out the center of or removing surface layers from the part and measuring the resulting dimensional changes. Nondestructive methods based on x-rays are also available.

2-2 PHYSICAL PROPERTIES

Physical properties other than strength are often of great importance and have to be satisfied by manufactured parts.

2-2-1 Tribological Properties

Tribology is the science, technology, and practice relating to interacting surfaces in relative motion. The term was coined in Britain in 1966 from the Greek (*tribein* = to rub), in recognition of the great importance of this interdisciplinary subject. It encompasses several fields.

Adhesion When two bodies are brought together into such intimate contact that atoms come within interatomic distances, strong bonds may develop; in the language of tribology, *adhesion* occurs, and it takes a measurable force to separate the two bodies. Adhesion between two solids may result in the formation of a strong joint (a pressure weld). This is desirable when the purpose is to make a composite structure, e.g., the nickel-clad copper used in U.S. coinage. It is undesirable when low friction and wear are to be secured.

Adhesion can be reduced by an appropriate choice of contacting materials. Generally, materials of greater hardness show less adhesion, and some materials

FIGURE 2-16
(*a*) The surface of materials differs from their bulk, showing evidence of prior processing and reactions with the atmosphere and other media. (*b*) Very few surfaces are truly smooth; most show peaks (asperities) and valleys.

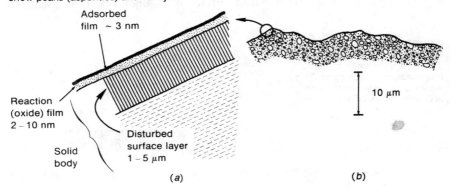

show inherently low adhesion (e.g., lead in contact with other metals, or PTFE in contact with most materials). Alternatively, a contaminant film may be interposed that prevents atomic bonding. Such contaminant films are provided by nature: Materials processed in the normal terrestrial atmosphere have surface films formed in contact with air. At the least, there are adsorbed films of gases and water vapor. On many surfaces, chemical reactions also occur: Most metals oxidize in air (Fig. 2-16a) and some polymers and ceramics undergo an irreversible change on contact with humid air. Thus, technical surfaces are never absolutely clean. Nevertheless, adhesion may still occur when relative sliding causes surface films to be broken through and when temperatures are high enough to cause migration (diffusion) of atoms from one body into the other.

Friction Mechanical components often slide against another body. The normal force P exerts a normal stress, which is usually called an *interface pressure* and is denoted p (instead of σ). The force required to move the body parallel to the surface is called a shear force F (Fig. 2-17); on dividing by the surface area A, a *shear stress* τ_i is obtained (the subscript i signifies the interface). By definition, the *coefficient of friction* μ is

$$\mu = \frac{F}{P} = \frac{\tau_i}{p} \tag{2-18}$$

On the micro scale, surfaces are not perfectly smooth but show hills (*asperities*) and valleys (Fig. 2-16b). Friction arises from the interaction of these asperities and from adhesion. In many applications it is necessary to minimize μ, either by the use of a lubricant, or by selecting materials that show inherently low friction, or both. Material pairs that show low adhesion usually—but not always—also give low friction. Manufacturing techniques can be directed to produce an internal structure in a component that is favorable for low friction (see Sec. 6-2-4). The surface texture (roughness and orientation) of the part, which is controlled by the manufacturing process, is also for prime importance.

Wear Economic losses due to wear are enormous. *Wear* is the progressive loss of substance from the operating surface of components. It is usually a

FIGURE 2-17
When two bodies are in contact, it takes a finite force to move them relative to each other. This allows us to walk, but it also accounts for much loss of energy.

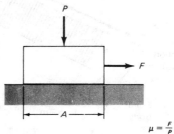

$$\mu = \frac{F}{P} = \frac{\tau_i}{p}$$

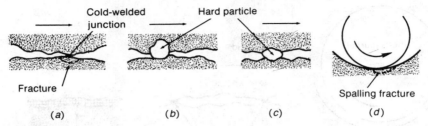

FIGURE 2-18
Wear is the progressive loss of material. It may be caused by: (*a*) the formation of adhesive junctions; (*b*) rubbing (abrasion) by a hard particle embedded in one of the mating surfaces; (*c*) abrasion by a hard particle trapped between surfaces; or (*d*) fatigue resulting from repeated loading.

consequence of the simultaneous action of several mechanisms, with one mechanism dominating. The most important ones are the following:

1 *Adhesive wear* occurs when a pressure-welded joint is stronger than one of the contacting bodies, and rips out a particle from that body (Fig. 2-18*a*).

2 *Abrasive wear* is caused by hard particles, whether they are within one of the contacting bodies (*two-body wear*, Fig. 2-18*b*) or are interposed between the two components (*three-body wear*, Fig. 2-18*c*).

3 *Fatigue wear* occurs when the repeated passage of a component over the surface of the other component leads to the separation of small particles from the surface, as in ball bearings (Fig. 2-18*d*).

4 *Chemical wear* is caused by chemical attack accelerated by the pressure and sliding prevailing in tribological contacts.

Numerous wear-evaluation techniques are available; they usually simulate, as closely as possible, the conditions encountered in service. Materials have been developed for high wear resistance. Alternatively, wear resistance can be increased by coating the surface with or transforming the surface into a material of greater wear resistance. Controlled, accelerated wear is intentionally induced in some manufacturing processes (Sec. 8-8).

Lubrication The purpose of *lubrication* is to reduce or, more accurately, to control both friction and wear. In addition to choosing material pairs that show low adhesion and friction, a separate substance (*lubricant*) is often interposed between the contacting surfaces. Lubricants may be grouped according to their mode of action:

1 Viscous fluids (such as mineral oils) introduced into a converging gap between moving surfaces (Fig. 2-19*a*) may build up a thick enough film to separate the two surfaces. Such *hydrodynamic lubrication* virtually eliminates wear, and friction is very low.

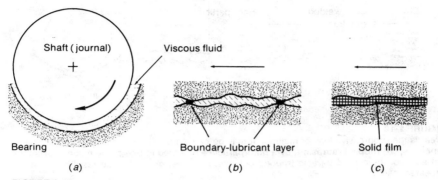

FIGURE 2-19
Friction and usually also wear may be reduced by (*a*) viscous fluids, (*b*) boundary lubricants attached to the surface by physical or chemical adsorption, or (*c*) solid films.

2 *Boundary lubricants* are organic substances (such as fatty acids) that adsorb on the surfaces of the contacting bodies and prevent adhesion, even when the fluid film thins out to the point where asperity contact takes place (Fig. 2-19*b*).

3 *EP* (extreme-pressure) *lubricants* are chemicals (usually organic materials with S, Cl, or P content) that react at elevated temperatures with metals to protect them from adhesion and rapid wear; often they also reduce friction.

4 *Solid lubricants* (such as graphite and molybdenum disulfide, MoS_2) separate the two surfaces with a layer of low shear strength (Fig. 2-18*c*). They lubricate even when sliding speeds are low or temperatures are high.

Lubrication is of critical importance in many manufacturing operations and in the service of mechanical devices. The successful operation of such devices demands very close control of dimensions and surface finish. This does not necessarily mean a very smooth finish; for example, operation of an internal combustion engine hinges on the controlled, crosshatched roughness produced on the cylinder bore.

2-2-2 Electronic Properties

While there is some relation between mechanical and tribological properties, electronic properties can be quite independent of either.

Electric current is conducted in most solids by the movement of electrons. In order to move, the electron must be given extra energy by the imposition of an electric field.

Metals can be visualized as consisting of positively charged centers (ions) bonded by freely moving electrons. Thus, metals are *conductors*, with resistivities on the order of 20×10^{-9} $\Omega \cdot m$. However, any crystal imperfections make passage of electrons more difficult, and maximum conductivity can be attained only if the manufacturing process sequence and the final condition of the part are closely controlled. Some materials become *superconductive* at temperatures close to absolute zero (up to 9 K in ductile materials, up to 22 K in intermetallics such as Nb_3Ge): their resistivity drops to zero. They are playing an increasing role in electromagnets, in electric power generation and distribution, and, potentially, in solid-state electronics.

Insulators are materials in which all—or virtually all—electrons are tied down in covalent, ionic, or molecular bonds. A large energy is required to break loose an electron (there is a large energy gap). Therefore, their resistivities are greater than $10^8 \ \Omega \cdot m$. They loose their insulating quality only at some critical field intensity, the *dielectric strength*.

Of great technical significance are solids that have a conductivity between that of insulators and conductors. They form the basis of the *semiconductor* industry and will be discussed in Chap. 10.

2-2-3 Magnetic Properties

Many materials are *ferromagnetic*: They contain *magnetic domains*. When these are readily reoriented under the influence of imposed magnetic fields, one speaks of *magnetically soft materials* (e.g., the core sheets in transformers or motors). In contrast, magnetically hard materials are difficult to remagnetize and *permanent magnets* retain the magnetic orientation imposed during manufacture (e.g., magnets of loudspeakers). Some materials can be magnetized repeatedly, opening opportunities for magnetic recording and data storage.

In all instances, not only composition but also manufacturing technologies must be closely controlled to obtain the desired properties.

2-2-4 Thermal Properties

Thermal properties such as the coefficient of expansion, specific heat, and latent heat of fusion and evaporation are important in many manufacturing processes and service situations, and their values may be found in handbooks.

In common with other thermal properties, *thermal conductivity* is an intrinsic, structure-independent material property. However, *heat transfer* in a structure often also depends, in addition to conduction through the structure, on the movement of some heated material, such as gas or other fluid (convective heat transfer), and on radiation. The purpose of manufacturing is often the production of a composite structure in which heat transfer by these means is either promoted or hindered.

Example 2-11

Internal combustion engines, in common with all heat engines, become more efficient at higher operating temperatures. However, temperature limits are set by lubricants and the materials of construction. Therefore, most engines are cooled with a water-based (aqueous) circulating fluid, from which heat is extracted with the aid of a sophisticated manufactured product, the radiator. In this, the coolant is pumped through parallel tubes from which heat is extracted through fins. Fins are designed and manufactured into often complex shapes so that air flowing over them removes heat most efficiently. Heat exchangers are vital to the operation of refrigerators, air conditioners, industrial and domestic furnaces, solar collectors, and heat sinks for computers; all of these products represent different manufacturing challenges. At the other end of the spectrum, heat transfer is minimized by insulating structures such as fiberglass mats, foamed plastics, and furnace refractories.

2-2-5 Optical Properties

Manufacturing processes are controlled to endow manufactured parts with desirable optical attributes, for both aesthetic reasons (appearance) and technical function.

The surface appearance of parts is controlled by manufacturing techniques to reflect light in a desirable manner. A very smooth finish reflects light at the same angle as the angle of incidence (*specular reflection*, as that given by a mirror-finish surface), whereas a rough surface reflects light randomly (*diffused reflection*, as given by a matte finish).

Some materials absorb light and are *opaque* (not transparent). Others, such as amorphous polymers, glasses, and ceramics are *transparent*. If, by appropriate manufacturing techniques, internal reflecting surfaces are created, the same materials become *translucent* (partially transparent) or opaque (see Sec. 6-5-1).

2-3 CHEMICAL PROPERTIES

Many manufactured structures are expected to survive for prolonged periods of time while being exposed to the atmosphere or other gases or liquids. Their deterioration by chemical or electrochemical action (*corrosion*) is governed primarily by the choice of materials, but is also affected by the method of manufacture.

The aim is usually that of avoiding harmful situations. For example, residual stresses could lead to accelerated corrosion and also to stress-corrosion cracking (Sec. 2-1-8); steel screws used for joining brass sheet would corrode; some stainless steels loose their corrosion resistance if slowly cooled from the welding temperature. On the positive side, steps can be taken to protect a structure from corrosion, for example by zinc coating a steel component.

Corrosion resistance may be undesirable in manufacturing when the function of a lubricant requires a chemical reaction to take place. For example, the corrosion resistance of stainless steel necessitates the use of special lubrication techniques in forming processes.

2-4 GEOMETRIC ATTRIBUTES

Design and manufacturing interact most directly in defining the shapes and dimensions of parts and structures.

2-4-1 Shape

The *shape* of a part is dictated, first of all, by its function. However, not all manufacturing processes are equally suitable for developing a given shape: often, the shape of the part can be changed—without affecting its function—so that it becomes easier to make by one technique or other. Shape also affects the complexity of machine motions and controls required for making the part.

When the movement of tool or workpiece is restricted to a single axis, one speaks of *one-axis* or *single-axis* (usually denoted as Z axis) *movement* or *control* (as, for example, when drilling a hole in a clamped workpiece, Fig. 2-20a). Table movement requires *two-axis* control (usually referred to as X and Y axes, Fig. 2-20b); programmed movement in the Z axis makes it into a *three-axis* machine. (When movement in the Z direction is simply on–off and proceeds at some preset rate, one usually speaks of two-and-a-half–axis control). Swiveling the tool (or table) would add the fourth and fifth axes (Fig. 2-20c). Every joint in the toolholder or table adds a further freedom (axis) of movement and permits more complex shapes to be made, but at the expense of more complex and expensive machinery and control. Thus, one of the purposes of design is to facilitate manufacture and assembly with minimum complexity. This is especially important with the trends toward automation and the use of robots. The human body has dozens of freedoms of movement and, while it might be possible to build machines and robots of similar versatility, *it is easier and cheaper to accommodate the limitations of machinery by appropriate design of the parts and assemblies the automated equipment will have to handle.*

There are some shape features that immediately set certain limitations:

1 Axial symmetry is, in many ways, the simplest because the shape can be generated by rotating the part or tool (around the Z axis).

2 Parts of nonrotational symmetry call for a minimum of two-axis control, and spatial curvature can be followed only with three- (or more) axis control.

FIGURE 2-20
Tools and workpieces may be moved and control may be exercised along (a) one; two; (b) three; or (c) several axes.

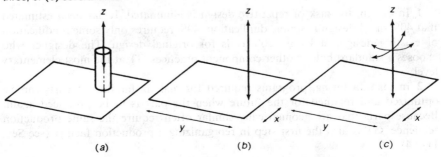

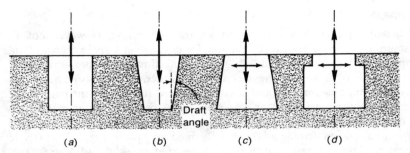

FIGURE 2-21
(*a*) A straight-walled pocket is often easy to produce but a (*b*) draft angle may be required to withdraw a tool. (*c*), (*d*) Undercut shapes require multiaxis control or complex tooling.

3 A surface in line with the tool movement (Fig. 2-21*a*) can be made with one-axis control although, if the tool is difficult to withdraw, a *draft angle* (Fig. 2-21*b*) may be necessary. Undercut shapes (Fig. 2-21*c* and *d*) require control in more than one axis.

With the spread of CAD/CAM, there is a much keener awareness of the need to *design with manufacturing in mind*. It is important not to fix the part configuration too early in the design process, because the most economical manufacturing process may thus be excluded. Once the optimum process is identified, *the part shape must be optimized for that process*. This approach is aided and made more economical by the application of group technology.

2-4-2 Group Technology

Group technology (GT) is a very broad concept; its essence is the recognition that many problems have similar features and, if these problems are solved together, great efficiency and economy result. In applying the concept to manufacturing, individual parts are analyzed in terms of commonalities of design features as well as manufacturing processes and process sequences. This way *families of parts* can be identified and economies are assured:

1 In design, the task of repetitive design is eliminated. It has been estimated that 40% of all design is simple duplication, 40% requires only some modification of existing design, and only 20% calls for original design. The designer who chooses a standard bolt or other component practices GT at the most elementary level.

2 In manufacturing, programs required for making families of parts can be optimized and retained for the future when the part is to be produced again. Because parts that are geometrically similar often require the same production sequence, GT is also the first step in reorganizing a production facility (see Sec. 11-2-4).

3 In production planning, cycle time estimation is accelerated, workpiece movement is rationalized, and process design is simplified. Cost estimation is facilitated too.

In recent years, the introduction of the computer has made GT particularly attractive, because programs relating to the design of standard elements such as solid and hollow cylinders, rectangular blocks, cones, etc., can be retained in memory and easily combined and modified for a large variety of part configurations. Similarly, process details can be filed away for later use, with modifications, if necessary.

The first step in GT is the *classification of parts* into families. Several approaches can be taken:

1 *Experience-based judgment.* This works only in the simplest cases. The part is classified into a family by visual judgment of its shape, and the classification is further refined from a knowledge of the usual production sequence. There is no assurance that such a sequence is actually the optimum one.

2 *Production-flow analysis* (PFA). Information relating to the sequence of operations in an existing plant is contained in routing sheets or routing cards. Thus the flow of parts through various operations can be easily extracted. Parts that are made by identical operations form a family. Good engineering judgement will tell whether parts on which some additional operations are performed should be included in the family. A critical examination may also reveal that some parts falling outside of a family could be made more economically by adopting the production sequence typical of the family. Parts that are made by the same processes but in different sequences may still logically be classified into the same family, but the flexibility of the production system will have to be greater to allow the return of the part to a previous operational position.

3 *Classification and coding.* This is a more formal exercise. There is no universally accepted system, and there will perhaps never be one. Some systems are more suitable for design, others for parts made by specific processes (casting, forging, machining, etc.), yet others aim at some universality. They all start from a classification of basic workpiece shapes (something similar is done in Fig. 12-1 although most of the commonly used systems are limited to a smaller variety of shapes). Part codes are usually made up of several (sometimes up to 30) digits which define various geometrical features as well as composition and requisite surface properties. Further digits may be added to define processes, process parameters, and processing sequences. Some computer-based systems facilitate coding by guiding the operator through the necessary steps in a conversational mode; others use the data base generated by CAD to help in assigning code numbers.

Classification can be powerful but is time-consuming and not free of problems. It can pay off in the design stage by identifying parts that could either be made identical or redesigned to fit into a larger family, and in the production stage by allowing the rational organization or reorganization of the existing plant and the planning of new production facilities.

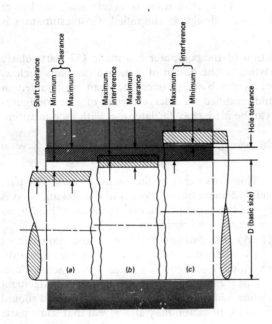

FIGURE 2-22

In the basic hole system the diameter of the hole is chosen from a table of preferred sizes and then the tolerances are applied to create (a) clearance, (b) transition, or (c) interference fits.

2-4-3 Dimensional Tolerances

No manufacturing process can make a part to exact dimensions (we will come back to this point in Sec. 12-2-3 after discussing various processes). Therefore, maximum and minimum *limits of dimensions* (length or angle) are specified with two goals in mind:

1 The limits must be set close enough to allow functioning of the assembled parts (including interchangeable parts).

2 The limits must be set as wide as functionally possible, because tighter limits usually call for expensive processes or process sequences. The single most important cause of excessive production costs is the specification of unnecessarily close dimensional limits.

The designer specifies dimensions and the *allowance*, i.e., the difference in dimensions necessary to ensure proper functioning of mating parts (the allowance is called also the *functional dimension* or *sum dimension*). This is best illustrated on the example of a shaft fitting into a hole (Fig. 2-22). The *basic size* of one of the mating parts is first defined, from tables of preferred sizes if possible at all, so that standard shafts or tools can be used. In principle, the basic size could be assigned to either the hole or the shaft. In practice, holes are often manufactured with some special tool (drill, reamer, punch) and are, furthermore, difficult to measure while the hole is being made; therefore, the *basic hole system* is generally used. The allowance (the minimum clearance or maximum interference) is then specified to satisfy functional requirements.

The position of the *tolerance zone* relative to the basic dimension defines the type of fit. *Clearance fits* allow sliding or rotation (Fig. 2-22a). *Transition fits* provide accurate location with slight clearance or interference (Fig. 2-22b). *Interference fits* ensure a negative clearance (interference) and are designed for rigidity and alignment, or even to develop a specified pressure (shrink pressure) on the shaft (Fig. 2-22c).

The next step is determination of the *tolerance*, that is, the permissible difference between maximum and minimum limits of size. Tolerance can be expressed with respect to the basic size as deviation in both upper and lower directions (*bilateral tolerancing*) or in only one direction, if the consequences of inaccuracy in that direction are less dangerous (*unilateral tolerancing*).

Experience has taught that, in most manufacturing processes, dimensional inaccuracies are proportional to the cube root of the absolute size (denoted D for diameter, in units of mm or in). The American National Standards Institute (ANSI) standard ANSI B4.1-1967, R1979 gives tables for 8 classes (comprising 34 subclasses) of fits, ranging from loose fit to force fit. The International Standards Organization (ISO) recommendation ISO R286-1980 and the corresponding ANSI R4.2-1978 are based on the tolerance unit i

$$i = 0.45D^{1/3} + 0.001D \qquad (2-19)$$

and the grade of tolerance is expressed as the standardized multiple of i (within the grades 5–16, the tightest tolerance IT5 implies a standard tolerance of $7i$, while the loosest, IT16, implies $1000i$). The actual value of tolerance may be obtained from tables. There are also computer-aided techniques for assigning dimensions and tolerances.

FIGURE 2-23
Example of a close-running fit, according to American National Standard preferred basic hole metric clearance G8/f7 (*ANSI B4.2-1978*).

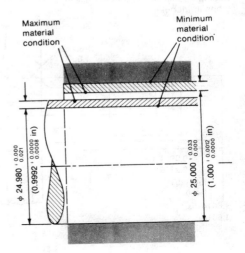

The example given in Fig. 2-23 is for an assembly which is suitable for running with a mineral oil (it is a hydrodynamic bearing). Capital letters show the position of the tolerance zone relative to the basic dimension of the hole and lowercase letters show it for the shaft.

2-4-4 Shape and Location Deviations

For a part to function properly with respect to other components, it is often necessary to place further restrictions on the location (position) of geometric features and on geometric properties such as concentricity, runout, straightness, flatness, parallelism, and perpendicularity (*geometric tolerancing*).

A simple example of a rotating hollow shaft is given in Fig. 2-24. It will run in journal bearings at its end and support a force-fitted flywheel at its center; appropriate tolerances are given for the A and C portions of the shaft. The nonmating surfaces B could have greatly relaxed tolerances, and dynamic balance could be attained by local removal of material at appropriate points, but it is usually preferable to apply tighter tolerances to B since it will be machined with the rest of the shaft anyway. Of equal importance is specification of hole concentricity: maintenance of diametral tolerances alone would not necessarily ensure a uniform wall thickness. Straightness must be specified too; otherwise the shaft would not fit or the assembly would be out of balance. The same considerations apply to parts not shown in Fig. 2-24: The journal bearings would need restrictions on OD (for fit into the machine frame), on ID (for ensuring the running clearance), and on concentricity for wall thickness variation (to ensure alignment). On a part of nonrotational symmetry, other qualities—such as flatness or freedom from bow—may be specified.

FIGURE 2-24
Example to show the functionally important dimensions to be held in manufacturing (dimensions in inches).

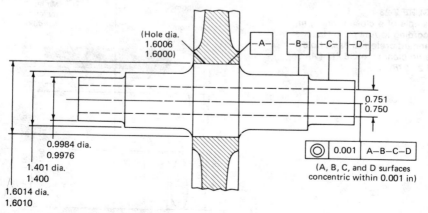

2-4-5 Surface Roughness and Waviness

Few surfaces are smooth and flat (or of cylindrical or other pure geometrical shape).

On the microscopic scale, surfaces exhibit *waviness* and *roughness*. The surface profile can be measured and recorded. For easier visualization, recordings are usually made with a larger gain on the vertical axis (Fig. 2-25). This gives a distorted image with sharp peaks and steep slopes; in reality the peaks (asperities) have gentle slopes of typically 5–20° inclination (as in Fig. 2-16b). The traces or, more frequently, the signal obtained from the profilometer may be processed electronically or, after digitization, in a computer, to derive various values for a quantitative characterization of the surface profile. Of the various measures given in ANSI B46-1-1978, the following are most frequently used:

1 R_t is the *maximum roughness height* (the height from maximum peak to deepest trough). It is important when the roughness is to be removed, for example, by polishing. Often a more meaningful figure is obtained by taking the average height difference between the 5 highest peaks and 5 deepest valleys within the sampling length (10-point height, R_z).

2 A line, drawn in such a way that the area filled with material equals the area of unfilled portions, defines the centerline or mean surface. The average deviation from this mean surface is called the *centerline average* (CLA) or *arithmetical average* (AA), denoted also as R_a

$$R_a = \frac{1}{l} \int_0^l |y|\, dl \quad \text{or} \quad R_a = \frac{y_1 + y_2 + y_3 + \cdots + y_n}{n} \tag{2-20}$$

3 The *root mean square* (rms) value R_q is frequently preferred in practice and also in the theory of contacting surfaces

$$R_q = \left[\frac{1}{l} \int_0^l y^2\, dl \right]^{1/2} \quad \text{or} \quad R_q = \left(\frac{y_1^2 + y_2^2 + \cdots + y_n^2}{n} \right)^{1/2} \tag{2-21}$$

R_q is closely related to R_a ($R_a = 1.11 R_q$ for a sine wave) and, for technical surfaces, the relationship between various values is fairly well defined (Table 2-2).

FIGURE 2-25
The roughness of technical surfaces can be revealed by various techniques; typical recordings are made with a larger magnification in the direction perpendicular to the surface.

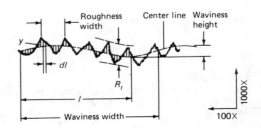

TABLE 2-2
APPROXIMATE RELATIONSHIP OF SURFACE
ROUGHNESS VALUES

Type of surface	RMS/CLA	R_t/R_a
Turned	1.1	4–5
Ground	1.2	7–14
Lapped	1.4	7–14
Random	1.25	8

4 *Skewness* expresses the distribution of roughness heights and is a quantitative measure of the "fullness" of the surface (Fig. 2-26). The Abbot curve shows the load-bearing area available when cuts are taken at various levels from the top of the profile.

Convenient units of measurement are the micrometer (μm) or nanometer (nm) and the microinch (μin).

$$1 \mu\text{in} = 0.025 \mu\text{m} = 25 \text{ nm} = 250 \text{ Å}$$

$$1 \mu\text{m} = 40 \mu\text{in}$$

The finer details of surface roughness are superimposed on larger-scale periodic or nonperiodic variations (waviness, Fig. 2-25). In measuring the surface

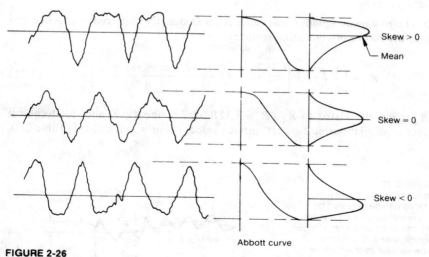

Abbott curve

FIGURE 2-26
For the same peak-to-valley roughness height, surfaces may have very different profiles, resulting in a skewing of the roughness-height distribution. (*After ANSI B46.1-1978, ASME, New York, 1978.*)

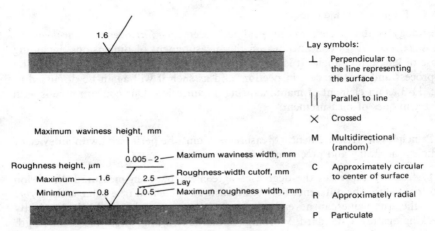

FIGURE 2-27
Characteristics of the surface finish are described by standard symbols (the example is given in SI units, with roughness in μm R_a).

roughness, the waviness is usually filtered out by electronic processing of the signal, although the allowable waviness is specified and measured (in units of mm or in) when it is functionally important.

On drawings, the *roughness limits* are given by a check mark written over the line to which the roughness designation applies (Fig. 2-27). A single roughness number indicates an upper limit, below which any roughness is acceptable; if a minimum roughness is required, two limits are shown. The waviness, when important, is limited by a number over the horizontal line of the check mark. Surfaces usually exhibit a topography characteristic of the finishing process (see Sec. 12-2-3). The characteristic *directionality* (*lay*) is indicated by a symbol placed under the check mark.

There is a close relationship between roughness and tolerances. A good rule of thumb is that the maximum roughness height R_t (and waviness, if any) should be about $\frac{1}{3}$ to $\frac{1}{2}$ of the tolerance, unless the fit is a forced fit and the surface roughness can be at least partially smoothed out in the fitting process. Remembering that $R_t = 10R_a$, a roughness value of 3.2 μm (125 μin) R_a would be too coarse for a tolerance of 0.025 mm (0.001 in), and a roughness of maximum 0.8 μm (32 μin) R_a should be specified for such a tight tolerance.

It must be recognized that the same numerical R_q or R_a values may be obtained on surfaces of greatly differing profiles, and that highly localized troughs add very little to the average values. Therefore, averages are often inadequate to describe surfaces for specific applications, and the problem of surface characterization remains a challenge. Nevertheless, the manufacturing process must be capable of providing a surface suitable for the intended function of the part, and further quantitative or descriptive terms can be and often are used to elaborate on the required finish.

2-4-6 Engineering Metrology

Metrology is the science of physical measurement. *Engineering metrology* (or *industrial metrology*) concentrates on the measurement of dimensions, including those of length and angle. It is of prime importance for the control of quality by in-process and post-process inspection, and as such it will again be discussed in Sec. 11-3 as an element of manufacturing organization. Our concern here is with the techniques of measurement.

Principles of Measurement   Measurement must be performed with a device of sufficient accuracy and precision.

1 *Accuracy* expresses the degree of agreement between measured dimension and true value. The difference between measured value and true value is the *error*; since the true value can never be known, the error can be established only by checking against a standard. Working standards (working gages) used for length measurement in the workplace are checked against reference standards which, ultimately, are checked against national standards. However, the SI unit of length, the meter, is now defined by the vacuum wavelength of the orange line of krypton 86, with a precision of 1 part in 10^9.

2 *Precision* is the degree of repeatability of measurement (Fig. 2-28). A rough rule of thumb is that precision should be ten times better than the tolerance; a better, statistical definition will be given in Sec. 11-3-3.

A good measuring instrument possesses a number of attributes:

1 *Sensitivity* is the smallest variation that the device can detect. It is called *resolution* when reading is digital or is made against a scale. A scale subdivided

FIGURE 2-28
Mechanical properties, dimensions, and other measured variables always show some dispersion. The distribution in (*a*) is accurate but imprecise; in (*b*) precise but inaccurate; only (*c*) is both accurate and precise.

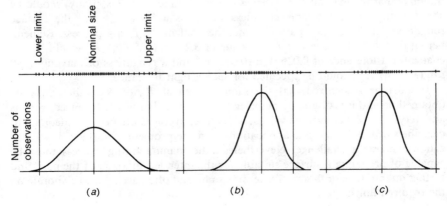

into increments smaller than the device can detect gives only spurious resolution; the accuracy of the device should be several times better than the smallest graduation of the readout.

2 *Linearity* affects readings over a specified measurement range. Even if an instrument is set (calibrated) against a standard at some point in the range, nonlinearity affects other points in the range.

3 *Repeatability* determines the possible highest precision that can be achieved under well-controlled conditions. The instrument must be capable of repeating readings to the same accuracy to which it can be read.

4 *Stability* expresses resistance to drift that would reduce both accuracy and precision and would necessitate frequent recalibration.

5 *Speed of response* is critical when a transient variable is to be measured, usually during production.

6 *Feasibility of automation* is important in many applications.

Repeated measurements are subject to statistical variations of two kinds:

1 *Assignable (systematic) errors* are measurable and often controllable. In addition to errors inherent in the device, temperature variation is the main source of systematic error. If tolerances are tight, the temperature of the part must be uniform and known so that an allowance can be made for thermal expansion; in post-production measurements this is best ensured by taking the part to a climate-controlled room and allowing it to equalize with the temperature of the measuring device.

2 *Random errors* stem from human error (inaccurate scale reading, improper setup, etc.) and from sources such as dust and rust. Again, a climate-controlled room helps with its filtered air and controlled humidity.

Measurements are often made with reference to a datum surface such as a flat, hole, or shaft. It must be chosen with due regard to the method of manufacture and inspection.

Example 2-12

An AISI 1020 steel shaft of 100.00-mm diameter is made by turning on a lathe. The part heated up to 70 °C during cutting. Can the dimension be measured to the nearest 0.01 mm without allowing for the temperature increase?

Unless otherwise stated, dimensions refer to room temperature (20 °C or 68 °F). The steel is a carbon steel of 0.2% C content. The coefficient of linear thermal expansion is 11.7 μm/m·C (MHDE, p. 1.51). Hence the diameter will increase by $(70 - 20)(0.100)(11.7) = 58.5 \mu$m, which is six times the desired precision of measurement.

Gages   In the narrower sense, the term *gage* refers to hardened steel, tungsten carbide, ceramic (glass), etc., bodies that are manufactured to close tolerances. They can be fixed or adjustable. Once set, the adjustable gage is also used as a

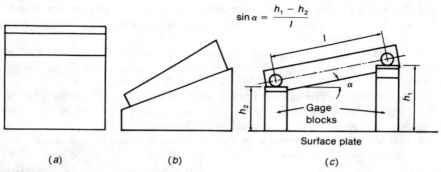

FIGURE 2-29
Hardened steel (*a*) gage blocks, (*b*) angle blocks, and (*c*) sine bars that are extensively used for comparative gaging purposes.

fixed gage. There are several types:

1 *Gage blocks* still are the primary length gages in most applications. They are made in sets that allow building any dimension by *wringing* (a sliding–twisting motion) of several blocks (Fig. 2-29*a*). Adsorbed moisture or oil films on the mating measuring surfaces have negligible thickness but provide sufficient adhesion to handle the built-up column as one unit; a twisting motion is again needed for separating the blocks. Gage blocks come in several grades. For a nominal dimension of 1 in, tolerances (expressed in microinches) are as follows: Grade 3 gages, used directly in production, $+8$, -4; Grade 2 sets, used as inspection and toolroom standards, $+4$, -2; Grade 1 laboratory gage blocks, for the calibration of other gages and indicating instruments, $+2$, -2; Grade 0.5 reference gages, used only in work of the highest precision, $+1$, -1.

2 *Angle blocks* (Fig. 2-29*b*) are constructed according to the same principles as gage blocks. *Sine bars* (Fig. 2-29*c*) are used in conjunction with gage blocks to create any angle.

3 Other length gages include *length bars* (measuring rods, Fig. 2-30*a*) and fixed (Fig. 2-30*b*) and adjustable (Fig. 2-30*c*) *gap gages*.

4 *Plug and ring gages* are used for the measurement of diameters (Fig. 2-31). They are usually of the GO–NOT GO type. The GO limit gage is the negative (the reverse replica) of the dimension at the *maximum material condition* (see Fig. 2-23), indicating that the mating parts can be assembled. The NOT GO limit gage is made to the dimension of the *minimum material condition* and rejects parts that are outside the tolerance. There are three problems with these gages; first, they themselves can be made only to certain tolerances, resulting in the rejection of good parts or passing of bad ones; second, they are subject to operator judgement; third, they give no information on the variations of part dimensions within the limits, and are thus of limited use for statistical production control. (Also, in the form shown, they violate Taylor's principle: Only the GO gage should be of

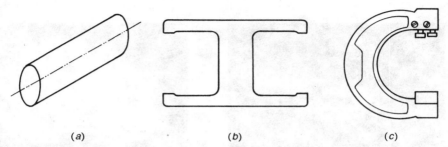

(a)   (b)   (c)

FIGURE 2-30
Comparative measurement of length dimensions is possible with (a) length bars, (b) fixed gap gages, or (c) adjustable gap gages.

full form to check *both* size and geometric features, whereas the NOT GO gage should check only one linear dimension.)

5 Multiple-diameter gages such as *thread plugs* (Fig. 2-32a) and rings, spline gages, etc., check the combined effect of several parameters. *Contour gages or templates* (including straightedges and radius gages, Fig. 2-32b) test the coincidence of shapes by visual observation or optical magnification. In the broader sense, *surface plates* also come in this category; they are often used for setting up other gaging elements and are made of some very stable material, such as granite, to specified flatness.

6 *Assembly gages* test not only dimensions but also alignment and coaxiality.

Graduated Measuring Devices   These allow the reading of a dimension against a scale. Some have a zero point, others read only relative displacement. Relative to fixed gages, their great advantage is that information on the *distribution of*

FIGURE 2-31
Diameters of holes can be checked with plug gages and diameters of bars with ring gages.

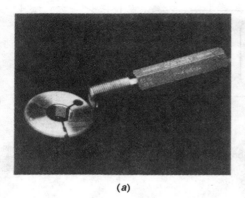

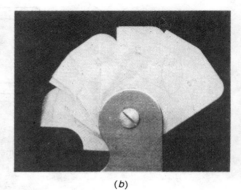

(a)   (b)

FIGURE 2-32
More complex configurations can be checked with (a) thread plugs and rings or (b) radius gages.

dimensions within a batch is obtained. For best results, Abbe's principle should be observed: The line of scale should coincide with the line of measurement.

1 *Line-graduated rules and tapes* limit reading to the nearest division.

2 The use of a *vernier* increases the sensitivity of *caliper gages* (Fig. 2-33a) to 25 μm (0.001 in) and that of *micrometers* (Fig. 2-33b) to 3 μm (0.0001 in). Abbe's principle is satisfied in measuring with a micrometer, but with caliper gages the line of measurement (between the jaws) is separated from the scale.

3 When two *diffraction gratings* (closely spaced parallel lines on a glass surface) are superimposed at a slight inclination, they produce interference fringes, the location of which depends on the relative position of the gratings (Fig. 2-34a). The number of fringes can be counted electronically, to give a sensitivity of 5 μm (0.0002 in).

4 *Linear digital transducers* (Fig. 2.34b) can be used to transmit pulses by electronic, photoelectric, or magnetic means, to a resolution of 4 μm (0.0002 in). A rotary pulse-generation encoder can be used for angular measurements and, with a rack-and-pinion or slide-wire movement, also for linear measurements.

5 *Numerical encoding disks* (Fig. 2-35) provide, with the appropriate interface, direct readout or, if desired, input to NC controls.

6 Solid-state electronic devices that convert light to an electrical signal (*photo-detector diodes* and *charge-coupled devices*, CCD) sense the presence or absence of light and, arranged in a linear array, offer resolutions of 3 μm (0.0001 in) or better when used alone or in TV cameras.

7 *Toolmaker's microscopes* are optical microscopes equipped with cross-slide stages driven by micrometers, eyepieces with cross hairs for length measurement, and protractor eyepieces for angle measurement. They can also be used for checking and measuring the form (shape) of parts.

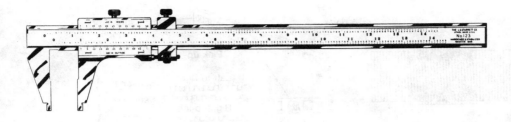

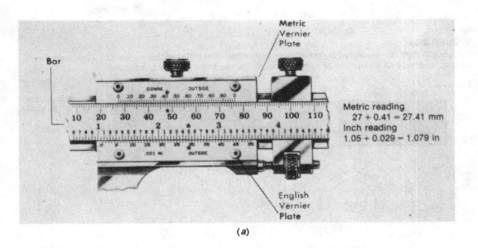

Metric reading
27 + 0.41 = 27.41 mm
Inch reading
1.05 + 0.029 = 1.079 in

(a)

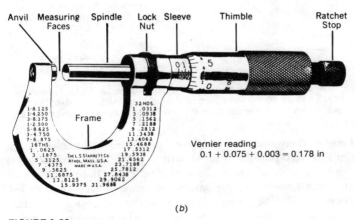

Vernier reading
0.1 + 0.075 + 0.003 = 0.178 in

(b)

FIGURE 2-33
A vernier allows reading to some fraction of the smallest division on the main scale. To read, the line of the vernier scale which coincides with a line on the major scale is read, and the reading is added to the basic reading on the major scale of the (a) vernier caliper or (b) vernier micrometer. There are no verniers on digital-readout instruments. (*Courtesy The L. S. Starrett Co., Athol, Mass.*)

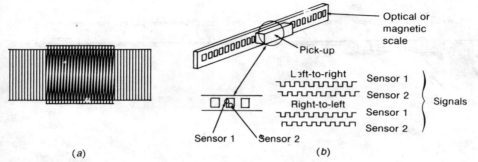

(a)

(b)

FIGURE 2-34

(a) Length may be measured by counting the number of interference fringes. (b) The direction of displacement of an optical or magnetic scale is sensed by two transducers.

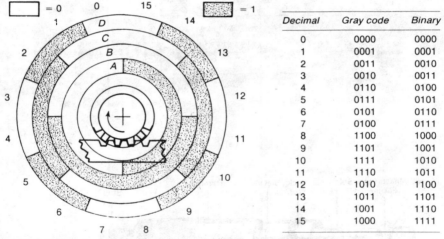

Decimal	Gray code	Binary
0	0000	0000
1	0001	0001
2	0011	0010
3	0010	0011
4	0110	0100
5	0111	0101
6	0101	0110
7	0100	0111
8	1100	1000
9	1101	1001
10	1111	1010
11	1110	1011
12	1010	1100
13	1011	1101
14	1001	1110
15	1000	1111

FIGURE 2-35

A numerical encoding disk, driven by a rack and pinion, provides a digital signal for control purposes. The transducer may be a photodiode, magnetic pickup, or electric contact. Ambiguous readings are avoided in this four-bit encoder by the use of the Gray code, which is then converted into binary code.

Comparative Length Measurement *Indicators* measure only the deviation from a zero position; the zero position is set up with a setting gage (master gage) that is chosen to give the nominal size of the part (Fig. 2-36). If an indicator of sufficient sensitivity is used and a suitable datum (reference surface) is provided, much relevant information can be obtained not just on length and its variation from part to part but also on runout, alignment, etc. The indicator can be of several kinds:

1 *Dial indicators* are purely mechanical devices that convert linear displacement into rotation (e.g., with a rack-and-pinion movement) and amplify it with a gear train to increase sensitivity to 1 μm (50 μin). Some gages have built-in

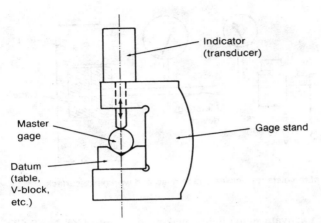

FIGURE 2-36
Dimensions may be read by gages which are equipped with an indicator or some form of position transducer.

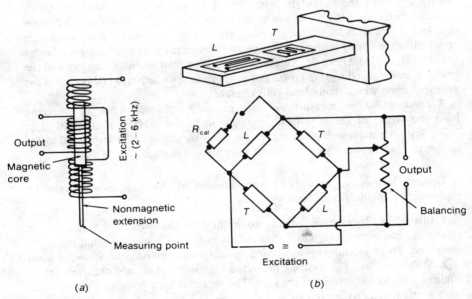

(a)   (b)

FIGURE 2-37
Displacements may be obtained from (a) the position of a differential transformer or (b) the deflection of a beam to which strain gages, connected into a Wheatstone bridge, have been attached.

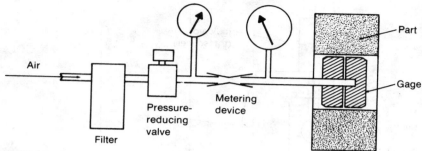

FIGURE 2-38
Air gages give a measure of the distance between gage head and workpiece surface.

electrical contacts which activate signal lights, making them into GO–NOT GO gages.

2 *Electronic gages* (transducers) transform mechanical movement into an electrical signal according to various principles. Frequently used is the differential transformer (Fig. 2-37a), the output of which is zero when the movable core is exactly centered; the output is proportional to displacement elsewhere. Other transducers measure the change in capacitance, yet others convert the deflection of a leaf spring into an electrical signal: strain gages (resistance wire loops) are attached to the spring with an adhesive and connected into a Wheatstone bridge circuit (Fig. 2-37b). Upon deformation, the resistance of the wire changes and the bridge is unbalanced to give an output proportional to deflection (this is also the principle upon which many load cells operate).

3 *Pneumatic gages* measure the back pressure generated when air emerging from the orifice of the gage head impinges upon the surface of the part (Fig. 2-38). Within a narrow dimensional range, pressure change is proportional to the size of the gap between gage head and workpiece surface.

Optical Devices Light waves have a number of characteristics that can be exploited for engineering metrology.

1 Instead of observing in a microscope, the magnified shape of a part can be projected onto a screen on which dimensions as well as angles and shapes can be measured. These instruments are called *optical projectors* or *optical comparators*.

2 A *light-sectioning microscope* projects a narrow band of light obliquely (at an angle of 45°) to the surface of the part. The reflected light gives an outline of the cross section, suitably magnified.

3 Visible light has wavelengths from 400 nm (violet end) to 760 nm (red end). When an optical flat (glass or fused quartz disk with parallel flats, true to within 50 nm) is placed at a slight angle to the workpiece surface and monochromatic light is beamed on it, light and dark bands (*interference fringes*) become visible to the eye (or a photodetector). The reason for this is that light rays from the

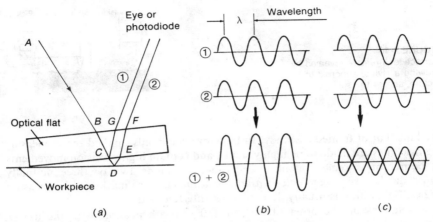

FIGURE 2-39
Flatness of a surface is obtained from (a) interference fringes produced with the aid of an optical flat. Light is split into two beams: (b) when in phase, they reinforce each other and a light band appears; (c) when out of phase, they cancel and a dark band appears.

monochromatic light source are reflected from both the bottom surface of the optical flat and the surface of the workpiece (Fig. 2-39a). The two reflected rays interact; the ray reflected from the workpiece surface travels a path that is longer by the distance CDE. If this distance equals a wavelength λ (or an integer multiple of it, $n\lambda$), the two rays reinforce each other (Fig. 2-39b) and the observer sees a light band. Conversely, if the distance is $\lambda/2$ (or $\lambda/2 + n\lambda$), the rays cancel and a dark band appears (Fig. 2-39c). Since DE is practically equal to CD, the fringes repeat every time the height between flat and workpiece surface changes by $\lambda/2$. By counting the number of fringes, the total distance (or the height of the workpiece from a reference plane) can be measured. (Note that refraction and phase changes are ignored here because they do not affect the argument.) Helium-neon lasers are increasingly used as the light source.

Interferometry is also useful for checking the flatness of surfaces: Fringes are straight, parallel, and evenly spaced when the surface is flat.

4 Highly collimated laser beams can be used for the noncontacting measurement of dimensions. In one approach, the workpiece is placed in the light path between source and photodetector. The beam sweeps at a preset rate, hence the length of time for which the light is cut off is a measure of the dimension. In another approach, the beam is split, and interference with the beam reflected from the surface of the part gives the distance on a digital display, to a resolution of 2.5 μm (0.0001 in) or better.

Measuring Machines The term *measuring machine* is used to denote structures built with extreme care to provide supports for transducers relative to a reference surface or reference axes. In a sense, a micrometer is also a measuring

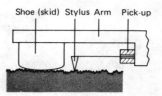

FIGURE 2-40
Surface features are revealed by
drawing a stylus, attached to a
pickup, across the surface.

machine, but of limited accuracy. In the more commonly applied sense, measuring machines are made to be highly stable, and contain high-precision movements to permit measurement along a single axis or along two or three mutually perpendicular axes (*coordinate measuring machines*). Some machines also measure angles. Calibration is usually done by laser interferometry.

Resolutions on the order of 250 nm (10 μin) are possible with the use of mechanical, electronic, or optical readout instruments. The readout is usually supplied in a digital form for processing by computer. With suitable drives for the measuring transducers, coordinate measuring machines are used extensively for the tracing of complex surfaces. For noncontact measurement, video-image processing and optical transducers are available.

Measuring machines can be used for layout prior to machining and for checking dimensions after machining. They can be linked to computers to perform automatic measurements, sometimes in conjunction with a computer-controlled production cell (flexible manufacturing cell, Sec. 11-2-4).

Surface-Roughness Measurement The most common surface-roughness measuring instrument is based on the principle of the record player (Fig. 2-40). An arm with a reference rest is drawn across the surface, while a stylus follows the finer surface details. The surface profile can be recorded (as in Fig. 2-25) and

FIGURE 2-41
Light-interference microscopy is used to observe deviations from a flat surface; in the example a scratch of 0.4-μm depth is revealed. (*From F. T. Farago, Handbook of Industrial Measurement, 2d ed., Industrial Press, New York, 1982.*)

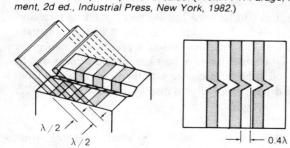

(*a*)   (*b*)

various roughness characteristics computed. Portable instruments used in the plant give a readout of R_q directly.

For plant use there are also collections of standard sample surfaces (*replica blocks*) available, with R_q marked for each sample. By drawing a fingernail across the sample and the production part, remarkably close estimates of R_q can be obtained.

Other inspection devices are based on the measurement of capacitance, optical interference (Fig. 2-41), diffraction, and air-pressure drop.

In some instances, surface attributes are difficult to quantify and then comparison specimens, chosen to represent acceptable and reject qualities, are used.

2-5 NONDESTRUCTIVE TESTING (NDT)

In critical components, the presence of cracks and other defects is checked by various *nondestructive testing* (NDT) techniques.

1 Surface defects are revealed by *liquid-penetrant inspection*. Penetrants (dyes) are applied as sprays or by immersion to a thoroughly cleaned and dried surface. After wiping off the excess, the penetrant trapped in defects is drawn out and made visible by an absorbent developer. Some dyes are fluorescent and make the defect highly visible in ultraviolet light.

2 *Magnetic particle inspection* is limited to ferromagnetic workpieces. When the workpiece is magnetized, cracks lying more or less perpendicular to the field interrupt the magnetic field and become visible when fine ferromagnetic particles are dusted onto the surface.

3 The workpiece does not need to be ferromagnetic for *eddy current inspection*. A probe supplied with a high-frequency current induces an electric field in the part; the field changes in the presence of surface or near-surface defects. These changes show up on instruments. The technique is noncontacting and is suitable for on-line inspection, measurement of the thickness of surface coatings, and changes in metallurgical condition.

4 *Ultrasonic inspection* is based on the observation that a beam of ultrasonic energy (high-frequency acoustic energy) passes through a solid structure with little loss but is partially reflected from internal surfaces; therefore, cracks and cavities show up on a VDT. Good coupling between the transducer and workpiece is ensured by a coupling fluid (couplant).

5 Internal defects as well as surface cracks change the absorption of penetrating radiation and can be revealed by *radiographic inspection* using x-rays, γ-rays, or neutrons.

6 *Electromagnetic sorting* is used to separate ferromagnetic components according to their hardness, composition, or compositional change in surface layers (such as occur in case hardening). Sorting is based on the effects of these variables on magnetic properties.

7 The *acoustic emission* technique is of great value for monitoring processes and machinery. Internal processes such as fracture, bulk plastic deformation, and surface processes such as shearing and sliding all result in the release of short bursts of elastic energy, which can be detected with transducers attached to the surface. Analysis of the emission spectrum gives valuable clues regarding the process and, in some instances, can be used for closed-loop control.

Interpretation of NDT readings requires considerable skill and judgement. Personal bias is minimized when computer graphics is drawn upon for displaying and interpreting signals. It is then possible to obtain a complete map of imperfections in a large workpiece.

2-6 MATERIALS SELECTION

We have now arrived at a point where it will be useful to summarize the many activities involved in the design of products (block 2 of Fig. 1-7) and recapitulate their interaction with manufacturing processing, with emphasis on the making of individual parts (components). The following steps are involved:

1 Determine the functions the part will have to satisfy, with due regard to operating conditions, safety aspects (including fail-safe characteristics), regulatory requirements, product liability implications, environmental impact (storage and disposal), packaging requirements, ease of maintenance, and service life.

2 Determine the configuration that will fulfill the requisite functions, and assign dimensions.

3 Analyze the design for loads and stresses, possible failure modes, and aspects of reliability. Consider the use of standard designs and components of known reliability.

4 Choose a material that satisfies all service criteria. Voluminous handbooks exist which show various properties of materials, usually classified according to composition. Since there are thousands of potential materials to be considered, such classifications are of little value unless some more generally applicable guidelines can be brought to bear on the problem. Such guidelines allow the designer to consider the broadest possible group or groups of materials without prematurely restricting the choice and thus limiting the possibilities of manufacture. Material choice is facilitated by computerized data bases which, if properly constructed, incorporate some of the logic necessary for making a sound decision.

5 Optimize the material choice by considering alternative materials. Depending on application, various factors become important: minimum cost for a given strength (load-bearing capacity), minimum weight for a given strength (strength-to-weight ratio) or stiffness (elastic modulus-to-weight ratio). Some feel for the wide range of possibilities may be gained from Table 2-3.

6 Assign the widest possible tolerances and roughest surface finish allowable for the given function.

7 Choose an appropriate process or process sequence, with due regard to the cost of processing and assembly and the number of parts to be produced. Savings can often be made by combining several parts (a subassembly) into a single part. Consider ease of production (producibility), inspection (inspectability), and test-

TABLE 2-3
PROPERTIES OF SOME ENGINEERING MATERIALS*

Material	Elastic modulus, GPa	Tensile strength, MPa	Fatigue strength, Mpa	Density, kg/m³	Specific energy,† MJ/kg	Cost, $/kg
0.15C steel	210	380	190	7860	38	0.5
Heat-treated steel	210	1800	560	7860	38	2.0
304 stainless steel	195	600	250	7900	50	3.0
Ti–6Al–4V alloy	110	900	500	4430	550	20.0
Cast iron (gray)	95	340	140	7200	30	0.7
Brass (30Zn)	110	310	140	8530	50	2.5
Al alloy (2024-T6)	70	420	140	2770	170	3.0
Mg alloy	45	250	90	1760	400	5.0
Zn alloy (die cast)	90	300	55	6600	70	1.5
Polyethylene, high dens.	(1)	30		960	130	1.3
Polypropylene	(1)	35		900	170	1.3
Nylon	(3)	85		1150	250	3.3
Reinforced concrete	30	400	20	2400	8	0.3
Plywood	12	15	6	500	2	0.5

*Compiled from various sources. Specific energy and cost are approximate and vary with the form in which the material is used. Prices are subject to large fluctuations.
†1 MJ/kg = 0.278 kWh/kg.

ing (testability). The cost of these functions can exceed, by a wide margin, the cost of the starting material. Establish acceptance and rejection criteria. Remember that some processes are not suitable for parts below or above certain sizes or of very thin or very thick walls, and that a process that may be economical for a few parts may be noncompetitive in mass production (Sec. 12-2).

8 Optimize the design by an iterative refinement of steps 2 to 6. Consider the total cost implications; while some material may fulfill the required function, it may also present substantial manufacturing difficulties.

9 At different times and in different places, some other considerations assume overriding significance. For example:

a The energy consumed in manufacturing varies greatly for various materials (Table 2-3). This is always an important consideration but, at times of energy shortages, it may become critical. Recycling often saves substantial energy; for example, aluminum can be remelted with only 5% of the energy used in the primary reduction of the metal.

b Many raw materials are found only in some parts of the world. Their supply may become critical in periods of upheaval, and substitution may then require different approaches to design and manufacture.

In this book we will limit ourselves to the manufacture of parts. The starting material is often the end result of prior operations. While many possible routes for such *primary processing* are available, even the few indicated in Fig. 2-42 show that the same starting material may often be obtained through a number of

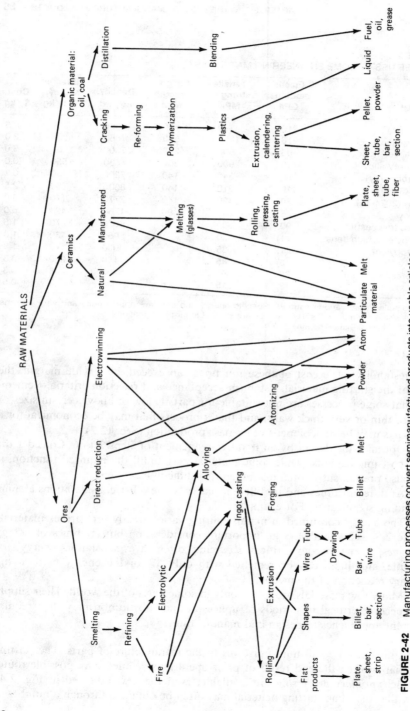

FIGURE 2-42 Manufacturing processes convert semimanufactured products into usable articles. Semimanufactured products may be obtained by a variety of processing techniques.

alternative routes—some of them much shorter than others—from raw material to semifabricated product. It would, however, be too hasty to conclude that the more complex processes are necessarily more expensive. Very often, economy is a matter of scale; thus, it is still possible to buy steel strip at a lower price than powder, partly because of the vast quantities produced in strip form.

Metals are still the most generally employed engineering materials and the growth of their production (and especially that of steel) has often been taken as an indicator of industrial development. With the increasing sophistication of many products and with the growth of plastics and microelectronics, these relationships are no longer valid, particularly in industrialized nations (the growth shown in Table 2-1 for the period 1972–1982 is unusually low because 1982 was a year of worldwide economic recession). Nevertheless, metals remain indispensable. Steel still represents an overwhelming portion of total metal production (Table 2-1), but other metals offer unique properties and some of them, notably magnesium and titanium, would become much more important if they could be extracted with a smaller energy outlay. Polymers (plastics) play an increasingly important role, a role that has been growing despite periodic shortages of oil. Not shown in Table 2-3, the various natural and man-made ceramics represent a vast source of engineering materials, many of them also used in the manufacturing industries.

There are several thousand engineering materials (metals, alloys, polymers, and ceramics) in everyday engineering use. Their behavior during processing in the liquid or solid state shows an enormous variability, and a cataloging of their manufacturing properties and of manufacturing processes would turn into a bewildering encyclopedia. There are, however, some basic principles one can identify that govern the behavior of a large majority of materials. By necessity, such rules will have to be generalizations, and exceptions to these rules will always be found. Nevertheless, these rules can give a broad, first indication of what materials and processes may possibly fulfill the requirements of the final product, and thus allow a more intelligent design of products and processes. These principles will be first identified for metals and their alloys and then extended to ceramics and polymers.

2-7   SUMMARY

Manufacturing is the production of durable and semidurable articles. The aim of process selection, development, and control is to satisfy all service requirements at minimum cost. The service attributes of a material of given composition greatly depend on the method of manufacture; the properties of a part may often be improved but, frequently, other properties are sacrificed. Processes and process sequences must be chosen and controlled to give an optimum combination of properties. For this reason, one of the important activities in manufacturing is the measurement of properties both during manufacture and in the finished product. Properties fall into several broad categories.

1 The response of materials to imposed stresses is characterized by mechanical properties determined under conditions designed to simulate loading in service. All these properties are structure-sensitive, i.e., they change—for a given material —with the microstructural features of the part. These features can be changed, in a controlled manner, by manufacturing techniques.

a Response to steady loading in tension is expressed by so-called static strength properties such as elastic modulus, yield strength, and tensile strength. Ductility, as expressed by elongation to fracture and reduction of area, is important in applications requiring toughness and makes processing by plastic deformation possible.

b Resistance to compressive loading is measured by the stress-strain curve in compression and by hardness.

c Materials of limited ductility are often better characterized by the modulus of rupture obtained in bending.

d Resistance to sudden loading and fast fracture is measured in impact and fracture-toughness testing and is greatly improved by manufacturing techniques that ensure internal cleanliness (freedom from inclusions).

e Resistance to repeated loading is established in fatigue tests. Fatigue strength is greatly increased by manufacturing techniques that minimize surface and internal defects and create compressive residual stresses on the surface.

f Prolonged service in the hot temperature regime calls for creep resistance and long stress-rupture life, properties that are again greatly improved by the absence of weakening internal features such as cracks and voids and by appropriate structural features induced by manufacturing techniques.

2 Other physical properties may assume great significance in various applications.

a Desirable tribological properties include controlled (not necessarily low) friction and wear, and may call for a change of these characteristics by means of lubrication. Most tribological attributes are sensitive to structure and surface topography (roughness) and thus amenable to control by manufacturing techniques.

b Properties such as electrical conductivity (resistivity) and magnetic and optical properties are structure-sensitive and are controlled by manufacturing techniques. Thermal properties are largely structure-insensitive but heat conduction can be controlled by the appropriate design and manufacture of components and assemblies.

3 Chemical properties such as corrosion resistance are vital in many applications and can be controlled by manufacturing techniques, including special treatments of the surface.

4 Manufactured parts must have closely defined shapes, dimensions, and surface roughness. Maintenance of specified tolerances is vital to the function of assemblies and makes interchangeability possible. Excessively tight tolerance and surface finish specifications do lead, however, to excessive manufacturing costs. Because of the importance of making parts right the first time, engineering metrology and, especially in-process measurement, assume a central role in manufacturing.

5 A powerful aid to quality improvement is nondestructive testing for the detection and quantification of surface and internal defects, residual stresses, and deviations from specified material conditions.

6 Design and manufacturing are (or should be) inseparably interwoven activities. Therefore, materials and material conditions must reflect an optimum combination of service and manufacturing properties.

PROBLEMS

2-1 The proposal is made that the component of Example 2-1 should be manufactured of Ti–6Al–4V titanium alloy or AISI 4340 steel (quenched and tempered at 425 °C). For both materials, recalculate (*a*) elongation under a load of 80 kN, (*b*) load for yielding, and (*c*) load for fracture.

2-2 A component may be represented as a bar of 400-mm length. It must carry a load of 80 kN in tension, with a safety factor of 2 (i.e., the stress must not exceed 50% of the YS). The bar may be made of 7075-T6 aluminum alloy, or the materials given in Prob. 2-1. Which of the materials will give the lowest-weight component?

2-3 Assume that the three bars, of compositions and diameters calculated in Prob. 2-2, are loaded until an elastic strain of 0.5% is obtained (check that no plastic deformation occurs). What is the elastic energy stored in each bar?

2-4 Continuing Example 2-3 calculate the contributions of the specimen and machine to the total elastic deflection. Assume that elastic extension outside the gage length of the specimen amounts to 50% of the extension in the gage length.

2-5 The specimen specified in Example 2-2 is tested on a machine of 20-kN capacity. Recording is made from the crosshead of the machine (as in Example 2-3). Would you expect the initial slope of the recording to be steeper for the smaller machine?

2-6 Steel wire used in steel-belted radial tires has YS = 2100 MPa. How much elastic extension is possible before permanent deformation sets in?

2-7 In the construction industry, either steel or reinforced concrete may be used in many applications where tensile stresses are generated (e.g., beams). What method of testing would you recommend to establish the safe tensile stress for each material?

2-8 A company specification calls for a steel component to have a minimum TS of 180 kpsi. Tension tests are conducted on selected samples, but all components are also subjected to Rockwell C hardness testing. What HRC is the minimum acceptable value?

2-9 If the components of Prob. 2-8 were too large to be tested in a Rockwell hardness tester, what method would one use to keep a running check on hardness?

2-10 A steel is to be used in the construction of an offshore drilling platform for the Beaufort Sea. What test would you recommend for quick checks on its susceptibility to brittle fracture?

2-11 Forgings made for aircraft applications must be free of laps, seams, and cracks. What NDT techniques would you consider if the material was: (*a*) steel, (*b*) aluminum alloy, or (*c*) titanium alloy?

2-12 What noncontacting technique would be suitable for automatic, 100% inspection of shafts produced to the tolerances shown in Fig. 2-23?

2-13 The shaft and bearing shown in Fig. 2-23 are checked for dimensions. What precision should the measuring device have, and what would be the smallest scale division on such an instrument?

2-14 Is tolerancing in Fig. 2-23 unilateral or bilateral? Why was this particular system of tolerancing chosen?

2-15 What kind and dimensions of GO–NOT GO gages would be used for the parts in Fig. 2-23? What kind of in-process measuring instruments would you consider for control?

2-16 Define allowance, tolerance, and fit.

2-17 Have the preferred sizes listed in standards been developed for reasons of technology, economy, or both? Explain.

2-18 A batch of components is returned to the manufacturer with the claim that all parts are outside tolerance, even though their dimensions are closely clustered. In-process, 100% inspection had been made with a pneumatic gage. Can the complaint be correct?

2-19 There is a change of 10 °C in room temperature; what will be its effect on the diameter of the shaft of Fig. 2-23?

2-20 A steel shaft of 3.00-in diameter is measured with a gap gage (C gage). The shaft is at 150 °F, the gage at room temperature. Calculate the magnitude of measurement error.

2-21 An aluminum block of 200-mm length is at 80 °C. Vernier calipers, made of steel, are used at 20 °C. Calculate the error.

2-22 The surface finish of a part is specified as 32 μin CLA maximum. The shop-type instrument gives readings in μin rms. Is the limit value greater or smaller than 32 μin?

2-23 A customer complains that a part is not usable because it was received with a scratch that destroys its appearance. Inspection records show that the part left the plant with a surface finish that satisfied the specified CLA roughness. Can the customer's claim be valid?

2-24 Make a sketch of the pattern of interference fringes produced when (*a*) a ball-bearing ball and (*b*) a roller-bearing roller are placed on an optical flat.

2-25 Calculate the $0.5T_m$ temperature for Zn, Ag, and Ni. Explain on this basis whether significant creep of these metals should be expected at 200 °C.

FURTHER READING

Bever, M. B. (ed.): *Encyclopedia of Materials Science and Technology*, (8 vols.), Pergamon, Oxford, 1986.

A General Introductory Texts on Materials

Ashby, M. F., and D. R. H. Jones: *Engineering Materials: An Introduction to their Properties and Applications*, Pergamon, Oxford, 1980.

Askeland, D. R.: *The Science and Engineering of Materials*, Brooks/Cole Engineering Division, Monterey, Calif., 1984.

_____ : *The Science and Engineering of Materials*, (alternate edition), PWS Engineering, Boston, Mass., 1985.

ASM Thesaurus of Metallurgical Terms, 3d ed., American Society for Metals, Metals Park, Ohio, 1979.

Beesley, C.: *Fundamentals of Engineering Materials*, Macmillan, London, 1984.

Clauser, H. R.: *Industrial and Engineering Materials*, McGraw-Hill, New York, 1975.

Cottrell, A. H.: *An Introduction to Metallurgy*, 2d ed., Arnold, London, 1975.

Flinn, R. A., and P. K. Trojan: *Engineering Materials and Their Applications*, 3d ed., Houghton Miffin, Boston, 1986.

Higgins, R. A.: *Properties of Engineering Materials*, Hodder and Stoughton, London, 1977.

Jacobs, J. A., and T. F. Kildruff: *Engineering Materials Technology*, Prentice-Hall, Englewood Cliffs, N.J., 1985.

John, V. B.: *Introduction to Engineering Materials*, 2d ed., Macmillan, London, 1983.

Keyser, C. A.: *Materials Science in Engineering*, 3d ed., Merrill, Columbus, Ohio, 1980.

Lewis, G.: *Properties of Engineering Materials*, Macmillan, London, 1981.

Murr, L. E. (ed.): *Industrial Materials Science and Engineering*, Dekker, New York, 1984.

Pascoe, K. J.: *Introduction to the Properties of Engineering Materials*, 3d ed., Van Nostrand Reinhold, New York, 1978.

Ralls, K. M., T. H. Courtney, and J. Wulff: *An Introduction to Materials Science and Engineering*, Wiley, New York, 1976.

Shackelford, J. F.: *Introduction to Materials Science for Engineers*, Macmillan, London, 1984.

Smallman, R. E.: *Modern Physical Metallurgy*, 4th ed., Butterworths, London, 1985.

Smith, C. O.: *The Science of Engineering Materials*, 2d ed., Prentice-Hall, Englewood Cliffs, N.J., 1977.

Thornton, P. A., and V. J. Colangelo: *Fundamentals of Engineering Materials*, Prentice-Hall, Englewood Cliffs, N.J., 1985.

Van Vlack, L. H.: *Materials for Engineering: Concepts and Applications*, Addison-Wesley, Reading, Mass., 1982.

_____ : *Elements of Materials Science and Engineering*, 5th ed., Addison-Wesley, Reading, Mass., 1985.

_____ and C. J. Osborn: *Study Aids for Introductory Material Courses*, Addison-Wesley, Reading, Mass., 1977.

Young, J. F., and R. S. Shane: *Materials and Processes*, 3d ed., (2 vols.), Dekker, New York, 1985.

B   Mechanical Behavior

Caddell, R. M.: *Deformation and Fracture of Solids*, Prentice-Hall, Englewood Cliffs, N.J., 1980.

Dieter, G. E., Jr.: *Mechanical Metallurgy*, 2d ed., McGraw-Hill, New York, 1976.

Felbeck, D. K. and A. G. Atkins: *Strength and Fracture of Engineering Solids*, Prentice-Hall, Englewood Cliffs, N.J., 1984.

Hertzberg, R. W.: *Deformation and Fracture Mechanics of Engineering Materials*, 2d ed., Wiley, New York, 1983.

Le May, I.: *Principles of Mechanical Metallurgy*, Elsevier, Amsterdam, 1981.

McClintock, F. A., and A. S. Argon: *Mechanical Behavior of Materials*, Addison-Wesley, Reading, Mass., 1966.

Meyers, M. A. and K. K. Chawla: *Mechanical Metallurgy: Principles and Applications*, Prentice-Hall, Englewood Cliffs, N.J., 1984.

C   Design

Andreasen, M. M., S. Kahler, and T. Lund: *Design for Assembly*, IFS (Publications) Ltd., Bedford, England/Springer, Berlin, 1983.

Begg, V.: *Developing Expert CAD Systems*, Unipub, New York, 1984.

Bjorke, O.: *Computer-Aided Tolerancing*, Tapir Publishers, Trondheim, 1978.

Bralla, J. G. (ed.): *Handbook of Product Design for Manufacturing: A Practical Guide to Low-Cost Production*, McGraw-Hill, New York, 1986.

Colangelo, V. J., and P. A. Thornton: *Engineering Aspects of Product Liability*, American Society for Metals, Metals Park, Ohio, 1981.

Dieter, G. E., Jr.: *Engineering Design: A Materials and Processing Approach*, McGraw-Hill, New York, 1983.

Flurscheim, C. H. (ed.): *Industrial Design in Engineering*, The Design Council/Springer, Berlin, 1983.

Foster, L. W.: *Geometrics: The Application of Geometric Tolerancing Techniques*, Addison-Wesley, Reading, Mass., 1983.

Gardan, Y., and M. Lucas: *Interactive Graphics in CAD*, Unipub, New York, 1984.

Hammer, W.: *Product Safety Management and Engineering*, Prentice-Hall, Englewood Cliffs, N.J., 1980.

Henley, E. J., and H. Kumamoto: *Designing for Reliability and Safety Control*, Prentice-Hall, Englewood Cliffs, N.J., 1985.

Kantowicz, B. H., and Sorkin, R. D.: *Human Factors: Understanding People-System Relationships*, Wiley, New York, 1983.

Kolb, J., and S. S. Ross: *Product Safety and Liability*, McGraw-Hill, New York, 1980.

Lange, J. C.: *Design Dimensioning with Computer Graphics Applications*, Dekker, New York, 1984.

Murrell, K. F. H.: *Ergonomics: Man and His Working Environment*, Chapman and Hall, London/Wiley, New York, 1979.

Niebel, B. W., and A. B. Draper: *Product Design and Process Engineering*, McGraw-Hill, New York, 1974.

Spotts, M. F.: *Dimensioning and Tolerancing for Quantity Production*, Prentice-Hall, Englewood Cliffs, N.J., 1983.

Trucks, H. E.: *Designing for Economical Production*, Society of Manufacturing Engineers, Dearborn, Mich., 1974.

Weinstein, A. S., A. D. Twerski, H. R. Piehler, and W. A. Donaher: *Products Liability and the Reasonably Safe Product*, Wiley, New York, 1978.

D Engineering Metrology

ASM: *Metals Handbook*, 8th ed., vol. 11: *Nondestructive Inspection and Quality Control*, 1976; 9th ed., vol. 8: *Mechanical Testing*, 1985; vol. 10: *Materials Characterization*, 1986, American Society for Metals, Metals Park, Ohio.

Batchelor, B. G., D. A. Hill, and D. C. Hodgson: *Automated Visual Inspection*, IFS (Publications) Ltd., Bedford, England, 1985.

Buck, O., and S. M. Wolf: *Nondestructive Evaluation: Application to Materials Processing*, American Society for Metals, Metals Park, Ohio, 1984.

Davis, H. E., G. E. Troxell, and G. F. W. Hauck: *The Testing of Engineering Materials*, 4th ed., Macmillan, New York, 1982.

Farago, F. T.: *Handbook of Dimensional Measurement*, 2d ed., Industrial Press, New York, 1982.

Hardt, D. E. (ed.): *Measurement and Control for Batch Manufacturing*, American Society of Mechanical Engineers, New York, 1982.

Hull, J. B., and V. B. John: *Non-Destructive Testing*, Macmillan, London, 1984.

Kennedy, C. W., and D. E. Andrews (eds.): *Inspection and Gaging*, 5th ed., Industrial Press, New York, 1977.

Lenk, J. D.: *Handbook of Controls and Instrumentation*, Prentice-Hall, Englewood Cliffs, N.J., 1980.

McMaster, R. C. (ed.): *Nondestructive Testing Handbook*, 2d ed., American Society for Nondestructive Testing, Columbus, Ohio, 1982.

Roth, E. S. (ed.): *Gaging, Practical Design and Application*, 2d ed., Society of Manufacturing Engineers, Dearborn, Mich., 1983.

Sydenham, P. H.: *Transducers in Measurement and Control*, 3d ed., Hilger, Bristol, 1985.

Thomas, G. G.: *Engineering Metrology*, Butterworths, London, 1974.

Warnecke, H. J., and W. Dutschke (eds.): *Metrology in Manufacturing Technology*, Springer, Berlin, 1984.

E Materials Selection

Aerospace Structural Metals Handbook (5 vols.), Mechanical Properties Data Center, Battelle, Columbus, Ohio, 1980.

ASM: *Metals Handbook*, 9th ed., *Properties and Selection*: vol. 1, *Irons and Steels*, 1978; vol. 2, *Nonferrous Alloys and Pure Metals*. 1979; vol. 3, *Stainless Steels, Tool Materials, and Special-Purpose Metals*, 1980, American Society for Metals, Metals Park, Ohio.

ASM: *Metals Handbook*, 8th ed., vol. 8: *Metallography, Structures and Phase Diagrams*, 1973; vol. 7: *Atlas of Microstructures of Industrial Alloys*, 1972; Vol. 10: *Failure Analysis and Prevention*, 1975; 9th ed., Vol. 9: *Metallography and Microstructures*, 1985, American Society for Metals, Metals Park, Ohio.

ASM: *ASM Metals Reference Book*, American Society for Metals, Metals Park, Ohio, 1981.

Boyer, H. E., and T. L. Gall (eds.): *Metals Handbook Desk Edition*, American Society for Metals, Metals Park, Ohio, 1984 (also *MetalSelector* software for IBM PC).

Brady, G. S., and H. R. Clauser: *Materials Handbook*, 12th ed., McGraw-Hill, New York, 1985.

Brandes, E. A.: *Smithells Metals Reference Book*, 6th ed., Butterworths, London, 1983.

Crane, F. A., and J. A. Charles: *Selection and Use of Engineering Materials*, Butterworths, London, 1984.

Ericsson, T. (ed.): *Computers in Materials Technology*, Pergamon, Oxford, 1981.

Farag, M. M.: *Materials and Process Selection in Engineering*, Applied Science Publishers, London, 1979.

Gibbons, R. C. (ed.): *Woldman's Engineering Alloys*, 6th ed., American Society for Metals, Metals Park, Ohio, 1979.

Hanley, D. P.: *Introduction to the Selection of Engineering Materials*, Van Nostrand Reinhold, New York, 1980.

Lynch, C. T. (ed.): *CRC Handbook of Materials Science* (3 vols.), CRC Press, Boca Raton, Fla., 1974, 1975, 1975.

Potts, D. L. and J. Arcuri: *International Metallic Materials Cross Reference*, 2d ed., General Electric Company, Schenectady, N.Y., 1983.

Summit, R., and A. Silker (eds.): *CRC Handbook of Materials Science*, vol. 4, CRC Press, Boca Raton, Fla., 1980.

Wick, C. and R. Veilleux (eds.): *Tool and Manufacturing Engineers Handbook*, 4th ed., vol. 3, *Materials, Finishing and Coating*, Society of Manufacturing Engineers, Dearborn, Mich., 1985.

F Materials Data Bases

Machinability Data Center, Metcut Research Associates, Inc., 3980 Rosslyn Drive, Cincinnati, Ohio 45209.

Mechanical Properties Data Center, Battelle Columbus Laboratories, 505 King Avenue, Columbus, Ohio 43201.

Metals. and Ceramics Information Center, Battelle Columbus Laboratories, 505 King Avenue, Columbus, Ohio 43201.

Metals Datafile (datafile on properties, composition, specification, design information) and *Metadex* (comprehensive on-line bibliographic data base for research and information). Available from American Society for Metals, Metals Park, Ohio 44073, or from SDC Information Services, 2500 Colorado Ave, Santa Monica, Calif. 90406.

Metals Information. A joint information service of American Society for Metals, Metals Park, Ohio 44073 and The Metals Society, 1 Carlton House Terrace, London, SW1Y 5DB.

Stahlschluessel (*Key to Steel*), 13th ed., Verlag Stahlschluessel, Marbach, 1983.

Unterweiser, P. (ed.): *Worldwide Guide to Equivalent Irons and Steels*, American Society for Metals, Metals Park, Ohio, 1979.

_____ : *Worldwide Guide to Equivalent Nonferrous Metals and Alloys*, American Society for Metals, Metals Park, Ohio, 1980.

G Journals

Journal of Materials Education
Journal of Nondestructive Evaluation
Journal of Tribology (*Trans. ASME*)
Materials Engineering
Materials Science and Engineering
Tribology International
Wear

CHAPTER 3

METAL CASTING

With few exceptions, the first step in the manufacture of metal components is casting. The essential steps are: melting a metal charge in a furnace; pouring the melt into a previously prepared mold; extraction of heat from the melt and solidification; removal and, possibly, treatment of the solidified part. The success of the casting process and the quality and service properties of the cast part all hinge upon the control of process parameters. Understanding the role of process variables requires a knowledge of some basic phenomena. These topics are usually covered in a first course on engineering materials; for those who have not had such exposure, the background material is discussed in the following section.

3-1 SOLIDIFICATION OF METALS

Solid metals are crystalline materials characterized by the metallic bond, reasonable strength and ductility, and good electrical conductivity. If their atoms, complete with electrons, are visualized as tiny spheres (of diameters around 0.2 nm or 2 Å), these spheres occupy strictly prescribed positions in space. The arrangement of points representing the center of atoms is called a *lattice*. Atoms vibrate about their lattice position; vibration is minimum at absolute zero. When the solid is heated, atoms vibrate at ever-increasing amplitudes; at a critical temperature—the melting point, T_m (Sec. 2-1-7)—the solid melts and turns into a liquid. The *long-range* crystalline order of the solid is largely lost, although some *short-range* order, extending to a few atoms, may exist. Thus, on melting, the *crystalline* solid changes into an *amorphous* liquid.

3-1-1 Pure Metals

We may follow the solidification of a *pure metal* by inserting a thermocouple into a melt contained in a small crucible and recording the change in temperature with

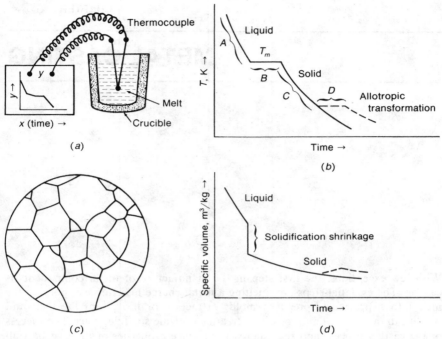

FIGURE 3-1
Solidification of a pure metal may be observed by (a) inserting a thermocouple and (b) recording the temperature as a function of time. (c) A micrograph of the resulting structure shows only grain boundaries. (d) The volume shrinks on solidification but may increase with an allotropic transformation.

time (Fig. 3-1a). If no heat is supplied, the melt gradually cools by releasing sensible heat or internal energy (A in Fig. 3-1b) until at T_m very small crystalline bodies, *nuclei*, form at several points in the melt. Temperature now remains constant while nuclei grow by the deposition of further atoms in the same crystallographic orientation, and the heat of fusion (B) is removed. When all melt is solidified, temperature drops again (Fig. 3-1b), and the solid releases its sensible heat energy (C).

The solidified body is *polycrystalline*, i.e., it consists of many randomly oriented crystals (usually called *grains*). Mechanical and other properties of a single crystal are *anisotropic*, i.e., a function of the direction of testing relative to lattice orientation. In contrast, a polycrystalline body consisting of a large number of randomly oriented grains is *isotropic* (has the same properties in all directions), with properties representing a mean of all crystallographic directions.

Since adjacent grains have different orientations, the *grain boundary* is a zone of disorder. To reveal grain boundaries, the solidified body may be cut, the surface ground, polished, and etched with a suitable reagent. Because of the higher chemical energy of atoms on the grain boundary, they are preferentially attacked and the groove appears as a dark line under the optical microscope (Fig. 3-1c).

In the liquid state, the randomly spaced, highly agitated atoms occupy much room, hence the *specific volume* (volume per unit mass) is large (Fig. 3-1d). During the cooling of the melt, thermal excitation becomes less violent, and specific volume drops gradually until the melting point is reached. Here the atoms occupy their lattice sites which are more closely spaced, and the specific volume drops substantially; the *solidification shrinkage* is typically 2.5–6.5%. This means that, if a casting free of cavities is to be produced, melt will have to be supplied to make up for solidification shrinkage. Diminishing thermal excitation during cooling in the solid state results in further shrinkage, as given by the thermal expansion coefficient. Typically, metals shrink about 1% per 1000 °C temperature

FIGURE 3-2
Lattice sites, slip planes, and slip directions in (*a*) face-centered cubic, (*b*) body-centered cubic, and (*c*) and (*d*) hexagonal close-packed structures.

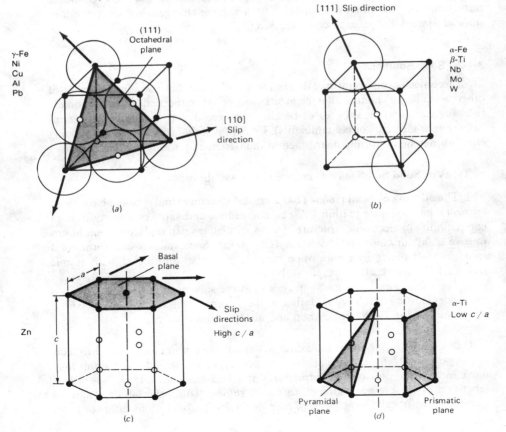

drop. Even in a fully solidified metal there will be some unoccupied atomic sites, *point-defects*, called *vacancies*.

The packing arrangement of atoms is characteristic of the metal and can be described by the *unit cell* (the smallest volume that fully defines the atomic arrangement). For practical engineering metals, three lattice types are of importance: *face-centered cubic* (fcc) with atoms at each corner and in the middle of the face of a cube (Fig. 3-2a); *body-centered cubic* (bcc) with atoms at each corner and in the middle of a cube (Fig. 3-2b); and *hexagonal close-packed* (hcp) with an atom at each corner and the center of the end face (*basal plane*) and at three sites in the middle of the body (Fig. 3-2c and d). Structure plays important roles in solidification and plastic deformation.

Some metals undergo, in the solid state, a change in crystal structure (*allotropic transformation*) at some critical temperature, releasing the latent heat of transformation (D in Fig. 3-1b). For convenience, different crystallographic forms of the same metal are denoted by Greek letters. Thus, on cooling, the bcc δ-iron changes to the fcc γ-iron at 1400 °C, which again changes to the bcc α-iron at 906 °C. The bcc β-titanium changes on cooling to hcp α-titanium at 880 °C. Allotropic transformations are often accompanied by a volume change which may result in sufficient internal stresses to cause cracking.

3-1-2 Solid Solutions

Most technically important metals are not pure metals but contain a number of other metallic or nonmetallic elements which are either added intentionally (*alloying elements*) or are present because they could not be removed economically (minor elements or contaminants). Under favorable conditions, the alloying element may be uniformly distributed in the base metal, forming a *solid solution*.

Types of Solid Solutions There are two possibilities:

1 The alloying element (*solute*) has a crystal structure similar to the base metal (*solvent*), has a similar (within 15%) atomic radius, and satisfies some criteria of compatibility in electronic structure. Then solute atoms can replace solvent atoms to give a *substitutional solid solution* (Fig. 3-3a). Some metals can form solid solutions over the entire composition range (e.g., copper and nickel, with atomic radii of 1.278 and 1.245 Å, respectively).

2 The solute atoms are much smaller than the solvent atoms and can fit into the spaces existing in the crystal lattice of the solvent metal, to form an *interstitial solid solution* (Fig. 3-3b, e.g., carbon and nitrogen atoms in iron).

Diffusion It is important to realize that atoms are not immovably tied to their lattice position. If, for example, there is a vacancy, one of the adjacent atoms may move in, and the previously occupied site now becomes vacant (Fig. 3-3c). By a repetition of these events, atoms can move, *diffuse* within the lattice. The case quoted above is called *vacancy diffusion* (or substitutional atom diffusion). An

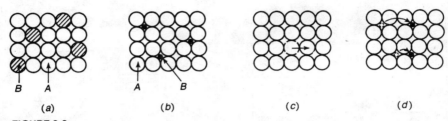

FIGURE 3-3
Alloying elements may be accommodated in (a) substitutional or (b) interstitial solid solutions.
Migration of atoms may take place by (c) vacancy (substitutional) or (d) interstitial diffusion.

interstitial solute atom can also move into an adjacent space between the solvent atoms by *interstitial diffusion* (Fig. 3-3d).

If the solute atoms are not distributed evenly in a solid solution, they will diffuse until concentration gradients are eliminated. According to Fick's first law, the *flux of atoms J* (the number of atoms passing through a plane of unit area, in unit time), in units of atoms/m$^2 \cdot$ s, is proportional to the concentration gradient ΔC (the change in concentration over a Δx distance)

$$J = -D \frac{\Delta C}{\Delta x} \tag{3-1}$$

where D is *diffusivity* or *diffusion coefficient*. The value of D is larger at higher temperatures

$$D = D_0 \exp\left(-\frac{Q}{RT}\right) \tag{3-2}$$

where D_0 is a constant for a given material pair, Q is the *activation energy* (the energy required to overcome the energy barrier involved in moving atoms through the lattice), and R is the gas constant (8.31 J/mol $\cdot$ K). Accordingly, diffusion is greatly accelerated by high temperatures. Diffusion is a most important mechanism not only in solidification but also in many other phases of manufacture.

Solidification of Solid Solutions   The events occurring during the solidification of solid solutions under equilibrium conditions may be followed by making up different melts of, say, copper and nickel, with the Ni content at 0, 50, and 100 weight percent (wt% or, in this text, simply %). The melts with 0% Ni (100% Cu) and 100% Ni are pure metals, and their cooling curves are the same as in Fig. 3-1b. The melt of $C_0 = 50\%$ Ni is different (Fig. 3-4). Solidification begins at 1315 °C with the formation of nuclei with 68% Ni content. Temperature drops gradually while alloy less rich in Ni solidifies onto the nuclei until, at 1270 °C, all melt disappears. If solidification was very slow and Cu atoms could diffuse into the already solidified crystals, the composition will be uniform everywhere at 50% Ni. At some intermediate temperature T_1, the alloy is in a *mushy state*: by

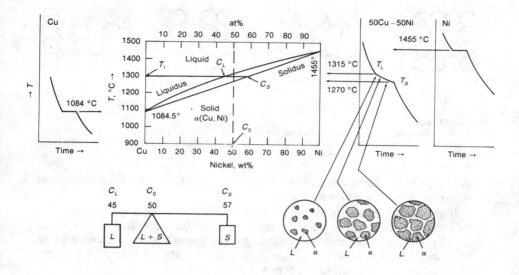

FIGURE 3-4
The copper–nickel equilibrium diagram shows complete solid solubility of the two elements in each other. Solidification of a solid solution takes place at gradually dropping temperatures and the proportion of solid and liquid phases may be found from the inverse lever arm rule.

drawing a horizontal *tie line*, we see that solid crystals (of composition C_S) coexist with a liquid (of composition C_L). Their relative quantities are given by the *inverse lever arm rule*: the weight fraction of solid S is proportional to the horizontal distance (lever arm) between the nominal composition C_0 and the composition of the liquid phase C_L:

$$\frac{S}{S+L} = \frac{C_0 - C_L}{C_S - C_L} 100 \ (\%) \tag{3-3a}$$

Similarly, the weight fraction of liquid L present is

$$\frac{L}{S+L} = \frac{C_S - C_0}{C_S - C_L} 100 \ (\%) \tag{3-3b}$$

Repeating the experiment at other concentrations, lines defining complete melting (the *liquidus*, T_L) and solidification (the *solidus*, T_S) are defined and an *equilibrium diagram* of temperature versus composition is obtained. Obviously, the quantity of solid is vanishingly small at all points on the liquidus, and solid crystals gradually grow during cooling to the solidus.

Because the solvent atoms are uniformly distributed in the solute, each grain in a polycrystalline body will appear *homogeneous* and will look like the grains of a pure metal (Fig. 3-1c). It is usual to denote solid solutions by a Greek lowercase letter.

Example 3-1

For an alloy containing 50% Cu and 50% Ni, calculate the amounts of solid S and liquid L present at 1300 °C. Show that the total nickel content in the solid and liquid phases adds up to 50%.

The equilibrium diagram (Fig. 3-4) shows that, for $C_0 = 50\%$ Ni, at 1300 °C a liquid of composition $C_L = 45\%$ Ni is in equilibrium with a solid of composition $C_S = 57\%$ Ni. We know that mass must be conserved, hence for a batch of 100-g mass,

$$L + S = 100 \text{ g}$$

We also know that the masses must balance, i.e., the total amount of Ni (C_0) must reside in the liquid and solid phases

$$C_L L + C_S S = 100 C_0$$

Substitution of $L = 100 - S$ results in Eq. (3-3a). Then

$$\frac{S}{S + L} = \frac{50 - 45}{57 - 45} 100 = 42\%$$

and

$$L = 100 - 42 = 58\%$$

Amount of nickel in 100-g alloy:

$$\text{solid} = (57)(0.42) = 23.9 \text{ g}$$

$$\text{liquid} = (45)(0.58) = \underline{26.1 \text{ g}}$$

$$\text{total} \qquad = 50.0 \text{ g or } 50.0\%$$

3-1-3 Eutectic Systems

Generally, elements that exhibit a greater than 15% difference in atomic radii or have a different crystal structure, are soluble in each other only up to a certain limit. When this limit of solid solubility is exceeded, the excess solute atoms are rejected into a second *phase* (phase means structurally homogeneous part of the system), which may again be a solid solution. The equilibrium diagram shows the temperatures and concentrations at which a given phase can exist. Thus, the equilibrium diagram is like a political map that reveals what phases to expect; therefore, it is also called a *phase diagram*.

An example is the silver–copper system (with atomic radii of 1.444 and 1.278 Å, respectively). The phase diagram (Fig. 3-5) shows that the maximum solubility

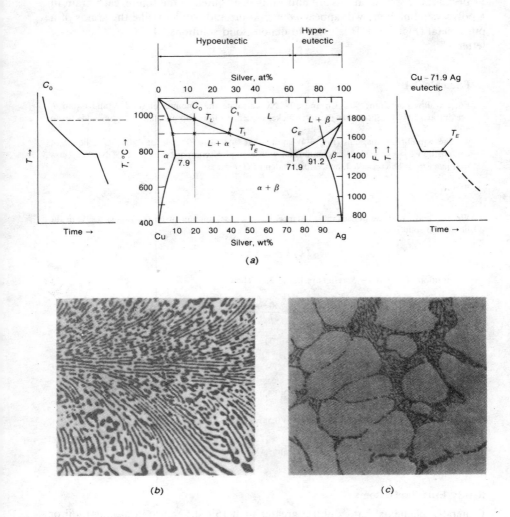

FIGURE 3-5
Limited solid solubility may result in (*a*) eutectic solidification. (*b*) The structure of a eutectic alloy is often lamellar. (*c*) A pro-eutectic alloy contains primary α grains surrounded by the eutectic. [(*b*) *Courtesy Dr. H. Kerr, University of Waterloo;* (*c*) *from Metals Handbook Desk Edition, American Society for Metals, 1985, p. 6.50. With permission.*]

of Ag in Cu is 7.9% and that of Cu in Ag is 8.8%. A solid of overall composition between these limits will consist of a two-phase mixture.

A unique point exists at 71.9% Ag. An alloy of this composition cools until it solidifies, like a pure metal, at a constant temperature T_E. However, T_E is below T_m of both copper and silver; therefore, this low-melting composition is called, from the Greek, the *eutectic composition*. The temperature of its solidification (or melting) is termed the *eutectic temperature*. An examination of the microstructure of the solidified eutectic shows that, within each crystal, two phases can be distinguished: one is a solid solution of 7.9% Ag in Cu, and can be conveniently called α solid solution, while the other, β solid solution, contains 91.2% Ag (i.e., it is a solid solution of 8.8% Cu in Ag). When the temperature drops below T_E, the mutual solubilities of Cu and Ag diminish, and the compositions of the α and β phases are given by the *solvus* lines. The two phases frequently appear as parallel plates, therefore, the eutectic shown in Fig. 3-5b is called *lamellar*. Because the *eutectic transformation* can occur only at a given composition and temperature, it is called an *invariant reaction*.

If the alloy contains 20% Ag, solidification begins at T_L with the formation of solid-solution nuclei, of approximately 6% Ag. At some intermediate temperature T_1, more α coexists with a liquid of C_1 composition. On reaching the eutectic temperature T_E, the remaining liquid is of the eutectic composition and solidifies as a eutectic. Thus, the microstructure consists of α solid-solution crystals embedded in the eutectic (Fig. 3-5c). Because the α crystals formed prior to eutectic solidification, we may also say that the structure consists of *pro-eutectic* α in a eutectic *matrix*. The eutectic itself is a two-phase structure, yet it is often regarded as a single phase, especially from the point of view of its effects on mechanical properties.

Example 3-2

Calculate the relative proportions of phases in a copper–silver alloy of eutectic composition, just below the eutectic temperature.

The inverse lever-arm rule, Eq. (3-3) can again be applied to find the proportion of α

$$\frac{\alpha}{\alpha + \beta} = \frac{C_\beta - C_E}{C_\beta - C_\alpha} = \frac{91.2 - 71.9}{91.2 - 7.9} = 23.2\%$$

the proportion of $\beta = 100 - 23.2 = 76.8\%$.

Example 3-3

Calculate the relative proportions of phases in a solidified copper–silver alloy of 20% Ag content just below the eutectic temperature. First calculate the total weight percent α and β, and then the relative proportions of pro-eutectic α and eutectic E.

The total weight percent α is, by analogy to Example 3-2,

$$\frac{\alpha}{\alpha + \beta} = \frac{C_\beta - C_0}{C_\beta - C_\alpha} = \frac{91.2 - 20}{91.2 - 7.9} = 85.5\%$$

and that of $\beta = 100 - 85.5 = 14.5\%$. This tells us little about the structure. To find the proportion of pro-eutectic α, the calculation is carried out for a temperature just above the eutectic.

$$\frac{\alpha}{\alpha + E} = \frac{C_E - C_0}{C_E - C_\alpha} = \frac{71.9 - 20.0}{71.9 - 7.9} = 81.1\%$$

The α crystals will be surrounded by $100 - 81.1 = 18.9\%$ matrix of eutectic composition (in which 23.2% is α, Example 3-2).

3-1-4 Other Systems

Phase diagrams of practical alloy systems may show further features. We may generalize the discussion by calling one of the metals A and the other B.

Peritectic Systems   When the melting points of two metals are greatly differ-ent, the invariant reaction is often of the *peritectic* type (Fig. 3-6a). An alloy of peritectic composition C_P begins to solidify with the formation of α solid-solution crystals. At the invariant temperature T_P all the remaining liquid must disappear and the entire solid must transform into a β solid solution. This can be achieved only by the circumferential diffusion of B atoms into the already solidified α crystals; hence the name *peritectic diffusion reaction*. In an alloy of C_0 composi-tion a two-phase structure results.

FIGURE 3-6
(*a*) Large differences in melting points often lead to peritectic solidification. (*b*) Many elements form intermetallic compounds ($A_m B_n$) or intermetallic phases (γ).

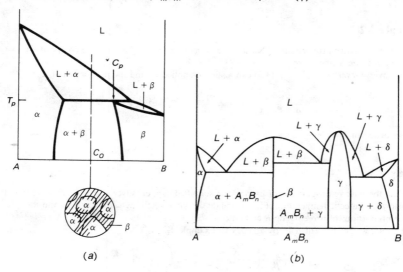

(*a*)   (*b*)

Intermetallic Phases In many alloy systems phases with distinct properties are formed. Their composition is characterized by a more or less fixed ratio of the two elements.

In *intermetallic compounds* the ratio is stoichiometric and can be denoted as $A_m B_n$. The two atomic species A and B occupy fixed sites in the unit cell. Bonds may have predominantly covalent characteristics, with electrons shared between atoms. Even though some electrical conductivity is retained, these compounds are often brittle, hard, and of a high melting point (Fig. 3-6b). Compounds may also form between metallic and nonmetallic elements. The most important example is Fe_3C in steels (Sec. 3-1-6).

In some cases the intermetallic can exist over an extended composition range and then one speaks of an *intermetallic phase* (Fig. 3-6b).

3-1-5 Solid-State Reactions

The solid alloy may undergo further changes as the temperature drops. We saw one such event in Fig. 3-5 where the phase boundary (solvus) of the *terminal solid solution* (i.e., solid solution at the end of the phase diagram) indicated that less of the solute species is kept in solution with decreasing temperature. The excess solute must then separate (*precipitate*) into a second phase, and this can be a powerful mechanism for controlling the properties of solid alloys.

As the temperature drops, the stability of various phases may change too and, at some critical temperature, transformations similar to those occurring in the liquid-to-solid transition may occur; to distinguish them from their counterparts, their names are formed by the ending "-oid." Thus, when a homogeneous solid solution decomposes into two phases, one speaks of *eutectoid* or *peritectoid* *transformation*.

3-1-6 The Iron–Iron Carbide System

Some transformations are best illustrated on the example of the Fe–Fe$_3$C system (solid lines, Fig. 3-7).

Above 2% C, the alloy is called a white cast iron. In the absence of other alloying elements and at fast cooling rates, solidification between 2% and 4.3% C begins with the rejection of γ solid-solution crystals and ends with the formation of a γ-Fe$_3$C eutectic matrix. At 4.3% C, solidification is eutectic (Fig. 3-8a); at higher carbon contents, Fe$_3$C is embedded in the eutectic matrix. Because of the industrial importance of different phases in the system, each phase is also given a name. Thus, the eutectic is called *ledeburite*, and is composed of *austenite* (the γ phase) and *cementite* (Fe$_3$C). Since the Fe$_3$C is formed during solidification, it is called *primary cementite*.

Steels are alloys of less than 2% carbon in iron. There is a peritectic reaction at 1495 °C which we may ignore for our purposes. At temperatures of practical importance, the C occupies interstitial sites, and at higher temperatures forms the

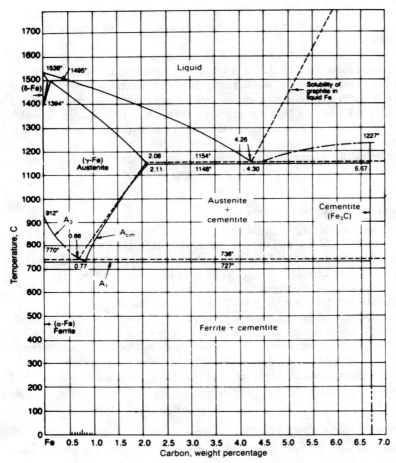

FIGURE 3-7
Carbon may occur in iron in the form of cementite (solid lines) or graphite (broken lines).
(*From Metals Handbook Desk Edition, American Society for Metals, 1985, p. 28.2. With permission.*)

fcc γ solid solution (austenite). With dropping temperatures, the solubility of C in austenite decreases and the excess over 0.8% is rejected in the form of Fe_3C (*secondary cementite*). Of greatest importance is the eutectoid decomposition of austenite, at 727 °C (also called the A_1 transformation temperature), into cementite and a solid solution of C in α-iron (*ferrite*). The eutectoid is of lamellar structure (Fig. 3-6L) but, because it is formed in the solid state, the short diffusion paths make the lamellae much finer than in eutectics. It is called *pearlite* because the Fe_3C lamellae cause a fresh fracture surface to shimmer in a pearl-like

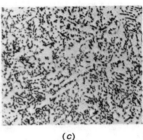

(a)   (b)   (c)

FIGURE 3-8
In the absence of graphite-stabilizing elements such as Si, carbon usually separates in the
form of cementite. (a) Primary cementite is found in the eutectic. (b) Eutectoid decomposition
of austenite results in pearlite in which cementite and ferrite platelets alternate. (c) By
spheroidizing heat treatment, the cementite can be brought into a spherical form. (*Courtesy
Dr. G. F. VanderVoort, Carpenter Technology Corporation. Also in Metals Handbook Desk
Edition, American Society for Metals, 1985, pp. 27.28, 35.37, 35.42.*)

manner. The solubility of C in ferrite decreases with dropping temperatures and
tertiary cementite is rejected.

(Note: The tertiary cementite usually precipitates onto the eutectoid cementite,
except in very low carbon steels in which it precipitates on the grain boundaries.)
Since ferrite and pearlite form by the solid-state decomposition of austenite, the
grain size is determined by the prior austenite grain size and the rate of cooling.

Example 3-4

Various countries adopt their own standards for the designation of steels. In the U.S. and Canada,
the American Iron and Steel Institute (AISI) and Society of Automotive Engineers (SAE)
four-digit system is widely accepted. The first two digits show what kind of alloying elements are
added to iron, and the last two digits give the carbon content in hundredths of percent. Thus, AISI
1040 is a plain carbon steel of 0.4% C content. (A complete description of the classification and
designation of steels is given in MHDE, pp. 4.1–4.19.) Calculate the metastable equilibrium phases
present in this steel at (a) 1000 °C and (b) room temperature.

(a) At 1000 °C, the structure is 100% austenite, with 0.4% C in interstitial solid solution in fcc
iron.

(b) If the proportion of the eutectoid is denoted as ED, the proportion of pro-eutectoid α at
room temperature can be obtained, as in Example 3-3, by taking a temperature just above the
transformation temperature:

$$\frac{\alpha}{\alpha + ED} = \frac{C_{ED} - C_0}{C_{ED} - C_\alpha} = \frac{0.77 - 0.40}{0.77 - 0.02}\,100 = 49.3\%$$

Eutectoid pearlite will constitute $100 - 49.3 = 50.7\%$ of the volume. The two phases will show up
in the same proportions in the microstructure (a two-dimensional slice through the body).

The proportions of ferrite (F) and cementite (CM) in the eutectic pearlite can be obtained, as in Example 3-2:

$$\frac{F}{F + CM} = \frac{C_{CM} - C_{ED}}{C_{CM} - C_F} = \frac{6.67 - 0.77}{6.67 - 0.02} 100 = 88.8\%$$

Thus, $100 - 88.8 = 11.2\%$ cementite platelets (by weight) are alternating with platelets of ferrite.

We could also calculate the total weight fraction of α present in both the pearlite and ferrite from the composition just below the eutectoid temperature:

$$\frac{F}{F + CM} = \frac{C_{CM} - C_0}{C_{CM} - C_F} = \frac{6.67 - 0.4}{6.67 - 0.02} 100 = 94.3\%$$

However, this will tell us little about the properties of the steel.

3-2 STRUCTURE-PROPERTY RELATIONSHIPS

The aim of most casting processes is to produce parts that not only have the correct size and shape, but also possess the best possible properties. For load-bearing components, this calls for high strength coupled with acceptable ductility, properties that are greatly influenced by structure.

3-2-1 Metals and Single-Phase Alloys

We saw in Sec. 2-1-1 that metals subjected to loading deform permanently at some critical stress. When the deformed specimen is observed under an optical microscope, deformation appears to have taken place by the slip of adjacent zones (Fig. 3-9a). At high magnifications each *slip zone* appears composed of many small steps, indicating that displacement must have taken place along *preferred slip planes* in each crystal (Fig. 3-9b).

Calculations can be made to show that slip by the massive movement of entire adjacent crystal zones would take much higher stresses than actually observed. Indeed, it is found that slip takes place by the movement of *line defects* (*dislocations*) along preferred slip planes in the crystal lattice: Shear stresses on these planes must reach a critical value before deformation can commence. This *critical shear stress* depends on the metal, crystal structure, and shear direction. In the simplest view, a dislocation could be regarded as an extra line or plane of atoms inserted into the structure (*edge dislocation*, Fig. 3-9c); thus, it is only necessary to dislodge this extra line of atoms along the slip plane instead of moving hundreds of thousands of atoms of a slip plane at the same time. Many of the deformation characteristics of metals can be interpreted by contemplating the ease with which these dislocations can move and by considering obstacles that may impede or arrest their movement.

One might expect that dislocation movement (slip) should be easier on planes that give the smoothest movement, the least bumpy ride: If atoms are visualized as touching spheres (Fig. 3-2), one finds that slip takes place most readily in the most *closely packed planes* in the *closest-packed crystallographic directions*.

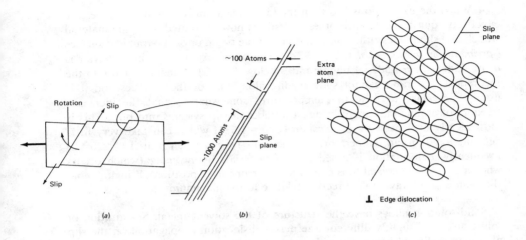

FIGURE 3-9
A single crystal subjected to tension deforms by (a) slip of adjacent zones. (b) At higher magnification, closely spaced slip planes are discovered. (c) Within each slip plane, there are numerous line defects, dislocations.

1 In the fcc structure (Fig. 3-2a) there are 4 equivalent closely packed planes (the {111} octahedral planes) with 3 equivalent slip directions ⟨110⟩, giving a total of 12 independent *slip systems* (i.e., combinations of slip planes and directions). If slip is limited on one plane because dislocations are arrested, there is always a likelihood that some other slip system will be oriented in the direction of the maximum deforming shear stress. Because of the close atomic packing, the critical shear stress is relatively low. Thus we can conclude that fcc metals should be readily deformable, essentially at all temperatures. Indeed, this is characteristic of Pb, Al, Cu, Ni, and γ-iron.

2 In the bcc structure, one cannot readily identify obviously close-packed planes, but a clearly closest-packed direction is found in the body diagonal (Fig. 3-2b). Therefore, these crystals slip in systems containing various planes that have the body diagonal ⟨111⟩ as a common slip direction, rather like a bunch of pencils that is deformed by sliding the pencils along their axes. This so-called *pencil slip* allows extensive deformation, for example in α-iron and β-titanium. However, because packing is not as close as in fcc metals, the critical shear stress may be higher.

3 The deformation of hcp structures is governed by the ratio of height-to-side dimensions, the *c/a ratio* (Fig. 3-2c). In an ideal hcp structure this ratio would be 1.633. Some metals show a larger ratio, i.e., the basal planes are more widely separated. Slip then occurs only in these planes along the three equivalent closest-packed directions (*basal slip*, Fig. 3-2c, as in zinc, c/a = 1.856).

4 When the c/a ratio is less than the ideal, the atoms of the basal planes are effectively squashed into each other and slip is now prevented here; the material will choose slip planes either along the side of the prism or on a pyramidal surface (*prismatic* or *pyramidal slip*, Fig. 3-2d). The prime example of this behavior is α-titanium ($c/a = 1.587$, at temperatures below 880 °C). A metal of close to the theoretical c/a ratio cannot slide readily along any of these planes and it is usually necessary to raise the temperatures somewhat so that the increased freedom of atomic movement brings a number of slip systems into play. This is most clearly evidenced by magnesium ($c/a = 1.624$) which can take very little deformation at room temperature but deforms readily when heated to 220 °C. Frequently, deformation is aided in hcp materials by *twinning* (which occurs when a part of the crystal flips over into a mirror-image position), bringing more slip planes into a favorable direction relative to the maximum shear stress.

Solid-solution alloys have the structure of the solvent metal. Substitution of solute atoms of slightly different size makes dislocation propagation on the slip planes more difficult; thus, strength increases without necessarily reducing ductility. Interstitial elements play a similar role in impeding dislocation mobility although they can also have an embrittling effect if they entirely block the movement of dislocations.

3-2-2 Two-Phase Materials

In considering the properties of two-phase structures, it is necessary to recognize that the presence of two phases immediately implies the existence of an *interface* between them. Any interface, even a grain boundary in a pure metal, is a site of many unsatisfied, broken interatomic bonds that add up to an excess energy, the *interfacial energy* γ. The magnitude of interfacial energy is larger when the mismatch between adjacent atomic groupings is greater. Thus, the interfacial energy between the vapor and solid γ_{SV} of a substance, say metal, is much larger than the interfacial energy between its liquid and solid phases γ_{SL}. The relative magnitudes of interfacial energies between two dissimilar materials are readily judged by placing a liquid drop of one on top of a flat, solid surface of the other. The liquid drop sits in place (hence the name *sessile drop*) but is free to change its

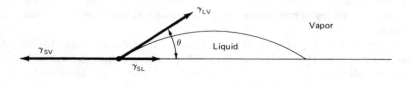

FIGURE 3-10
A liquid wets the solid surface when the angle θ is small.

shape until the *surface tensions* establish a force equilibrium (Fig. 3-10):

$$\gamma_{SV} = \gamma_{SL} + \gamma_{LV} \cos \theta \qquad (3\text{-}4)$$

When $\theta < 90°$, the surface is *wetted* and the drop spreads out; when $\theta > 90°$, the surface is *not wetted* and the liquid forms, in the limit, a spherical droplet.

Wetting, then, is an indication of relative surface energies and, through these, a measure of the strength of the interface. Wetting is a sign of reasonable match between the atomic lattices of the contacting phases, and one can expect a wetted interface to resist stresses that might pull the phases apart. A nonwetted interface, on the other hand, can behave as a preexisting crack and thus impair mechanical properties (Sec. 2-1-5). It should be remembered that surface tension is indeed a surface effect; therefore, even the minutest amounts of a contaminant, segregated at the interface, can substantially reduce wetting.

With these preliminaries in mind, one would intuitively expect that the properties of a two-phase structure must depend on a number of factors, such as the properties, quantity, distribution, shape, and size of individual phases and the nature of interfaces between them. Some basically different situations can be envisaged (Fig. 3-11):

1 Both phases are ductile and wet each other; they behave like a homogeneous body, dislocations pass freely through both phases, and properties can be estimated from the relative volumes of the two phases.

2 One of the phases is ductile, the other is brittle and wetted. In this case, the relative quantity, shape (*morphology*), and location of the brittle phase become dominant, because dislocations will pass freely through the ductile phase but will be blocked by the brittle phase.

  a If the brittle phase is the matrix (i.e., the other phase is embedded in it), the resulting structure will be brittle; it may have a high compressive strength, but tensile strength and ductility are low because cracks initiated in the brittle phase will easily propagate through the material (an example of this is white cast iron).

  b If the matrix is ductile, but coarse plates (*lamellae*) or needles (*aciculae*) of the brittle phase weave through it, the brittle phase causes stress concentrations on loading, cracks propagate through the brittle phase, and the structure will be both weak and of low ductility, and it will have low fracture toughness.

  c Finer plates confined within a grain and systematically aligned into a lamellar structure will substantially strengthen the matrix but usually at the expense of ductility, since the brittle plates fracture during deformation and the stress raisers terminate plastic deformation in the ductile matrix. An example of this is shown by pearlitic carbides in Fig. 3-12.

  d If the plate-like structure is extremely fine and closely spaced so that the hard phase acts as a barrier to dislocation mobility, the composite structure may show high strength coupled with reasonable ductility, because rapid strain hardening delays necking (see Sec. 4-1-1).

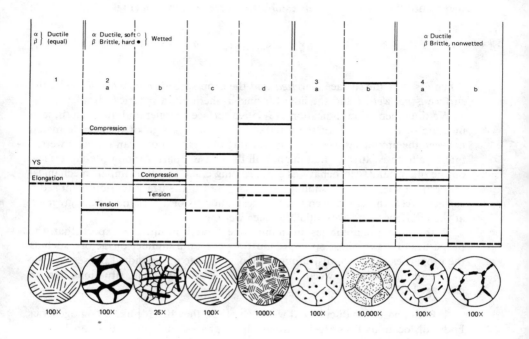

FIGURE 3-11
Properties of two-phase structures depend on the properties, wetting, shape, size, and distribution of the two phases. Note the different magnifications for the schematic microstructures.

FIGURE 3-12
Ductility is highly sensitive to second-phase particles, especially if they are of unfavorable shape and of low ductility. (*After T. Gladman, B. Holmes, and L. D. McIvor, in Effect of Second-Phase Particles on the Mechanical Properties of Steel, Iron and Steel Institute, London, 1971. With permission of The Metals Society.*)

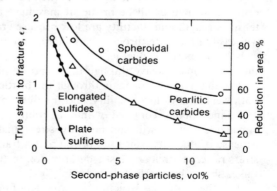

3 The hard constituent is wetted by the soft phase and is in a roughly spheroidal form.

a If the hard particles are relatively coarse and widely spaced, they have only a minor effect on the strength of the material. Ductility is impaired, but not excessively (spheroidal carbides in Fig. 3-12) because dislocations circumnavigate such large blocks of harder material, and the notch effect is minimal because of the large notch radius.

b When the particle size becomes small enough to arrest or slow down dislocations, the structure will be strengthened and ductility reduced, depending on the total quantity and spacing of hard particles. Great strengthening can be obtained by loading the ductile matrix with masses of particles (until the matrix becomes little more than a ductile cement), but then ductility will greatly suffer.

4 The hard phase is not wetted by the matrix; the interface between them acts as a premade crack.

a When the nonwetted particles are inside ductile grains, they can be relatively harmless, although any cracks that form can coalesce readily, greatly impairing ductility. If the base material is relatively brittle, strength is reduced too because the interface acts as a notch.

b Nonwetted particles are most harmful when located on the grain boundary which tends to be less ductile in any event.

3-2-3 Ternary and Polycomponent Alloys

In practice, very few truly *pure metals* or *binary alloys* (i.e., alloys formed by two atomic species) are used. Even if only on the order of a few parts per million, contaminants are always present, while intentional additions may reach such high proportions that it becomes difficult to classify a material according to its base metal. Such *ternary*, *quaternary*, etc., alloys (*polycomponent systems*) still exhibit distinct phases found in binary alloys. Thus, one can find solid solutions, eutectics, peritectics, intermetallics, and their various combinations. The behavior and properties of polycomponent systems can be derived by analogy to two-phase materials, especially if the phase diagram is known or at least a section of the phase diagram (a *pseudobinary diagram*, for constant percentages of the other alloying elements) has been established. Great strides have been made in using the computer to predict what phases to expect in polycomponent alloys, at any temperature and composition, on the basis of binary phase diagrams and a data bank of thermodynamic data.

3-2-4 Inclusions

The term *inclusion* is used to describe foreign particles in a metallic structure. They find their way into the alloy usually during melting (for example, from the furnace lining, contaminations of the charge, or even as a result of reaction—usually oxidation—with the surrounding atmosphere), or during pouring. As with all second-phase particles, their effect depends greatly on whether they are wetted by the matrix or not (Sec. 3-2-2).

If inclusions are wetted, strong, and perhaps even ductile, and are dispersed inside the grains in an approximately globular or fibrous form, they are harmless and sometimes even useful. Arranged along grain boundaries they are likely to be harmful, unless they are extremely small and well distributed. Brittle plates and, particularly, films (such as are formed by aluminum oxide) are detrimental, as are low-strength inclusions in elongated or platelike forms. Strength may not be greatly affected, but ductility (Fig. 2-4a; also sulfides in Fig. 3-12), fatigue strength, and fracture toughness suffer. Hence modern casting technology aims at producing clean metals, free from inclusions.

Nonwetted inclusions are almost always harmful, reducing the strength, ductility, and fatigue and impact properties of the material. If gases are present, they tend to congregate on the interface between inclusion and matrix and can build up such high pressures that a bubble (blister) is formed on inclusions close to the surface, particularly if the part is in high-temperature service or is heated during manufacture. Even inside the body, gases segregating on nonwetted interfaces aggravate the crack effect and lead to embrittlement (e.g., hydrogen embrittlement in steel, Sec. 9-3-2).

3-2-5   Effects of Grain Size

In a polycrystalline material, at relatively low temperatures, individual grains can deform only by the propagation of dislocations. Grain boundaries represent defects in the structure and are thus sources of dislocations. At the same time, grain boundaries also present obstacles to dislocation propagation; therefore, it is generally found that the strength of a material increases with decreasing *grain size* (Fig. 3-13) according to the Hall–Petch relationship*

$$\sigma = \sigma_0 + k_y d^{-1/2} \tag{3-5}$$

where d is the average grain size, and σ_0 and k_y are material constants. In some metals an even better agreement is found between strength and the *size of subgrains* (relatively strain-free subunits within the larger grain, arranged with a slight crystallographic misorientation).

It should be noted that Eq. (3-5) holds only when dislocation propagation is the primary mechanism of deformation. This is true of deformation in the *cold* temperature range (Fig. 2-11), i.e., typically below $0.5T_m$ on the homologous temperature scale. At higher temperatures, in the *hot* temperature range, other deformation modes are possible at certain strain rates $\dot{\epsilon}$. At high strain rates, dislocation movement still dominates (although other mechanisms come into play

---

*E. O. Hall, *Proc. Phys. Soc. London* **64B**:747 (1951); N. J. Petch, *J. Iron Steel Inst. London* **173**:25 (1953).

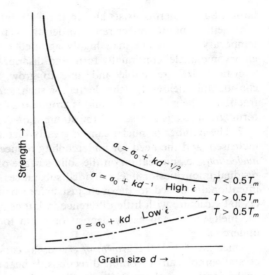

FIGURE 3-13
The strength of metals greatly increases with diminishing grain size, except when deformation takes place at high homologous temperatures and low strain rates, with a massive diffusion of atoms.

too; Sec. 4-1-4) and strength decreases with increasing grain size (Fig. 3-13). However, at the very low strain rates typical of creep (Sec. 2-1-7) and temperatures close to T_m, there is time for substantial diffusion to take place. The part may deform by the sliding of grains as complete blocks relative to each other, or by the reshaping of individual grains in the loading direction. Both processes are easier if grain size is small; hence, creep strength drops with decreasing grain size, and a large grain size is preferable for materials destined for high-temperature service.

3-3 STRUCTURE AND PROPERTIES OF CASTINGS

It is clear from the above discussion that the properties of a solidified alloy are dependent not just on composition but also on grain size and the shape and distribution of phases. These factors may be controlled and modified in the course of solidification.

3-3-1 Nucleation and Growth of Grains

In Sec. 3-1 we gave a highly simplified account of solidification. In reality, the processes of nucleation and growth are more complex.

Nucleation and Growth There are two ways in which nuclei can form:

1 *Homogeneous nucleation* occurs only in very clean melts. The nucleus is formed by the ordering of atoms into positions corresponding to the crystal lattice. Such ordering exists also in the melt but only over short distances. Below the melting point longer-range ordering is possible but much of it is only temporary. Atoms are in a highly agitated condition at this temperature and embryonic nuclei continually form and disappear: Only nuclei that have reached a critical size are stable and able to grow, and then only at temperatures considerably below T_m (the degree of such *undercooling* can be expressed as a fraction of the melting point and is around $0.2T_m$ in pure metals); since few nuclei form, grain size is coarse and the strength of the solidified casting is low.

2 The number of nuclei can be greatly increased, grain size reduced, strength increased, and the need for undercooling reduced or eliminated by *heterogeneous nucleation*, i.e., nucleation on the solid surface of *nucleating agents* which may be residual impurities or finely divided substances (often intermetallic compounds), intentionally added to the melt just before pouring. If they have a compatible crystal structure with little difference in lattice spacing and if they are wetted by the melt, atoms can easily deposit on them to form crystals at less than 5 °C undercooling.

Once nucleated, the *growth of crystals* occurs in all directions but is faster in crystallographically favorable directions. If heat extraction is omnidirectional and nuclei form throughout the melt, the resultant structure will consist of *equiaxed crystals* (of roughly equal dimensions in all directions).

Grain Size Nucleation and growth of grains occur simultaneously, but at different rates (Fig. 3-14). Nucleation rate is maximum at substantial undercooling, whereas growth rate peaks close to the solidus temperature. Therefore, grain size depends on residence time at a given temperature, which in turn depends on *cooling rate*. At low cooling rates, there is time for the few nuclei formed to grow and the structure will be coarse-grained. At high rates of cooling, a high nucleation rate gives many sites on which growth can take place and grain size

FIGURE 3-14
During solidification, growth of grains is fastest at temperatures close to the solidus, but more grains are nucleated at lower temperatures.

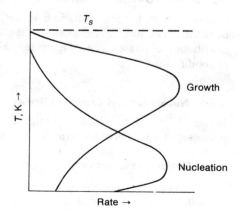

will be small. At extremely high rates of cooling crystallization may be suppressed and a noncrystalline (*amorphous*) body obtained (Sec. 6-2-1).

Nuclei that preexist in the melt (either as homogeneous or heterogeneous nuclei) are dissolved by overheating the melt. Therefore, with increasing superheat, grain size increases too. The magnitude of *superheat* is usually expressed as the difference between melt temperature and liquidus temperature

$$\text{Superheat} = T_{\text{melt}} - T_L \qquad (3\text{-}6a)$$

There is not much information, but it is reasonable to assume that, for materials of different melting points, the effects of superheat may be rationalized by relating it to the melting point

$$\text{Superheat} = \frac{T_{\text{melt}} - T_L}{T_L} \qquad (3\text{-}6b)$$

Freshly formed crystals are extremely weak and easily broken to provide more nuclei. Thus grain size is refined by thermal currents or mechanical agitation of the solidifying melt, provided that the superheat is low and crystal fragments are not remelted but survive as nuclei (*grain multiplication*).

3-3-2 Solidification of Melts

When a melt is poured into a colder mold, metal in contact with the mold solidifies in the form of roughly equiaxed fine grains, because cooling rates are high (*chill zone*) and the mold wall induces heterogeneous nucleation. The latent heat of fusion released during solidification slows down the rate of solidification and the course of further solidification depends on the type of alloy being cast.

Pure Metals Solidification proceeds by the growth of a few favorably oriented nuclei, in the direction of heat extraction. This leads to the often observed *columnar structure* (Fig. 3-15a) throughout the bulk of the casting. Because of the preferred growth direction of these large grains, the casting will have very anisotropic properties. (In a larger casting the central zone cools very slowly and heat extraction is almost omnidirectional; if solidification proceeds by heterogeneous nucleation, this results in an equiaxed structure, of grains much coarser than on the surface).

Since most metals shrink on solidification (Fig. 3-1c), the liquid meniscus gradually drops and, if there is no supply of fresh liquid, a *shrinkage cavity* remains. A cavity of the geometry shown in Fig. 3-15a is called a *pipe* and always forms when a pure metal solidifies.

Eutectics Eutectics, like pure metals, solidify at a constant (invariant) temperature (Fig. 3-5) and the solidification front is more or less plane (Fig. 3-15a). Within each grain, there are several groups, *eutectic cells* or *colonies*. The

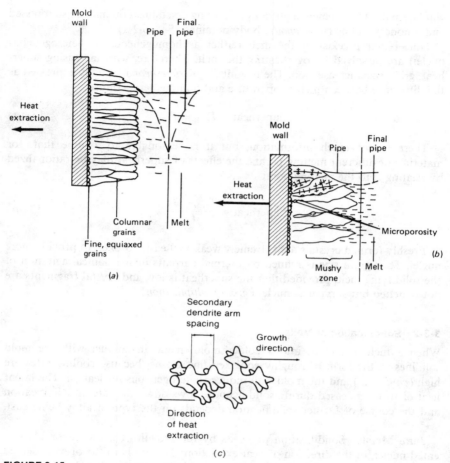

FIGURE 3-15
Solidification proceeds with (a) the growth of columnar grains in pure metals but (b) with the growth of dendrites in solid solutions. (c) Dendrites grow in crystallographically favorable directions.

properties of the casting can be influenced by various means:

1 Rapid cooling reduces cell size and, in lamellar eutectics, also interlamellar spacing and thus the strength of the casting.

2 Nucleating agents promote the formation of fine equiaxed eutectic grains of superior mechanical properties.

3 The lamellar structure is only one of the possible forms of the eutectic. In certain instances the "natural" morphology of the eutectic may be changed by *modification*, with marked changes in properties. For example, the platelets of a lamellar eutectic may be changed into spheres (spheroidal structure) or rod-like particles. Such structures have distinctly different properties; typically, a spheroidal eutectic has higher ductility than a lamellar one (see Fig. 3-11).

Solid Solutions Solid solutions solidify over the freezing range $T_L - T_S$ (Fig. 3-4) and this has significant effects on the structure. Crystals again grow in the direction of heat extraction (Fig. 3-15b) but solidification begins with a leaner solid solution (one that contains less than the nominal amount of alloying element) whereas the remaining solid is enriched. This, coupled with local undercooling in the liquid, leads to the formation of a branched crystal skeleton. This resembles a tree (Fig. 3-15c), and is, therefore, called *dendrite* (from the Greek *dendron* = tree). When the melt finally solidifies, each grain contains one or more complete dendrites (*cellular dendritic structure*). Dendrite arms are initially very weak and can be easily broken by thermal and/or mechanical agitation to give, at low superheat, many nuclei and thus a fine grain size. At higher cooling rates or in the presence of nucleating agents, grains are refined and, more importantly for mechanical properties, the *secondary dendrite arm spacing* is also reduced.

The intricate network of dendrite arms makes free movement of the remaining liquid difficult and spaces formed between arms may be starved of the fluid necessary to make up for solidification shrinkage. Consequently, *microporosity*—characterized by the presence of holes with ragged edges—is typical of solid solutions. Such holes represent inclusions of zero strength and, because of the notch effect, are harmful for strength and ductility. In alloys the total shrinkage is similar to that of the constituent metals, but the pipe is much smaller and a large proportion of total shrinkage is in a *distributed* form (Fig. 3-16b). An alloy with a wider solidification range (long freezing range, $T_L - T_S$) is more prone to porosity. The freezing range may be normalized in relation to T_L

$$\text{Freezing range} = \frac{T_L - T_S}{T_L} \tag{3-7}$$

Other Systems Properties and porosity in a binary alloy system may be predicted with fair accuracy by considering the phase diagram. For example, in the eutectic system of Fig. 3-16, microporosity increases from A to B until the solubility limit of the α solid solution is reached, declines toward the eutectic composition, to rise again to the β solid solution. Strength (characterized here by the yield strength) rises by solid-solution alloying and changes little with the appearance of the eutectic. Ductility may rise or fall with solid-solution alloying; the effect of the eutectic depends greatly on its morphology. Lamellar eutectics tend to be less ductile than spheroidal ones (Fig. 3-11).

Occasionally an alloying element cannot be dissolved even in the liquid metal; instead, it exists as a separate liquid phase (*limited liquid solubility* or *total immiscibility*). A prime example is lead which is practically insoluble in many metals. Its effect on properties depends on its distribution. Since it is soft, it can

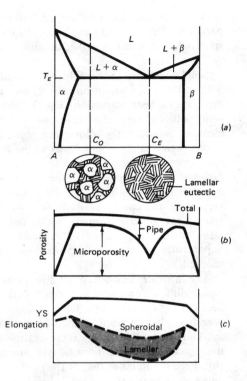

FIGURE 3-16
(a) In a eutectic system, (b) microporosity dominates when the structure consists primarily of a solid solution, and (c) the mechanical properties reflect the effects of structure.

act as a useful lubricant if trapped in interdendritic spaces within grains or in a globular form on grain boundaries. However, because of its low melting point, it destroys hot strength when it segregates on grain boundaries.

3-3-3 Nonequilibrium Solidification

We assumed until now that cooling conditions during solidification allow the attainment of complete equilibrium. This is seldom the case because cooling rates in most casting processes are relatively fast (on the order of a fraction of a degree to a few degrees per second). Therefore, cast structures typically show a number of nonequilibrium microstructural features.

Microsegregation (Coring) Diffusion processes are, in general, too slow to establish equilibrium and, particularly when the freezing range is wide, concentration gradients due to lack of diffusion are found in each grain of a solid solution.

In a system like that shown in Fig. 3-17, solidification of a melt of composition C_0 begins with the rejection of α solid-solution crystals of composition C_1. On

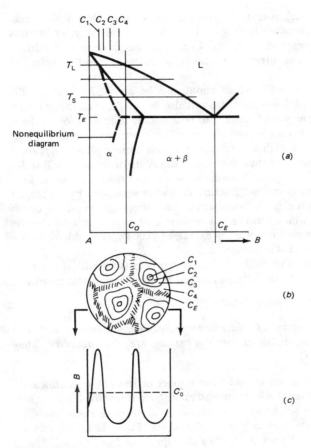

FIGURE 3-17
At usual rates of cooling, (a) a solid solution of C_0 composition will show (b) coring, with (c) a higher concentration of the alloying element on the grain boundaries.

further cooling, the crystals not only grow but, according to the equilibrium phase diagram, their composition would also have to become enriched in the B element as dictated by the solidus. If time is insufficient to allow B atoms to diffuse into the already solidified core, the core remains leaner in the alloying element. The excess B atoms are retained in the melt, and solidification does not end when the T_S equilibrium solidus temperature is reached; instead it continues by the gradual deposition of richer and richer layers. The nonequilibrium solidus shown as a broken line in Fig. 3-17 represents the average composition of the solid. It is even possible that a liquid phase remains until the eutectic temperature is reached, when the remaining liquid finally solidifies along the grain boundaries as a eutectic.

Because the centers (cores) of crystals grown during nonequilibrium solidification have a different composition (Fig. 3-17b), it is usual to refer to this phenomenon as *microsegregation* or *coring*. Coring is also evident in dendrites, with the alloying element concentration increasing from the core to the surface of each dendrite arm.

If the B element of Fig. 3-17 (which could also be an A_mB_n intermetallic compound) is brittle, the solidified structure will also be brittle, even though from the equilibrium diagram one would judge it to be a ductile solid solution of the α phase. Such an alloy will also suffer from *hot shortness* when heated above the T_E eutectic temperature: even though according to the equilibrium phase diagram it is nominally a solid solution and thus should be readily deformable, it will suffer fracture by separation at the grain boundaries where the nonequilibrium, low-melting eutectic is present. Hot-short fracture is readily identified by its ragged appearance as it follows the grain boundaries. Sometimes the presence of an unsuspected contaminant which forms a low-melting eutectic may totally destroy ductility. A striking example is sulfur in excess of 0.004% (i.e., 40 parts per million) in nickel or high-nickel superalloys.

Microsegregation may be undesirable for a number of reasons and, if it cannot be prevented, may be partially or fully eliminated by subsequent heat treatment (see Sec. 3-5-2).

Macrosegregation Instances of *macrosegregation*, i.e., compositional differences extending over long distances within a casting, are often observed. They are basically of three kinds:

1 So-called *normal segregation* occurs when a more or less plane solidification front (as in Fig. 3-15a) drives the lower-melting constituent toward the center. A section taken through the solidified cross section will show a lower alloying-element concentration on the surface than in the center (Fig. 3-18a). If gases are liberated during solidification, they drive the richer fluid out from the solidifying zone and contribute to the segregation of alloying elements to the center.

2 *Inverse segregation* is typical of solid-solution alloys with a dendritic solidification pattern (Fig. 3-15b). Since dendrite arms form first and have a lower alloying-element concentration, the interdendritic spaces formed by solidification shrinkage must be filled in by a liquid of higher solute concentration. This liquid flows back in a direction opposite to the growth direction of dendrites; hence, the surface has a higher than average alloying element concentration (Fig. 3-18b).

3 If insoluble compounds, inclusions, or metals immiscible in the liquid have a density that is greatly different from the melt, they rise or sink to give *gravity segregation* (Fig. 3-18c). One of the attractions of manufacturing in space is that, in the weightless environment, unusual alloys—consisting of metals of greatly different densities—may be solidified without segregation.

These forms of macrosegregation, if undesirable, must be prevented during solidification because diffusion distances are too large to equalize the composition in subsequent heat treatment.

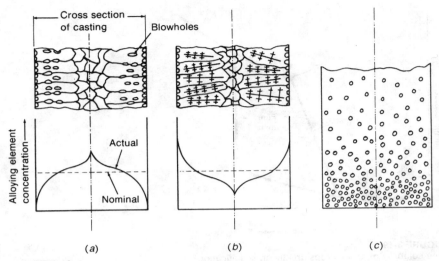

FIGURE 3-18
Solidification of almost pure metals leads to (a) "normal" macrosegregation, especially in the presence of gas evolution. (b) Dendritic solidification leads to "inverse" segregation. (c) High-density constituents that do not dissolve in the melt separate by gravity segregation.

3-3-4 Gases

Gases normally exist in a molecular form but, at higher temperatures and in contact with metal, a significant portion may dissociate into the atomic form and enter into the metal. They can be accommodated interstitially in the relatively loose, nonordered structure of melts. Thus, above the melting point, solubility of gases may be high (Fig. 3-19). Solubility drops steeply as the melt solidifies. Some gas may be trapped in the solid in the atomic form, but much is rejected at the solid-liquid interface to combine into molecules. These molecules coalesce into gas bubbles which rise in the melt or, if trapped during solidification, cause *gas porosity* (*pinholes* or larger *blowholes*) in the structure. In contrast to microporosity resulting from the freezing pattern, gas pores are generally round and, if they contain a neutral or reducing gas, they have a clean, bright surface. They too can be regarded as inclusions of zero strength, but their larger radius makes them less damaging to mechanical properties. They can also cause blistering, as discussed in conjunction with nonwetted inclusions.

Not all gases are equally soluble in all metals. Hydrogen is soluble in practically all metals because of the small size of its atoms. It may be introduced into the melt by dissociation of water from the air, the charge, or combustion products. It is particularly troublesome for aluminum and magnesium alloys. In contrast to hydrogen, nitrogen is soluble in iron but not in nonferrous metals. Noble gases (of which argon is technically most significant) are completely insoluble.

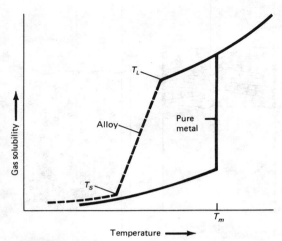

FIGURE 3-19
The solubility of gases drops greatly on solidification.

The solubility S of any one gas in the melt increases (or decreases) with the square root of the *partial vapor pressure* p_g of that gas over the melt (Sievert's law)

$$S = k\sqrt{p_g} \tag{3-8}$$

where k is the equilibrium constant. It follows that the concentration of any gas in the melt can be reduced by either reducing the overall gas pressure (by drawing a vacuum) or by bubbling a nonsoluble *scavenging* gas through the melt just before pouring. Because the partial pressure of the offending gas is zero in the scavenging gas bubbles, the offending gas is drawn out of solution into the rising scavenging gas and is removed.

Example 3-5

The maximum equilibrium solubility of hydrogen in liquid magnesium is 26 cm³ H/100 g; this drops to 18 cm³ H/100 g upon solidification. What would be the porosity if liquid Mg saturated with H were allowed to solidify?

The volume of H rejected upon solidification is $26 - 18 = 8$ cm³.

The volume of Mg is 100 g/(2.4 g/cm³) = 41.6 cm³.

The solid will look like Swiss cheese, and will have a total volume of $8 + 41.6 = 49.6$ cm³, of which $8/49.6 = 0.16$ or 16 vol.% will be pores.

Example 3-6

When the partial pressure of hydrogen is 1 atm above a melt of aluminum, the equilibrium solubility of the gas is 0.7 cm³/100 g Al. Maximum solubility in solid Al is 0.04 cm³/100 g. What partial vapor pressure p_{H_2} should be maintained over the melt if a pore-free casting is to be obtained?

Substituting into Eq. (3-8) for the liquid, the equilibrium constant is obtained

$$0.7 \text{ cm}^3/100 \text{ g} = k\sqrt{1 \text{ atm}}$$

$$k = 0.7 \left(\text{cm}^3/100 \text{ g}\sqrt{\text{atm}}\right)$$

For the solid

$$0.04 = 0.7\sqrt{p_{H_2}}$$

$$p_{H_2} = (0.04/0.4)^2 = 0.0033 \text{ atm}$$

Note that a very low partial vapor pressure must be maintained.

3-4 CASTING PROPERTIES

Solidification characteristics combine with fluid properties to determine the suitability of various alloys for casting.

3-4-1 Viscosity

The pouring of the melt into a mold is essentially a problem in fluid flow, and as such, it is greatly affected by the resistance exerted by the fluid against flow. This resistance can be measured as a shear stress τ. If a fluid film of h thickness is sheared between two flat parallel plates, one of which moves at v velocity, the shear stress τ is the force per unit area acting on these plates (Fig. 3-20a)

$$\tau = \eta \frac{dv}{dh} = \eta\dot{\gamma} \tag{3.9}$$

where $\dot{\gamma}$ is the shear strain rate and η is the dynamic viscosity (in units of $N \cdot s/m^2$).

The laws governing the flow of substances are the subject of *rheology* (from the Greek *rheos* = current, flow). Many fluids (e.g., mineral oils used in machines) exhibit *Newtonian viscosity*, independent of $\dot{\gamma}$ (Fig. 3-20b, line A). Fluids in which solids are suspended shear readily at low strain rates, but the solid particles obstruct flow at high strain rates (*dilatant fluids*, line B). Substances in which the particles or molecules can orient themselves in the direction of flow shear readily at high strain rates (*pseudoplastic flow*, line C). An important group of materials begins to deform only after some minimum initial shear stress is applied, and then continue to shear in a viscous manner (*Bingham solids*, line D).

Above T_m (or T_L), most metals behave as Newtonian fluids whose viscosity is a function of free volume and, therefore, drops with superheat. While information is sketchy, one might generalize by saying that viscosity is a function of composition and of superheat as expressed on the homologous temperature scale (Eq. (3-6b)). However, the nature of phases that are present is also important. For example, in eutectic systems one may find (Fig. 3-20c) that the viscosity changes

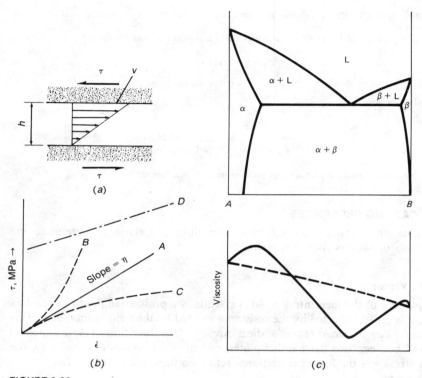

FIGURE 3-20
Shearing of a fluid requires (a) a shear stress which increases with increasing fluid viscosity. (b) Fluids may exhibit Newtonian (A), dilatant (B), pseudoplastic (C), or Bingham behavior (D). (c) In a eutectic system, viscosity may vary greatly.

linearly with alloy composition, but it could also show marked variations with phase boundaries (maximum viscosity at the limit of solid solubility, minimum at the eutectic composition).

Between T_L and T_S, the presence of the solid phase introduces non-Newtonian effects. An *apparent viscosity*, which is a function of the quantity and structure of the solid phase, can be defined for a constant $\dot{\gamma}$. Equiaxed crystals hardly affect viscosity up to about 60% concentration by volume. Dendrites increase the apparent viscosity greatly, except when shear rates are large enough to break up the dendrites; then viscosity is low, similar to that found with equiaxed crystals.

3-4-2 Surface Effects

When the melt has to flow through small (typically, below 5 mm) channels, surface tension (Eq. (3-4)) becomes significant. A high surface tension makes it impossible to fill sharp corners.

On exposure to the atmosphere, the surface of many melts becomes rapidly coated with an oxide film, and the nature of this film greatly influences casting behavior. Thus, the extremely dense and tenacious oxide of aluminum (Al_2O_3) makes it flow as if it were inside a rather tough bag, and alloying elements that modify the oxide greatly affect the casting behavior of aluminum alloys. Aluminum as an alloying element in other metals usually oxidizes preferentially; an aluminum oxide skin forms which has the effect of increasing surface tension.

3-4-3 Fluidity

When a mold is filled, heat is extracted and solidification begins while flow is taking place. Therefore, *mold filling* depends on many factors, the exact effects of which may not be known. To characterize materials under complex conditions, it is customary to develop *technological tests* that allow a quantitative comparison of materials, *but only if the test conditions are carefully specified.*

FIGURE 3-21
Fluidity is a technicological property which is determined by pouring into a (*a*) spiral or (*b*) plate mold or (*c*) by pulling a vacuum.

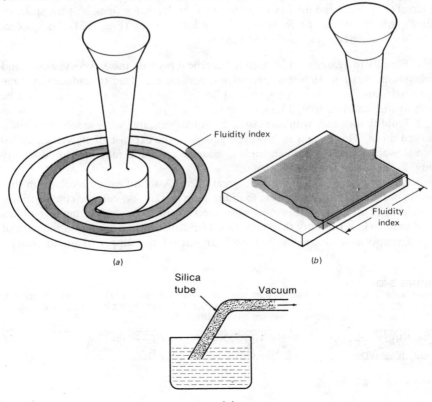

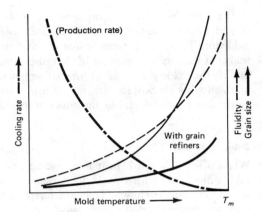

FIGURE 3-22
Mold temperature is a powerful factor in determining production rates, attainable shape complexity, and mechanical properties.

The mold-filling ability of a metal is described as *fluidity*. It is a system property that is a function not only of the metal but also of the mold. Typically, a long spiral-shaped or thin plate-like cavity is made in the mold material of interest (Fig. 3-21) and fluidity is quoted as a *fluidity index* (length of the spiral or plate). Alternatively, the length of fill under vacuum (Fig. 3-21c) is quoted. Fluidity is affected by a number of factors:

1 Fluidity increases with increasing superheat because this lowers viscosity and delays solidification. However, excessive superheat may lead to undesirably large grain size, and may also be impractical because the melting furnace may not be able to withstand such high temperatures.

2 Fluidity increases with increasing mold temperature, because solidification is slowed down. This benefit is, however, gained at the expense of slower solidification (lower cooling rate) which leads to coarser grain and may limit productivity (Fig. 3-22).

3 The type of solidification has a great effect. A solidification mechanism that allows orderly freezing, such as is found in pure metals and eutectic compositions, is helpful (Fig. 3-23a). However, pure metals with their higher melting points tend to have lower fluidity than eutectics. Dendrite arms growing into the path of liquid supply slow down the flow and can cut off the supply of liquid entirely

FIGURE 3-23
When a melt solidifies in a channel, communication with hot metal is (a) kept open longer with the frontal solidification of pure metals and eutectics than with (b) the dendritic solidification of solid solutions. (c) Debris of crystals chokes off the flow.

(a)   (b)   (c)

(Fig. 3-23b); therefore, the fluidity of alloys with a long freezing range is generally low. However, if the fluid is forced to flow (by a large gravity head or by externally applied pressure), dendrites are ripped off and broken up, and fluidity increases greatly. Flow stops when broken crystals freeze to form a "plug" at the meniscus (Fig. 3-23c).

4 Surface tension and the presence of oxide films have an effect.

5 Mold material and mold dressing affect fluidity by influencing heat extraction and wetting of the mold surface.

In a rather loose sense, one also speaks of the *castability* of a metal. This term incorporates, in addition to the technological concept of fluidity, aspects that define the ease of producing a casting under average foundry conditions. Thus, an alloy is regarded as highly castable when it not only has a high fluidity but is also relatively insensitive to accidental changes in process conditions, is more tolerant to the design of the fluid supply system, is less sensitive to wall thickness variations, and, in general, will produce castings of acceptable quality with less skill.

3-5 CHANGING PROPERTIES AFTER CASTING

In Sec. 3-8 we will see how conditions can be varied during melting and casting for controlling the properties of the finished casting. Even with the best control, it may still not be possible to attain the desired properties and further treatment may then be imposed on the already solidified casting.

3-5-1 Application of Pressure

Undesirable features (large grain size, porosity) of cast structures can be removed entirely by hot working but in doing so the shape of the casting is changed too. Hot working will be discussed in conjunction with bulk deformation processes (Sec. 4-1-4). Here we limit ourselves· to discussing *high-temperature isostatic pressing* (HIP) which improves properties without substantially changing the shape or dimension of castings.

The finished casting is placed into a well-insulated furnace which is then loaded into a specially constructed pressure vessel. A neutral gas such as argon is used for pressurizing the system. Pressures up to 200 MPa (30 kpsi) and temperatures up to 2000 °C (3630 °F) are possible, but 100 MPa (14.5 kpsi) and 1250 °C (2280 °F) are more typical of HIPing the superalloy and titanium alloy castings used in jet engines. The metal becomes soft enough for internal porosity to be closed under the imposed pressure and, if the surface of pores is clean, adhesion and solid-state welding result at the high temperatures employed. Since pressure is applied isostatically (from all directions), shape change is negligible but the properties of the casting—especially ductility, stress rupture, and fatigue properties—greatly improve.

3-5-2 Annealing

Annealing is the process of heating a material to some elevated temperature, holding at that temperature, and cooling back to room temperature. The rate of heating and cooling may have to be controlled. Undesirable reactions—in particular, oxidation—may be significant at high temperatures and, if this is objectionable, annealing is carried out in vacuum or in an inert or reducing gas atmosphere. Annealing may serve several purposes.

We already discussed stress-relief annealing in Sec. 2-1-8; it is usually performed below the temperature at which recrystallization or phase transformation would occur. Annealing of cold-worked materials will be discussed in Sec. 4-1-3.

In some applications, the compositional variations typical of microsegregation (Sec. 3-3-3) are objectionable. The casting may be subjected to *homogenization* by heating to just below the solidus temperature (T_S) and holding it, usually for several hours, until diffusion equalizes the concentration of the alloying element throughout the casting. The distances over which diffusion must take place (*diffusion paths*) are large in material of coarse grain or large dendrite spacing, and holding times may have to be increased to several days. Even so, some elements may diffuse too slowly to achieve complete homogenization.

3-5-3 Precipitation Hardening

In alloy systems where the solubility of an alloying element changes with temperature, there are opportunities for influencing mechanical properties by various heat treatments, especially if the excess solute is rejected in the form of an intermetallic compound. An example is shown in Fig. 3-24, with the composition of the alloy marked as C_0.

1 When the alloy cools slowly through the solvus, the rejected solute atoms combine with solvent atoms to form large, stable second-phase $A_m B_n$ particles (Fig. 3-24, Annealed). The relatively few and large particles have little effect on strength or ductility.

2 The second-phase particles can be completely dissolved by heating the alloy into the homogeneous α solid-solution temperature range. Such *solution treatment* is completed faster at temperatures closer to the solidus T_S but it must be kept in mind that, if solidification was nonequilibrium, a low-melting eutectic may be present and the casting may distort or, in extreme cases, it may even fall apart. Also, high solution-treatment temperatures may allow some grains to grow at the expense of others, and the coarse-grained structure will have lower strength.

Since diffusion of atoms and precipitation of the second phase takes time, the process can be suppressed by rapid *quenching*, usually in water. The resulting solid solution (Fig. 3-24, Quenched) is *metastable*, i.e., it reverts to the stable two-phase structure if conditions are favorable for diffusion.

a In many alloys precipitation proceeds at room temperature, and then one speaks of *natural aging*. In other alloys diffusion is too slow at room temperature

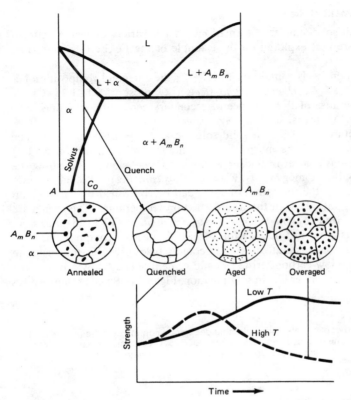

FIGURE 3-24
After solidification, some alloys may be substantially strengthened by precipitation-hardening heat treatment.

and is accelerated by holding at some elevated temperature; this is called *artificial aging*. The result of either process is that the second phase separates, *precipitates*. At the relatively low temperatures employed, diffusion is limited to short distances, precipitation begins at many sites, and the precipitated particles are small (Fig. 3-24, Aged), often not visible in an optical microscope even at high magnifications. As shown in Fig. 3-11, strength greatly increases without undue loss of ductility; thus, this heat treatment is also termed *precipitation hardening*.

b At higher temperatures further diffusion leads to a coarsening of fewer particles and the strength of such *overaged* structure (Fig. 3-24, Overaged) declines. Such overaging may also occur in service if the material is exposed to too high temperatures.

3-5-4   Heat Treatment of Steel

Alloy systems with solid-state transformations offer a variety of heat treatment possibilities that are best explored on the example of the Fe–Fe$_3$C system (Fig. 3-7).

In practice, carbon steels contain up to 1.7% C; the eutectoid composition lies at approximately 0.8% C. *Hypereutectoid steels* (i.e., those of 0.8–1.7% C) are hard but brittle because of the presence of secondary cementite, and thus have limited application. Most cast steels are *hypoeutectoid* (i.e., contain less than 0.8% C) and their structure consists of α solid solution and a pearlite eutectoid. The distribution of phases and the morphology of the eutectoid depend on cooling history: The pearlite is coarser in thicker sections that cool slower. Reheating the solid casting into the homogeneous γ range (*austenitizing*) takes all C into solution; subsequently, the formation of Fe$_3$C can be controlled by choosing an appropriate cooling rate through the transformation temperature. The events are graphically summarized in *time-temperature-transformation* (TTT) diagrams, an example of which is given in Fig. 3-25 for a steel of eutectoid composition.

Above 723 °C, stable austenite exists. On cooling below 723 °C, the austenite decomposes into the low-carbon ferrite (α) and cementite (Fe$_3$C) by a process of nucleation and growth (rather like in solidification, Fig. 3-14). Since both processes

FIGURE 3-25
Depending on cooling rates, the eutectoid decomposition of austenite may result in a variety of structures. The example shown is for steel of 0.8% C content; the hardness of the resultant structures is given, approximately, on the right-hand ordinate.

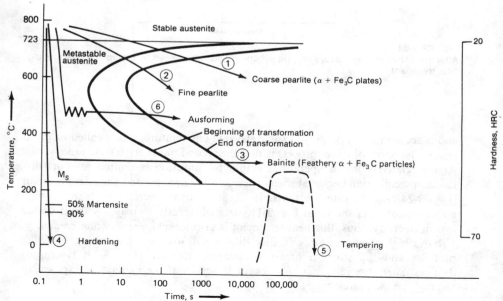

take time, transformation begins and is completed only after a certain time has elapsed, and this time is a function of temperature, giving the characteristic C-shaped curves in the TTT diagram.

1 On slow cooling (Fig. 3-25, line 1) few nuclei form just below the transformation temperature and are able to grow rapidly; the structure will consist of coarse cementite platelets in a ferrite matrix. Such coarse *lamellar pearlite* is relatively soft but not very ductile. The heat treatment consisting of austenitization followed by cooling on air is called a *normalizing anneal*. The cementite may be changed to a spheroidal form by holding the steel just below the eutectoid temperature or by repeatedly heating and cooling just above and below this temperature. *Spheroidized* steels (Figs. 3-8c) have lower strength and higher ductility.

2 On faster cooling (Fig. 3-25, line 2) transformation is somewhat delayed and a metastable austenite exists until the time and temperature of transformation is reached. A great many nuclei form and the structure will consist of much finer but still lamellar pearlite.

3 If the steel is cooled very rapidly and then held at an intermediate temperature, say around 300 °C, the nose of the transformation curve is missed and transformation occurs *isothermally*, along line 3 in Fig. 3-25, with the formation of *bainite*. In this, the lack of diffusion time makes the carbide particles appear as extremely fine spheroids in a matrix of α solid solution. As expected from Fig. 3-11, such a structure possesses a desirable combination of strength and ductility. Isothermal transformation at around 400 C results in the formation of very fine pearlite (*patenting heat treatment*).

4 When cooling is again fast enough to miss the nose of the curve entirely (Fig. 3-25, line 4), but is now taken to room temperature (*quenching*), separation of the carbide phase is suppressed but transformation of iron from the fcc to the bcc form cannot be prevented. Transformation starts at the temperature marked M_s in Fig. 3-25 and is completed at M_f before room temperature is reached. Carbon atoms are retained in a supersaturated solid solution, distorting the bcc structure into a body-centered tetragonal lattice. This highly stressed structure (*martensite*) is very hard and brittle.

5 Ductility can be restored by reheating the martensite (Fig. 3-25, line 5) so that carbide can precipitate in a very fine form. The strength and hardness of such *tempered martensite* is somewhat lower but the sacrifice is well justified by the increased ductility and toughness. Heating at yet higher temperature causes *overtempering*: The Fe_3C particles coarsen and the hardness drops. (It should be noted that there are also martensites in other alloy systems, but not all martensites are necessarily hard.)

The phases appearing upon solidification can be modified by the addition of alloying elements (usually several elements) to a carbon steel. Low-alloy steels still form a characteristic martensitic structure upon quenching but the *critical cooling rates* are reduced (the nose of the curve in Fig. 3-25 is shifted to the right), thus allowing heat treatment of thicker sections. Manganese is a solid-solution element which is also effective in this respect, thus increasing the *hardenability* of the steel (i.e., the depth to which hardening is obtained upon quenching). It also combines with sulfur in the form of inclusions, thus preventing the formation of iron sulfide which would cause hot shortness. Some alloying elements increase the stability of the austenite (e.g., the solid solution alloying element nickel), others that of ferrite (e.g., chromium). Alloying-element concentrations can be raised to the level where austenite is retained at room temperature in the stable form (austenitic stainless steels). In yet other steels alloying elements such as chromium, vanadium, and molybdenum are introduced that combine with carbon to form very stable carbides, so that the steels retain their hardness at temperatures where the martensite would be overtempered.

Example 3-7

A part made of AISI 1040 steel is to be subjected to heat treatment for higher strength. We saw in Example 3-4 that there is sufficient carbon to form 50.7% pearlite; hence, martensite will form on quenching. First, the casting must be heated into the austenitic temperature range. From Fig. 3-7, all traces of ferrite and pearlite disappear at 780 °C (actually, the transition temperature is 793 °C on heating (MHDE, p. 28.11)). In practice, one heats to a somewhat higher temperature, say, by 50 °C. Excessive temperatures cause coarsening of the austenite with some deterioration in final properties. Hence, we austenitize at 830–855 °C (MHDE, p. 28.13) and quench. Because the nose of the transformation diagram for this steel is similar to that in Fig. 3-25, water quench will be necessary. This could cause distortion. Furthermore, the depth to which martensite forms is also limited because, inside the part, cooling rates will not be high enough to avoid pearlitic transformation. (If through-hardening is required, an alloy steel must be chosen.) The as-quenched hardness will be high, about HV 700. Tempering reduces hardness, but imparts ductility (toughness). From MHDE, p. 4.21:

Tempering Temperature, °C	TS, MPa	YS, MPa	Elong., % in 2 in	R.A., %	HB
205	780	590	19	48	260
425	760	550	21	54	240
540	720	490	26	57	210
650	635	435	29	65	190

3-5-5 Surface Treatment of Steel

In many applications—such as gears, shafts, rolling-mill rolls, components subjected to wear—it is desirable to have a high hardness on the surface combined with great toughness throughout the body of the part. This aim may be achieved by three fundamentally different approaches:

1 The steel has sufficient carbon and alloying-element concentration to form martensite upon quenching. *Surface hardening* is then possible by first heat treating the component to obtain the toughness required in the core, followed by rapid heating (e.g., by induction, flame, or laser beam) and immediate quenching of the surface layer. Alternatively, a composition is chosen that will, upon controlled quenching from the austenitic temperature, transform into martensite on the surface but into pearlite in the core (*shell hardening*).

2 The steel has a low (typically 0.2%) carbon content and the surface is made hardenable by diffusing carbon into the surface—from a gas atmosphere, a liquid, or a solid pack—in the austenitic temperature range. Upon quenching, the carbon-enriched surface layer or *case* transforms into martensite, while the core remains tough. Hence the term *case hardening* or *carburizing*.

3 The steel is hardened by diffusing nitrogen into the surface below the A_1 transformation temperature (*nitriding*). Even though the treatment is carried out at elevated temperatures, no quenching is required. The danger of distortion is reduced or completely avoided by the injection of N atoms into the surface (see Sec. 9-5).

3-6 CASTING ALLOYS

With the exception of metals and alloys that are produced directly by powder metallurgy or electrolytic techniques, all metals and alloys must first go through the melting and casting stage (Fig. 2-42). It is, however, usual to distinguish between two broad classes:

1 *Wrought alloys* possess sufficient ductility to permit hot and/or cold plastic deformation (these will be discussed in Chaps. 4 and 5). They represent some 85% of all alloys produced, and are cast into simple shapes suitable for further working.

2 *Casting alloys* are selected for their good castability (such as the eutectics) or are materials of a structure that cannot tolerate any deformation (for example,

TABLE 3-1
SHIPMENTS OF CASTINGS* (UNITED STATES)

	Thousand tonnes	
Type	1972	1984
Gray iron	14 000	7 300
Malleable iron	860	330
Ductile iron	1 830	2 380
Steel	1 450	870
Aluminum alloys	850	830
Zinc alloys	460	260
Copper and brass	345	285
Magnesium alloys	21	10
Tin (+ solder)	7(+ 21)	5(+ 13)
Lead (+ solder)	7(+ 64)	20(+ 20)

*1972 data compiled from *Metal Statistics 1974,* American Metal Market, Fairchild Publications, Inc., New York, 1974. 1984 data from American Foundrymen's Society, Des Plaines, Ill.

TABLE 3-2 PROPERTIES OF SELECTED CASTING ALLOYS*

Alloy			Preferred casting method	Liquidus (solidus) °C	Shrinkage allowance,‡ %	Mechanical Properties†			
Name	ASTM No.	Typical composition, w%				TS, MPa	YS, MPa	Elongation (2 in), %	Hardness,§ BHN
Ferrous:									
Cast steel	60–30	≤ 0.25C	Expendable mold		1.5–2	420	210	24	< 180
	175–145		Expendable mold		1.5–2	1200	1000	6	360
Gray iron	20	3.5C, 2.4Si, 0.4P, 0.1S	Expendable mold	1180	1	150	(570)¶	(< 1)	160
	60	2.7C, 2.0Si, 0.1P, 0.1S, 0.8Mn	Expendable mold	1290	1	420		(< 1)	270
Malleable iron	A47	2.5C, 1.4Si, 0.05P, 0.1S, 0.4Mn	Expendable mold		1	350	220	10	< 150
Ductile iron	60-40-18	3.5C, 2.4Si, 0.1P, 0.03S, 0.8Mn	Expendable mold		0.8–1	420	320	15	160
Stainless steel	CF8	0.08C, 19Cr, 9Ni	Expendable mold		2.5	500	240	45	
Cu based:									
Tin bronze	C90500	10Sn, 2Zn	All	999(854)		320	150	40	80
Leaded red brass	C83600	5Sn, 5Pb, 5Zn	All	1010(854)	0.8–1.8	250	105	32	62
High-lead tin bronze	C93500	10Sn, 10Pb	All	926(760)	1–2	220	90	30	60
Leaded yellow brass	C85400	1Sn, 3Pb, 29Zn	All	940(925)	0.8–1.5	230	100	27	55
Al based:									
	108	3Si, 4Cu	Sand	627(521)	1.5	150	100	2.5	55
	D132	9Si, 3.5Cu, 0.1Mg, 0.8Ni, 3Zn	Permanent mold	582(520)	1	250	195	(1)	55
	A380	8Si, 3.5Cu	Die	593(538)	0.6	300	170	(2)	105
	A413	12Si	Die	577		295	145	(2.5)	
Mg based:									
	AZ 91B	9Al, 0.7Zn, 0.2Mn	All	596(468)	1.5 (die 0.6)	280	135	5	70
	EZ 33A	2.7Zn, 0.5Zr, 3 rare earths	Sand and permanent mold	643(543)	1.2	160	110	3	50
Zn based:									
	AC41A	4Al, 1.25Cu, 0.04Mg	Die	387(381)	0.3–0.6	285		10	82
	∠A12	11Al, 1Cu, 0.03Mg	Sand and permanent mold	432(377)	1.3	300	200	2	94
Pb based:									
Babbit		5Sn, 10Sb	Bearings	256(240)		70		5	19
Sb lead		9Sb	Die	265(252)		50		17	15
Sn based:									
Babbit	B102	4.5Sb, 4.5Cu	Die	371(223)		65		2	17
Pewter	B560	7.5Sb, 1Cu	Permanent mold	295(244)					24

*Data compiled from *Metals Handbook*, 9th ed., vol. 1, 1978; vol. 2, 1979; vol. 3, 1980, American Society for Metals, Metals Park, Ohio.
†Minimum properties in the as-cast condition, except malleable and nodular cast iron (annealed) and D132 and EZ 33A (precipitation hardened). To convert MPa into 1000 psi, divide by 7.
‡Patternmakers' allowance.
§Load: 3000 kg for ferrous, 500 kg for nonferrous materials.
¶Compressive strength.

alloys with unfavorably distributed hard or nonductile phases or with high proportions of intermetallic compounds and other hard constituents). These are cast directly into the final shape. There are, of course, overlaps between the two groups, and the same material—because of its attractive service properties—may be produced in both wrought and cast forms.

The total quantities cast into shape have gradually declined in the industrially developed nations, but their value has gone up considerably because of the increased complexity and highly improved quality of products. Ferrous castings still represent the largest tonnages (Table 3-1) but nonferrous castings contribute much of the value. It should be noted that not all the quantities shown in Table 3-1 were produced from *primary* (*virgin*) *metal*. Scrap generated in foundries and other metalworking plants (*primary scrap*) is, in most cases, completely recycled. In addition, *secondary scrap*—obtained from breaking up used machinery and even from garbage dumps—constitutes a significant source of raw materials.

Some properties of the most popular alloys are given in Table 3-2; general properties are discussed in the following.

3-6-1 Ferrous Materials

In its most familiar form, the iron–carbon diagram actually shows the equilibrium phases in the iron–carbide (Fe–Fe$_3$C) system (solid lines in Fig. 3-7). The compound Fe$_3$C is, however, metastable, and under certain conditions can revert to the more stable carbon (*graphite*) form. Alternatively, melt composition and solidification conditions may be controlled to allow the carbon to separate in the form of graphite during solidification, at a somewhat higher eutectic temperature (Fig. 3-7, broken lines). Thus, several families of materials can be derived from the iron–carbon system.

Cast Steels   In carbon steels (up to 1.7% C) the carbon is always in the form of Fe$_3$C. Their high melting point and, above 0.15% C, the long freezing range makes steels less suitable for casting purposes. However, they are ductile and have a high strength and fatigue resistance which, as discussed in Sec. 3-5-4, can be further increased by heat treatment and alloying. Below 0.2% C quenching does not materially increase hardness and such steels are used in the as-cast or annealed condition. Many steels can be readily welded to build up components of unusually large sizes or complexity. Hence they have important applications, primarily in the railroad industry (wheels, truck frames, couplers); construction- and mining-equipment manufacture (track shoes, axle housings, hoist drums, buckets and bucket teeth, grinding balls); metalworking machinery (rolling mill, press, and hammer housings); oil-field and chemical plant components (valve bodies, impellers, drill-rig parts).

Because of poor fluidity, wall thickness must be fairly large and castings tend to be of larger size (almost half of all castings are in the 200–500-kg weight range). Stainless steels are indispensable in various applications, but their high melting point and long freezing range present substantial technological challenges.

White Cast Irons As discussed in Sec. 3-1-6, cast irons contain in excess of 2% C. The form in which the carbon solidifies depends on cooling rates as well as composition. Control is exerted primarily by the total C and Si (and also P) content, and their combined effect can be expressed by the *carbon equivalent* C.E.:

$$C.E.(\%) = C\% + \frac{Si\% + P\%}{3} \tag{3.10}$$

At C.E. < 3 and fast cooling (small section thickness; say, below 6 mm in sand casting), the entire cross section will solidify with a white microstructure, i.e., with all carbon in the form of Fe_3C; even lower C.E. must be maintained for thicker walls. The presence of primary cementite in the eutectic makes these white iron castings hard and brittle; hence, their use is limited for wear-resistant parts such as grinding balls, liners for ore-crushing mills, and some agricultural machinery parts. They are virtually unmachinable except by grinding.

Malleable Iron The Fe_3C of white iron may be converted to the stable graphite by an annealing treatment. The castings are heated to 850–1000 °C for 50 h. In this first stage the primary cementite decomposes into graphite (*temper carbon*) and austenite. The castings are allowed to cool slowly through the eutectoid decomposition temperature, and are held around 700 °C for 25 h (second stage), to allow austenite to decompose into temper carbon and ferrite. The carbon appears in the form of irregular aggregates (Fig. 3-26a) embedded in a ferrite matrix: If some C remains in the austenite, the matrix will be pearlitic. Either way, the relatively large graphite particles do not substantially impair strength or ductility and a material similar to steel—but of lower melting point, higher fluidity, hence much better castability—is obtained. Since the casting must be white to begin with, only thin-walled products (max. 40 mm) can be cast. There are many applications in the automotive and agricultural equipment industries (housings, yokes, wheel hubs) and for general fittings. The presence of graphite imparts good machinability.

Gray Iron At relatively high C.E. and slower cooling rates, there is time for the iron to solidify in the stable form, with the carbon separating in the form of *graphite flakes* (Fig. 3-26b), making the fracture surface dull gray; hence the name *gray iron*. The formation of graphite counterbalances much of the solidification shrinkage of the iron, thus ensuring soundness and relative freedom from solidification porosity.

The graphite flakes reduce ductility to practically nil, and Young's modulus is also lower than that of pure iron (it ranges from 70–150 GPa). The size, shape, and distribution of flakes can be controlled to give castings of low to high

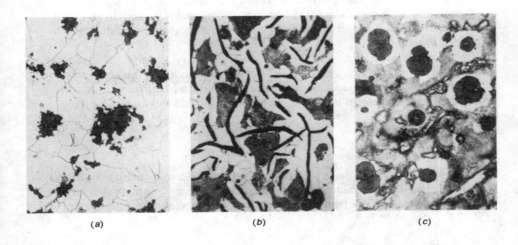

(a)   (b)   (c)

FIGURE 3-26
Carbon is present in different forms of graphite in (a) malleable iron (ferritic), (b) gray iron
(50% ferrite, 50% pearlite), and (c) nodular iron (50% ferrite, 50% pearlite). (*From Metals
Handbook Desk Edition, American Society for Metals, 1985, p. 27.26. With permission.*)

strength (Table 3-2); properties are always better in compression because the
graphite flakes act as incipient crack sites in tension. Cooling too fast results in a
mottled structure in which primary carbide is also present; rapidly cooled surface
layers may be entirely white, creating problems in machining a material of
otherwise excellent machinability.

    Its low cost makes gray iron the preferred choice in all fields when ductility
and high strength are not needed (weights; frames; motor, gear, and pump
housings; pipe fittings). Its high damping capacity is an advantage for machine
tool bases. High fluidity and good tribological properties have made it the
traditional engine block material.

    Nodular Iron   A ductile cast iron is obtained when the graphite is brought into
a less detrimental, globular form upon solidification. This is achieved by adding
to the melt, just prior to casting or during pouring, a small amount of magnesium
or cerium (introduced in the form of a ferroalloy) which, through a mechanism
that is only partially understood, causes the graphite to separate into well-defined,
roughly spherical (or elongated, *nodular*) particles, distributed in the α-iron or
pearlite matrix (Fig. 3-26c). Since Mg promotes the formation of cementite, the
iron is then inoculated with silicon. *Nodular* (or *ductile* or *spheroidal*) *cast iron*
combines the good castability and machinability of gray iron with some of the
ductility of steel. For the control of properties, castings are usually heat treated.

Nodular iron has an extremely wide range of applicability, from automotive crankshafts and hypoid gears to pump housings, rolling mill rolls, and, in general, for parts subjected to impact loading or requiring a high elastic modulus ($E = 150–175$ GPa).

The matrix of cast irons may be produced with varying levels of carbon content; therefore, cast irons may be heat treated just as steels are. Also, cast irons can be alloyed for enhanced mechanical or chemical (corrosion-resistant) properties. Grain refinement is possible with calcium silicide and other nucleating agents.

3-6-2 Nonferrous Materials

The most important alloy groups are discussed here in order of increasing melting point which also indicates increasing cost and difficulty of melting and superheating them to the appropriate temperature. Properties of selected alloys are given in Table 3-2.

Tin-Based Alloys Of the widely used metals, tin has the lowest melting point (232 °C). It is highly corrosion resistant and nontoxic, but its low strength precludes its use as a construction material.

An important application is for bearing surfaces where its low shear strength and low adhesion to other metals assures low friction coefficients even when lubrication fails. It must be backed by a stronger material or, if the bearing layer is to be thicker, it must be strengthened by creating a duplex structure in which a hard compound is dispersed in the soft tin matrix. This is achieved by adding Sb to form the hard intermetallic compound SbSn, in the shape of small, hard, cubic crystals (*cuboids*). These tend to rise to the surface of the melt; the addition of some copper improves the situation by forming a copper–tin intermetallic which solidifies as a spatial network of needles, thus trapping the cuboids.

Old pewter contained lead, but modern pewter is free of lead (Table 3-2), and is suitable for decorative items and tankards.

Lead-Based Alloys Lead too has a low melting point (327 °C) and good corrosion resistance, but it is toxic and its use is limited to applications where human contact is avoided. Large sand or permanent-mold castings are used as x-ray and γ-ray shields.

The low strength of lead and its low solubility in other metals qualifies it as a bearing material, although of somewhat lower quality than tin. Strengthening is again obtained by alloying, usually with tin and antimony, so that the SbSn cuboids are dispersed in a matrix of ternary Sn–Pb–Sb eutectic. These ternary alloys are not only hard but also possess a high fluidity imparted by the presence of tin; therefore they were also used as type metal that gave a clear, clean typeface.

Antimonial or calcium lead is extensively used for cast lead–acid battery grids. Great economy is ensured by recycling most of the used batteries (a majority of all lead used in the metallic form is recycled).

Zinc-Based Alloys Zinc is the only low-melting (419 °C) metal widely used as a structural casting material. Its major weakness is low creep strength. Also, its corrosion resistance is low in the presence of contaminants such as Cd, Sn, and Pb which leads to intergranular corrosion. However, its high fluidity and low melting point make it eminently suitable for casting into steel dies. Strengthening is obtained by solid-solution alloying with approximately 4% Al and 1–2% Cu (the eutectic is at 5% Al in the Zn–Al system). By the use of 99.99%-pure zinc and careful control of contaminants, good corrosion resistance is secured, making these alloys highly competitive (even with plastics) for thin-walled parts of intricate shape, such as instrument housings and automotive components and trim; for the latter application, the excellent response to chromium plating is an advantage. Alloys with 11% Al offer high strength combined with good fluidity.

Magnesium-Based Alloys The melting point of magnesium is substantially higher (649 °C) but still low enough to allow casting by all techniques. Its low density and reasonable strength, coupled with corrosion resistance (except in marine environments) make it very attractive for structural applications including air-cooled automotive engine blocks, transmission housings, and wheels. The major barrier is its cost. Casting alloys are solid-solution strengthened with up to 10% Al (the eutectic composition with 32% Al is too brittle to be practical), and some precipitation hardening may be obtained by adding Mn, Zr, or Zn. Fluidity is quite adequate because the oxide is not dense and does not hinder flow. The Zn–Zr–rare earth alloys are suitable for service up to 260 °C. Grains are refined by adding Zr.

Aluminum-Based Alloys Melting at an only slightly higher temperature (660 °C), almost as light and considerably cheaper than magnesium, aluminum and its alloys represent (beside nodular iron) the fastest-growing segment of the casting industry. The corrosion resistance of aluminum is excellent (except to alkali) and its strength is readily improved through solid-solution and precipitation-hardening mechanisms. Increased recycling of secondary scrap has reduced the total energy consumed in making aluminum parts.

Pure aluminum is used for domestic utensils. High-conductivity, 99.6%-pure aluminum is pressure die cast into squirrel-cage rotors for fractional horsepower motors and is used, as a permanent mold casting, for larger motors too.

The oxide film on the melt surface is dense and tough and, as already mentioned, reduces fluidity. The ease of casting is greatly affected by the influence of alloying elements on this oxide film. Silicon is most beneficial, making silicon alloys the most castable aluminum alloys. The eutectic composition (around 12% Si, with a melting point of 577 °C) is, of course, the most favorable. Its properties are greatly improved by refining (modifying) the eutectic structure through the addition of a small quantity of sodium (more recently, strontium) to the melt just prior to pouring, whereupon the eutectic silicon is modified and separates in the form of fine rods instead of coarse flakes. Hypereutectic alloys contain the very hard and brittle silicon in a pro-eutectic form and are thus extremely wear-resistant. The structure is refined by the addition of 0.01% P to form aluminum–phosphide nuclei. A 17% Si alloy has been used as an engine block without cast-iron liners.

The hard and brittle silicon of the eutectic limits the ductility of the alloy. Therefore, in the most popular casting alloys some Si is replaced with Cu which increases strength by solid-solution strengthening and also opens the door to precipitation hardening (Sec. 3-5-3). Castability is still high, especially in die casting where the dendrites are broken by the applied pressure.

Magnesium is a useful solid-solution strengthening element but creates problems typical of a long freezing range.

Hydrogen would lead to porosity (Examples 3-5 and 3-6) and scavenging with argon or chlorine just prior to pouring is practiced. Grain refinement is achieved by adding Ti–B alloy nucleating agents.

The use of aluminum alloy castings is very wide ranging and is always worth considering, especially when high strength-to-weight ratio and corrosion resistance are desired. Typical applications include automotive transmission cases, pistons, engine blocks, and some aircraft components.

Copper-Based Alloys The melting point of copper (1083 °C) is too high for steel dies (unless protected by heavy coatings) but other casting methods are practiced because the metal has attractive color, good corrosion resistance, and high electrical conductivity.

The majority of castings are made of alloys that combine good fluidity with reasonably high strength. Because copper alloys have been around for such a long time, many of them were given proprietary names and sometimes misleading designations (some brasses are commonly called bronzes). Extensive past use of copper alloys now allows a substantial part (up to 35%) of the total consumption to be covered from secondary scrap. The technologically important features of alloys can be deduced from their composition and phase diagrams.

There are few copper alloy systems with useful eutectics; therefore, most casting alloys are hardened by adding solid-solution elements, up to the limit of solubility for maximum strength without undue embrittlement by excessive intermetallic particle content. Whenever the solidification range is wide, fluidity suffers.

Tin bronzes (Cu–Sn alloys) have a very long freezing range, and fluidity is increased by adding phosphorous which forms a low-melting ternary eutectic (*phosphor bronzes*). The addition of zinc with its low vapor pressure also increases fluidity. An 88Cu–10Sn–2Zn alloy still has high strength, making it suitable for gears, bearings, and pump parts. Lead is often added, primarily to improve machinability, but it also benefits fluidity. The 85Cu–5Sn–5Pb–5Zn alloy is the most castable of all and is extensively used for water fittings, fixtures, pump bodies, and general castings. The high lead content in the 80Cu–10Sn–10Pb alloy reduces strength but makes the alloy suitable for bearing applications.

Aluminum bronzes have a short freezing range but the oxide reduces fluidity. They do give, however, high strength, especially in the heat-treated condition (with iron and other precipitation-hardening additions). Their excellent corrosion resistance makes them favorites for marine applications, worm gears, valves, and nonsparking tools.

Brasses are Cu–Zn alloys. They have a short freezing range, and most casting alloys also contain lead which improves fluidity. They are especially suitable for fittings, plumbing fixtures, and other smaller parts.

Nickel- and Cobalt-Based Alloys   The high melting points of nickel (1435 °C) and cobalt (1495 °C) and their corrosion resistance make them eminently suitable for many critical applications. Their strength and, particularly, hot strength can be greatly increased with solid-solution and precipitation-hardening alloying elements. Some of these superalloys have such a high second-phase content that they are not deformable, and these cast superalloys can outperform other materials in high-temperature applications, particularly as gas turbine (jet engine) parts.

High-Temperature Materials   Some of the higher-melting alloys are used in only very specific cases to produce castings.

Titanium (melting point 1670 °C) can be alloyed to give high elevated-temperature strength combined with low weight (high strength-to-weight ratio) and corrosion resistance. These alloys are thus used in chemical plants and in subsonic and, especially, in supersonic aircraft construction. The great affinity of Ti to oxygen, the high melting point, and the low fluidity place great demands on the skills of the foundry worker. Properties of castings can be greatly improved by HIPing.

By definition, *refractory metals* are resistant to heat and are difficult to melt. The most important ones are: molybdenum, Mo (melting point 2610 °C); niobium, Nb (also called columbium, Cb, 2470 °C); and tungsten, W (3410 °C). They oxidize extremely rapidly; therefore, special (vacuum arc or electron beam) melting and casting techniques are needed. They are indispensable in some applications, such as rocket motor nozzles.

3-7 MELTING AND POURING

Thus far we established the basic principles bearing on casting; we may now proceed to discuss actual techniques.

3-7-1 Melting

The first step is to prepare a melt of the correct composition. Figure 3-27a shows, schematically, the major elements of the system:

1 A *charge* is made up to yield, upon melting, the alloy of specified composition. It is seldom necessary, practicable, or even desirable to make up a charge entirely of metals obtained from ores (primary or virgin metals). Alloying elements of much higher melting point than the base metal would be slow in dissolving and would require excessive overheating, therefore, they are added in

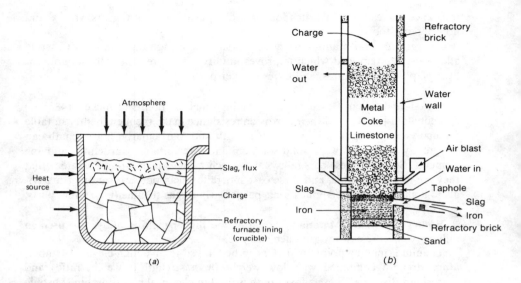

FIGURE 3-27
(*a*) Elements of a melting system are shown schematically; the furnace may have very different configurations. (*b*) Water-cooled cupola for melting cast iron.

the form of a *master metal* (*temper alloy*, *hardener*), which contains a higher concentration of the alloying element in the base metal. For economy of operation, it is most important that as much scrap as possible should be added. The aim is that of producing a melt of the composition specified by relevant standards while holding contaminants below the allowed maximum levels, and accomplish all this at the lowest possible cost. Computer programs are available to facilitate this task. Smaller plants often find it more economical to use prealloyed material cast into ingots purchased from specialized companies (*ingoted melting practice*).

    2 The starting materials are charged into a *furnace* which contains the melt and provides a source of heat.

    a The furnace is lined with a material of substantially higher melting point than the metal while also minimizing contamination of the melt by inclusions or dissolved elements. The lining is chosen so that its oxide is not reduced by the metal; otherwise the liner would be quickly attacked (e.g., Al is never melted in steel). The material may range from iron (for lead) to graphite or refractory crucibles, and to refractory-lined furnace structures (a *refractory* is a ceramic of high temperature resistance). Alternatively, the melt may be contained by maintaining a chilled outer zone that forms its own container.

    b *Heat* is provided externally (for example, by electric, gas, or oil heating), internally (as by electric induction) or, only for cast iron, by mixing the fuel with the charge itself. Cast iron is usually melted semicontinuously in a vertical shaft furnace (*cupola*, Fig. 3-27*b*); lining of the cupola with a refractory is being abandoned in favor of water-cooled steel jackets. The charge is mixed with coke and some minerals (primarily limestone, $CaCO_3$), and hot air is blown through the column. Coke burns to give heat and is also a source of carbon for the cast iron. The liquid metal is tapped at the bottom, separately from the slag which is formed by the limestone with nonmetallic contaminants and metal oxides. In the *duplex process*, the liquid metal is tapped into an electric holding furnace where alloying and superheating is also practiced.

3 An inevitable factor is the presence of an atmosphere. This may be air which, with its humidity and various pollutants, is a source of N, H, and O gas absorption; it could be a *protective atmosphere* (such as argon gas); or even vacuum, produced at some expense. Combustion products including H_2O and H are also present in oil- and gas-fired furnaces. When the charge is mixed with the fuel (such as coke in the cupola), reactions of the fuel and its combustion products with the melt are inevitable. Thus, interactions with the atmosphere may range from simple dissolution of gases in the melt to reactions such as oxidation or, in the presence of reducing agents, reduction, and even carbon enrichment.

4 The charge is covered or mixed with *fluxes*, various (usually inorganic) compounds that can be spread on the surface or mixed into the metal to react with the melt. They melt and react with contaminants and nonmetallic elements and inclusions; the resulting *slag* floats to the surface of the melt. These may also isolate the melt from the atmosphere and reduce vapor losses of metals of low vapor pressure. Metal lost in the slag and losses due to oxidation or evaporation also represent a financial loss and the aim is their minimization, except when selective loss of contaminants is desired.

Many foundry operations generate fumes, gases, and dust. Techniques are available to minimize and even eliminate *environmental pollution*, and some of the most difficult and unpleasant jobs are now performed by robots and other mechanical devices.

3-7-2 Pouring

When the melt reaches the desired temperature and composition, it is *tapped*. A stationary furnace is tapped by breaking through a refractory plug placed in a hole close to the bottom of the furnace. As their name implies, tilting furnaces are tapped by tilting. Lower melting-point metals can be pumped or syphoned out of the furnace.

The melt may be transferred directly to the mold or tapped into a *ladle* (a refractory-lined vessel) which is then taken to the mold; metal is dispensed through a bottom orifice or by tilting the ladle. In some instances, melt is distributed from a central melting facility to several plants located at some distance.

There may be mismatch between the rates of melting and using material, and then *holding furnaces* are employed in which some treatment or alloying of the melt may also take place.

It is at the *pouring* stage where temperatures are finally adjusted. Highly volatile alloying elements that—because of their high vapor pressure—would be lost during melting, may be introduced (e.g., Mg into Al melts). Elements for deoxidation may also be added. In general, the aim is to keep the metal flowing free from turbulence that would cause entrapment of oxides and slag. Pouring rates and the quantity poured must be controlled too. Automated pouring, where economically possible, gives the most reproducible results.

3-7-3 Quality Assurance

Great strides have been made in improving the quality of castings, and many of these improvements are related to the melting and pouring stage.

1 Composition limits used to be held by careful attention to charge makeup. Rapid analytical methods, including high-speed spectrography now provide analyses for the important elements within minutes so that adjustments can be made, to each charge, before pouring.

2 We saw in Sec. 3-2-4 that inclusions can greatly impair mechanical properties, particularly impact properties (Sec. 2-1-5) and fatigue resistance (Sec. 2-1-6). Thus, the reliability of castings can be increased and the range of applications broadened by the introduction of techniques aimed at reducing the number of inclusions or, if this is not practicable, changing inclusion morphology and distribution to minimize harmful effects (these techniques are, of course, also used for wrought alloys). Many possibilities exist:

a *Holding* the metal at a constant temperature allows lighter inclusions to separate. An active flux helps to gather up inclusions. Alternatively, the melt is passed through *filters* (especially for aluminum alloys). *Electroslag refining* (related to electroslag welding, Sec. 9-3-5) is applicable to steels and superalloys, and involves remelting of an electrode, previously cast to the specified composition, by drawing an arc submerged in the slag.

b *Purging* (scavenging) of the melt with a gas (Sec. 3-3-4) reduces gas contents. Reactions with the melt may also occur, as when using chlorine gas to remove hydrogen from magnesium or aluminum melts. Sometimes these effects are a byproduct of another operation; e.g., in the oxygen blowing of steel the aim is to remove carbon, but nitrogen and hydrogen concentrations are also reduced. This is true also of the argon–oxygen blowing of stainless steel. Oxygen is later reduced by deoxidation.

c Many inclusions are the products of *unwanted reactions* (usually oxidation) with a gas. Cleanliness is improved by reducing the partial vapor pressure (Eq. (3-8)) by the application of vacuum during melting (vacuum induction melting, consumable-electrode remelting), after melting is complete (vacuum degassing in the furnace or ladle), or during pouring.

d Gases can form inclusions of zero strength (Sec. 3-3-4) and are, in most instances, unwanted. In addition to purging and the application of vacuum, there is the option of tying up gases in a solid reaction product. Thus, copper alloys are *deoxidized* with phosphorus; steels are deoxidized with aluminum, silicon, manganese, or calcium. The reaction products remain in the casting, and process controls aim to distribute them in the least harmful form. Because of its importance to wrought steels, we will return to deoxidation in Sec. 4-2-1.

Alloys (especially steels) of a given composition are commercially produced with various levels of impurities and inclusions, with the cleaner—and more expensive—versions used in more critical applications.

3-8 CASTING PROCESSES

The number of casting processes is very large. From the technological point of view, they can be grouped into two broad categories: casting into expendable and into permanent molds. The *expendable mold* is used only once and must be broken up to free the solidified casting, whereas a *permanent* mold is expected to last up to several hundreds or thousands of castings and must be of such a construction as to release the solidified casting.

An alternative classification is based on the purpose of casting:

1 *Ingot, slab, and billet casting.* The metal is a wrought alloy; in preparation for rolling, extrusion, or forging, it is cast into a simple shape suitable for further working. As mentioned, some 85% of all metals is processed this way, in specialized plants. It is essentially a primary manufacturing activity and will be discussed here only to the extent necessary in preparation for the discussion of Chaps. 4 and 5.

2 *Remelt ingots.* These are simple shapes, cast from melts of closely controlled and analyzed composition, for easy transport and loading into the furnaces of secondary manufacturers.

3 *Shape casting.* The melt is cast into the final shape which needs only cleaning and/or machining to produce a finished part. This is a typical secondary manufacturing process and will be the focus of our discussion.

3-8-1 Ingot Casting

Cast bodies of circular, octagonal, or round-cornered square cross sections are called *ingots* when their diameter or side dimension is about 200 mm or greater, and are called *billets* when smaller. Bodies of rectangular cross section are generally called *slabs*. Almost universally, they are poured into permanent molds by a variety of techniques:

1 *Ingot molds,* usually of iron or steel, are used for static casting of all alloys (Fig. 3-28a). Solidification begins from the mold walls and proceeds toward the center, giving rise to the typical solidification patterns shown in Fig. 3-15. Better quality (better surface, less slag and gas entrapment) is often obtained by bottom pouring (broken lines in Fig. 3-28a). Piping is avoided by feeding molten metal, either from the ladle, or from a hot metal reservoir contained within a

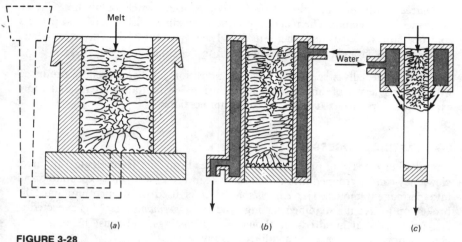

FIGURE 3-28
Metals destined for further working by plastic deformation are cast into simple shapes in (*a*) permanent molds or (*b*) water-cooled molds, or (*c*) are subjected to semicontinuous casting (continuous casting is similar in principle).

refractory-lined extension of the mold. The metal may be kept hot even longer and the depth of pipe reduced by placing an insulating collar or an exothermic compound (a compound that ignites on contact with the hot metal and produces heat) on top of the melt (*hot top*). This gives a higher yield of sound metal.

2 *Water-cooled molds* (Fig. 3-28*b*) are employed mostly in the casting of copper-base alloys.

The smallest ingots cast by the first two techniques may be 25–50-mm thick in nonferrous metals and 150–200-mm thick in steels, and may range up to 20 tons in weight in nonferrous metals and 300 tons in steels.

3 *Continuous casting* processes (Fig. 3-28*c*) are used for casting the vast majority of slabs and billets in aluminum and copper alloys and now also in steels. The solidification zone is localized in the water-cooled die when casting nonferrous alloys, but extends to great distances beyond the mold when casting ferrous alloys. The casting is withdrawn gradually as solidification progresses. The emerging casting may or may not be further cooled with water sprays. The process may be interrupted periodically to allow removal of an ingot (*semicontinuous casting*, mostly for nonferrous metals), or it may go on almost indefinitely (*continuous casting*, also called *strand casting*, mostly for steels), in which case the slab or bar is cut up during its movement with a flying saw or torch. Smaller cross sections (wirebar and strip), if cast with an acceptable surface quality, may be fed directly into a rolling mill, creating a completely continuous process, mostly for aluminum and copper alloys.

A mold wash (ceramic powder for steel casting) or a lubricant-type parting compound (often containing graphite or molybdenum disulfide) prevents adhesion and welding of the melt to the mold. In the continuous casting of steel, further protection is obtained by oscillating the mold. Contact with the mold may be entirely eliminated and the surface quality greatly improved in the casting of aluminum by containing the molten zone with air pressure or with an electromagnetic field.

3-8-2 Casting of Shapes

When the casting process aims at producing a component of complex shape, a mold is prepared with a cavity that defines the shape of the component, with due allowance for shrinkage after solidification. A sound casting is obtained if the melt is brought to the cavity in an orderly fashion and solidifies in a planned manner.

Fluid Flow The fluid supply system of a mold is designed in accordance with principles of fluid flow. Ideally, flow should be laminar (smooth) but, in practice, turbulence cannot be entirely avoided. However, turbulence must be kept at a minimum to avoid erosion of the mold and entrapment of slag, mold material, and gases. The mold system must be filled with metal under positive pressure, so that no gas is aspirated (sucked in) anywhere.

The fluid supply systems have some common features for all shape casting processes (Fig. 3-29):

1 The *pouring basin* is a receptacle large enough to accommodate the stream of metal, and is often shaped to ensure smooth flow of the melt. At the surface (level 1, Fig. 3-29), hydrostatic pressure is nil and potential energy is maximum. *Dross* (oxides and other inclusions rising to the surface) may be held back with a *skimmer*, and heavier inclusions with a *weir*.

FIGURE 3-29
Orderly distribution of the melt to a mold cavity requires a well-designed running and gating system.

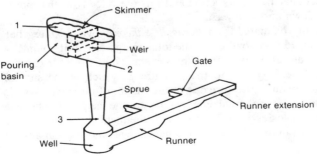

2 The fluid is transported down into the mold by the *sprue*. In doing so, potential energy is converted into kinetic energy and velocity increases. Because the mass flow is constant, the stream will pull away from the sprue walls as the velocity increases and this will draw unwanted air into the mold. To avoid aspiration of air, a positive pressure differential must be maintained throughout, and for this the sprue must be tapered downward (from level 2 to level 3, Fig. 3-29). A *well* of large cross section is provided at the base of the sprue; the sudden slowing of flow dissipates kinetic energy and helps to drop out inclusions, scum, and various refractory materials that may have been washed in with the fluid stream. Ceramic or wire mesh filters are sometimes placed at the sprue base to filter our dross and other large inclusions.

3 The melt is distributed through *runners* which are of larger cross section and often streamlined to slow down and smooth out the flow, and are designed to provide approximately uniform flow rates to various parts of the cavity.

4 The runners are connected by *gates* to the mold cavity. At the junction to the cavity, these gates are much reduced in thickness (*in-gates*) not only to allow easy separation from the solidified casting but also to choke the flow of metal and ensure quiet entrance into the cavity. The runner is often extended beyond the last gate to serve as a trap for inclusions carried into the runner by the first metal.

Castings are usually gated at the side, as shown in Figs. 3-29 and 3-31. Some parts are suitable for top gating, with the sprue serving also as a riser. Feeding from the bottom gives the quietest flow, although the top of the casting is then filled by the coldest metal. A particularly favorable situation exists when the metal is drawn up into a heated mold by vacuum (*suction casting*) because the mold fills without splashing and there are few, if any, mold gases to dissipate.

The dimensions of the various parts of the fluid distribution system may be calculated in an approximate manner by considering that melts are incompressible. Therefore, the *flow rate* (the volume passing through any given cross section in unit time) in any part of the system obeys the *equation of continuity*:

$$A_0 v_0 = A_1 v_1 \quad \text{etc.} \tag{3-11}$$

where A is the cross-sectional area of a section and v is the velocity at that point. If $A_1 > A_0$, flow slows down and vice versa.

The velocity may be approximated from *Bernoulli's theorem* which states that, under steady, well-developed flow conditions, the total energy of a unit volume of material must be a constant at every part of the system. There are four components of energy: pressure energy due to the pressure p, which is the sum of external and hydrostatic pressure (pressure due to the weight of the melt); kinetic energy due to the velocity v; potential energy due to the height h above a reference plane; and energy losses f due to friction in the melt (this term may be taken to include energy lost in turbulence, directional change, and friction against the mold walls). Thus, total energy per unit volume is

$$p_0 + \frac{\rho v_0^2}{2} + \rho g h_0 = p_1 + \frac{\rho v_1^2}{2} + \rho g h_1 + f \qquad (3\text{-}12)$$

where ρ is density.

If velocities are too low, freezing occurs before the mold is completely filled; if they are too high, the mold will be washed out and inclusions will be swept into the mold cavity.

Positioning and dimensioning of the runner system vitally influence the soundness of the casting, because they determine the supply rates of the melt to various parts of the cavity, and thus also influence the solidification pattern. The subject has been greatly developed in recent years, both experimentally and theoretically. Computer programs are usually based on a blend of basic theory (in the simplest cases, Eqs. (3-11) and (3-12)) and experience, and aid in the gating of castings in an interactive mode.

Example 3-8

Assuming negligible frictional losses, use Eqs. (3-11) and (3-12) to show that the areas of the top and bottom of the sprue must obey the following relation to avoid aspiration

$$\frac{A_2}{A_3} = \sqrt{\frac{h_3}{h_2}}$$

Let the pressure at the top of the sprue (Fig. 3-29, level 2) be p_2; at the bottom (level 3) p_3. To avoid aspiration, $p_3 > p_2$; for purposes of this example, take $p_3 = p_2$.

From Eq. (3-12)

$$\frac{\rho v_1^2}{2} + \rho g h_1 = \frac{\rho v_2^2}{2} + \rho g h_2 \qquad (3\text{-}12')$$

The cross-sectional area of the pouring basin is very large; hence,

$$v_1 = 0 \quad \text{at} \quad h_1 = 0, \quad \text{and} \quad v_2 = \sqrt{2 g h_2} \qquad (a)$$

Similarly,

$$v_3 = \sqrt{2 g h_3} \qquad (b)$$

From Eq. (3-11),

$$v_2 = \frac{A_3/A_2}{v_3} \qquad (c)$$

Substituting (c) into (a) and equating to (b) gives the relationship

$$\frac{A_2}{A_3} = \frac{v_3}{v_2} = \sqrt{\frac{h_3}{h_2}}$$

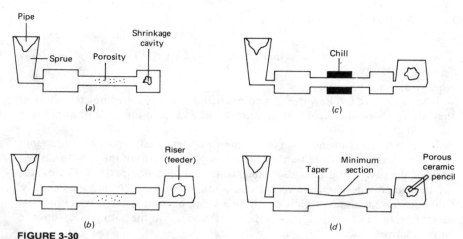

FIGURE 3-30
(a) A casting may show shrinkage cavities and microporosity. (b) Feeder heads or risers, removed after solidification, provide hot metal. Microporosity may be eliminated with directional solidification by (c) incorporating a metal chill into the mold or (d) tapering the thinnest section.

Heat Extraction and Solidification Once the melt enters the mold, heat is extracted through mold walls and solidification begins. If no special measures are taken, heat is extracted all around, so that solidification occurs *progressively* from all surfaces inward.

Solidification time t_s is, as might be expected, directly proportional to volume (which governs heat content) and inversely proportional to surface area (over which heat extraction occurs). It can be shown that, for a large variety of shapes and sizes, the relationship is quadratic (*Chvorinov's rule*)*

$$t_s \propto (V/A)^2 \tag{3-13}$$

thus, chunky portions of the casting freeze last. Therefore, progressive solidification can lead to early freezing of thinner sections, denying access of liquid to thicker parts and leading to porosity and the formation of shrinkage cavities (Fig. 3-30a). Remedies may take different forms:

1 *Risers* (*feeder heads*) provide a reservoir of molten metal. Made with a high V/A ratio (Fig. 3-30b), they solidify last and feed enough liquid to heavy sections of the casting to make up for shrinkage before and during solidification. An actual example is shown in Fig. 3-31; it will be noted that the risers are placed between runners and casting, so that they are filled last and contain the hottest metal (*live* or *hot risers*). At other times, as in Fig. 3-30, a riser must be placed at

---

*N. Chvorinov, *Proc. Inst. Br. Foundrymen* **32**:229 (1938–1939).

FIGURE 3-31
An example of a cored gray-iron casting showing sprue, runners, gates, and risers. Note the strainer configuration at the base of the sprue. (*Courtesy Massey-Ferguson Brantford Foundry, Brantford, Ontario.*)

the end of the casting (*dead* or *cold riser*). To ensure uninterrupted feeding, the junction between riser and casting can reach, in the casting of steel, 70–90% of the cross section to be fed.

Risers may be *open* to the atmosphere and then exothermic compounds may be placed on them; the *blind riser* shown in Fig. 3-30 loses less heat but a porous ceramic *pencil core* must be inserted to equalize pressure in the shrinkage cavity. While often indispensable, risers reduce the yield and increase the amount of scrap to be recycled.

2 Porosity in a thin section can be avoided by initiating freezing in that section and moving the solidification front toward thicker sections, i.e., by changing

FIGURE 3-32
Jet engines may be operated at higher temperatures by changing from (*a*) equiaxed, polycrystalline turbine blades produced by conventional casting to (*b*) columnar-grain and (*c*) single-crystal blades produced by directional solidification. (*Courtesy United Technologies, Pratt & Whitney.*)

progressive solidification to *directional solidification*. In expendable molds, this can be aided by placing metal inserts (*chills*) into the refractory mold at points where maximum cooling is desired (Fig. 3-30c). In permanent molds localized cooling is achieved by placement of cooling fins or pins on external surfaces, or even by air or water cooling passages in the mold. In addition, it is necessary that liquid should be supplied to compensate for solidification shrinkage. A temperature gradient of minimum 1.5 °C/cm is required to ensure good feeding and avoid microporosity.* Porosity may also be eliminated by tapering (Fig. 3-30) although this requires extra material (*padding*).

Heat extraction and the mode of solidification affect not only the soundness of the casting but also the structure and grain size of the solidified metal. Indeed, one of the powerful ways of improving the properties of a casting is by controlling the grain size. We saw in Sec. 3-3-1 that this usually entails a small grain size, and nucleating agents together with controlled cooling rates are employed for the purpose. There are, however, exceptions: In high-temperature, creep-resistant applications a coarse grain is preferable (Fig. 3-13). Hence, the properties of turbine blades improved when the polycrystalline structure (Fig. 3-32a) was replaced by oriented grains (Fig. 3-32b) and then by a single grain (Fig. 3-32c). For the latter, the mold is placed on a water-cooled base and is

* M. C. Flemings, *Solidification Processing*, McGraw-Hill, 1974.

slowly withdrawn from a heated enclosure; solidification occurs strictly direction-ally. Yet further improvements are possible by the directional solidification of eutectic structures, so that fibers of a hard, strong phase (such as an intermetallic compound) become oriented in the direction of loading and provide integral reinforcement.

Mold Design With the above principles in mind, mold design proceeds by the following steps:

1 The volume and weight of the casting are determined.

2 Based on volume and geometric configuration (long and narrow, or blocky, or of nonuniform cross sections, etc.), the size and number of risers are de-termined.

3 On the basis of empirical relationships, the optimum pouring time is determined.

4 The feeder system is designed to feed the mold, in the allowed time, in the smoothest possible manner.

Computer programs are helpful in many ways. Relatively simple programs, suitable for microcomputers, take the mold designer through the steps outlined above, using simple theory and a great deal of empirical data.

Mathematical models, based on analytical or numerical methods, have ad-vanced to the point where mold filling can be observed on a screen, the gating and risering system can be designed, freezing times, and even microstructure and properties can be predicted to some extent. The influence of changes in casting conditions can be evaluated without extensive experimentation. Such programs require a more powerful computer and are justified only for high-volume or high-value production.

The solidification of melts can also be studied in the laboratory by the use of simulating materials (e.g., organic solutions whose crystallization may be observed in a transparent plastic mold).

Classification of Shape-Casting Processes There is a bewildering variety of old-established and newer (and sometimes, proprietary) processes; however, all can be classified according to whether the mold, cores, and patterns are expenda-ble or permanent (Fig. 3-33). It should be noted that, to aid process selection, characteristics of various casting processes are summarized in Table 12-2.

3-8-3 Expendable-Mold, Permanent-Pattern Casting

Expendable molds are prepared by consolidating a refractory material (*sand*, which may be silica or other refractory powder) around a *pattern* that defines the shape of the cavity and also incorporates, in most instances, the gates, runners, sprue, and risers required to fill the mold (Fig. 3-34).

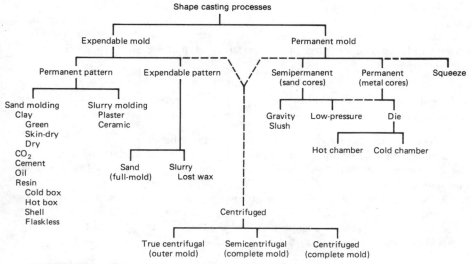

FIGURE 3-33
Classification of shape-casting processes.

FIGURE 3-34
Some characteristic elements of a cope-and-drag sand mold. To avoid the "dead" feeder head (riser), the mold could be fed from the right, making material in the "live" riser the hottest.

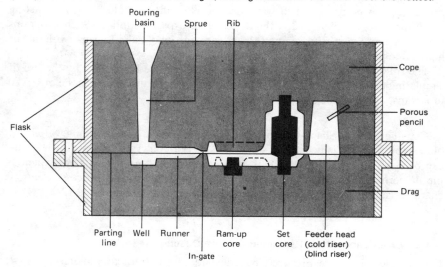

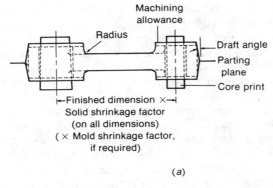

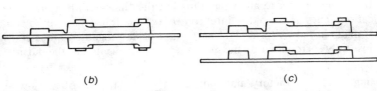

FIGURE 3-35
A pattern must allow for (*a*) solid shrinkage and easy removal from the mold. For faster production, it is fastened onto (*b*) a match plate or (*c*) cope-and-drag plates.

Patterns Patterns differ from the finished part in some important respects. All dimensions are increased to account for the contraction (solid shrinkage) of the casting from the solidus to room temperature (not to be confused with solidification shrinkage). There are patternmaker's rules that are longer by the *shrinkage allowance*; in CAD/CAM the allowance is preprogrammed. If the casting is to be machined, an appropriate thickness (*machining allowance*) is added.

Because permanent patterns are to be used repeatedly, they are made of wood or, for greater durability and dimensional stability, of a metal or strong plastic. The pattern must be easily removed from the consolidated mold; for this, molds will have to be made in two halves. Accordingly, a *parting plane* is selected that conveniently divides the shape into two parts (Fig. 3-35*a*). Surfaces parallel to the direction of withdrawal are given a draft (Fig. 3-35*a*) to allow removal of the pattern without damaging the mold.

Cavities, undercuts, and recesses in the cast shape must be formed by the insertion of *cores*. Thus, greater complexity of shape is attainable, but at a higher cost. For the accurate location of cores, the pattern provides nesting holes (*core prints*).

The simplest pattern for producing the shape shown in Fig. 3-35 would be in *one piece* (but usually split along the parting plane), and gates, runners, and risers would be added during molding (*loose pattern molding*). This makes molding slow and labor intensive. For higher productivity, elements of the feeding system are incorporated into the pattern, split along the parting line. The two halves are either mounted to the two surfaces of the same plate (*match plate*, Fig. 3-35*b*) or to separate upper and lower plates (*cope half* and *drag half*, Fig. 3-35*c*). Production rates are further increased if several pieces are molded and cast simultaneously in the same mold; for this, *multipiece pattern plates* are prepared.

Large parts of fairly simple configuration are often molded by hand, using skeleton patterns or, if the part is of rotational symmetry, by rotating a cross-section board (a *sweep pattern*) in the sand.

The refractory will have to be contained around the pattern, and the container is traditionally called a *flask*. When split into two pieces to accommodate the upper and lower halves of the pattern, one speaks of *cope* and *drag halves*, respectively. Very large molds may be formed into a pit in the ground.

Cores, like the mold itself, are made of refractory materials. However, their bond strength must be greater to allow handling but they must still be removable after solidification. They are molded into *core boxes* made of wood or metal. Cores may be made in halves (or several parts) and pasted together. Since cores are often almost fully surrounded by the melt, they must be vented to the outside.

Sand Casting Of all refractory materials, silica sand (SiO_2) is of the lowest cost and, if quality (composition and contaminants) is carefully controlled, it is satisfactory for quite high casting temperatures, including that of steel. Some other refractories such as zircon ($ZrSiO_4$), chromite ($FeCr_2O_4$), or olivine (($MgFe)_2SiO_4$) are used for special purposes. Sand in itself flows freely and must be bonded temporarily. The *bond* must be strong enough to withstand the pressure of and erosion by the melt, yet it must be sufficiently weakened by the heat of the metal to allow shrinkage of the casting and, finally, removal of the sand without damage to the solid casting. However, the bond must not destroy the permeability of the sand so that gases—present in the melt or produced by the heat of the melt in the binder itself—can escape. The quality of the sand is routinely tested in the sand laboratory for properties such as grain size; compressive, shear, and tensile strength; hardness; permeability; and compactability (decrease in height under a specified load).

Processes are often described according to the bonding agent (*binder*) used with the sand:

1 *Green sand molds* are the cheapest because they are bonded with clay. Clay is a hydrated aluminosilicate with a layered structure (Sec. 6-3-3). It is fairly strong but brittle in the dry state. It becomes readily deformable when water is added: water adsorbs on the platelets and allows their movement relative to each other. Some sands already contain the required few percent clay, but superior qualities are usually obtained when a quality clay (e.g., 6–8% bentonite) is added to pure quartz sand. With 2–3% water and thorough mixing (*mulling*), a readily transportable and moldable sand mix is obtained. When left in the damp condition, one speaks of a green sand mold. A great advantage is that used sand is readily reclaimed.

2 The clay-bonded sand may be partially dried around the cavity to improve the surface quality of the casting and reduce pinhole defects that may develop as a result of steam generation (*skin-dried sand mold*), or the entire mold may be dried out (*dry-sand mold*). Dry sand gives better surface finish but, because of its greater strength, may cause tearing in hot-short materials. It also places greater demands on energy, floor space, and equipment.

3 The *CO_2 process* uses a silica gel as the bonding agent. The sand is mixed with 3–5% water glass ($Na_2O \cdot xSiO_2 + nH_2O$), a liquid. On completion of molding, CO_2 is bubbled through to form the reaction products Na_2CO_3 and a gel of $xSiO_2 \cdot nH_2O$ composition. This gives a firmer sand mold with less wall movement and, therefore, larger, more accurate castings can be made.

4 Hydration of cement results in the formation of a gel of great strength; hence 10–15% cement is used occasionally as a binder, mostly for large steel castings molded in a pit. The sand is hard to break away from the finished casting.

5 *Oil sands* consist of sand mixed with a drying-type vegetable oil (such as linseed oil) and some cereal flour. These oils are unsaturated hydrocarbons (with double and/or triple bonds in the carbon chain) and form a polymer on heating to temperatures around 230 °C. Thus, the sand is bonded with what could be regarded as a flour-filled polymer, acquiring high strength, and is thus also suitable for cores.

6 *Resin-bonded sands* are bonded with thermosetting resins (polymers, Sec. 7-6-3). The resin-coated sand may be blown into a core box heated to 200–250 °C (*hot-box method*) or the resin may be cured with an airborne catalyst (*cold-box method*). Originally developed for single-piece, strong cores, resins are increasingly used for molding too.

7 Sand is firmly lodged in place if air is removed. This phenomenon is exploited in the *vacuum-molding process*. Patterns have small holes in them so that a thin, heated thermoplastic polymer (ethylene vinyl acetate) sheet can be tightly drawn over their surface by vacuum. Clean, unbonded sand is then applied in a flask, the surface of the flask is sealed, and vacuum is drawn on the sand. The vacuum is now released on the pattern, the pattern is removed, the mold is assembled, and the metal is poured. The polymer sheet burns up and, once the casting is solidified, the vacuum is released and the loose sand falls off.

Bonded sand is compacted by various techniques, chosen according to production rates and the total number of parts produced:

1 For only a few parts, the sand may be shoveled into the flask around a one-piece pattern and rammed by hand. It requires high skill to produce a mold of uniform packing.

2 For mass production, the sand is conveyed to the molding station and dropped, blown, or slung (by having it flow over a fast-rotating wheel) onto patterns surrounded by flasks. The sand is compacted in the flask by mechanical

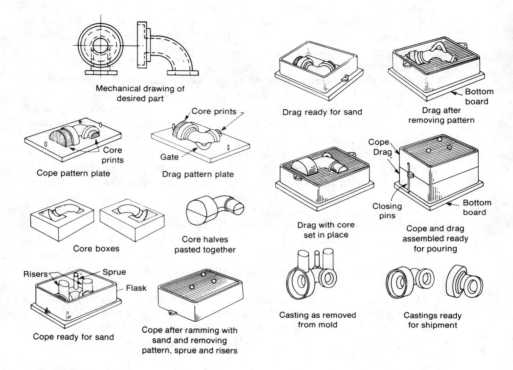

FIGURE 3-36
A typical sand molding sequence. (*From Steel Castings Handbook, 5th ed., Steel Founder's Society of America, Des Plaines, Ill., 1980. With permission.*)

means, such as jolting or squeezing. A typical sequence of molding operations is given in Fig. 3-36.

3 The flask may be evacuated and then a large valve opened through which sand is drawn in; permeability tends to be much more uniform than when compacting with static pressure. Similar good results are obtained with dynamic compaction using the pressure wave from the detonation of natural gas above the flask.

4 When pressures are high enough (around 7 MPa or 1000 psi), a properly bonded sand acquires enough strength to maintain the integrity of the mold without a supporting flask (*flaskless molding*). Production can reach 250–750 molds/h, stacked end-to-end, in a single production line. The high strength of molds also minimizes wall movement during casting and solidification, and gives castings of higher accuracy.

In all but flaskless molding weights must be placed on the cope; otherwise, the metallostatic pressure exerted by the melt would lift up the cope and a breakout would occur.

It is customary to use a finer sand in the vicinity of the mold surface. Facing materials (coal or graphite) can be added to generate gases on contact with the hot metal, reducing metal penetration and adhesion to the sand (burning of the sand), thus giving a better surface finish, free of defects. Alternatively, various refractory materials can be suspended in a liquid and applied as a coating to the mold and core surfaces.

Cores greatly increase the variety of shapes that can be cast. If their weight cannot be supported, cores are placed on *chaplets* (small, often perforated metal supports that will melt into the casting alloy).

Shell Molding *Shell molding* is a variant of the resin-bonded sand technique. The pattern must be made of metal and heated to 200–260 °C. After coating with a parting agent, it is placed on top of a box that contains sand, coated with a heat-curing resin. On inverting the box, the sand settles on the pattern and a thin shell cures in situ, faithfully reproducing details of the pattern. Once the shell is thick enough, the box is turned back, whereupon excess, unbonded sand drops back into the box. The shell is stripped, combined with the other half, placed in a flask, and backed with some inert material such as steel shot to provide support. The greater strength of the mold often allows forming integral cores with the mold. Compared to green-sand casting, the surface finish and tolerances of parts are closer, floor space and sand quantity in circulation are reduced, but recycling of the sand is more expensive.

Shell molding is eminently suitable for cores. Sand is blown into heated molds; because the cores are hollow, they give good venting.

Slurry Molding Instead of compacting the sand by force, a finer-grained refractory may be made into a *slurry* with water and poured around the pattern. A smoother surface finish is obtained and, if so chosen, the refractory may be more heat-resistant than the bonded-sand variety. Since the shrinkage of mold and casting can be closely controlled, one often speaks of precision casting.

Plaster molding relies on the well-known ability of a plaster of paris slurry (gypsum, $CaSO_4 \cdot 2H_2O$) to flow around all details of a pattern. Various inorganic fillers may be added to improve strength and permeability. After a rather complex baking sequence, the mold is assembled and the metal poured. In a patented variant (*Antioch process*), steam pressure treatment is applied to produce intergranular air passages. Since gypsum is destroyed at 1200 °C, it is not suitable for ferrous castings. For other materials, plaster molding gives very good surface finish and tight tolerances.

Ceramic-mold casting is suitable for all materials, because the slurry is now made up of selected refractory powders, such as zircon ($ZrSiO_4$), alumina (Al_2O_3), or fused silica (SiO_2), with various patented bonding agents. The fine-grained ceramic slurry is applied as a thin facing to the pattern and is backed up with lower-cost fire clay. The mold is fired at 1000 °C and the melt is poured while the mold is still hot. The higher cost of these processes (such as the *Shaw process*) is well justified by their success in producing quality castings of highly alloyed, high melting-point metals. Precision is obtained by taking account of dimensional changes at each stage of mold making and casting, often by CAD/CAM. Large constructional parts as well as forging and casting dies can be cast to final shape and dimensions, often without the need for subsequent surface finishing.

3-8-4 Expendable-Mold, Expendable-Pattern Casting

An *expendable pattern* must be made of a material that can be either melted out before pouring or burned up during casting. This way, the pattern may be left in the mold, and there is no need for parting planes, draft angles, or even cores. Shape limitations are few, and the only criterion is that the refractory can be shaken out or otherwise removed from all cavities and intricate details of the finished casting (the refractory is sometimes left in cavities of sculptures).

Expendable patterns are made by injecting the pattern material into the cavity of a pattern mold. Thus, there is the requirement that the pattern must be extracted from the mold, however, more complex shapes are easily produced by assembling the pattern from several simpler shapes. In some instances, a rubber (usually silicone rubber) mold can be used. Shrinkage of the pattern material must be taken into account, and a draft must be provided if the mold cavity is deep. Processes are usually named after the pattern material.

Investment Casting   *Investment casting*, also called the *lost-wax process*, was already used in ancient Egypt and China, but has found widespread industrial application only since the Second World War, with the need to produce precision parts in high-temperature materials for jet engines. It is capable of producing the most complex shapes, because the pattern is made of wax (sometimes of a plastic such as polystyrene, or even of frozen mercury) complete with feeding system, and the refractory slurry is poured around this.

Wax patterns are readily produced in large quantities by injection molding into metal dies. Individual patterns are assembled with wax sprues, runners, and gates into a so-called tree, simply by local melting of the wax, using a hot knife or blade between the two mating surfaces. Two approaches are then practical:

1 In *solid investment*, the tree is precoated by dipping in a refractory slurry, dusted with refractory sand, and placed in a flask where a thick, coarser refractory slurry is poured around it. When the slurry has gelled by drawing off excess water, the mold is dried in an oven in an upside-down position to allow the wax to run out. Before casting, the mold is fired at 700–1000 °C; this imparts strength to the mold, eliminates the danger of gas formation from water during casting, increases the fluidity of the melt that will be poured, and ensures a good surface finish as well as close dimensional tolerances.

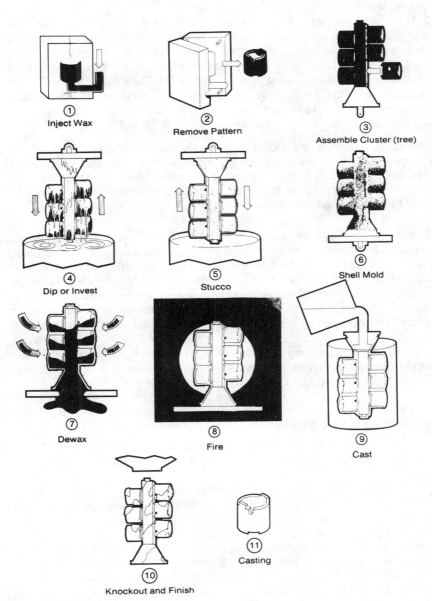

① Inject Wax

② Remove Pattern

③ Assemble Cluster (tree)

④ Dip or Invest

⑤ Stucco

⑥ Shell Mold

⑦ Dewax

⑧ Fire

⑨ Cast

⑩ Knockout and Finish

⑪ Casting

FIGURE 3-37
Typical sequence of ceramic shell molding. (*Investment Shell Casting Institute, Dallas, Tex.*)

2 The cost of the mold may be reduced and the rate of production increased by dispensing with the solid mold. In the *ceramic shell-molding process* (Fig. 3-37) the tree is prepared as before, but is then covered with refractory in a fluidized bed. (When air is blown from the bottom of a container partially filled with powder, the powder is suspended in the air and flows like a fluid.) Several layers of gradually coarsening coats are applied to reach sufficient thickness. The repetitive operation can be entrusted to a robot. The shell mold is dried and fired, if necessary supported by a granular material, and the metal is cast.

Significant improvements are obtained with vacuum pouring. The mold is placed over the melt (with a nozzle reaching into the melt); vacuum is then drawn on the mold so that the melt rises smoothly to fill the cavity. The sprue and runner are made large enough to prevent solidification in them, and the liquid is allowed to flow back into the melt. Yields rise to 85–95%.

Evaporative Casting An interesting variant of sand casting utilizes an expendable mold with an expendable pattern made of expanded polystyrene foam, similar to that used in cups for hot beverages. Very complex shapes can be built up and runners, etc., can be attached with rubber cement or hot-melt resins. Such *evaporative casting*, also called *lost-foam* or *full-mold process*, again allows great freedom in shapes, without draft, because the pattern is left in the mold to evaporate and burn up during casting. The plastic foam is firm but would be damaged by high compaction pressures. It is first coated with a refractory wash, then a slurry is poured around it. Loose sand may be held by vacuum, or dry sand is simply compacted by vibration and weighted down during pouring. Alternatively, steel shot may be kept in place by a magnetic field.

3-8-5 Permanent-Mold Casting

While in the processes described above the mold was destroyed after the solidification of the casting, the mold is reused repeatedly in permanent-mold casting processes.

The mold material must have a sufficiently high melting point to withstand erosion by the liquid metal at pouring temperatures, a high enough strength not to deform in repeated use, and a high thermal-fatigue resistance to resist premature crazing (the formation of thermal fatigue cracks) that would leave objectionable marks on the finished casting. Finally, and ideally, it would also have low adhesion (Sec. 2-2-1) to the melt to prevent welding of the part to the mold.

The mold material may be cast iron, although alloy steels are the most widely used. For casting higher-melting alloys (brasses and ferrous materials), the mold steel must contain large proportions of stable carbides so that strength is retained at higher temperatures. More recently, refractory metal alloys, particularly the precipitation-hardenable molybdenum alloy TZM (0.015% C, 0.5% Ti, 0.08% Zr), have found increasing application. Graphite molds can also be used for steel, although only for relatively simple shapes.

Coatings or dressings, composed of refractory powder in a suspending medium, are applied to the die surface to protect the die and reduce heat transfer. Graphite, silicone, and other films (*parting compounds*) reduce adhesion and facilitate ejection. Refractory coatings are sometimes built up to thicker layers for the purpose of reducing temperature fluctuations on the mold surface. Coatings are an important element in the system; their uniform application is most important, and programmable robots are well suited for such repetitive task.

All permanent molds (dies) have some common features:

1 A prime requirement is that the solidified casting be readily removable from the die cavity. Therefore, shapes that can be produced are more limited than when casting in an expendable mold, although great complexity is allowable when the mold is made in several parts (for example, for office-machine, sewing-machine, chain-saw, etc., housings, automotive carburetor bodies, etc.).

2 Internal cavities are formed with the aid of fixed or movable *metal cores*. Undercut shapes require that cores be made in several interlocking parts which are withdrawn in a fixed sequence. When casting with gravity or low-pressure feed, sand cores may be inserted into the permanent mold (*semipermanent molds*).

3 The metal die affords means of producing composite castings by the accurate location of inserts (threaded inserts, heating elements, etc.). Appropriate nests are provided in the mold.

4 *Ejector pins* are necessary to remove the solidified casting, particularly if the process is mechanized. The number and location of ejector pins must be chosen to prevent objectionable surface marks or distortion of the casting. Early ejection is important when the casting would shrink onto bosses, but enough time must be allowed for solidification.

5 The mold material is not permeable, therefore, *vents* must be provided to avoid trapping gases. Clearances along parting planes and around ejector pins may also serve as vents.

6 The permanent mold works as a heat exchanger. At the start of a production run, the mold must be preheated to the desired temperature (typically 150–200 °C for Zn, 250–275 °C for Mg, 225–330 °C for Al, and 300–700 °C for Cu alloys). During steady-state production, heat given off by solidification is removed by means of radiating pins or fins, or internal water cooling channels. Evaporation of water from mold dressings, prior to closing the mold, also cools the mold face.

Close control of die temperatures and mold dressing composition and application allow casting with thinner walls, even though at the expense of slower solidification and a prolonged casting cycle (Fig. 3-22). Nevertheless, solidification rates are much higher than in refractory molds; therefore, output rates are high and grain size is small. However, the supply of melt to thicker sections of the casting may be cut off prematurely. It is imperative, therefore, that proper feeding be provided; even so, porosity can be and usually is higher than in similar castings made in expendable molds. The permanent mold is always stronger than the solidifying casting; therefore, casting alloys prone to hot-shortness (those of long freezing range, or containing a low-melting matrix) are avoided. There are several variants of permanent-mold casting, distinguished by the method and pressure of feeding.

Gravity-Feed Permanent-Mold Casting The process is usually referred to simply as *permanent-mold casting* or *gravity die casting*. It builds on the same principles as expendable-mold casting, except that the mold is made of an appropriate permanent material. The casting machine is basically a bed that supports the stationary and movable mold halves. The halves may be hinged as pages in a book (*book mold*). Manually operated machines are equipped with a long handle and clamps; mechanized machines have hydraulic actuators (Fig. 3-38a). In conjunction with split metal cores or collapsible sand cores, the process is very versatile.

The process is widely used for aluminum alloys (e.g., internal combustion engine pistons cast in a mold equipped with multipiece movable cores) as well as magnesium and copper alloys. Smaller cast iron and steel castings can also be made. The mold is protected by a lubricant-type coating for aluminum and magnesium alloys, and with ceramic coatings of up to 1-mm (0.040-in) thickness for copper-based alloys and gray iron.

FIGURE 3-38
Permanent molds may be filled under (*a*) gravity, or with the application of (*b*) low pressure, or high pressure in (*c*) hot-chamber or (*d*) cold-chamber die casting.

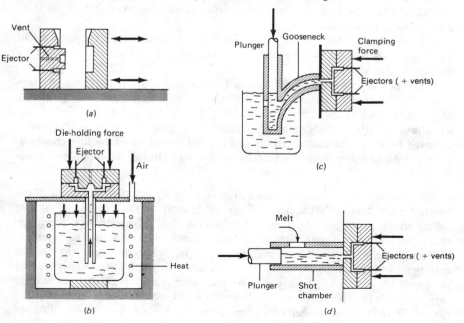

The *slush casting* process is a variant of permanent-mold casting. It is used mostly for nonstructural, decorative products such as hollow lamp bases, candle sticks, and statuettes. The mold is filled with the melt and, after a short time, inverted to drain off most of the melt, leaving behind a hollow casting with a good outer surface but very rough inner surface.

Low-Pressure Permanent-Mold Casting   The mold is situated right above the melting or holding furnace, and metal is fed by air pressure through the bottom gate into the mold cavity (Fig. 3-38*b*), ensuring smooth filling. Solidification is directed from the top downward. Air pressure is released as soon as the cavity is filled with solid metal; thus, material losses are minimized. Thin coatings ensure an acceptable surface quality. The die halves must be held together under sufficient force to resist the force generated by fluid pressure in the cavity. In a variant of the process, smoother die filling is obtained by drawing vacuum on the mold.

The process finds widest application to aluminum alloys. A variant is used for casting steel into graphite molds.

Die Casting   The term *die casting* is used to denote processes in which the mold cavity is filled under moderate to high pressures, thus forcing the metal into intricate details of the cavity. The die halves are held together by a correspondingly high force; therefore, die-casting machines resemble hydraulic presses of two-, three-, or four-column construction. The die-holding force is exerted by the press movement but, for greater rigidity, the dies are usually locked with toggle clamps while the melt is forced into the die—slowly at first and then at increasing speeds—by a separate plunger. Machines are usually rated by the die-holding force. Shot sizes range from a few grams to tens of kilograms, and production rates of up to 1000 shots/h can be achieved in the smaller sizes. There are two basic variants:

1 In the *hot-chamber process* the liquid metal is transferred to the mold directly from the holding pot by a submerged pump (cylinder and plunger, Fig. 3-38*c*) at pressures up to 40 MPa (6000 psi). Pumps constructed of steel are used for zinc alloys and magnesium alloys. There are still some problems with the durability of ceramic pumps needed for aluminum alloys.

In a variant of the process, *direct injection* of Zn alloys takes place through a heated manifold and mininozzles, so that gates and runners are eliminated and the yield is raised.

2 The *cold-chamber process* uses separate melting facilities. Melt of a quantity sufficient for one shot is individually transferred, often automatically, to the cylinder (*shot chamber*) from which the plunger squirts it into the die cavity (Fig. 3-38*d*) at pressures up to 150 MPa (20 kpsi). The alloys that can be handled are limited only by the mold material. The process has long been established for zinc, magnesium, and aluminum alloys, and is finding increasing use for brass. Occasionally, steel is cast in TZM dies.

Success depends on control of the die casting process. In the past, this was the task of the die-casting machine operator. More recently, a better understanding of the role of process variables has led to extensive instrumentation so that measured variables can be fed into a microprocessor which then performs control. A number of variables are of importance: melt temperature, dissolved gas content, die temperature and its distribution, plunger velocity and its variation during the stroke, chamber pressure, cavity pressure, and gas composition. In a machine of given capacity, the pressure exerted drops as the rate of metal delivery (pumping rate) increases. From experiment and theory, much is now known about the optimum pumping rates and pressures needed to make good, thin-walled castings. From the experimentally determined machine capability, the size of the runners and gates that gives the optimum velocity can be calculated: Excessive velocity would result in mold erosion and gas entrapment due to turbulence; too slow filling in incomplete parts (misruns) and cold shuts. (Gate velocities can reach 40 m/s in casting zinc alloys.)

Most alloys are solid solutions and, even though the solidification pattern (Fig. 3-23*b*) would choke off the flow, the applied pressures are high enough to rip off dendrites, and intricate cavities can be filled. With proper control, it has been possible to reduce wall thicknesses to the point where the entire cross section exhibits the fine structure typical of cast surfaces, thus increasing the strength of the product. With decreasing wall thicknesses, elastic deflections of the die and press become significant, and forces on each press column are measured and equalized.

Since solidification is too rapid for the proper feeding of heavier sections, porosity is a major problem in solid-solution alloys. Microporosity due to dendritic solidification can be minimized or completely eliminated if the injection pressures are high enough or if other measures are taken:

1 In a variant of the process, injection pressures are kept at a medium level until the casting is almost solidified; at this point pressures are raised to 150 MPa (20 kpsi) to consolidate the still mushy material.

2 In another version, a premeasured amount of melt is loaded into a die, allowed to cool below the liquidus temperature, and then the die is closed while solidification is completed. Such *squeeze-casting* or *melt forging* uses typical forging dies; it represents a transition between die casting and hot forging, and gives highly refined grain structures. It is also used for making aluminum alloy parts such as diesel-engine pistons reinforced with Al_2O_3 fibers.

3 Yet another possibility exists in exposing the melt to shear (e.g., by a rotor submerged in the melt) during solidification. The structure is greatly modified by the breakdown of the dendritic pattern. The partially solidified but still fluid melt can be injected into a die or allowed to solidify for later remelting (*rheocasting*).

The other source of porosity is gas entrapped in the casting. The difficulty lies in the need for the rapid expulsion of air from the die cavity and in the presence of air itself. One solution is based on the evacuation of the cavity, the other on the observation that in some alloys gas pores contain only nitrogen. If the cavity is purged prior to casting with pure oxygen, the gas reacts with the melt to form finely dispersed oxide particles. This process, called *pore-free die casting*, produces better castings, although porosity from solidification or hydrogen entrapment is still possible.

In all die-casting processes it is essential to apply a die lubricant to the mold surfaces between each shot. The lubricant is usually graphite or MoS_2 in an oily carrier, which is then dispersed in water. Evaporation of the water aids cooling.

The repetitive tasks involved in opening and closing the die, removing and quenching the casting, and insertion into a trim press can be performed by robots. Some of the newer die-casting machines use an indexing turret to move the part through flash removal and minor machining steps.

3-8-6 Centrifugal Casting

When a mold is set in rotation during pouring, the melt is thrown out by the centrifugal force under sufficient pressure to assure better die filling. Solidification progresses from the outer surface inward; thus, porosity is greatly reduced and, since inclusions tend to have a lower density, they segregate toward the center (which in axially symmetric parts is often machined out anyway). Forced movement by shearing the melt results in grain refinement. Centrifuging can be applied

FIGURE 3-39
Good filling and fine grain can be obtained by rotating the mold in (*a*) centrifugal, (*b*) semicentrifugal, or (*c*) centrifuged casting. [(*b*) *and* (*c*) *G. E. Schmidt, Jr., Massachusetts Institute of Technology.*]

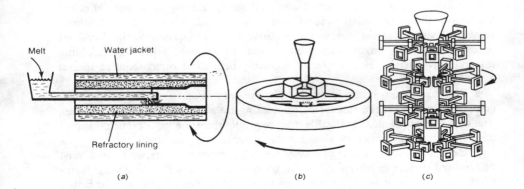

(*a*)	(*b*)	(*c*)

to all casting processes if the mold is strong enough to withstand rotation. It is customary to distinguish between various processes according to the shape of the mold (Fig. 3-39).

1 True *centrifugal casting* employs molds of rotational symmetry, essentially tubes, made of steel (protected with a refractory mold wash or even with a green- or dry-sand lining) or of graphite. The melt is poured while the mold rotates, resulting in a hollow product such as a tube or ring (Fig. 3-39*a*). By controlling flow rates and moving the pouring orifice along the axis, long and large tubes of very uniform quality and wall thickness can be cast. If desired, the outer contour of the casting can be varied, while the inside remains cylindrical. Surface quality is good outside but can be poor inside.

2 Centrifuging may be applied to the production of castings in molds prepared by any of the techniques discussed under expendable and permanent molds. When only one piece of approximately rotational symmetry (e.g., a wheel with spokes and central hub) is cast, the term *semicentrifugal casting* is applied (Fig. 3-39*b*).

3 When odd-shaped parts are placed around a central sprue in a balanced manner (e.g., by investment molding), the term *centrifuged casting* is usual (Fig. 3-39*c*). Jewelry is centrifugally cast (*spin cast*) by placing the investment mold at the end of a rotating arm and placing the crucible, containing the molten metal, next to it. Centrifugal force transfers the melt and ensures good mold filling.

3-9 FINISHING PROCESSES

The solidified casting must be subjected to a number of auxiliary operations before it can be used.

3-9-1 Cleaning and Finishing

1 When casting is performed in expendable molds, the first step is freeing the casting from the mold. For green and dry sand, *shaking* is a most effective procedure; the clay-bonded sand is then recycled and, with suitable additions, reused. This is one of the reasons for the survival of sand casting as the dominant process for making larger parts even in mass production. With other molding materials reclamation is a matter of economy because it often requires special equipment or processes.

2 Residual sand is removed by *shot blasting* (*airless blasting*). Round particles (*shot*) are hurled on the surface of the casting by a fast-rotating paddle wheel. The shot is made of steel, malleable iron, or white iron for harder castings, and of mild steel, bronze, copper, or glass for the softer nonferrous materials.

3 Dependent on the casting process, the gates, runners, risers, and sprue must be removed (before or after shotblasting), together with any fin (*flash*) that forms when melt flows out into gaps between two mold halves or at cores. In brittle materials, this is simply done by breaking the excess material off; in more ductile materials, sawing or grinding becomes necessary. Robots are often employed in these operations, especially in die casting.

4 The entire surface is cleaned by various processes, including shot blasting; tumbling in a dry or wet medium of some refractory material or steel shot; or chemical pickling.

5 If each part has sufficient value, any defects detected may sometimes be repaired by welding without jeopardizing the function of the finished part. Otherwise, defective parts are rejected and remelted.

6 The finished casting is sometimes subjected to HIP to reduce porosity, or to various heat treatments to modify mechanical properties or reduce residual stresses (Sec. 3-5).

3-9-2 Quality Assurance

Quality control and inspection at all stages of production are vital to the success of the modern foundry.

Many (but not all) of the casting defects can be detected by visual inspection, and may be classified with the aid of atlases showing typical defects. Some of the most frequently encountered defects are briefly discussed here.

1 Improper gating and risering or poor molding practices may cause sand erosion and embedment (*sand skin*, *scab*) in expendable-mold casting. In the extreme case, the melt penetrates into the sand and causes the metal to fuse into a mass. When sand drops off the surface, an *expansion scab* is formed.

2 Poorly controlled sand compaction may cause the dimensional tolerance limits to be exceeded by allowing too much movement of the mold walls (*swell*) in expendable-mold casting.

3 *Shifting* of mold halves and particularly of cores is a common cause of exceeding tolerance specifications.

4 The melt may break out from the flask if the mold is inadequately weighted or the mold wall is too thin. An incipient *breakout* shows up as a thin fin at the parting plane.

5 Insufficient cavity filling (*short run*, *misrun*) may occur with any process and can be caused by inadequate metal supply, improperly designed gating, or too low mold or melt temperatures.

6 Too low melt temperature can also lead to uneven die filling and visible folds (*cold laps*). *Cold shuts* form when two metal streams meet without complete fusion; this can be particularly troublesome when the metal flows inside an oxide cover, as do alloys that contain aluminum.

7 Surface or subsurface *pinholes* and *blowholes* are caused by gas liberated from the melt or formed as a result of metal–mold reactions.

8 We already discussed the causes and prevention of shrinkage cavities and porosity. They may become exposed when gates and risers are removed.

9 Shrinkage on cooling from the solidus temperature requires that the mold should give. If this does not happen, the casting will be larger than expected, will have residual stresses, or will become distorted (thinner sections remaining longer). In the extreme case, when the material is hot-short, fracture occurs in the section that solidifies last.

3-9-3 Inspection

Visual inspection has always been practiced, sometimes aided by the use of penetrant dye and magnetic particle techniques. The greatest advances in the quality of castings have been made by the extensive use of nondestructive testing techniques, including x-ray, ultrasonic, and eddy-current inspection for internal soundness (Sec. 2-5). While these measures can be costly, they have made the casting process competitive in applications hitherto reserved for forgings. Some of the prime examples are aircraft-quality parts and the crankshaft and, to a lesser degree, connecting rod of the internal combustion engine. Nondestructive techniques are particularly important in detecting internal defects, whether they be due to solidification shrinkage, internal hot tearing, or gas porosity.

3-10 DESIGN FOR CASTING

While the skill of the foundryworker may make it possible to cast a very large variety of shapes with relatively thin walls and sudden changes in direction and cross section, design practices that violate the rules of sound casting principles increase the cost of production, make production possible only at the expense of accepting the high cost of large rejection rates or, in the extreme case, make production impossible (it may turn out that the part should not be made as a casting at all).

Prototypes of products are often assembled from machined components in preparation for casting design, and the possibility of making a subassembly as a single casting should be contemplated.

There are several books available that detail the usual design practices for castings; however, it is possible to indicate the major points in a general manner, with reference to the principles discussed hitherto.

1 As with all design, it is essential that preconceived ideas about the final shape of the part should not intrude at early stages of design. Instead, the basic functions of the part must be considered, and alternative shapes that satisfy these functions must be kept under consideration as long as possible. The final design will have to be optimized for the chosen casting process. The use of weldments and of cast inserts should be kept in mind.

2 Stress levels should be kept consistent with the strength readily achievable in casting alloys. If the material is brittle (e.g., gray and, particularly, white cast iron), one must remember that the tensile strength is much lower than the compressive strength. In many materials the as-cast structure is likely to contain porosity, inclusions, and unfavorably located second phases, some of them resulting from nonequilibrium solidification. Therefore, the tensile and, particularly, fatigue strengths are lower than in a thoroughly worked material.

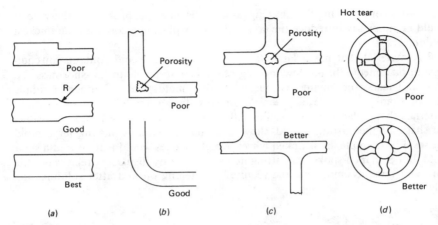

FIGURE 3-40
Design features help to alleviate casting problems due to (a) wall thickness variation, (b) hot spot in corner, (c) hot spot at cross ribs, and (d) hot tearing.

3 Tolerances, surface roughness, and minimum and maximum wall thickness should be consistent with the capabilities of the process. While some processes such as investment casting can produce parts of very good surface finish and close tolerances, the cost may not be justifiable. The capabilities of various processes are discussed in Sec. 12-2.

4 The shape of the casting must ensure that the pattern can be removed from an expendable mold (except in expendable-pattern techniques) and that the casting can be released from a permanent mold. The location of the parting plane must be considered, because this will determine the need for and location of cores and draft angles. Cores and the extra material required by the draft add to production cost. Therefore, draft should be chosen to contribute to load bearing, and it should not interfere with positive clamping and location in subsequent machining. Locations of drilled holes should be strengthened by bosses, preferably shaped so as to make the drill enter perpendicular to the cast surface.

5 The shape of the casting should allow orderly, directional solidification by moving the solidification front from the remotest parts toward the feeding end, and should not close off access of melt to thicker sections. Otherwise, shrinkage cavities and microporosity result.

a When wall-thickness variations are unavoidable, transition must be made by generous radii (Fig. 3-40a). Small radii or sharp corners act as stress raisers in the finished casting, create turbulence during pouring, and prevent proper feeding during solidification.

b Localized heavy cross sections—such as result when applying the appropriate radius only to the inside surface of a corner or when two ribs cross each other—create hot spots where the melt solidifies only after adjacent zones have frozen; therefore, shrinkage cavities form. Applying a radius to the outer surface

(Fig. 3-40*b*) or offsetting the ribs (Fig. 3-40*c*) alleviates the problem. Otherwise, it would be necessary to reduce the cross section by placing a core into the thickest section.

6 A material that has a large solidification shrinkage and contains an equilibrium or nonequilibrium low-melting phase is susceptible to hot-shortness. To avoid hot tears, such materials must be cast in molds that are either of simple enough shape not to develop tensile stresses during solidification or in mold materials that collapse or give sufficiently to allow shrinkage. Whenever possible, the shape of the casting should allow deformation without moving large mold masses. For example, in a spoked wheel straight spokes would tear even in a sand mold, but S-shaped spokes can straighten somewhat by displacing relatively little sand and thus accommodate the required shortening on and after solidification (Fig. 3-40*d*).

3-11 SUMMARY

Solidification processes are involved in producing the vast majority of metallic materials. Almost all wrought materials are cast before deformation, and a significant proportion of metals is cast directly into a shape. Cast components are used for applications as diverse as machine-tool bases, automotive engine blocks and crankshafts, turbine blades, plumbing fittings, and decorative hardware. With the appropriate molding process, parts of a complexity unmatched by any other process may be cast.

A sound casting results only if the limitations imposed by the solidification process are recognized:

1 For solidification to occur, the mold must be colder than T_m of the metal. Problems of fluid flow and of heat transfer limit the minimum attainable wall thickness, especially if the alloy solidifies with the growth of dendrites which choke off fluid flow.

2 Gating and risering techniques must ensure smooth, complete filling of the die cavity followed by orderly solidification, with a liquid metal supply sufficient to feed the pipes that would otherwise form.

3 Heat transfer must be locally controlled to prevent starvation of late-solidifying portions of the casting and to minimize porosity.

4 Grain-size control is one of the most powerful means of improving mechanical properties. One usually aims at a fine grain, but exceptionally (for high-temperature creep resistance) at coarse grain, in the limit, by directional solidification of a single dendrite.

5 Mechanical properties, particularly ductility and fatigue and impact strength may be greatly improved by melting, pouring, and casting techniques that reduce the number and size of inclusions and bring inclusions into the least harmful shape and location.

6 The properties of castings can be improved by subjecting them to pressure treatment during or subsequent to solidification.

7 Desirable properties may also be imparted by heat treating the solidified casting to remove residual stresses (stress-relief anneal), homogenize the structure (homogenization), increase strength (solution treatment and aging of precipitation-hardenable alloys, and quenching and tempering of steel), or develop the optimum structure (annealing of white iron to convert it into malleable iron, and annealing of ductile cast iron).

PROBLEMS

3-1 Apply a material balance to Example 3-2 to show that 28.1% Cu is in the alloy.

3-2 Apply a material balance to Example 3-3 to show that 80% Cu is in the alloy.

3-3 Draw a hypothetical phase diagram containing eutectic, peritectic, intermetallic compound, and eutectoid transformations.

3-4 Would you expect significant coring in alloys of (*a*) Cu–5Sn; (*b*) Cu–30Zn; (*c*) Al–5Mg?

3-5 Under otherwise identical conditions, would a melt heated to $1.2T_L$ give a finer-grained casting than one heated to $1.1T_L$?

3-6 Make judgements on the changes in fluidity expected on moving in the Ag–Cu system from the Ag to the Cu end.

3-7 Describe the temperatures and sequence of precipitation-hardening treatment in a Ag–6Cu alloy. Would you expect significant hardening?

3-8 Explain the difference between precipitation hardening in an aluminum alloy and martensite formation in steel.

3-9 Plot the property data given in Example 3-7 as a function of tempering temperature. Connect the points with continuous curves; note the more rapid decline of strength properties at the higher temperatures. Check whether hardness is indeed three times TS.

3-10 By taking σ_0 and k_y in Eq. (3-5) arbitrarily equal to unity, calculate and plot a generalized σ versus d curve (with d varying from 10^{-2} to 10^2).

3-11 Show that Bernoulli's equation (Eq. (3-12)) is dimensionally homogeneous and indeed represents energy per unit volume.

3-12 Solid-solution alloys of wide freezing range are generally regarded as poor casting alloys, yet they are extensively used in die casting. Provide an explanation.

3-13 Define the following terms, using appropriate sketches: pattern, core, core box, core print, flash, pouring basin, sprue, well, runner, gate, in-gate, hot riser, cold riser, feeding head, dead riser, live riser.

3-14 Define green-sand and dry-sand mold. Which is stronger, and which is more suitable for casting hot-short materials?

3-15 Look up the Al–Si equilibrium diagram and select compositions with 2% and 12% Si. Which alloy is (*a*) more ductile in the equilibrium condition, (*b*) more prone to coring, (*c*) more susceptible to hot-shortness, (*d*) of higher fluidity, (*e*) easier to feed, and (*f*) more prone to microporosity? Support your considerations with sketches, and then (*g*) state which alloy is more favorable for casting.

3-16 The part (Fig. P3-16) shown is to be cast of 10% Sn bronze, at the rate of 100 parts per month. Review Fig. 12-5 for surface finish. To find an appropriate casting process, consider all, then reject those that are (*a*) technically inadmissible or (*b*) technically feasible but too expensive for the purpose, and (*c*) identify the most

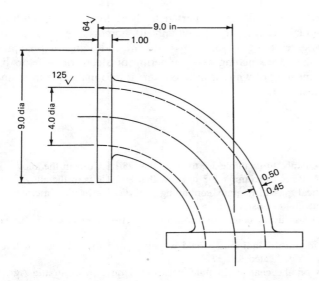

economical one (make common-sense assumptions about costs, reinforced by data from Table 12-2).

3-17 Liquid steel with 0.01 wt% oxygen is to be strand cast. Assume no solubility of O in the solid and that the O will be rejected in the form of CO. (*a*) Calculate the amount of gas released in units of $cm^3/100$ g metal. (*b*) Recalculate as a percentage of the total (gas + metal) volume. (*c*) If the steel is continuously cast into a 100 mm × 100 mm square strand at 1.5 m/min, calculate the feed rate of 10-mm-diam aluminum wire used to deoxidize the steel.

3-18 A 60Cu–40Zn brass melt is poured into a sand mold. The metal level in the pouring basin is 200 mm above the centerline of the runner which is taken as the zero level. The cross section of the runner is 10 mm × 10 mm. Calculate, from Bernoulli's theorem, the velocity and rate of flow at the entry to the mold, ignoring friction losses (the pouring basin is so large that the velocity in it can be taken as zero).

3-19 Al–4Cu alloy castings are solution treated in a continuous, belt-type furnace at 545 °C for 15 min. The furnace has three heating zones, each capable of holding temperature within ±10 °C. Lately many castings have suffered severe distortion, sagging in the furnace. Suggest a reason or reasons for the problem.

3-20 The part shown in the figure for Example 12-6 will be sand cast of steel. (*a*) Using data on machining allowances, draft angles, and radii (e.g., from *Metals Handbook*, 9th ed., vol. 1), design and draw the cast shape. (*b*) Taking shrinkage into consideration, make a drawing of the pattern. (*c*) For greater productivity, eight rings will be cast in each mold. Sketch a possible layout complete with sprue, runners, and gates; provide for streamlined flow; ensure equal feeding rates to each ring by taking the runner cross sections proportional to the flow rates at various points. (*d*) A cylindrical riser is to be used with a maximum volume to surface area ratio. Using Chvorinov's approximation, determine the size of the riser.

3-21 The part of Prob. 3-16 is machined at the flange face. Utilizing data on machining allowances, drafts, shrinkage, etc. (from the literature, e.g., *Metals Handbook*, 9th ed., vol. 1), choose an appropriate parting line and design a properly dimensioned pattern and core box.

3-22 A wheel is cast of low-carbon steel with straight spokes (as in Fig. 3-40, "poor"). The length of each spoke is 100 mm. The mold is of a refractory material, unyielding, and changes its dimensions insignificantly during heating or cooling. The spoke cools from 1100 to 900 °C in 10 min. Calculate (*a*) the strain, if the thermal expansion coefficient is 23×10^{-6} per °C; (*b*) the average tensile strain rate; (*c*) the flow stress at 1000 °C (from Table 4-2). (*d*) Assuming that the material at this temperature behaves like an ideal elastic-plastic body, and Young's modulus is 60% of the room-temperature value, determine whether shrinkage will be accommodated by the development of elastic (residual) stresses or by plastic deformation. (*e*) Subject the above problem statement to a detailed critique regarding the validity of the simplifying assumptions.

3-23 With the aid of Chvorinov's rule, calculate the relative solidification times for castings of identical volumes and of the following shapes: (*a*) sphere of diameter d_s; (*b*) cylinder with $h/d = 1$; (*c*) cylinder with $h/d = 10$; (*d*) cube; (*e*) right rectangular prism with $h/a = 10$; (*f*) flat plate of the same length as (*e*) but of $\frac{1}{3}$ the thickness. (*g*) Plot the results to illustrate the effect of shape changes.

3-24 If the part discussed in Prob. 3-20 were to be die cast in an aluminum alloy, what size machine would be required?

FURTHER READING

A Detailed Process Descriptions

Allsop, D. F., and D. Kennedy: *Pressure Die Casting—Part II: The Technology of the Casting and the Die*, Pergamon, Oxford, 1983.

ASM: *Metals Handbook*, 8th ed., vol. 5, *Forging and Casting*, 1970. American Society for Metals, Metals Park, Ohio.

Kaye, A., and A. Street, *Die Casting Metallurgy*, Butterworths Scientific, London, 1982.

Mikelonis, P. J. (ed.): *Foundry Technology, Source Book*, American Society for Metals, Metals Park, Ohio, 1983.

Rauch, A. H. (ed.): *Source Book on Ductile Iron*, American Society for Metals, Metals Park, Ohio, 1977.

Romanoff, R.: *Centrifugal Casting*, TAB Books, Blue Ridge Summit, Penn., 1981.

Street, A.: *The Diecasting Book*, Portcullis Press, Redhill, Surrey, 1977.

Upton, B.: *Pressure Die Casting—Part I: Metals-Machines-Furnaces*, Pergamon, Oxford, 1982.

B Textbooks

Beeley, P. R.: *Foundry Technology*, Butterworths, London, 1972.

Chalmers, B.: *Principles of Solidification*, Wiley, New York, 1964.

Davies, G. J.: *Solidification and Casting*, Applied Science Publishers, Barking, Essex, 1973.

Flemings, M. C.: *Solidification Processing*, McGraw-Hill, New York, 1974.

Heine, R. W., C. R. Loper, and C. Rosenthal: *Principles of Metal Casting*, 2d ed., McGraw-Hill, New York, 1967.

Kondic, V.: *Metallurgical Principles of Founding*, Arnold, London, 1968.

Sylvia, J. G.: *Cast Metals Technology*, Addison-Wesley, Reading, Mass., 1972.

Szekely, J.: *Fluid Flow Phenomena in Metals Processing*, Academic Press, New York, 1979.

C Design of Castings

Walton, Ch. F., and T. J. Opar (eds.): *Iron Castings Handbook*, 3d ed., Iron Castings Society, Des Plaines, Ill., 1981.

Wieser, P. F. (ed.): *Steel Castings Handbook*, 5th ed., Steel Founder's Society of America, Des Plaines, Ill., 1980.

D Miscellaneous

Analysis of Casting Defects, American Foundrymen's Society, Des Plaines, Ill., 1974.

Brody, H. D., and D. Apelian (eds.): *Modeling of Casting and Welding Processes*, Metallurgical Society of AIME, Warrendale, Penn., 1981.

Dantzig, J. A., and J. T. Berry (eds.): *Modeling of Casting and Welding Processes, II*, Metallurgical Society of AIME, Warrendale, Penn., 1984.

Miller, R. K.: *Robots in Industry: Applications for Foundries*, SEAI Institute, Madison, Ga., 1982.

Minkoff, I.: *The Physical Metallurgy of Cast Iron*, Wiley, New York, 1983.

Rowley, M. T. (ed.): *International Atlas of Casting Defects*, American Foundrymen's Society, Des Plaines, Ill., 1974.

E Journals

British Foundryman
Casting Engineering and Foundry World
Foundry Management and Technology
Foundry Trade Journal
Modern Casting
Transactions of the American Foundrymen's Society

CHAPTER 4

BULK DEFORMATION PROCESSES

In Sec. 3-6 we mentioned that some 90% of all metals is cast into ingots, slabs, or billets for further working by plastic deformation. Plastic deformation implies that the shape of the workpiece is changed without a change in volume or melting of the material. It is, obviously, essential that the material should be able to undergo plastic deformation without fracture but, because all deformation occurs in the solid state, die filling will not be as easy as it was in casting. Therefore, in the design of metalworking processes it will be necessary to consider not only the laws governing material flow (because they define whether the desired configuration can be obtained) but also pressures, forces, and power requirements (because they determine the loading of tools and equipment).

The success of processes depends on interactions between material properties and process conditions, and the principles to be discussed here have universal applicability. However, for practical reasons, it is usual to divide metalworking processes into two groups. In *bulk deformation processes* the thickness, diameter, or other major dimension of the workpiece is substantially changed. In *sheet-metalworking processes* thickness change is incidental; furthermore, the sheet—which is the starting material—is the product of a bulk deformation process, namely rolling. For these reasons, sheet-metalworking processes will be discussed separately in Chap. 5, and an examination of material properties that are of primary importance to sheet metalworking will also be held over. Here our concern is with bulk deformation processes which account for large production quantities (Table 4-1).

TABLE 4-1
SHIPMENTS OF WROUGHT PRODUCTS* (UNITED STATES)

Alloy group	Thousand tonnes	
	1972	1982
Steel		
Sections, rails	5 900	3 500
Plate	7 300	4 200
Hot-rolled sheet and strip	14 200	9 600
Cold-rolled sheet and strip	17 800	13 500
Galvanized sheet	4 900	4 900
Tinplate	5 000	3 000
Hot-rolled bar	11 800	4 800
Cold-finished bar	1 600	1 000
Wire	2 300	1 000
Tube, pipe	6 900	4 500
Forgings	1 200	600
Copper and brass	2 600	2 100
Aluminum	4 100	4 500
Lead (incl. battery)	480	340
Zinc	45	38
Magnesium	16	14

*Compiled from *Metal Statistics 1974 and 1984*, American Metals Market, Fairchild Publications Inc., New York, 1974 and 1984.

4-1 MATERIAL PROPERTIES

In Secs. 2-1 and 2-2 we already discussed many properties of solid materials, but we did so with reference to the properties required in the service of the manufactured part. Now we need to reexamine these properties with emphasis on their relevance to deformation processing.

4-1-1 Flow Stress in Cold Working

For metalworking calculations, the stress required to deform the workpiece material must be known. By definition, this stress is a true stress (Eq. (2-1)). Hence the engineering stress (Eq. (2-3)) conventionally calculated from the tension test is of little use. To derive a true stress from the forces measured in the tension test, the instantaneous cross-sectional area A of the test piece must be calculated using the principle of constancy of volume. (During elastic deformation interatomic distances change and thus the volume also changes, as indicated by a Poisson ratio of about 0.3; the plastic component of deformation is due to the movement of atoms with little change in vacancy density; hence, the volume remains constant and Poisson's ratio is 0.5.) As long as elongation is uniform over the gage length (Fig. 2-2c), the cross-sectional area can be calculated from the recorded instantaneous length l

$$A = A_0 \frac{l_0}{l} = \frac{V}{l} \tag{4-1}$$

Once necking begins, the minimum diameter—which is the only diameter of relevance—is unknown and no further points can be calculated.

Our interest is in permanent deformation which begins at the point of yielding; hence, the true stress is usually calculated from initial yielding to necking. Each calculated point defines the stress that must be applied to keep the material deforming, flowing; hence, we call it the *flow stress* σ_f

$$\sigma_f = \frac{P}{A} \tag{4-2}$$

where P is the instantaneous force. We could—and sometimes do—plot flow stress as a function of engineering tensile strain e_t (Eq. (2-4)). However, for calculating purposes the *true strain* ϵ (also called *natural* or *logarithmic strain*) is needed. By definition, it is obtained as the natural logarithm of the ratio of instantaneous length l to original length l_0

$$\epsilon = \ln\frac{l}{l_0} = \ln\frac{A_0}{A} \tag{4-3}$$

The data derived from the tensile force-displacement curve may now be plotted to define the true-stress–true-strain curve (Fig. 4-1a) between the limits of initial yielding and necking. For comparison, the engineering-stress–true-strain curve is shown in broken lines. (There is one point that can be calculated, even if only approximately, beyond necking: the fracture force P_f is available, and the corresponding minimum cross-sectional area A_f (Fig. 2-2d) can be measured on the broken specimen. For reasons to be explained later, the true stress thus calculated is somewhat high.)

FIGURE 4-1
In the cold-working temperature range (a) many materials obey the power law of strain hardening as shown by (b) the linear log-log plot of flow stress versus true strain.

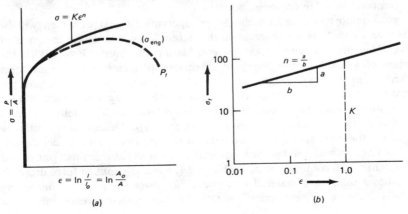

Flow stress curves of many materials have been determined, and one could build an atlas of such curves. However, a more condensed—and for calculating purposes more convenient—record can be kept. When σ_f is replotted against ϵ on log-log paper, a straight line frequently results (Fig. 4-1b), indicating that σ_f must be a power function of ϵ

$$\sigma_f = K\epsilon^n \qquad (4\text{-}4)$$

where K is the *strength coefficient* and n is the *strain-hardening exponent*. From the log-log plot, K is the stress at a strain of unity, and n is the slope of the line, measured on a linear scale.

A problem with the tension test is that necking limits the uniform strain that can be obtained. Many metalworking processes involve heavy deformation, hence the compression test (Sec. 2-1-2) is more useful. The true stress is really an interface pressure. It is obtained (Eq. (2-12)) by invoking constancy of volume (Eq. (2-11)) and, if friction effects can be neglected, may be regarded as the flow stress. True strain ϵ can be calculated from

$$\epsilon = \ln \frac{h}{h_0} = \ln \frac{A_0}{A} \qquad (4\text{-}5a)$$

The calculation yields a negative number; for convenience (and because, from the metallurgical point of view, compressive and tensile deformations have the same effects), the convention is usually ignored and true strain is taken as the natural logarithm of the ratio of the larger value to the smaller value

$$\epsilon = \ln \frac{h_0}{h} = \ln \frac{A}{A_0} \qquad (4\text{-}5b)$$

From a log-log plot of σ_f versus ϵ, the K and n values can be extracted. Indeed, most published data (including those that will be given in Tables 4-2 and 4-3) have been determined in compression tests.

Of great importance for stretching-type operations (most of which will be discussed in Chap. 5) is the observation that strain hardening delays the onset of necking. This may be understood by considering the events involved in the formation of a neck. In the course of extension, an incipient neck may form anywhere along the gage length, at a point of inhomogeneity, i.e., where the material is, for any reason, weaker (because of a surface irregularity, inclusion, or a large grain of weak orientation). If the n value is high, localized deformation in the incipient neck raises σ_f at this point (Eq. (4-4)). Deformation will now continue in other, less strain-hardened parts of the specimen, until hardening can no longer keep up with the loss of load-bearing capacity; at this time, one of the necks stabilizes and continues to neck (Fig. 4-2a) while the applied force drops. It can be shown that, for a material that obeys the power law of strain hardening (Eq. (4-4)), the n value is numerically identical to the uniform (prenecking) strain

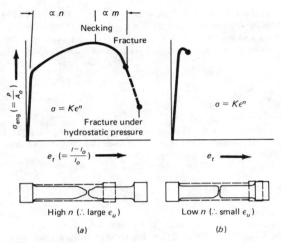

FIGURE 4-2
(a) A high strain-hardening rate, as expressed by a high n value, results in large uniform (pre-necking) elongation; post-necking deformation increases with increasing strain-rate sensitivity or high m value, and fracture is delayed under hydrostatic pressure. (b) A material of low n necks early and, if m is low, fractures soon.

expressed as true strain ϵ_u; therefore, a material of low n necks soon after initial yielding (Figs. 4-2a and b).

Example 4-1

From the force-displacement curve given in Example 2-2, calculate the flow stress of the material at several points. Plot to obtain the K and n values.

To obtain the instantaneous cross-sectional area, the volume of the specimen is calculated for the length l_0. $V = (6.35)(6.38)(25.0) = 1013 \text{ mm}^3$. The instantaneous area A is obtained from Eq. (4-1), flow stress from Eq. (4-2), and true strain from Eq. (4-3). (Note that $l = l_0 + \Delta l$ and is always obtained by drawing a line, from the point of interest, parallel to the elastic line). Data are tabulated:

Δl, mm	l, mm	A, mm²	P, N	σ, N/mm²	ϵ
2	27	37.52	9100	243	0.077
4	29	34.93	11200	321	0.148
6	31	32.68	12600	386	0.215
8	33	30.70	13500	440	0.278
10	35	29.82	14000	469	0.336
12.5	37.5	27.83	14200	510	0.405
...	...	(2.85)(3.5)	9300	932	1.40

(Note that the true stress is always higher than the engineering stress.)

The last point in the table was calculated from the fracture area: $\sigma_f = P_f/A_f = 9300/(2.85)(3.5)$ = 932 MPa and from strain based on the fracture strain $\epsilon = \ln(A_0/A_1) = \ln(40.5/9.98) = 1.4$. The plot of points on log-log paper defines a straight line; thus, the material obeys Eq. (4-4). $K = 760$ MPa, and $n = 0.45$ (quite high but not unreasonable since the material is a solid solution of ductile elements).

Example 4-2

Check whether $n = \epsilon_u$ for the material of Example 2-2.

By the definition of Eq. (2-9a), uniform strain can be expressed as natural strain $\epsilon_u = \ln(l_u/l_0)$ = $\ln(37.5/25.0) = 0.405$. This is less than $n = 0.45$. Good agreement between the two values can be expected for steels; agreement is often less good for nonferrous materials.

Example 4-3

Find the K and n values for the steel of Example 2-4.

When we plotted the true-stress–compressive-strain curve in Example 2-4, we already had all the relevant data. Only the true strain has to be calculated from Eq. (4-5b) (results are entered in the table of Example 2-4). From the log-log plot, $K = 800$ MPa and $n = 0.13$. Note that the strain-hardening capacity, while not as high as for the Cu–Ni alloy of Example 4-1, is still quite substantial for this interstitial solid solution of C in Fe. The test material was in the slightly cold-drawn condition, hence n is less than it would be for the same steel in the annealed condition. (Note also that the engineering and natural strains are very similar at low reductions but the numerical value of natural strain becomes progressively larger with increasing reductions.)

Example 4-4

As far as their effect on strain hardening is concerned, extension of a bar from $l_0 = 1$ unit length to $l = 2$ units should be the same as compressing a bar of the same material from $h_0 = 2$ units to $h = 1$ unit height. Calculate the corresponding strains.

$$\text{Engineering tensile strain (Eq. (2-4}b\text{))} \qquad e_t = 100(2-1)/1 = 100\%$$
$$\text{compressive strain (Eq. (2-13))} \qquad e_c = 100(2-1)/2 = 50\%$$
$$\text{Natural strain, tension (Eq. (4-3))} \qquad \epsilon = \ln(2/1) = 0.69$$
$$\text{compression (Eq. (4-5}a\text{))} \qquad \epsilon = \ln(1/2) = -0.69$$

Note that it is very misleading to quote engineering strain without specifying whether it is tensile or compressive because, to calculate tensile strain, the *change in dimensions* is divided by the *smaller* dimension, whereas for compressive strain it is divided by the *larger* dimension. The absolute value of natural strain is the same for tension and compression, correctly indicating that the two deformations are equivalent in their effects on the material.

4-1-2   Effects of Cold Working

It is obvious from Fig. 4-1 that an ever-increasing true stress is needed for the continuing deformation of the metal. Because this is a direct consequence of working or straining, one speaks of *work hardening* or *strain hardening*. The reason for it is to be found in the mechanism of plastic deformation.

We saw in Sec. 3-2-1 that crystalline metals deform by slip and, on the atomic scale, by the propagation and multiplication of dislocations (Fig. 3-9). Slip occurs on close-packed planes in close-packed directions (Fig. 3-2); inspection of Fig. 3-2 will show that there are a number of equivalent *slip systems* in each crystal structure. As deformation proceeds, dislocations may move on several systems. It takes a higher stress to move a succession of dislocations on the same plane, and a yet higher stress is needed to move them once dislocations propagating on different planes become entangled. This higher stress leads to the increase in flow stress. Distortion of the crystal lattice by foreign atoms increases strain hardening; therefore, solid solutions have a higher n value: Since this gives a larger prenecking strain, solid solutions have good ductility.

A material subjected to cold working, for example, by rolling or drawing, strain hardens too. Dislocation density increases and, when a tension test is performed on this strain-hardened material, a higher stress will be needed to initiate and maintain plastic deformation; thus, the YS rises. The TS rises too, although not as rapidly as the YS, and the TS/YS ratio approaches unity (Fig. 4-3). However, the ductility of the material—as expressed by total elongation and reduction of area—drops because of the higher initial dislocation density. Similarly, K rises and n drops. The microstructure changes too: Crystals (grains) become elongated in the direction of major deformation. These changes are summarized in Fig. 4-4a. The material may also develop directional properties, which are of paramount importance in sheet metalworking, (hence will be discussed in Sec. 5-1-3).

FIGURE 4-3
Tension tests conducted on previously worked material show that cold working increases strength and reduces ductility.

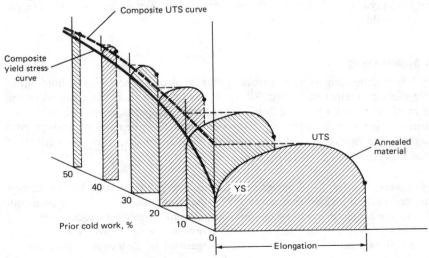

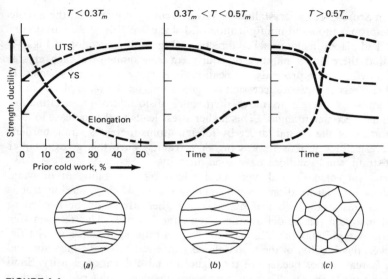

FIGURE 4-4
(a) The effects of prior cold work are (b) partially removed by recovery and (c) the original, soft condition is fully reestablished by recrystallization.

Strain hardening is important in manufacturing for two reasons. First, many cold-worked materials retain a reasonable level of ductility; therefore, cold working is a useful approach to the production of higher-strength materials. Second, many products are made by a succession of cold-working steps; the increased flow stress can generate excessive tool pressures, and the reduced ductility may lead to fracture of the workpiece. It is then necessary to remove the effects of cold working by an appropriate heat treatment.

4-1-3 Annealing

We defined annealing as heat treatment that involves heating to (and holding) at some elevated temperature (Sec. 3-5-2). When its purpose is the removal of the effects of cold working in the finished product, one speaks simply of *annealing*. When the purpose is softening of a workpiece for purposes of further cold working, one speaks of *process anneal*. Fundamentally, these are identical processes.

Recovery In Sec. 2-1-7 we introduced the concept of the homologous temperature scale and indicated that above $0.5T_m$ the strength of many materials drops, because the larger thermal excursions allow atoms to move to vacant sites and thus change places with relative ease. Before this temperature is reached, increased atomic mobility allows the rearrangement of dislocations into regular

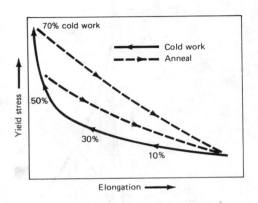

arrays (typically, at temperatures of $0.3–0.5T_m$). Given enough time, such *recovery* restores some of the original softness without changing the visible grain structure (Fig. 4-4b).

Ductility drops rapidly with even a small degree of cold work (Fig. 4-4a and solid line in Fig. 4-5). In some metals recovery increases ductility without greatly affecting strength (broken line in Fig. 4-5); therefore, *recovery anneal* is a useful method for producing a material of higher strength yet reasonable ductility.

Recrystallization Above $0.5T_m$ atoms can move, diffuse to form new, relatively dislocation-free nuclei which grow until all the cold-worked structure is recrystallized. Diffusion is greatly time- and temperature-dependent (Fig. 4-4c). An equiaxed structure normally results, with a grain size that is a function of prior cold work, annealing temperature, and time.

The driving force for *recrystallization* is provided by the increased energy content resulting from the higher dislocation density induced by cold working. Therefore, recrystallization begins at a lower temperature with increasing *prior cold work* (Fig. 4-6). High temperatures lead to grain growth and thus coarser grains but, for any given temperature, grain size diminishes with increasing cold work, because more nuclei form. There is, of course, no recrystallization possible if cold work is zero, and the original grain size is retained. However, slightly increased dislocation densities resulting from very slight (say 2–4%) cold work encourage the formation of only a few nuclei which can then grow to a large size. Such *critical cold work* is usually undesirable because of the poor mechanical properties of coarse-grained structures (Sec. 3-2-5). Very fine grain, obtained by annealing a heavily cold-worked metal, can give material of high strength yet reasonable ductility.

The temperature of $0.5T_m$ should be taken only as a very rough guide, since even minor amounts of alloying elements can substantially delay the formation of new grains and thus raise the recrystallization temperature. In alloys specifically

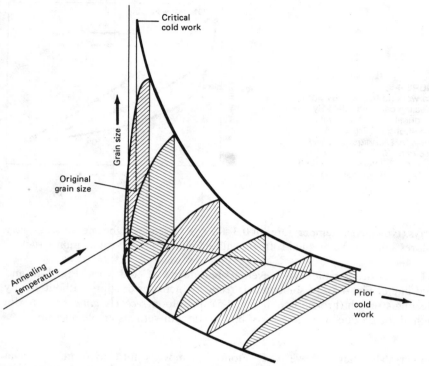

FIGURE 4-6
Recrystallization begins at lower temperatures and recrystallized grain size decreases with increasing prior cold work.

designed for high-temperature service, such as the superalloys, heavy alloying pushes the onset of recrystallization to around $0.8T_m$.

When a metal is held at temperature for a prolonged time, larger grains—which have a smaller surface area per unit volume and hence a lower surface energy—grow at the expense of smaller grains. Such *grain growth* is, in general, undesirable because strength and, if excessive grain growth occurs, even ductility suffers.

Recovery and recrystallization are collectively termed *softening processes* or *restoration processes*.

Example 4-5

The effects of cold working and of the grain size of an annealed material are well demonstrated in the properties of 70Cu–30Zn brass. Because of its high ductility, this was the traditional material for cartridge cases; hence, it is called *cartridge brass*, but it is used for many other purposes, mostly in the sheet form. It is supplied in various rolled "tempers," or annealed to a specified grain size.

From *Metals Handbook* (9th ed., vol. 2, 1979, p. 324):

Temper	Rolling reduction, %	TS, MPa	YS, MPa	Elong., %, 50 mm	Hardness
		Cold rolled			
H01 ($\frac{1}{4}$ hard)	10.9	370	275	43	55*
H02 ($\frac{1}{2}$ hard)	20.7	425	360	23	70*
H04 (hard)	37.1	525	435	8	82*
H08 (spring)	60.5	650	...	3	91*
		Annealed			
OS100		300	75	68	54†
OS050		325	105	62	64†
OS025		350	130	55	72†
OS015		365	150	54	78†

*HRB.
†HRF.

The designation OS indicates that annealing produced a prescribed average grain size, expressed in units of μm (thus OS025 indicates an average grain size of 25 μm).

4-1-4 Hot Working

We have noted that temperatures above $0.5T_m$ greatly facilitate the diffusion of atoms. This means that an arrested dislocation has the option of climbing, and can thus move into another, unobstructed atomic plane. Therefore, if deformation itself takes place at such elevated temperatures, many dislocations can immediately disappear; in fact, one finds that softening processes work simultaneously with dislocation propagation. Material resulting from such *hot working* has a much lower dislocation density and, therefore, is less strain hardened than cold-worked material.

In practice, hot working is conducted at higher temperatures, where softening processes are fast, but not at such high temperatures that there would be danger of incipient melting (typically between $0.7T_m$ and $0.9T_m$).

Mechanisms of Hot Working Since $0.5T_m$ is also the temperature of recrystallization, it is often said that hot working is conducted above the recrystallization temperature. However, recrystallization *during* hot working (*dynamic recrystallization*) is by no means universal; in many materials dynamic recovery takes place during working, resulting in quite low flow stresses. Recrystallization may still occur on holding at or cooling from the hot-working temperature. Therefore, the distinctive mark of hot working is not a recrystallized structure, but the simultaneous occurrence of dislocation propagation and softening processes, with or without recrystallization during working. The dominant mechanism depends on temperature, strain rate, and grain size, and may be conveniently shown on deformation mechanism maps. In general, the recrystallized structure becomes finer with lower deformation temperature and faster cooling rates, and material of superior properties is often obtained by controlling the finishing temperature.

Flow Stress in Hot Working   Since all softening processes require the movement of atoms, the time available for these processes is critical. This means that in hot working there is substantial *strain-rate sensitivity*. We already observed that strain rate should not be confused with deformation velocity. In its simplest definition, strain rate is the instantaneous deformation velocity divided by the instantaneous length or height of the workpiece (Eq. (2-17)). For compressive deformation (Fig. 2-5)

$$\dot{\epsilon} = \frac{v}{h} \tag{4-6}$$

Again, $\dot{\epsilon}$ is expressed in units of $1 \ s^{-1}$.

To find the flow stress of a metal, specimens are heated to a constant temperature and then compressed (or tested in tension) at a constant strain rate, on machines in which the crosshead velocity changes in a programmed manner so as to keep $\dot{\epsilon}$ (Eq. (4-6)) constant. From recordings of force versus displacement, stress–strain curves are plotted which may show a number of trends (Fig. 4-7):

1 After an initial peak, flow stress drops with increasing strain. Such strain softening is usually a sign of dynamic recrystallization.

2 The stress–strain curve may be fairly flat after initial yielding, indicating that strain hardening and softening processes roughly balance each other.

3 At yet higher strain rates, stresses increase with increasing strain, indicating that softening processes could not keep pace with strain hardening.

To a first approximation, hot working can be regarded as though it was governed purely by strain rate. Then flow stress values for a given strain may be extracted from the true-stress–true-strain curves (Fig. 4-7a) and replotted as a function of strain rate on a log-log scale (Fig. 4-7b). In the majority of instances, the line thus defined will be straight, indicating that hot-working flow stress is a power function of strain rate $\dot{\epsilon}$

$$\sigma_f = C\dot{\epsilon}^m \tag{4-7}$$

where C is a *strength coefficient*, and m is the *strain-rate-sensitivity exponent*. The value of C is found at a strain rate of unity, and m is the slope of the line, again measured on a linear scale (Fig. 4-7b). Evidently, different C and m values will be found for different strains. Both C and m also change with temperature: Increasing temperature usually increases strain-rate sensitivity and thus m, but always decreases the flow stress and thus C. (Note: The full form of Eq. (4-7)

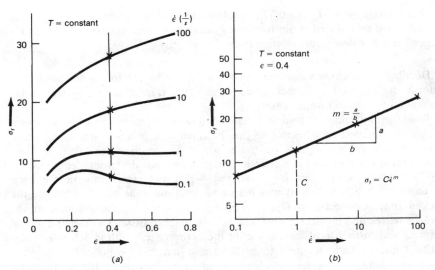

FIGURE 4-7
Hot working proceeds with simultaneous hardening and softening. (*a*) The flow stress is sensitive to strain rate and (*b*) for a given temperature and strain, it is often a power function of strain rate.

would have $\dot{\epsilon}/\dot{\epsilon}_0$ in it; with $\dot{\epsilon}_0 = 1$, the universally used form of Eq. (4-7) is obtained).

For purposes of calculation, experimentally determined C and m values (for example, from Tables 4-2 and 4-3) or flow stress curves must be used. It is worth noting, however, that time and temperature are equivalent in their effects on softening. Therefore, it is sometimes possible to express all hot-working flow stress values with a single curve that is a function of a *velocity-* (or *strain-rate-*) *modified temperature*.

In discussing cold-working flow stresses (Eq. (4-4)) we made the tacit assumption that strain-rate effects could be ignored (i.e., $m = 0$). This is not entirely true; a fuller description of the response of metals would include both strain and strain rate. Strain-rate sensitivity increases with increasing homologous temperatures, and increases rather suddenly when the hot-working temperature is reached. Typical values of the strain-rate sensitivity exponent are

Cold working	$-0.05 < m < 0.05$
Hot working	$0.05 < m < 0.3$
Superplasticity	$0.3 < m < 0.7$
Newtonian fluid	$m = 1$

Deformation at $0.3T_m$ to $0.5T_m$ is often denoted as *warm working*, and is characterized by reduced strain hardening, increased strain-rate sensitivity, and a somewhat lower flow stress relative to cold working.

Ductility A high m value means that markedly higher forces are needed to deform the material at higher strain rates. This translates into greater total elongation for the following reason:

When, in the course of testing in tension, a neck begins to form, this incipient neck is the smallest cross section of the specimen and, in a non-strain-rate-sensitive material, it would also be the weakest. However, deformation is now momentarily concentrated in the neck, the instantaneous deforming length in Eq. (2-17) suddenly drops (see Figs. 2-2 and 4-16), and strain rate in the neck becomes much higher than it was before necking, whereas it drops outside the necked zone. Consequently (Eq. (4-7)), the flow stress of the material in the neck increases, and the neck resists further deformation. Instead, adjacent material deforms and further locations neck until the entire gage length is deformed (Fig. 2-12). Thus, we find that next to n (the strain-hardening exponent), a high m value also indicates greater possible elongation by increasing the postnecking strain (Fig. 4-2a). This will be important in stretching-type operations (Chaps. 5 and 7).

Superplasticity In some extremely fine-grained materials, most often alloys with a two-phase *microduplex structure*, high-temperature deformation takes place by extensive grain-boundary sliding and accompanying diffusion (essentially, by entire grains sliding past each other) or by mass diffusion which reshapes entire grains. Deforming forces can be very low and, as long as strain rates are kept within the limits that allow these deformation mechanisms to prevail (Fig. 4-8), the superplastic behavior is maintained and very large elongation values (up to several hundreds and even thousands of percent) are readily obtained. Thus,

FIGURE 4-8
Some very-fine-grained materials exhibit superplasticity, with very high m values within a limited strain-rate range. Arrows indicate effects of decreasing grain size or increasing temperature.

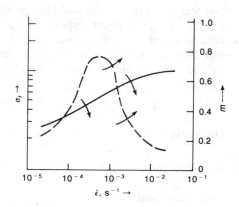

techniques developed for the forming of polymers (Sec. 7-7-4) can be applied to these metals.

After cooling from the superplastic temperature, many alloys develop substantial strength. However, the same mechanisms that allow superplastic deformation also account for the poor creep resistance of fine-grained materials (Sec. 3-2-5). Therefore, superplastically deformed parts may be made suitable for high-temperature service by a high-temperature anneal. The coarse grains thus formed have relatively little grain-boundary area and offer greater resistance to creep at low strain rates (Fig. 3-13). This process sequence is the basis of *Gatorizing*®,* a patented process for making superalloy turbine blades.

Example 4-6

The flow stress of metals is to be determined by compressing 20-mm-high cylinders at constant strain rates. Calculate the press speed needed for compression to 60% reduction in height at $\dot{\epsilon} = 5 \text{ s}^{-1}$

From Eq. (4-6), the press must slow down as the height diminishes to the 8-mm final height.

Height, mm	Press speed, mm/s
20	100
16	80
12	60
8	40

Example 4-7

Calculate C and m for the material shown in Fig. 4-7.

Assuming that σ_f is given in MPa, $C = 11.8$ MPa (remember to read on the log scale) and $m = 7.5/17 = 0.44$ (remember to read on the linear scale). This high m value indicates a superplastic material.

4-1-5 Interactions between Deformation and Structure

Up to now we tacitly assumed that the workpiece was homogeneous. This can be far from reality, and the interactions of deformation processes with structural features can be exploited to control the service properties of materials.

Destruction of Cast Structure The structure of cast ingots or billets shows a number of undesirable features. Grains and dendrite arm spacing within grains tend to be large, thus strength is low; columnar grains (Fig. 3-15) may be oriented in unfavorable directions, further reducing strength and ductility in some directions. Concentration gradients usually exist, as evidenced by microsegregation

---

*Registered trademark of Pratt & Whitney, United Technologies.

(coring, Fig. 3-17) and macrosegregation (Fig. 3-18). Microporosity, typical of dendritic solidification (Fig. 3-15b) is often present and there may even be gross piping (Fig. 3-15a). Pinholes and blowholes may remain as a result of gas evolution during solidification (Fig. 3-18a; Sec. 3-3-4).

Hot working is the most powerful method for eliminating harmful features because:

1 The forced movement of atoms favors the equalization of composition and thus accelerates homogenization.

2 Pores are compressed until their walls touch; upon further extension, the high prevailing pressures and temperatures lead to adhesion and solid-state welding, effectively eliminating the pore as a defect (at least if its walls were originally free of contaminants).

3 Any oxide or other internal contaminant films are greatly extended and—*if brittle*—broken up into small particles around which pressure welding can take place. Thus, even though such inclusions remain in the material, they may be rendered harmless from the point of view of mechanical properties. This is also true of oxide films that may be present on the surfaces of pores and pipes. Because intermetallic compounds are generally brittle they may also be broken up. However, any large-size *ductile* inclusions will be stretched out and could considerably impair properties. Heavy oxides and slag inclusions found in pipes prevent welding and cause laminations in the rolled product. Also, cracks oriented in the direction of force application are likely to open up rather than heal.

4 Cast ingots are usually subjected to a sequence of hot working steps (*passes*), and recrystallization during or in between passes replaces the coarse cast grain with a fine, equiaxed structure of much better mechanical properties.

5 The more or less randomly distributed inclusions and second-phase particles become aligned and, to some extent, oriented in the direction of major deformation. This *mechanical fibering* gives rise to *anisotropy*, i.e., a variation of properties with testing direction, quite independent of any directionality which may be due to crystal structure (Sec. 5-1-3). Typically, in the direction of fiber orientation the properties of the matrix dominate, and strength and ductility are high (Fig. 4-9a). When the material is loaded (during testing or in service) in the transverse direction, inclusions serve as effective stress raisers (Fig. 4-9b). Therefore the so-called short-transverse properties (such as strength and, even more so, impact strength, fatigue strength, and ductility) suffer (Fig. 4-9c).

6 Fibering can be revealed by deep *macroetching* (as opposed to the lighter microetching used to reveal the grain structure). The fibered structure developed in earlier passes is distorted on subsequent working, and etching a cross section to reveal *flow lines* is a most useful tool in studying material flow. Even in the absence of inclusions or second-phase particles, flow lines show up when homogenization is not perfect and traces of microsegregation remain. This is true of steels in which the large phosphorus atoms remain segregated even after heavy

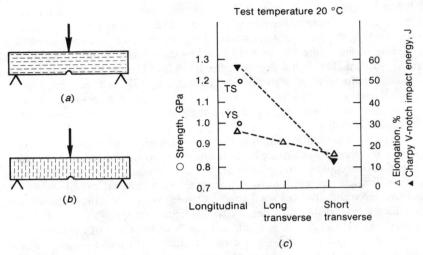

FIGURE 4-9
Fibering due to alignment of second-phase particles, inclusions, and segregation leads to directional properties, which (*a*) and (*b*) are revealed in impact testing and (*c*) are most clearly evident in ductility and impact energy. (*Data for Inconel 718 taken from Forging Design Handbook, American Society for Metals, 1972, p. 14.*)

hot working, thus outlining the flow lines upon macroetching (as in Figs. 4-18*d* and 4-31).

Example 4-8

A brittle inclusion embedded in a bar has a width of 5 mm, length of 12 mm, and thickness of 1 mm. It is oriented in the direction of elongation (elongation may be the result of any metalworking process that causes elongation by reducing the height of the workpiece). If the bar is reduced by 90% in thickness without a change in width, and if the inclusion breaks up without any change in thickness, over what length will the fragments be distributed?

A 90% reduction in thickness results in a strip with a height equal to $(100 - 90) = 10\%$ of the original thickness. From constancy of volume, $l_0 w_0 h_0 = l_1 w_1 h_1$. If $w_0 = w_1$, the extension is $l_1/l_0 = h_0/h_1 = 100/10$ or 10-fold. Thus fragments of the inclusion of 12-mm original length will be scattered over a length of 120 mm. In between, over a distance of $120 - 12 = 108$ mm (or 90%), the bar material will contact and reweld.

Thermomechanical Processing Because plastic deformation involves the movement of atoms, it accelerates all processes that rely on diffusion or transformation. We saw that dislocations, multiplied and entangled during deformation, provide sites for recrystallization. They also provide sites for the nucleation of precipitate particles, thus increasing the number and decreasing the size of precipitates. Many possibilities exist of which only the major ones will be discussed here.

1 When a steel is alloyed so that the metastable austenite can exist for some reasonable time (the nose of the curve is pushed to the right in Fig. 3-25), there is time for working the metastable austenite. For this, the steel is austenitized and then rapidly cooled some 100 to 200 °C below the transformation temperature where it is worked. The high dislocation density induced in the austenite results in a substantial refinement of transformation products, and such *ausformed steels* have high strength. If the austenite is worked at lower temperatures (*low-temperature thermomechanical* working, Fig. 3-25, line 6), strength increases further but at the expense of ductility.

2 Precipitation-hardening materials (Sec. 3-5-3), such as aluminum and nickel alloys, may be worked while heated into the homogeneous solid-solution temperature range. Upon cooling, precipitates are refined because they begin to form at sites of dislocation concentrations. Alternatively, the material may be solution treated and the supersaturated solid solution (which is quite ductile) cold worked to introduce a high dislocation density which refines the precipitates on subsequent aging. Great increase in strength may result without loss of ductility.

3 An already aged material—and even a tempered martensite—may be cold worked to take advantage of the great strengthening resulting from dislocation pileup against finely distributed obstacles. Ductility usually suffers.

4 Further possibilities exist if the material undergoes an allotropic transformation. A material heated to the vicinity of the transformation temperature often shows low strength and high ductility, although not to the same degree as a superplastic material. This is exploited in the hot working of titanium and its alloys around the α-to-β transformation temperature. Typically, grains are also refined, and the morphology of transformation products may change too. For example, pearlite formed while working steel at the transformation temperature is much refined and can become globular (spheroidal). Steel that is worked right through the transformation temperature may show unusually high strength and reasonable ductility (*controlled hot working*).

4-2 WROUGHT ALLOYS

In all plastic metalworking processes, the workpiece shape is formed from the solid metal or alloy by displacing material from unwanted locations into positions required by the part shape. This demands that the material should possess a property rather vaguely described as ductility—that is, the ability to sustain substantial plastic deformation without fracture. As we shall see (Sec. 4-3-3), process variables enter into the picture and, at this point, all we can say is that wrought alloys must possess a minimum ductility commensurate with the contemplated process.

This requirement is amply satisfied by all pure metals that have a sufficient number of slip systems (Sec. 3-2-1) and also by most solid-solution alloys of the same metals. Two-phase and multiphase (Sec. 3-2-3) materials are deformable if they meet certain minimum requirements. There must be no liquid or brittle phase on the grain boundaries or across several grains (thus, gray iron, white cast iron, or a hypereutectic Al–Si alloy cannot be cold worked). Excessive amounts of brittle constituent are not permissible even in a ductile matrix, especially if the brittle constituent is also coarse or lamellar. The greater the quantity of brittle constituents and the lower the ductility of the matrix, the more important it is that the material should be free of other weakening features such as inclusions, voids, or grain-boundary contaminants.

Steels represent the largest segment of wrought products (Table 4-1) and, in line with the system adopted in Chap. 3 for casting alloys, ferrous materials will be discussed first, followed by nonferrous materials. Relevant properties of selected alloys are given in Tables 4-2 and 4-3.

4-2-1 Carbon Steels

We already discussed, in Sec. 3-6-1, steels as casting alloys. We mentioned that steels (as well as other metals) are normally deoxidized in order to avoid gas porosity. This, however, need not be the case for castings destined for metalworking, and several classes of steel can be distinguished according to deoxidation practice.

1 The so-called *rimmed steels* are not deoxidized. Carbon reacts in the melt with oxygen to form carbon monoxide according to the reaction $2C + O_2 = 2CO$. Since CO is a reducing gas, the large blowholes formed have clean surfaces and, at the high temperatures and pressures prevailing in hot working, weld without trace. An advantage of the large number of gas bubbles formed during solidification is that piping is virtually eliminated. The blowholes are prominent at some distance below the ingot surface and help to move contaminants toward the center, imparting a strong normal segregation pattern (Fig. 3-18a) which persists through all processing steps. The ingot surface (*rim*) is particularly clean and low in carbon. The clean surface is an advantage in many applications, and sheet up to 0.25% C is often produced in this form.

2 Gas evolution is suppressed to some extent when a *cap* (a metal plug) is placed on the ingot (*capped steels*), thus retaining some of the surface cleanliness but achieving greater structural homogeneity than in a rimmed steel. *Semikilled* steels are partially deoxidized and are suitable for applications where great structural uniformity is not required, as in many steels used for construction purposes.

3 The most demanding applications call for *killed steel* in which gas reaction is prevented (*killed*) by the addition of aluminum, silicon, etc. Segregation is virtually absent, properties are uniform throughout, and grain size can be controlled in the finished product. However, proper feeding must be assured to prevent piping.

A further distinction can be made according to carbon content. *Low-carbon steels* (below 0.15% C) contain too little carbon to benefit from hardening and are used in the hot-worked or, for maximum ductility, in the annealed condition, primarily in the form of sheet and wire. Steels of below 0.25% C (often referred to

TABLE 4-2 MANUFACTURING PROPERTIES OF STEELS AND COPPER-BASED ALLOYS*
(Annealed condition)

Designation and composition, %	Liquidus/solidus, °C	Hot-working Usual temp., °C	Flow stress,† MPa at °C	Flow stress,† MPa C	Flow stress,† MPa m	Work-ability‖	Cold-working Flow stress,‡ MPa K	Cold-working Flow stress,‡ MPa n	$\sigma_{0.2}$, MPa	TS, MPa	Elongation, %	q R.A., %	Annealing temp.,§ °C
Steels:													
1008 (0.08C), sheet		< 1250	1000	100	0.1	A	600	0.25	180	320	40	70	850–900(F)
1015 (0.15C), bar		< 1250	800	150	0.1	A	620	0.18	300	450	35	70	850–900(F)
			1000	120	0.1								
1045 (0.45C)		< 1150	1200	50	0.17	A	950	0.12	410	700	22	45	790–870(F)
			800	180	0.07								
~8620 (0.2C, 1Mn, 0.4Ni, 0.5Cr, 0.4Mo)			1000	120	0.13	A							
			1000	120	0.1								
D2 tool steel (1.5C, 12Cr, 1Mo)		900–1080	1000	190	0.13	B	1300	0.3	350	620	30	60	880(F)
H13 tool steel (0.4C, 5Cr 1.5Mo, 1V)			1000	80	0.26	B							
302 SS (18Cr, 9Ni) (austenitic)	1420/1400	930–1200	1000	170	0.1	B	1300	0.3	250	600	55	65	1010–1120(Q)
410 SS (13Cr) (martensitic)	1530/1480	870–1150	1000	140	0.08	C	960	0.1	280	520	30	65	650–800
Copper-base alloys:													
Cu (99.94%)	1083/1065	750–950	600	130	0.06	A	450	0.33	70	220	50	78	375–650
				(48)	(0.17)								
			900	41	0.2								
Cartridge brass (30Zn)	955/915	725–850	600	100	0.24	A	500	0.41	100	310	65	75	425–750
			800	48	0.15								
Muntz metal (40Zn)	905/900	625–800	600	38	0.3	A	800	0.5	120	380	45	70	425–600
			800	20	0.24								
Leaded brass (1Pb, 39Zn)	900/855	625–800	600	58	0.14	A	800	0.33	130	340	50	55	425–600
			800	14	0.20								
Phosphor bronze (5Sn)	1050/950		700	160	0.35	C	720	0.46	150	340	57		480–675
Aluminum bronze (5Al)	1060/1050	815–870				A			170	400	65		425–750

*Compiled from various sources; most flow stress data from T. Altan and F. W. Boulger, *Trans. ASME, Ser. B, J. Eng. Ind.* **95**:1009 (1973).
†Hot-working flow stress is for a strain of $\epsilon = 0.5$. To convert to 1000 psi, divide calculated stresses by 7.
‡Cold-working flow stress is for moderate strain rates, around $\dot{\epsilon} \approx 1 \text{ s}^{-1}$. To convert to 1000 psi, divide stresses by 7.
§Furnace cooling is indicated by F, quenching by Q.
‖Relative ratings, with A the best, corresponding to absence of cracking in hot rolling and forging.

TABLE 4-3 MANUFACTURING PROPERTIES OF VARIOUS NONFERROUS ALLOYS[a]
(Annealed condition, except 6061-T6)

Designation and composition, %	Liquidus/ solidus, °C	Hot-working Usual temp., °C	Hot-working Flow Stress,[b] MPa at °C	Hot-working Flow Stress,[b] MPa C	Hot-working Flow Stress,[b] MPa m	Work-ability[f]	Cold-working Flow stress,[c] MPa K	Cold-working Flow stress,[c] MPa n	Cold-working $\sigma_{0.2}$, MPa	Cold-working TS,[d] MPa	Cold-working Elongation,[d] %	R.A.,[g] %	Annealing temp.,[e] °C
Light metals:													
1100 Al (99%)	657/643	250–550	300 / 500	60 / 14	0.08 / 0.22	A	140	0.25	35	90	35		340
Mn alloy (1Mn)	649/648	290–540	400 / 500	35 / 36	0.13 / 0.12	A			100	130	14		370
~2017 Al(3.5Cu, 0.5Mg, 0.5Mn)	635/510	260–480	400 / 500	90 / 36	0.12 / 0.12	B	380	0.15	100	180	20		415(F)
5052 Al(2.5Mg)	650/590	260–510	480 / 400	35 / 50	0.13 / 0.16	A	210	0.13	90	190	25		340
6061-0(1Mg, 0.6Si, 0.3Cu)	652/582	300–550	400 / 500	50 / 37	0.16 / 0.17	A	220	0.16	55	125	25	65	415(F)
6061-T6	NA[g]	NA	NA	NA	NA	NA	450	0.03	275	310	8	45	
~7075 Al(6Zn, 2Mg, 1Cu)	640/475	260–455	450	40	0.13	B	400	0.17	100	230	16		415
Low-melting metals:													
Sn (99.8%)	232	100–200	100	10	0.1	A				15	45	100	150
Pb (99.7%)	327	20–200	75	260	0.1	A				12	35	100	20–200
Zn (0.08% Pb)	417	120–275	225	40	0.1	A				130/170	65/50		100
High-temperature alloys:													
Ni (99.4Ni + Co)	1446/1435	650–1250	1150	~140	0.2	A			140	440	45	65	650–760
Hastelloy X (47Ni, 9Mo, 22Cr, 18Fe, 1.5Co, 0.6W)	1290	980–1200				C			360	770	42		1175
Ti (99%)	1660	750–1000	600 / 900	200 / 38	0.11 / 0.25	C			480	620	20		590–730
Ti–6Al–4V	1660/1600	790–1000	600 / 900	550 / 140	0.08 / 0.4	C			900	950	12		700–825
Zirconium (99.8%)	1852	600–1000	900	50	0.25	A			210	340	35		500–800
Uranium (99.8%)	1132	~700	700	110	0.1				190	380	4	10	

[a] Empty spaces indicate unavailability of data. Compiled from various sources; most flow stress data from T. Altan and F. W. Boulger, *Trans. ASME, Ser. B, J. Eng. Ind.* **95**:1009 (1973).

[b] Hot-working flow stress is for a strain of $\epsilon = 0.5$. To convert to 1000 psi, divide calculated stresses by 7.

[c] Cold-working flow stress is for moderate strain rates, around $\dot{\epsilon} = 1 \text{ s}^{-1}$. To convert to 1000 psi, divide stresses by 7.

[d] Where two values are given, the first is longitudinal, the second transverse.

[e] Furnace cooling is indicated by F.

[f] Relative ratings, with A the best, corresponding to absence of cracking in hot rolling and forging.

[g] NA Not applicable to the -T6 temper.

simply as mild steels) have somewhat higher strength but are still easy to weld and are widely employed for structural purposes as hot-rolled bars, sections, and plate. *Medium-carbon steels* (0.25–0.55% C) are often heat treated (quenched and tempered) after manufacturing by hot or cold metalworking. *High-carbon steels* (0.55–1.0% C) find applications as springs and wear-resistant parts. Steel wire of 0.8% C is often subjected to isothermal transformation heat treatment (*patenting*, just above line 3 in Fig. 3-25) that produces extremely fine pearlite. The wire is then further drawn, cold, into wire used in wire ropes, musical instruments, and steel-belted automotive tires.

Steels that are to be cold worked are usually annealed, and those of higher carbon content are spheroidized to ensure maximum ductility.

A special combination of properties is obtained when the surface of a wrought low-carbon steel part is carburized (Sec. 3-5-5). After heat treatment, the part such as a gear will have a hard, wear-resistant surface and a tough core.

4-2-2 Alloy Steels

For many applications, carbon steels cannot provide the required combination of properties and then the more expensive alloy steels will be specified.

1 Relatively small amounts of alloying elements allow heat treatment of thicker sections (Sec. 3-5-4). Higher alloying element concentrations, in combination with higher carbon content, raise the hardness and hot hardness of tool and die steels by introducing temperature-resistant carbides (such as WC, VC, and chromium carbides). Such *alloy steels* are most readily worked in the annealed condition, even though increasing carbide content increases the forming forces and die wear and reduces ductility. Therefore, these materials are usually hot worked, since in the austenitic temperature range their flow strength is not much higher than that of carbon steels.

2 Most *stainless steels* can be hot worked if proper precautions are taken. Those containing both nickel and chromium are among the most cold-formable materials because of their high strain-hardening rate.

3 The temperature at which the martensitic transformation starts (M_s) can be depressed by various alloying elements, and a metastable austenitic structure can be retained at room temperature. When such a material is subjected to deformation, the greater mobility of atoms during deformation initiates the transformation to martensite. Therefore, in the course of tension testing, an incipient neck is stabilized by the transformation of austenite into the much stronger martensite, and the onset of localized necking is delayed until the entire volume of the specimen is transformed. This provides, then, a third means (besides increasing n and m values) of increasing the ductility of a metal. Some stainless steels, as well as *transformation-induced plasticity* (TRIP) steels offer great strength with unusual ductility. The latter steels benefit from a high dislocation density induced in the metastable austenite.

4 We mentioned that not all martensites are hard. If the carbon content is very low, as in *maraging steels*, the martensite will be soft and readily worked, but subsequently can be greatly strengthened by the precipitation of intermetallic compounds (such as Ni_3Ti or Ni_3Mo) at the numerous sites of high dislocation density induced by cold working.

4-2-3 Nonferrous Metals

As in Chap. 3, nonferrous materials will be discussed in order of increasing melting point, not according to their relative importance.

Tin Alloys The low strength of tin makes it unsuitable as a structural material, except for foil and collapsible tubes (but then it is used for its corrosion resistance). Of the tin alloys, modern pewter is readily deformed, mostly into decorative products.

Lead Alloys Even though lead has low strength, its corrosion resistance amply justifies its use in the forms of sheet, tube, and cable sheathing. It can be strengthened by a number of elements (As, Sn, Bi, Te, and Cu). Antimonial lead with 6–7% Sb is widely used as flashing and roofing in the building industry. It also serves as an excellent sound, vibration, and radiation absorber. A Pb–Ca–Sn alloy is used, in the form of expanded sheet, for electrical storage batteries.

Zinc alloys Pure zinc has wide use as the material for drawn battery cans, corrugated roofing, and weather stripping (usually with 1% Cu for the last two applications). Because of its hexagonal structure, it is cold worked above 20 °C.

A eutectoid transformation in the zinc–aluminum system allows commercial production of extremely fine-grained material that exhibits superplasticity (Sec. 4-1-4). Binary allows with 22% Al and further-alloyed variants can be deformed at elevated temperatures almost like plastics and attain a substantial strength at room temperature. They have been used for prototype work and instrument cabinets where considerable detail of design is to be reproduced.

Magnesium Alloys The hexagonal structure of magnesium makes it rather brittle at room temperature, but it is worked readily at only slightly elevated temperatures, typically above 220 °C. Such low temperatures create no tool or lubrication problems and yield the benefit of great ease of forming. Both solid-solution alloying and precipitation hardening are exploited to obtain material of greater strength. Its low density combines with high strength to give high strength-to-weight ratios, desirable for aerospace and automotive applications.

Aluminum Alloys The fastest-growing segment of the metalworking industry has been the working of aluminum alloys. An fcc material, aluminum is readily deformable at all temperatures. With the aid of solid-solution and precipitation-hardening mechanisms, materials of great strength can be produced with an often unsurpassed strength-to-weight ratio. Aluminum alloys have been the main construction material for aircraft and are beginning to make larger inroads into the construction of land vehicles as bumpers, wheels, and some body components. Corrosion resistance and light weight make them attractive for a great many household, food industry, container, marine, and chemical plant applications. Equivalent electrical conductivity may be obtained at a cost often below that of copper and large quantities are used in high-voltage power lines, busbars, and motor windings.

The metallurgical condition is called temper and is designated by a letter, followed by numbers. Most alloys are formed in the annealed (O) condition. Non-heat-treatable alloys acquire useful strength through cold working (H1 condition) although at the expense of ductility; a second digit describes the degree of hardening (e.g., H12 = quarter-hard; H14 = half-hard; H18 = hard). The H2 temper (strain hardened and partially annealed) gives higher ductility for a given strength (Fig. 4-5). Heat-treatable alloys may be worked in the annealed condition, then subjected to solution treatment followed by natural aging (T4) or artificial aging (T6 condition). Even greater strength may be obtained by cold working a solution heat-treated material, since on subsequent natural aging (T3) or artificial aging (T8) the precipitates become extremely fine and well distributed (Sec. 4-1-5).

High-purity aluminum is an excellent conductor. Commercial purity (1100) aluminum is extensively used for foils, cooking utensils, etc. The solid-solution manganese alloy has higher strength and still adequate ductility, and is a general-purpose material for sheet-metal articles. The solid-solution Al–Mg alloys have excellent corrosion resistance and are hence suitable for automotive trim and marine applications. The 2000 series alloys acquire great strength and reasonable ductility in the age-hardened condition and, together with the 7000 series, are the primary materials of aircraft construction, although the lighter aluminum–lithium alloys can offer yet higher strength-to-weight ratios (each percent Li reduces density by 3% and increases the elastic modulus by 6%). Some alloys, such as Al–Li and Al–Cu–Zr are superplastic.

Copper-Based Alloys Copper is one of the most ductile materials, and it has the highest electrical conductivity after silver. Its high thermal conductivity and easy joining by soldering and brazing methods make it the main constructional material for electrical wiring, automotive radiators, and household water systems. Its strength can be increased without great loss in electrical conductivity by small amounts of Ag, Ca, or Be.

The solid-solution alloys with zinc (brasses) are the most widely used. As its name implies, *cartridge brass* (an α brass) is extremely ductile and is suitable for the heaviest deformations, whereas the α + β brasses (such as *Muntz metal*) are less ductile but are tolerant of contaminants, and their good machinability can be further improved by the addition of lead. Lead can be added also to α brasses but then they are not hot workable because the lead is located on grain boundaries.

Tin bronzes are normally deoxidized with phosphorus which forms a low-melting ternary eutectic. They are thus hot-short unless homogenization is ensured prior to hot working. Aluminum bronzes are readily hot worked, as are the nickel alloys (*cupronickel*) and ternary alloys (such as *nickel silver*, a Cu–Ni–Zn alloy).

The warm glow of copper and the infinite gradations of yellows of brass and bronze have appealed to humans over millenia; their aesthetic appeal is often enhanced by corrosion products (*patina*).

Nickel-Based Alloys   Nickel in its pure form is readily deformable, at both elevated and room temperatures. Some of its alloys, particularly those with copper, present no manufacturing problem. Nickel-based superalloys are heavily alloyed with both solid-solution and precipitation-hardening elements to give high creep strength at elevated temperatures. This makes them difficult to work because the hot-working temperature range is very narrow and close to the solidus. Sophisticated melting and pouring techniques are used to exclude contaminants and gases, and a thorough knowledge of the metallurgy of the alloys and of processing technologies is needed to prevent cracking during hot working.

High-Temperature Alloys   Hexagonal titanium, stable at room temperature, has low ductility and requires frequent process anneals. The bcc form (over 880 °C) is most ductile. For control of finished properties, alloys are often worked just below the transformation temperature but at higher strain rates they have relatively high strength. Therefore, isothermal forging is frequently employed; since cooling is of no concern, the high strain-rate sensitivity and ductility of the material can be exploited by working at very low strain rates and correspondingly low stresses.

Because of their corrosion resistance, titanium and its alloys—in the forms of tubing and sheet—are extensively used in chemical applications. Heat-treated titanium alloys of high strength-to-weight ratios have become indispensable for critical aircraft components, including the compressor stages of jet engines.

The refractory metal alloys (molybdenum, tungsten, and niobium) readily form a volatile oxide at high temperatures; hence they must be processed in vacuum or protective atmosphere. Tungsten is used extensively in the form of wire in incandescent lamps. Bars are compacted by powder metallurgy, and are first worked hot and then at gradually lower temperatures, as the ductile-to-brittle transformation temperature drops with increasing deformation. Recent developments in refractory metal alloys were spurred by space-age technology which demanded materials that would function at very high temperatures. An offshoot of this progress is the molybdenum-based die material TZM which is often used in the cold-worked condition; some effects of cold work are retained to about 1000 °C.

4-3   PRINCIPLES OF DEFORMATION PROCESSING

There is a great variety of plastic deformation processes but some principles can be applied to all of them. Without an understanding of these principles, no process can be intelligently designed or controlled.

4-3-1 Classification

Processes may be classified according to several points of view, all of which may be valid under certain conditions.

Temperature of Deformation We saw that material properties are a function of temperature, with $0.5T_m$ as a rough dividing line between hot and cold behavior. In everyday usage, distinction is made relative to room temperature.

1 *Hot working* simply refers to the working of preheated material. Usual hot-working temperatures are given in Tables 4-2 and 4-3. Since for most technical materials (except tin and lead) $0.5T_m$ is above room temperature, the everyday definition is also correct.

Hot working offers several advantages: flow stresses are low, hence forces and power requirements are relatively low, and even very large workpieces can be deformed with equipment of reasonable size. Ductility is high, hence large deformations (in excess of 99% reduction) can be taken (usually in a succession of passes) and complex part shapes can be generated. The cast structure can be destroyed (Sec. 4-1-5), in general, by a deformation equivalent to 75% reduction in height or area, although reductions of 90% (10:1 ratio) may be needed if highest properties are to be attained.

There are also disadvantages. It takes energy to heat the workpiece to the elevated temperature. Most materials oxidize and the oxides of some metals (e.g., scale on steel) can impair surface finish. Variations in finishing temperatures lead to fairly wide dimensional tolerances and also to a less well-defined set of properties in the as-hot-worked condition.

Hot working may be carried out by:

Nonisothermal forming. The deforming tool must be several times stronger than the workpiece, and this usually means that the tool must be kept much colder. *Cooling* of the surface layers of the workpiece has several disadvantages: variable cooling introduces further variations in properties; cooling of thin sections limits the minimum wall thickness attainable; periodic contact with the hot workpiece exposes the tooling to thermal cycling which leads to thermal fatigue (Sec. 2-1-7).

Isothermal forming. Some of the above problems disappear when the tool is at the same temperature as the workpiece. However, it can be a problem to find an appropriate tool material; the high temperature also increases the difficulties of lubrication.

Controlled hot working is usually conducted nonisothermally and is used to impart desirable properties (Sec. 4-1-4).

2 *Cold working*, in the everyday sense, means working at room temperature, although the work of deformation can raise temperatures to 100–200 °C. Cold working usually follows hot working. Scale and other surface films are normally removed by chemical etching (*pickling*) or shot blasting.

Cold working has several advantages. In the absence of cooling and oxidation, tighter tolerances and better surface finish can be obtained and thinner walls are also possible. The final properties of the workpiece can be closely controlled and, if desired, the high strength obtained during cold working can be retained or, if high ductility is needed, grain size can be controlled to advantage in annealing (Sec. 4-1-3). Lubrication is, in general, somewhat easier.

There are also drawbacks. For most technological materials, room temperature is below $0.5T_m$; therefore, flow stresses are high and hence tool pressures, deformation forces, and power requirements are high too. The ductility of many materials is also limited, thus limiting the complexity of shapes that can be readily produced.

3 *Warm working* combines some of the advantages of both hot and cold working, especially in the warm working of steel (typically between 650 and 700 °C). Temperatures are low enough to avoid scaling, thus producing a good surface finish, yet they are high enough to reduce flow stress and thus allow the forming of parts that would generate excessive die pressures in cold working. The elevated temperature results in substantial strain-rate sensitivity, and the flow stress (Eq. (4-7)) remains low only if strain rates are kept low.

Purpose of Deformation A further useful distinction may be made according to the purpose of deformation.

1 If the process aims at destroying the cast structure by successive deformation steps and the resulting semifabricated product is destined for further shaping or forming, it is customary to speak of *primary processes* (as shown in Fig. 2-42). They are usually conducted hot and on the large scale, in specially constructed plants.

2 *Secondary processes* take the products of some primary process and further transform them into a finished part; as indicated in Fig. 1-11, they are at the focus of our discussion. Secondary processes include specific variants of the bulk deformation processes (Fig. 4-10) and all of the sheet-metalworking processes to be discussed in Chap. 5.

Analysis From the point of view of understanding and analyzing bulk deformation processes, it is useful to make a different distinction:

1 In *steady-state processes* all parts of the workpiece are subjected to the same mode of deformation (Fig. 4-10). Thus, once the situation is analyzed for the deformation zone, the analysis remains valid for the duration of the process.

2 In *non-steady-state processes* the geometry of the part changes continually and the analysis must be repeated for various points in time, from the starting condition to the end of the stroke of the deforming machine. Some processes have a transitional character; for example, deformation is non-steady-state at the beginning and end of extrusion, but acquires steady-state characteristics while the greater part of a billet is extruded.

The same principles apply whether deformation is conducted hot or cold, on a large or small scale.

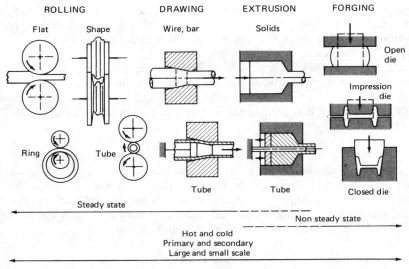

FIGURE 4-10
For purposes of analysis, bulk deformation processes are best classified according to whether deformation is steady state or non-steady state.

4-3-2 Pressures and Forces

Plastic deformation is conducted with the aid of tools (dies). All die materials have limited strength; therefore, a primary concern is the magnitude of pressures that are developed in the course of deformation; if these are too large, the process is not feasible. The calculation of interface pressures and deformation forces is the primary preoccupation of books dealing with plastic deformation processes. Very often the emphasis is on the relative accuracy of various competing theories. For our purpose it is much more important that any estimate of the pressures and forces should be truly relevant. The simple approach presented here will be accurate to within $\pm 20\%$; more sophisticated theories could improve the accuracy by a few percent, but in choosing tooling and equipment an appropriate safety factor must be allowed anyway. In order to get a meaningful estimate of pressures and forces, four points must be observed: (1) the stress state must be analyzed: (2) a relevant flow stress must be found: (3) the effects of friction must be judged: (4) inhomogeneous deformation must be taken into account.

Yield Criteria The stress state is, in the general case, triaxial—that is, stresses act in all directions. Analysis is simplified if the coordinate system is oriented in such a way that shear stresses disappear and only three normal stresses act. These are then called *principal stresses* and are denoted σ_1, σ_2, and σ_3 (Fig. 4-11a).

For plastic flow to occur, the combination of stresses must satisfy the *yield criterion*. Yield criteria have been formulated to describe the beginning of plastic

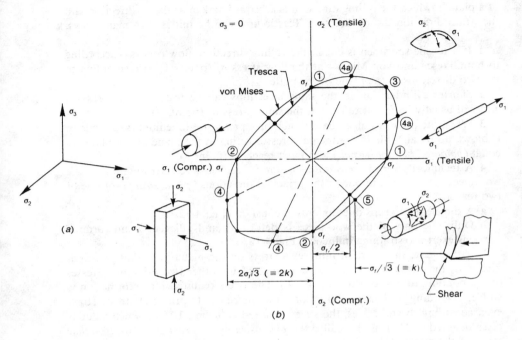

FIGURE 4-11
(a) The coordinate system may be rotated to obtain only principal stresses. (b) Under plane-stress conditions, some of the important stress states may be shown on the Tresca yield hexagon and von Mises yield ellipse. (*From J. A. Schey, Tribology in Metalworking: Friction, Lubrication and Wear, American Society for Metals, 1983, p. 12. With permission.*)

deformation by relating the principal stresses to the tensile or compressive yield strength of the material (Sec. 2-1-1). Our concern is with large plastic deformations; therefore, we will use the flow stress σ_f (thus we should really speak of flow criteria, however, the term *yield criterion* is so widely entrenched that we will retain it for our purpose). For metals, two criteria are frequently used.

The yield criterion due to Tresca can be written as

$$\frac{\sigma_{max} - \sigma_{min}}{2} = \frac{\sigma_f}{2} \tag{4-8}$$

where σ_{max} is the most positive and σ_{min} the most negative stress.

The yield criterion according to von Mises is

$$(\sigma_1 - \sigma_2)^2 + (\sigma_2 - \sigma_3)^2 + (\sigma_3 - \sigma_1)^2 = 2\sigma_f^2 \tag{4-9}$$

The significance of yield criteria is best illustrated by examining a simplified stress state in which $\sigma_3 = 0$ (*plane stress*). For ease of visualization, one may think of a plate in which the rolling direction is arbitrarily taken as the σ_1 direction and the width direction the σ_2 direction. Plastic flow can be initiated in many ways:

1 If a tensile specimen is cut in the rolling direction, flow occurs—according to both Tresca and von Mises—at the flow stress σ_f (points 1, corresponding to the two directions in the plane of the plate).

2 Shorter cylinders cut in the same directions can be tested in compression and will usually be found to flow at the same stress σ_f (points 2).

3 When the plate is bulged by a punch or a pressurized medium (as a balloon is blown up by air), the two principal stresses acting in the plane of the plate are equal (*balanced biaxial tension*) and must reach σ_f (point 3).

4 A technically very important condition is reached when deformation of the workpiece is prevented in one of the principal directions (*plane strain*) for one of two reasons:

a A die element keeps one dimension constant (Fig. 4-12a).

b Only one part of the workpiece is deformed, and adjacent nondeforming portions exert a restraining influence (Fig. 4-12b).

In either case, the restraint imposes a stress on the material in that principal direction; the stress is the arithmetic average of the other two principal stresses (corresponding to points 4 in Fig. 4-11b). The stress required for deformation is still σ_f according to Tresca who ignores the intermediate principal stress. However, according to von Mises, the stress required is higher, $1.15\sigma_f$, which value is often denoted as $2k$. It is also called the *plane-strain flow stress* or *constrained flow stress* of the material. Plane strain may also be imposed in tension, points 4a.

FIGURE 4-12
Deformation in one direction is often prevented and a plane-strain condition established by the restraint given either by (*a*) die elements or (*b*) nondeforming parts of the workpiece adjacent to the deformation zone. (*From J. A. Schey, as Fig. 4-11, p. 13.*)

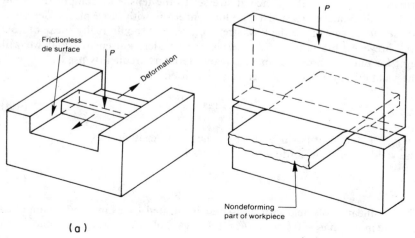

(a)

(b)

5 If a cylinder is cut out and twisted (torsion), the two principal stresses on the surface of the cylinder are of equal magnitude but of opposing sign (points 5 in Fig. 4-11b). This is a condition of *pure shear*, and flow occurs at the shear flow stress τ_f, which is equal to $0.5\sigma_f$ according to Tresca and $0.577\sigma_f$ according to von Mises. The shear flow stress according to von Mises is often denoted as k. The important point is that when, in the course of deformation by compression, a transverse stress of opposing sign (a tensile stress) is imposed, the stress required for compression will decrease. Thus, this offers a powerful mechanism for reducing die pressures.

6 A special condition is reached when all three principal stresses are equal in magnitude (*hydrostatic stress state*). An inspection of the yield criteria (Eqs. (4-8) and (4-9)) will show that superimposition of a hydrostatic stress simply shifts all principal stresses by the same amount; thus, there is no change in the yield criterion.

It will be noted that, for certain stress states, von Mises predicts a critical stress that is 15% higher than the uniaxial flow stress. Not all materials obey the von Mises criterion, but to be on the safe side, we will always use $1.15\sigma_f$ ($= 2k$) as the flow stress in plane strain.

The Relevant Flow Stress In all calculations, the stress sufficient to maintain plastic deformation, σ_f must be taken for the temperatures, strains, and strain rates prevailing in the process. It cannot be sufficiently emphasized that our interest is not just in initiating but also in maintaining plastic flow. Thus, the yield strength found in many handbooks is of little use; the flow stress traverses the true-stress–true-strain curve (Fig. 4-1a or 4-7a) within the strain limits defined by the starting material and the end strain.

1 In cold working it can be assumed that the power law, Eq. (4-4) holds and, whenever available, the K and n values should be used (a selection is given in Tables 4-2 and 4-3). For a non-steady-state process, such as forging, the *instantaneous flow stress* σ_f at the point of interest is taken; if we are concerned with the maximum force, which occurs at the end of deformation, the flow stress corresponding to the final strain must be calculated (Fig. 4-13a). In a steady-state process, such as rolling or wire drawing, the workpiece strain hardens as it passes through the deformation zone. To simplify calculations, a *mean flow stress* σ_{fm} is used. This is found by integration of Eq. (4-4) between the limits of strain. For an annealed material:

$$\sigma_{fm} = \frac{K}{\epsilon}\left[\frac{\epsilon^{n+1}}{n+1}\right] \qquad (4\text{-}10)$$

Alternatively, the flow stress curve is plotted and the mean is found by visual averaging (Fig. 4-13b).

A problem arises when the K and n values are not known. If equipment is available, a compression test (Sec. 2-1-2) can be conducted rapidly. Otherwise, the only guide could be the TS, from Tables 4-2 and 4-3 or other source. Paradoxi-

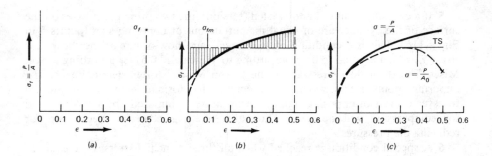

FIGURE 4-13
The relevant flow stress is calculated (a) at the strain of interest in non-steady-state processes, (b) as a mean flow stress in steady-state processes or (c), in the absence of K and n values, from the TS.

cally, the basically nonsensical method of calculating the TS (Sec. 2-1-1) happens to give a reasonable approximation of the mean true flow stress σ_{fm}, at least for strains on the order of the necking strain (Fig. 4-13c). Since at the point of necking $n = \varepsilon_u$, some reasonable correction can be made for smaller or larger strains (see Example 4-9).

2 For hot working, the flow stress can be calculated from the power function, Eq. (4-7), with the appropriate C and m values (Tables 4-2 and 4-3). If these values are not available for various strains, one has to assume that the flow stress remains constant throughout deformation (as in the curve for a strain rate of $1/s$ in Fig. 4-7a; in Tables 4-2 and 4-3 the values are given for a strain of $\varepsilon = 0.5$). If no C and m data are available, one is obliged to make a compression test. It is quite inadmissible to use hot-strength values determined in conventional, slow tension tests, because they often represent only a fraction of the true flow stress prevailing at the much higher (typically, $1–1000\text{-s}^{-1}$) strain rates attained in deformation processes. Extrapolation from low strain rates to high strain rates is hazardous because m may also change with strain rate (see Fig. 4-8).

It should be noted that the constants used in flow stress calculations are also a function of the starting condition of the material. Data given in Tables 4-2 and 4-3 are representative values for annealed material. Every effort was made to use reliable data, and the two C and m values entered for copper show the worst of the extreme variations occasionally found in published data.

Example 4-9

In Example 2-2 we found TS = 351 MPa for an annealed 80Cu–20Ni alloy. In Example 4-1 we determined $K = 760$ MPa and $n = 0.45$. If we had not made the calculations of Example 4-1 (or if only the TS of a material is available from the literature), would the TS give any guidance on flow stress?

From the plot given for Example 4-1, a true stress of 350 MPa corresponds to a true strain of about 0.17. Since $\varepsilon = \ln(l/l_0) = 0.17$, $l/l_0 = \exp 0.17 = 1.185$, and the corresponding tensile strain is $e_t = (l - l_0)/l_0 = (1.185 - 1.0)/1.0 = 18.5\%$. To convert to compressive strain, regard $l = h_0$

and $l_0 = h$; then $e_c = (h_0 - h)/h_0 = (1.185 - 1.0)/1.185 = 15.6\%$. Thus, the TS is a reasonable estimate of σ_f for a small compressive deformation, but would be too low if this heavily strain-hardening material were to be worked to a higher strain.

Effects of Friction Because deformation is most frequently conducted by bringing the workpiece into contact with a tool or die, friction between the two contacting bodies is unavoidable. We already examined friction in conjunction with situations existing in machinery elements (e.g., in bearings) in Sec. 2-2-1. We found that friction can be described by a coefficient of friction μ; because of its importance, we reproduce here Eq. (2-18)

$$\mu = \frac{F}{P} = \frac{\tau_i}{p} \qquad (2\text{-}18)$$

When the interface pressure p is low relative to the flow stress σ_f of the contacting materials (as it would be in a bearing), Eq. (2-18) holds: With increasing pressure p the interface shear stress τ_i increases linearly (Fig. 4-14a), and μ could assume any constant value.

In plastic deformation processes one of the contacting materials (the workpiece) deforms and in doing so also slides against the harder surface (the tool or die). A frictional stress τ_i is again generated, but this time there is a limit to μ, because the material will choose a deformation pattern that minimizes the energy of deformation. If friction is high, the interface shear stress τ_i will reach, in the limit, the shear flow stress τ_f of the workpiece material (Fig. 4-14a). At this point the workpiece refuses to slide on the tool surface; instead, it deforms by shearing inside the body (Fig. 4-14b). Since $\tau_f = 0.5\sigma_f$ (Fig. 4-11b), it is often said that the maximum value of $\mu = 0.5$. This statement is true only when $p = \sigma_f$; at higher p, the maximum value of μ is lower (Fig. 4-14b). In general, it is much more accurate to say that the coefficient of friction becomes meaningless when $\tau_i = \tau_f$, since there is no relative sliding at the interface. This is described as *sticking friction*, even though the workpiece does not actually stick to the die surface.

Because of the conceptual difficulties introduced by the coefficient of friction, it is often preferable to use the actual value of τ_i, especially when interface pressures are very high, and we shall follow this practice for extrusion calculations. Alternatively, the interface shear stress can be denoted as a fraction of the shear flow stress

$$\tau_i = m^* \tau_f = m^* \frac{\sigma_f}{2} \quad \left(\text{or} = m^* \frac{\sigma_f}{\sqrt{3}} \right) \qquad (4\text{-}11)$$

where m^* is the *frictional shear factor* (the literature uses m but, because of possible confusion with m in Eq. (4-7), we add the asterisk). For a perfect lubricant, $m^* = 0$; for sticking friction, $m^* = 1$.

Friction increases pressures and forces and could easily limit the attainable reduction. With a few exceptions noted later, every effort is made to reduce friction by applying a suitable *lubricant* (Table 4-4). A good lubricant accomplishes much more: It separates die and workpiece surfaces and thus prevents

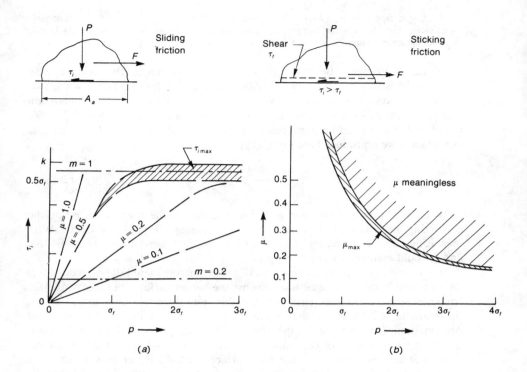

FIGURE 4-14
(a) The interface shear stress can never exceed the shear flow stress of the material and (b) the maximum possible coefficient of friction decreases when interface pressures exceed the flow stress of the material. (*From J. A. Schey, as Fig. 4-11, p. 15.*)

adhesion with its undesirable side effects of tool pickup, workpiece damage, and die wear; it reduces die wear due to abrasion and other mechanisms; it controls the surface finish of the part produced; and it cools the system in cold working and helps to prevent heat loss (or removes heat at a controlled rate) in hot working. The lubricant must not be toxic or allergenic, it must be easy to apply and remove, and residues must not interfere with subsequent operations or cause corrosion.

Example 4-10

Sticking friction sets in where $\mu p = \tau_f$. Calculate the values of μ for various interface pressures p.

It is convenient to express p as a multiple of σ_f. If the Tresca yield criterion is used, $\tau_f = 0.5\sigma_f$ and, by definition, Eq. (2-18)

$$\mu_{max} = \frac{\tau_i}{p} = \frac{\tau_i}{x\sigma_f} = \frac{\sigma_f}{2x\sigma_f}$$

TABLE 4-4 TYPICAL LUBRICANTS* AND FRICTION COEFFICIENTS IN PLASTIC DEFORMATION

Workpiece material	Working	Forging Lubricant	μ	Extrusion† lubricant	Wire drawing Lubricant	μ	Rolling Lubricant	μ	Sheet metalworking Lubricant	μ
Sn, Pb, Zn alloys		FO-MO	0.05	FO or soap	FO	0.05	FA-MO or MO-EM	0.05 / 0.1	FO-MO	0.05
Mg alloys	Hot or warm	GR and/or MoS₂	0.1–0.2	None			MO-FA-EM	0.2	GR in MO or dry soap	0.1–0.2
Al alloys	Hot	GR or MoS₂	0.1–0.2	None			MO-FA-EM	0.2		
	Cold	FA-MO or dry soap	0.1 / 0.1	Lanolin or soap on PH	FA-MO-EM, FA-MO	0.1 / 0.03	1–5% FA in MO(1–3)	0.03	FO, lanolin, or FA-MO-EM	0.05–0.1
Cu alloys	Hot	GR	0.1–0.2	None (or GR)			MO-EM	0.2		
	Cold	Dry soap, wax, or tallow	0.1	Dry soap or wax or tallow	FO-soap-EM, MO	0.1 / 0.03	MO-EM	0.1	FO-soap-EM or FO-soap	0.05–0.1
Steels	Hot	GR	0.1–0.2	GL (100–300), GR			None or GR-EM	ST‡ / 0.2	GR	0.2
	Cold	EP-MO or soap on PH	0.1 / 0.05	Soap on PH	Dry soap or soap on PH	0.05 / 0.03	10% FO-EM	0.05	EP-MO, EM, soap, or polymer	0.05–0.1
Stainless steel, Ni and alloys	Hot	GR	0.1–0.2	GL (100–300)			None	ST‡	GR	0.2
	Cold	CL-MO or soap on PH	0.1 / 0.05	CL-MO or soap on PH	Soap on PH or CL-MO	0.03 / 0.05	FO-CL-EM or CL-MO	0.1 / 0.05	CL-MO, soap, or polymer	0.1
Ti alloys	Hot	GL or GR	0.2	GL (100–300)					GR, GL,	0.2
	Cold	Soap or MO	0.1	Soap on PH	Polymer	0.1	MO	0.1	Soap, or polymer	0.1

*Some more frequently used lubricants (hyphenation indicates that several components are used in the lubricant):
CL = chlorinated paraffin.
EM = emulsion; the listed lubricating ingredients are finely distributed in water.
EP = "extreme-pressure" compounds (containing S, Cl, and P).
FA = fatty acids and alcohols, e.g., oleic acid, stearic acid, stearyl alcohol.
FO = fatty oils, e.g., palm oil and synthetic palm oil.
GL = glass (viscosity at working temperature in units of poise).
GR = graphite; usually in a water-base carrier fluid.
MO = mineral oil (viscosity in parentheses, in units of centipoise at 40 °C).
PH = phosphate (or similar) surface conversion, providing keying of lubricant.

†Friction coefficients are misleading for extrusion and are therefore not quoted here.
‡The symbol ST indicates sticking friction.

Source: Data extracted from J. A. Schey: *Tribology in Metalworking: Friction, Lubrication, and Wear*, American Society for Metals, Metals Park, Ohio, 1983.

hence, when

$$x = 1 \quad 2 \quad 4 \quad 8$$
$$\mu_{max} = 0.5 \quad 0.25 \quad 0.125 \quad 0.062$$

The points are plotted in Fig. 4-14b. The von Mises criterion gives $\tau_f = 0.577\tau_f$ and thus slightly higher μ_{max} values.

Example 4-11

From Eqs. (2-18) and (4-11)

$$\tau_i = \mu p = m^* \tau_f$$

If p is again expressed as an x multiple of σ_f, calculate the equivalent μ and m^* values.

$$\text{When} \quad p = \sigma_f \quad 2\sigma_f \quad 4\sigma_f \quad 8\sigma_f$$
$$\text{then} \quad m^* = 2\mu \quad 4\mu \quad 8\mu \quad 16\mu$$

There is thus no simple relationship between the two parameters used to describe friction. Furthermore, the above-calculated relations fail to hold when partial sticking sets in.

Inhomogeneous Deformation   There is another important source of high interface pressures and forces which has nothing to do with interface friction, and, therefore, is not affected by lubrication. It can be best understood from the example of *indentation* of a semiinfinite body with a narrow anvil. Inspection of Fig. 4-15a will show that a small tool cannot possibly deform the entire bulk of a large (semiinfinite) workpiece. It can be observed experimentally that when the tool penetrates, localized indentation with highly *inhomogeneous material flow* takes place.

This can occur by the mechanism shown in Fig. 4-15a: a part of the workpiece (1) immediately under the indentor remains immobile relative to the indentor and moves with it as though it would be an extension of the indentor itself. This rigid wedge then pushes two triangular wedges (2) aside, which in turn push up two outer wedges (3), thereby forming a hump corresponding to the volume displaced by the indentor. The rest of the workpiece (4) is only elastically loaded. The difficulty of moving the material purely locally—*against the restraint given by the surrounding elastic material*—raises the required interface pressure.

In many bulk deformation processes a workpiece of finite thickness is deformed simultaneously from two sides (Fig. 4-15b). The effects of inhomogeneous deformation then depend on how far the two deformation zones are separated, and this is most usefully expressed by the h/L ratio, that is, the ratio of height to contact length. It is found from both theory and experiment that when $h/L > 8.7$, the two deformation zones are entirely separated; the material between these zones is only elastically deformed and exerts the same restraining effect as though it were of infinite thickness. At lower h/L ratios the two wedges cooperate (Fig. 4-15b)

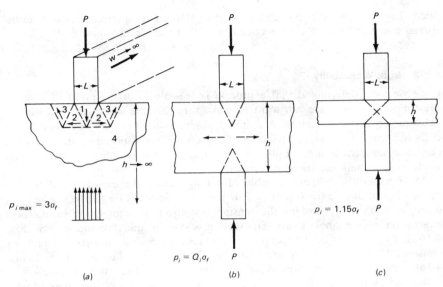

FIGURE 4-15
Deformation is (a) highly inhomogeneous when indenting a semiinfinite body; (b) at high h/L values, deformation is still inhomogeneous; (c) only at $h/L = 1$ is homogeneity approacned.

and the pressure drops. As might be expected, at a ratio of $h/L = 1$ the two deformation zones fully cooperate (Fig. 4-15c) and the material flows at a minimum pressure. If the h/L ratio were to diminish further, deformation would be homogeneous, but the effects of friction would now increase die pressures.

Inspection of Fig. 4-15b indicates that the two wedges penetrating from top and bottom tend to pull the workpiece apart; in other words, inhomogeneous deformation generates *secondary tensile stresses* (i.e., stresses that are not externally imposed but are generated by the process of deformation itself). Several consequences are possible:

1 Internal fracture may occur in the workpiece during deformation.

2 A residual stress pattern (internal stresses) may be set up that may cause subsequent deformation (warping) of the workpiece, particularly on heating.

3 Surface residual tensile stresses can combine with other effects to cause delayed failures (e.g., stress-corrosion cracking in the presence of a corrosive medium).

In general, therefore, the aim of process development is to make deformation as homogeneous as possible. If harmful residual stresses remain, a stress-relief heat treatment is given (Sec. 2-1-8).

We have seen that compressive residual stresses concentrated in a thin surface layer greatly improve the fatigue-resistance of the workpiece in service (Sec. 2-1-8). Highly inhomogeneous compressive deformation is then purposely applied. The surface compressive stresses are balanced by internal tensile stresses, spread over such a large cross-sectional area that their level is harmless.

4-3-3 Bulk Workability

Once we have determined that a process is feasible from the point of view of pressures and forces, we will want to make sure that the workpiece will survive deformation without fracture. A material of given ductility may fare very differently in various processes, depending on the conditions imposed on it. Therefore, our main concern is not simply ductility, but a more complex property called *workability* in bulk metalworking operations.

Most materials that are capable of taking plastic deformation fail by the mechanism of ductile fracture (Fig. 2-4b), which is induced by tensile stresses. An example of this is found in the tension test itself. In the region of uniform elongation, only a single stress acts (the stress state is uniaxial, and $\sigma_1 = \sigma_f$, Fig. 4-16a); the diameter of the specimen decreases only because the volume remains constant. However, at the point of necking, the stress state becomes triaxial (Fig. 4-16b), because deformation is now limited to the neck and contraction is restrained by the nondeforming portions adjacent to the neck. It is these triaxial tensile stresses that open up preexisting weak spots and generate cavities which finally merge and lead to fracture (Sec. 2-1-1). Thus, the technological concept of bulk workability has two components:

1 The basic ductility of the material allows it to deform to some extent, without fracture, even in the presence of tensile stresses. Therefore, reduction in area measured in the tension test (Eq. (2-10)) is a useful (but not universally applicable) measure of basic ductility; it is essentially a measure of resistance to void formation. Other possible measures are the number of turns to fracture in a torsion test, or the reduction in height in upsetting tests designed to generate high secondary tensile stresses. Upsetting with sticking friction at the end face causes severe barreling and thus surface cracking in a material of low ductility (Fig.

FIGURE 4-16
In a tension test, the stress state is (a) uniaxial during uniform extension but (b) becomes triaxial in the neck zone.

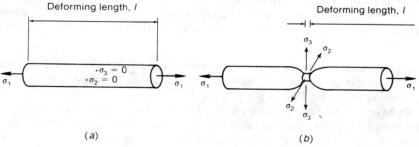

(a) (b)

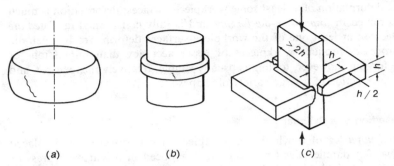

(a) (b) (c)

FIGURE 4-17
Workability may be evaluated in (a) compression with sticking friction, (b) upsetting a collared specimen, or (c) partial-width indentation.

4-17a). Tensile stresses are higher on a collared specimen (Fig. 4-17b) or in a partial-width indentation test (Fig. 4-17c).

2 The stress state induced by the process modifies ductility. If the process maintains compressive stresses in all parts of the deforming workpiece, cavity formation cannot begin and ductile fracture does not occur. Thus, if a tension test is conducted with the apparatus submerged in a high-pressure container, so that equal compressive stresses (a hydrostatic pressure) are applied, cavity formation is delayed and the neck reduces further before fracture (Fig. 4-2a); this immediately shows that ductility as a material property cannot be separated from the prevailing stress state. (At very heavy deformations, ductility of the material may be exhausted and then brittle, shear-type fracture may develop.) If, however, the process allows tensile stresses to develop, cavity formation can begin and will, finally, lead to fracture.

At what point this fracture should occur is predicted by *workability criteria*, none of which has proven to be universally applicable. The most useful one has been formulated by Cockroft and Latham* who state that, for a given metal, the work done by the highest local tensile stress must reach a critical value. It follows that when the development of secondary tensile stresses can be suppressed, deformation can be taken much further, just as fracture in the tension test is delayed by hydrostatic pressure. Therefore, one of the important aims of process design is to increase the *hydrostatic pressure component* $\sigma_H = (\sigma_1 + \sigma_2 + \sigma_3)/3$ in the system.

4-4 FORGING

Forging processes are among the most important manufacturing techniques. As shown in Fig. 4-10, three broad groups can be distinguished: *open-die forging*

*M. G. Cockroft and D. J. Latham, *J. Inst. Met.*, **96**:33–39, (1968).

allows free deformation of at least some workpiece surfaces; deformation is much more constrained in *impression-die forging* and is fully constrained in *closed-die forging*. Because at least one of the workpiece surfaces deforms freely, open-die forging processes produce workpieces of lesser accuracy than impression- or closed-die forging; however, tooling is usually simple, relatively inexpensive, and allows the production of a large variety of shapes.

4-4-1 Upsetting of a Cylinder

In the *axial upsetting* of a cylinder, a workpiece of cylindrical shape is placed between two flat parallel dies (*platens*) and is reduced in height by a press or hammer force applied to the platens. Upsetting is a very versatile process, practiced hot or cold. The end products range from huge, 150-ton or larger steel rotors for power-generation stations to minute components. Frequently, a head is upset at the end of a part, in special-purpose mechanized (automated) machines, producing vast numbers of nails, screws, bolts, pins, and similar components.

Frictionless Upsetting Let us assume that, by the application of some very good lubricant, we succeed in reducing friction to virtually zero. If we divide the cylinder into many small elements, each element now deforms equally, in other words, deformation is *homogeneous*. The cylinder becomes shorter and, to preserve constancy of volume (Eq. (2-2)), it assumes a greater diameter, but still remains a true cylinder (Fig. 4-18a). Because upsetting is a non-steady-state process, a complete analysis requires the calculation of variables at several points during the press stroke. Here we take the example of the final point, where forces will be highest. In calculations of a repetitive kind, it is best to follow a set sequence of operations, as shown here step by step.

Step 1: In practice, only one of the final dimensions is defined. If, for example, a cylinder of h_0 height and d_0 diameter is upset to a final height of h_1, the final diameter d_1 can be calculated from constancy of volume

$$A_1 = A_0 \frac{h_0}{h_1} \tag{4-12a}$$

and

$$d_1 = \sqrt{\frac{4A_1}{\pi}} \tag{4-12b}$$

Step 2: The compressive engineering strain is needed only for conversational purposes. It is usually calculated from the height change (Eq. (2-13)) but, because the volume remains constant, cross-sectional areas can be used equally well:

$$e_c = \frac{A_1 - A_0}{A_1} = \frac{h_0 - h_1}{h_0} \tag{2-13'}$$

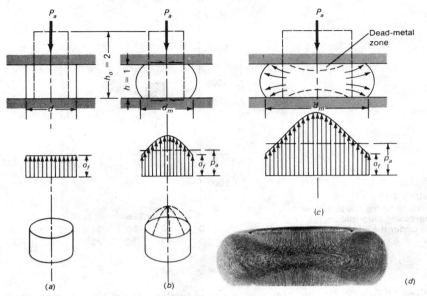

FIGURE 4-18
Interface pressures are (*a*) equal to the flow stress in frictionless compression but (*b*) friction generates a friction hill which (*c*) is larger for a large *d/h* ratio. (*d*) Sticking at the end face leads to folding-over of the sides.

Step 3: For purposes of calculating the flow stress, the true strain is obtained from Eq. (4-5*b*)

$$\epsilon = \ln \frac{h_0}{h_1} \tag{4-5b}$$

In hot working, the strain rate is also needed

$$\dot{\epsilon} = \ln \frac{v}{h} \tag{4-6}$$

Step 4: We are now ready to calculate the relevant flow stress. In cold working, it is given by Eq. (4-4)

$$\sigma_f = K\epsilon^n \tag{4-4}$$

In hot working, it is obtained from Eq. (4-7)

$$\sigma_f = C\dot{\epsilon}^m \tag{4-7}$$

(Note that strain rate must always be expressed in units of reciprocal seconds, s^{-1}.)

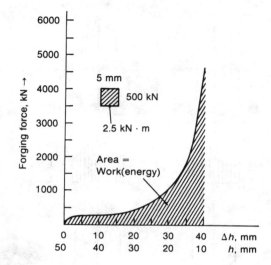

FIGURE 4-19
Upsetting forces rise steeply with progressing deformation; the area under the curve represents work.

Step 5: To calculate die pressures, we need to check the effects of stress state, friction, and inhomogeneity of deformation (Sec. 4-3-2). The stress state is uniaxial; hence, the flow stress is σ_f. The platen overlaps the workpiece; hence, no indentation effect is possible and we need not worry about any h/L ratio. Since we assumed that friction is absent, the interface pressure p_a (where the subscript refers to axial symmetry) is simply the uniaxial flow stress σ_f. This is the pressure that the tooling will have to withstand (Fig. 4-18a).

Step 6: The press force P_a is simply the interface pressure multiplied by the area over which the pressure acts

$$P_a = p_a A_1 \tag{4-13}$$

This defines the size of press needed.

Step 7: For some forging equipment, it is necessary to know also the total energy expended in deforming the workpiece. This can be obtained by repeating the calculations for the press force P_a at various points in the stroke. Thus, the force-displacement curve (Fig. 4-19) is defined. Force rises rapidly because the area A_1 also increases rapidly in the course of upsetting. The area under the curve has the dimensions of work (work = force × distance). Thus, the work or energy E_a to be delivered by the press or hammer can be obtained by graphical integration of this area.

The energy absorbed by the workpiece is converted into heat. In the absence of cooling, the adiabatic temperature rise ΔT would be

$$\Delta T = \frac{E_a}{V \rho c} \tag{4-14}$$

where V is the volume, ρ is the density, and c is the specific heat (more correctly, heat content per unit volume) of the workpiece. In practice, deformation takes place in finite time, and some of the heat is lost through conduction into the dies, and by radiation and convection into the surrounding atmosphere. Therefore, the actual temperature rise is less but can, nevertheless, be significant. In hot working it may raise the temperature above the solidus to cause hot-shortness, and in cold working it may result in lubrication breakdown.

Upsetting with Sliding Friction   In practice it is highly unlikely that zero friction can be attained even with the best lubricant. Die pressures are high and deformation of the cylinder requires that its end faces should slide on the tool surfaces; thus, a measurable frictional shear stress τ_i is always present. This shear stress opposes the free expansion of the end faces, with two consequences (Fig. 4-18b):

1 The cylinder assumes a *barrel* shape. We may ignore this in calculating the new diameter (Step 1) simply by calculating a mean diameter d_m from constancy of volume (Eq. (4-12b)).

2 In order to overcome the frictional stress, a higher and higher normal pressure must be exerted as we move toward the center of the cylinder. At the free edge, the pressure equals σ_f and rises from here like a hill does. The greater the friction (expressed as a coefficient of friction μ or frictional shear factor m^*), the steeper the *friction hill* will be. Therefore, in Step 5 of our calculations, we need to find the maximum stress $p_{a\,max}$; this is most simply done with the use of m^*

$$p_{a\,max} = \sigma_f \left(1 + \frac{m^*}{\sqrt{3}} \frac{d_1}{h_1}\right) \tag{4-15}$$

This will determine the maximum stress to which the tool is exposed.

In Step 5, only the average interface pressure p_a is needed for calculating the forces. A comparison of Fig. 4-18b and 4-18c will show that, for the same magnitude of friction, a cylinder of the same height but of larger diameter gives rise to a taller friction hill and, therefore, higher p_a. The average interface pressure p_a is conveniently expressed as a multiple of the uniaxial flow stress σ_f. The *pressure-multiplying factor* Q_a (where the subscript signifies axial symmetry) must take into account both the effects of friction (μ or m^*) and workpiece geometry (the d/h ratio, which characterizes the squatness of the cylinder). Without deriving the appropriate formulae, the relevant multiplying factors can be calculated, if m^* is used, from

$$p_a = \sigma_f Q_a = \sigma_f \left(1 + \frac{m^*}{3\sqrt{3}} \frac{d_1}{h_1}\right) \tag{4-16}$$

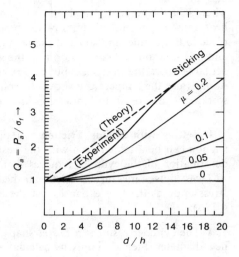

FIGURE 4-20
Average pressures in upsetting a cylinder increase with increasing friction and squatness of the cylinder. (*After J. A. Schey, T. R. Venner, and S. L. Takomana, J. Mech. Work. Tech.* **6**:23– 33 (1982). *With permission of Elsevier Science Publishers.*)

Alternatively, if μ is used, Q_a can be taken from Fig. 4-20 and

$$p_a = \sigma_f Q_a \qquad (4\text{-}17)$$

Upsetting with Sticking Friction In the extreme case, when the platen surface is rough and no lubricant is used, the interface shear stress τ_i may reach or exceed the shear flow stress τ_f of the workpiece material (Sec. 4-3-2) and movement of the end face is totally arrested. All deformation now takes place by internal shear in the cylinder; material adjacent to the platens does not move (*dead-metal zones*) and the sides of the cylinder fold over (Fig. 4-18*d*).

This is an example of inhomogeneous deformation that actually lowers interface pressure. Because the outer fibers of the cylinder are deformed by shearing superimposed on compression, the interface pressure remains low (Fig. 4-11) and the pressure-multiplying factor remains close to unity as long as $d/h < 2$. Simple theory cannot cope with this complexity and the limiting values of the pressure-multiplying factor given in Fig. 4-20 have been determined experimentally.

Limitations For production purposes it is important not only that the material can be deformed with feasible pressures and forces, but also that deformation should be uniform and *free of defects*. Several limits must be observed.

1 A very slender cylinder may *buckle* instead of upsetting uniformly. Therefore, it is advisable to limit the h_0/d_0 ratio to 2 when friction is high (Fig. 4-21*a*). When friction is very low, h_0/d_0 should be less than 1.5 to prevent skewing of the billet.

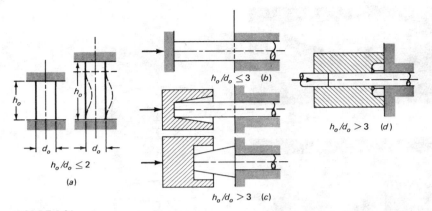

FIGURE 4-21
The initial height-to-diameter ratio is (a) limited by buckling and may be increased by (b) upsetting the end of a gripped bar, (c) limiting the deflection in conical cavities, or (d) expanding the bar into the cavity of a cold-header tool.

2 Very often upsetting is conducted in a *heading* operation, i.e., only the end of a cylindrical workpiece is upset. The longer part of the workpiece—firmly clamped in die halves—becomes fixed, and the increased resistance to buckling allows somewhat greater free lengths (Fig. 4-21b).

3 An even longer length can be upset when deflection of the workpiece is limited in *progressive upsetting* into conical and cylindrical shapes (Fig. 4-21c).

4 In so-called *cold headers* the long overhanging part of the wire or bar is supported in the bore of a die (Fig. 4-21d) and the head is produced by a punch that moves the material into the space made available in this die. Because the workpiece is guided on both of its ends, buckling is prevented and larger heads can be formed in a single stroke.

A second group of defects involves actual fracture of the workpiece. If deformation is truly homogeneous (Fig. 4-18a), most ductile materials can take relatively large strain in upsetting before their ductility is exhausted and fracture occurs by shearing at 45° to the application of the compressive stress (Fig. 4-22a).

In practice, the presence of friction leads to barreling (Figs. 4-18b to 4-18d). It is readily seen that material in the bulge is not directly compressed; instead, it is deformed indirectly, by the radial pushing action of the centrally located material. This expanding action creates circumferential as well axial secondary tensile stresses on the free (barreled) surface which may cause *cracking* (Fig. 4-22b). The direction of cracks depends on the relative magnitudes of the secondary tensile stresses (Fig. 4-22c and d). Since barreling is the primary culprit, improved lubrication (which reduces friction and thus barreling) may alleviate the problem.

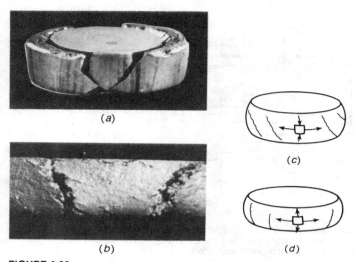

FIGURE 4-22
Fracture may occur by (a) exhausting ductility in cold working or by (b) intergranular fracture in hot working. (c), (d) The direction of cracks depends on the relative magnitudes of secondary tensile stresses generated by bulging.

It is quite common that one has to accept a limited deformation in a single stroke. Reheating in hot working and process anneals in cold working will allow further deformation.

Example 4-12

An AISI 1045 steel billet of $d_0 = 50$ mm and $h_0 = 50$ mm is hot upset, at 1000 °C, to a height of $h_1 = 10$ mm, on a hydraulic press operating at $v = 80$ mm/s. No lubricant is used. Calculate the press force and the energy expenditure.

To obtain the press force it would be sufficient to calculate for the final height only; however, the force is needed at several points of the press stroke if energy is to be determined too. It is best to setup a table.

Point No.	h, mm	A, mm²	$\dot{\varepsilon}$, s⁻¹	σ_f, MPa	d, mm	d/h	Q_a	p_a, MPa	P_a, kN
0	50	1963	1.6	...	50.0	1.0			
1	40	2454	2.0	131	55.9	1.4	1.0	131	320
2	30	3272	2.7	136	64.5	2.2	1.1	150	490
3	20	4909	4.0	144	79.1	4.0	1.5	216	1060
4	15	6545	5.3	150	91.3	6.1	2.0	300	1965
5	10	9827	8.0	157	111.8	11.2	3.0	470	4630

In setting up the table, the following were considered: The volume is $(50^2\pi/4)(50) = 98175$ mm^3. Instantaneous area is obtained from Eq. (4-12a), diameter from Eq. (4-12b). Since we deal with hot working, only strain rate is needed from Eq. (4-6). For flow stress, $C = 120$ MPa, $m = 0.13$ from Table 4-2. The factor Q_a is taken from Fig. 4-20 for sticking friction. Equation (4-17) gives p_a, and Eq. (4-13) gives P_a.

The results are plotted in Fig. 4-19. To obtain the energy requirement, the area under the force/displacement curve is integrated. One square corresponds to (500 kN) (5 mm) = 2500 N·m; the total area is about 14.5 squares or 36250 N·m (= 26700 lb·ft or 320000 lb·in).

4-4-2 Forging of Rectangular Workpieces

Two fundamentally distinct cases are to be distinguished: forging with *overhanging platens* and forging an *overhanging workpiece*.

Upsetting with Overhanging Platens When a rectangular slab is upset between two platens that are larger than the workpiece in all directions (Fig. 4-23a), the situation resembles that of the axial upsetting of a cylinder, at least as far as stresses and deformation in the cross section are concerned. Overall, however, the situation is different, especially if one of the dimensions of the slab is much greater than the other. The material will always flow in the *direction of least resistance*. Frictional resistance is proportional to the distance over which sliding takes place. Therefore, flow in the longer direction (which we shall call the *width direction w*), is much restricted and the condition of plane strain (Fig. 4-12a) is

FIGURE 4-23
In upsetting a rectangular workpiece (a) material flows in the direction of least resistance, marked L; (b) the friction hill now has a ridge shape.

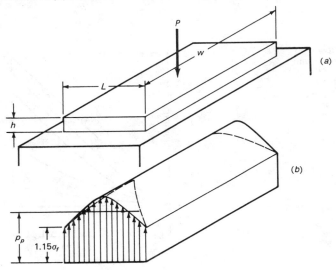

approximated. Most material flow takes place in the short direction, and for purposes of analysis we shall call this the *contact length L* between workpiece and tool surface; evidently, in upsetting with overhanging anvils, L increases as compression progresses.

Since the two platens are parallel, material flows away from the centerline where no material flow takes place and is, therefore, called the *neutral line* (the line that divides the flow directions).

In the process of calculations, Steps 1–4 are the same as in upsetting a cylinder. However, there are differences in Step 5.

First, the material will now begin to flow on reaching the plane-strain flow stress $1.15\sigma_f$ (points 4 in Fig. 4-11b). Second, the shape of the friction hill will now resemble a mountain ridge (Fig. 4-23b). The cross section of the friction hill can still be calculated by analogy to the axial upsetting case, as long as it is clearly understood that the friction hill is defined by that dimension of the workpiece which is measured *in the direction of major material flow*, i.e., the contact length L. Then, the friction hill will be higher for any given μ or m^* and for a larger L/h ratio. By analogy to the upsetting of a cylinder (Eq. (4-15)), the peak of the friction hill will be

$$p_{p\,\max} = 1.15\sigma_f\left(1 + \frac{m^*}{2}\frac{L}{h}\right) \qquad (4\text{-}18)$$

This peaks develops at the neutral line.

The friction hill resembled a single-pole tent in axial upsetting (Fig. 4-18b) but it is more like a ridge tent in the forging of a slab; therefore, the average pressure p_p (where the subscript refers to plane strain) is now

$$p_p = 1.15\sigma_f\left(1 + \frac{m^*}{4}\frac{L}{h}\right) \qquad (4\text{-}19)$$

FIGURE 4-24
Average pressures in upsetting a rectangular slab increase with friction and L/h ratio. (*After J. F. W. Bishop, Quart. J. Mech. Appl. Math.* **9**:236–246 (1956). With permission of Pergamon Press.)

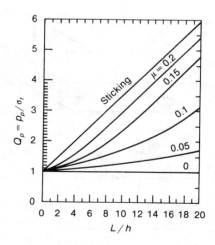

When friction is expressed as μ, the pressure-multiplying factor Q_p is taken from Fig. 4-24 and the average pressure is

$$p_p = 1.15\sigma_f Q_p \qquad (4\text{-}20)$$

In Step 6, the force P_p at any one point in the press or hammer stroke is again obtained by multiplying p_p by the area A_1 over which this pressure acts

$$P_p = p_p A_1 = p_p L w \qquad (4\text{-}21)$$

Rectangular workpieces are frequently forged, and elements of more complex forgings can often be regarded as rectangular ones.

Example 4-13

A 302 stainless steel pin is to be produced from a square wire. One end is flattened, and the center is pinched (as shown in the illustration). Calculate die pressures and forces, assuming that no lubricant is used.

The flattening operation may be considered as upsetting a rectangular workpiece. The 4-in length of the pin increases very little during flattening, and this dimension must be regarded as the width w during plane-strain compression (Fig. 4-23a). Most of the material gets displaced in the width of the pin; in terms of analysis, this becomes L.

Step 1: Since there is little change in the 4-in (width) dimension, $h_0 L_0 = h_1 L_1$ or $(0.25)(0.25) = (0.075) L_1$; thus $L_1 = 0.83$ in (the approximate value of 0.8 in marked in the illustration allows for some growth of the 4-in dimension).

Step 2: From Eq. (2-13a): $e_c = (0.25 - 0.075)/0.25 = 0.7$ (or 70%)

Step 3: From Eq. (4-5b): $\epsilon = \ln(0.25/0.075) = \ln 3.34 = 1.21$

For cold working, strain-rate sensitivity can be ignored.

Step 4: At the end of the stroke, the flow stress σ_f (from Table 4-2 after conversion): $\sigma_f = 190(1.21)^{0.3} = 200$ kpsi.

Step 5: From Fig. 4-24, for $L/h = 10.6$ and sticking friction, $Q_p = 3.7$ and, from Eq. (4-20), $p_p = 1.15(200)(3.7) = 850$ kpsi. This is too high for any tool material, and a suitable lubricant (Table 4-4) should be used. Assuming that the coefficient of friction is reduced to $\mu = 0.1$, from Fig. 4-11, $Q_p = 1.9$ and $p_p = 1.15(200)(1.9) = 437$ kpsi, which is still high but feasible (Sec. 4-6-1).

Step 6: The upsetting force, from Eq. (4-21): $P_p = 437(0.8)(4) = 1400$ klb or 700 tonf. Note the large size of press needed for this seemingly minor operation even with the application of a

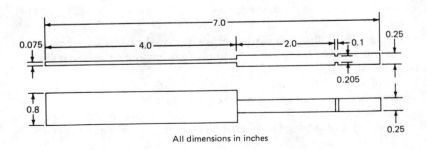

All dimensions in inches

lubricant. Die pressures and forces could be reduced by flattening in two steps, with an intermediate anneal.

Forging an Overhanging Workpiece   A very different situation exists when the platen is narrow (and is usually called an *anvil*). Since the forged part now overhangs the anvil, we cannot expect the entire mass of the workpiece to be deformed, and deformation can become inhomogeneous even within the work zone. To judge the degree of inhomogeneity, we must return to Fig. 4-15. When the workpiece is wide, deformation is again in plane strain (Fig. 4-12b). Major flow occurs in the direction of the short dimension of the anvil; hence, this now becomes L for purposes of analysis. Three distinct possibilities exist:

1 When $h/L > 8.7$ (Fig. 4-15b), the situation is the same as in indenting a semiinfinite body (Fig. 4-15a). It can be shown that the pressure required for indentation $p_{i\,max}$ is approximately three times the uniaxial flow stress σ_f of the material

$$p_{i\,max} = \sigma_f Q_{i\,max} = 3\sigma_f \qquad (4\text{-}22)$$

It will be recognized that while physically the situation shown in Fig. 4-15a appears very different from a hardness test (Fig. 2-7), the strain state is actually very similar. In hardness testing the specimen is, for all intents and purposes, infinite in the width, length, and thickness directions; thus, the indentor has to push the material out, as in Fig. 4-15a. Therefore, the *indentation hardness* of a material is approximately three times its uniaxial (compressive) flow strength. Since the highly localized deformation causes rather severe strain hardening, the indentation hardness is three times the mean flow stress σ_{fm} prevailing in the shear zones and, for reasons mentioned in Sec. 4-3-2, the TS is a good approximation of this mean value. It is for this reason that the indentation hardness is often taken as $3 \times$ TS (remember that hardness is quoted in kg/mm^2). For a strain-hardening material, better agreement is obtained when hardness is taken as three times the flow stress at 7% cold work.

2 When $8.7 > h/L > 1$, the two deformation zones gradually interact, requiring less and less force to maintain plastic deformation (Fig. 4-15b). Therefore, the pressure-multiplying factor also diminishes, and can be taken from Fig. 4-25. The indentation pressure is

$$p_i = 1.15\sigma_f Q_i \qquad (4\text{-}23)$$

It should be remembered that penetration of the two wedges sets up secondary tensile stresses which, at $h/L > 2$, can lead to internal fracture (*centerburst*) in a less ductile material.

3 At a ratio of $h/L = 1$ the two deformation zones fully cooperate (Fig. 4-15c) and the material flows at a minimum pressure (at $1.15\sigma_f$).

4 When $h/L < 1$ (or, more conveniently, $L/h > 1$), friction becomes significant and the pressure-multiplying factor must be obtained according to Eq. (4-19)

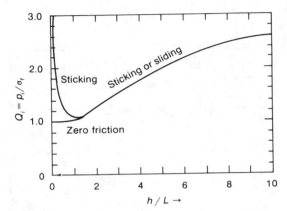

FIGURE 4-25
Pressures needed to indent a workpiece increase with h/L but are independent of friction. (*After R. Hill, The Mathematical Theory of Plasticity, Clarendon Press, Oxford, 1950. With permission.*)

or (4-20). The friction hill is chopped on its sides and the pressure-multiplying factor drops when $w/L < 8$; it drops to Q_a when $w/L = 1$.

Example 4-14

The pin in Example 4-13 is pinched at the center. Consider the geometry of the operation. The resemblance to Fig. 4-15b is obvious; $L = 0.1$ in, $w = 0.25$ in.

Step 3: $\epsilon = \ln(0.25/0.205) = 0.2$.
Step 4: $\sigma_f = 190(0.2)^{0.3} = 118$ kpsi.
Step 5: $h/L = 0.205/0.1 = 2$, thus deformation is indeed inhomogeneous. From Fig. 4-25, $Q_i = 1.5$. Thus $p_i = (1.15)(118)(1.5) = 204$ kpsi. Note that, because of the inhomogeneity of deformation, there is uncertainty about the proper value of strain and flow stress. However, the error is usually within practically permissible limits.

4-4-3 Open-Die Forging

In addition to upsetting and indentation, open-die forging employs various other processes, all of which can be analyzed by analogy to the processes discussed in Secs. 4-4-1 and 4-4-2. A great variety of shapes can be produced with relatively simple dies, although often through a complex sequence of deformation steps: The simplicity of tooling is gained at the expense of the complexity of process control.

Cogging The surface area of a rectangular workpiece can be very large, resulting in an impracticably high total force; therefore, it is customary to deform only one part of a large workpiece at a time. Properly sequenced individual *bites* gradually reduce the height of the entire length of the workpiece by the process of *cogging* or *drawing out* (Fig. 4-26). Successive bites must be spaced close enough to produce an even surface, but too short a bite ($b < h_0/3$) will just fold the

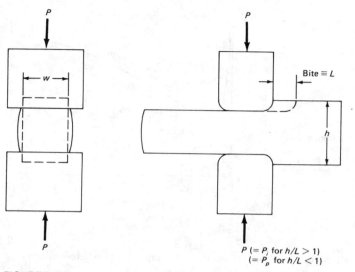

FIGURE 4-26
Bars may be reduced in height by a sequence of strokes in the process of cogging or drawing out.

material down instead of deforming the entire cross section. The h_0/w ratio should be less than 1.5; otherwise, tilting and barreling may occur, just as it does in the axial upsetting of a cylinder.

Drawing out is sometimes used as a substitute for rolling when the quantities to be deformed are small or the material is prone to hot cracking. Parts of substantial degree of complexity can be forged by a planned sequence of open-die forging steps. The part is often held in mechanical arms (*manipulators*), the motions of which must be closely coordinated with anvil movements; hence, computer control has rapidly spread in the industry.

The calculation of stresses and forces follows the principles described in Sec. 4-4-2. The contact length L is again measured in the direction of major material flow, and is thus equal to the bite (Fig. 4-26). To obtain an appropriate pressure-multiplying factor, the h/L ratio must be found. When its value is greater than unity, inhomogeneous deformation prevails and the interface pressure is found from Fig. 4-25; when its value is below unity, friction predominates and Fig. 4-24 (or Eq. (4-19)) should be used. Plane-strain conditions are approximated only when $w/L > 10$. For narrower pieces the multiplying factor is smaller.

Fullering and Edging Many parts have thick and thin sections and it is then necessary to redistribute material. Forging between flat anvils is inefficient because some material moves in the width direction (*spreads*) and, when L/h is large, pressures and forces are high. Forging with *inclined surfaces* solves these

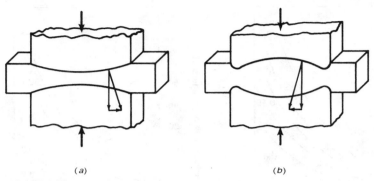

(a)   (b)

FIGURE 4-27
Material may be moved (a) away or (b) toward the center by inclined die surfaces, in the processes of fullering and edging, respectively.

problems because there is a pressure component acting in the direction of material flow (Fig. 4-27). This has two effects. First, it counteracts frictional retardation (when $\tan \alpha = \mu$, the effect of friction is neutralized) and thus lowers the die pressure. Second, it moves the material perpendicular to the direction of load application. The effect can be exploited to move material away from the center (*fullering*, Fig. 4-27a) or toward the center (*edging*, Fig. 4-27b). Repeated strokes, with the workpiece rotated around its axis between strokes, allow substantial material redistribution.

Ring Upsetting When a ring is compressed between flat platens with zero friction, it expands as though it would be a solid cylinder. Friction resists expansion, hence the hole expands less and, at higher friction, it becomes actually smaller. Therefore, the *ring-compression test* has become a favorite method of lubricant evaluation. Rings of OD:ID:height = 6:3:2 ratio are normally used. Lesser contraction of the ID indicates a better lubricant for upsetting operations. Approximate values of μ and m^* may be obtained from the curves of Fig. 4-28.

Example 4-15

Aluminum alloy rings of 30.0-mm OD, 15.0-mm ID, and 10.0-mm height (6:3:2 ratio) were compressed at a press speed of 50 mm/s, some with stearic acid (a solid at room temperature; may be deposited from an organic solvent or melted above 60 °C), others with mineral spirits (a paint thinner, sometimes used as a very light lubricant). Rings were reduced to a height of 5 mm; the internal diameter was 15.5 mm with stearic acid and 10.5 mm with the mineral spirits.

   The diameter of 15.5 mm corresponds to $(15-15.5)/15 = -3\%$ *decrease* in internal diameter. In Fig. 4-28a, a horizontal line is drawn at this value; a vertical line is drawn at $(10 - 5)/10 = 50\%$ reduction in height; the two lines intersect at $\mu = 0.05$ for stearic acid. Repeating for $(15 - 10.5)/15 = 30\%$ *decrease* in internal diameter, $\mu = 0.2$ for mineral spirits.

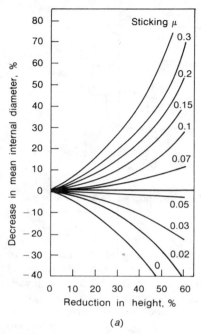

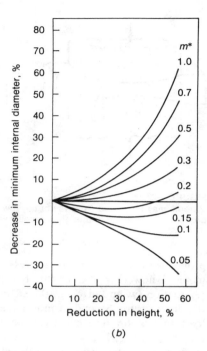

(a)   (b)

FIGURE 4-28
In ring upsetting (a) the coefficient of friction may be found from experimental calibration curves and (b) the interface shear factor may be derived from theory. ((a) After A. T. Male and M. G. Cockroft, J. Inst. Metals **93**:38– 46 (1964– 1965); With permission of The Metals Society. (b) after G. D. Lahoti, V. Nagpal, and T. Altan, Trans. ASME, J. Eng. Ind. **100**:413– 420 (1978). With permission of American Society of Mechanical Engineers.)

Piercing   Impressions or holes are made in a workpiece by piercing. Several variants of the process are used.

1 When *piercing in a container*, the workpiece is supported at its base and around its sides (Fig. 4-29a). Therefore, the workpiece behaves as a semiinfinite body, and the punch pressure is at least $3\sigma_f$ (Eq. (4-22)). When the punch penetrates to significant depths in a strain-hardening material, pressure rises to $4\sigma_{fm}-5\sigma_{fm}$. The material displaced by the punch flows back in a direction opposite to punch movement. Friction on the punch and container surfaces should be minimized, otherwise the piercing pressure will rise further.

2 When the workpiece is *unconstrained* (Fig. 4-29b), the deformation pattern depends on the ratio of workpiece diameter d_0 to punch diameter D_p. When $d_0/D_p > 3$, the workpiece behaves as a semiinfinite body and Eq. (4-22) applies. At lower d_0/D_p ratios (Fig. 4-29b) complex deformation takes place and pressures drop roughly linearly to reach the value of the uniaxial flow stress at

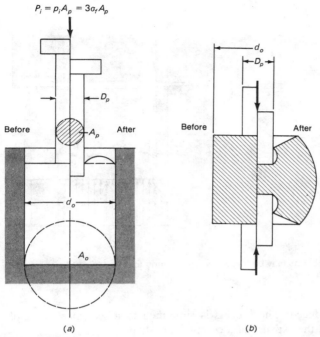

$$P_i = p_i A_p = 3\sigma_f A_p$$

FIGURE 4-29
Billets may be pierced (*a*) in containers or (*b*) with two opposing punches.

$$d_0/D_p = 1$$

$$p_{\text{pierce}} = \sigma_f \frac{d_0}{D_p} \qquad (4\text{-}24)$$

A cylindrical workpiece can be pierced with two punches from opposing ends to prepare a through-hole; the remaining web is removed in a separate operation.

The most frequent application of piercing is to the indentation of the heads of screws and bolts. Since this is done mostly cold and in a container, pressures on the indenting tool can become excessive. Another limitation is imposed by cracking, resulting from either secondary tensile stresses set up by the expansion of an unrestrained head, or from exhaustion of ductility in the prior heading operation.

4-4-4 Impression-Die Forging

More complex shapes cannot be formed with great accuracy by open-die forging techniques. Specially prepared dies are required that contain the negative shape of

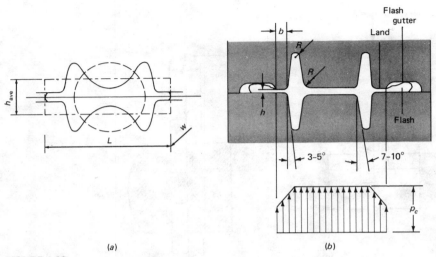

(a)   (b)

FIGURE 4-30
Parts with thin webs and tall ribs may be forged by (*a*) blocking followed by (*b*) finishing in a die provided with a flash gutter.

the forging to be produced: The process is simplified to a sequence of simple compression strokes at the expense of a complex die shape.

In one variant of the process (Fig. 4-30) the shape is obtained by filling out the die cavity defined by the upper and lower die halves. Excess material is allowed to escape into the *flash*; since the die is not fully closed, it is properly called an *impression die*. The term *closed-die* is, nevertheless, often applied, and the term *drop forging* is sometimes used to denote forging conducted upon a hammer; however, this distinction has no particular technical merit.

Material Flow The first concern is that the material must completely fill the die without defects of material flow, such as could occur when parts of the workpiece material are pinched, folded down, or sheared through. Therefore, the shape of the component must be redesigned to promote smooth material flow.

1 *Machining allowances* are applied to all surfaces that will be subsequently machined for better surface finish or closer dimensional tolerances.

2 A *parting line* is chosen with proper consideration of the fiber structure of the finished forging. Fibers (Sec. 4-1-5) should follow the contour of the forging as far as possible (Fig. 4-31), because this ensures greatest toughness, fatigue strength, and ductility. At the parting line the fibers are unavoidably cut through when the flash is trimmed; therefore, the parting line is best placed where minimum stresses arise in the service of the forging (Fig. 4-32).

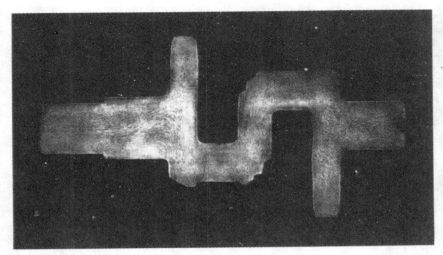

FIGURE 4-31
Grain flow (flow lines) in a forged steel workpiece can be revealed by macroetching. (*Courtesy Forging Industry Association, Cleveland, Ohio.*)

3 *Corners* in the die must be given appropriate radii to facilitate smooth material flow, and *fillets* must be radiused (Fig. 4-32*b*) to prevent stress concentrations that would reduce die life.

4 The cavity walls are given sufficient draft to allow removal of the forging from the die cavity. In nonisothermal forging the *internal draft* is greater than the *external draft* (Fig. 4-30) because the forging would shrink onto bosses of the die prior to its removal from the die.

5 Die pressures increase with increasing d/h or L/h ratios. Thus, the thickness of *webs* (Fig. 4-32*c*) may have to be increased to keep die pressures within

FIGURE 4-32
Parting lines and draft angles must be chosen to give sound material flow and allow removal of the forging from the die.

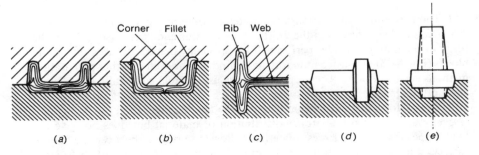

(a)	(b)	(c)	(d)	(e)

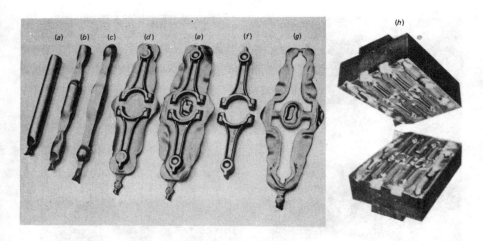

FIGURE 4-33

Hammer forging two connecting rods: (*a*) bar stock; after (*b*) fullering, (*c*) "rolling," (*d*) blocking, (*e*) finishing, (*f*) trimming; (*g*) the flash; and (*h*) the forging dies. (*Courtesy Forging Industry Association, Cleveland, Ohio.*)

acceptable limits, or lubrication has to be improved, or cooling in the die minimized.

6 Complex shapes, zero draft, and undercuts are possible when the die is constructed of several movable pieces (*segmented dies*).

The art of *die design* aims at determining the minimum number of steps that lead from the starting material (usually a round or rectangular bar) to the finished shape.

Die Design A complex shape cannot be filled without defects (and excessive die wear) simply by forging the starting bar into the finished die cavity. Various intermediate steps become necessary:

1 The first aim is to distribute the material correctly, so that little change in cross-sectional area occurs in the finishing die. To this end, *free-forging* (open-die forging) operations may be performed, on specially shaped surfaces in the die blocks (Fig. 4-33), or on separate forging equipment, or even by other preparation methods such as rolling. These operations are usually related to fullering, edging, and upsetting.

2 The preform may be brought closer to the final configuration by forging in a *blocker die*, which ensures proper distribution of material but does not give the final shape (Fig. 4-30*a*). Excess material is allowed to run out between the flat die surfaces and this *flash* is sometimes removed (*trimmed*) prior to further forging.

3 Final shape is imparted in the *finishing die*. The excess material is again allowed to escape into a flash, which must now be thin to aid die filling and produce close tolerances. As a general rule, flash thickness $h = 0.015(A)^{1/2}$ (mm), where A is the projected area of the forging (mm^2). A thin flash running out between parallel die surfaces would lead to very large L/h ratios and thus to high die pressures (Fig. 4-24). Therefore, L is reduced by cutting a *flash gutter* (Fig. 4-30b); this allows free flow of the flash and limits the minimum flash thickness to only a small width, the *flash land* (generally, the land is $3h$ to $5h$ wide). The flash is trimmed either hot or cold, in a separate die resembling a blanking die (Sec. 5-2).

A forging die designer will be able to judge die filling from long experience. However, modeling can be of great help. Plasticine specimens can be deformed in transparent plastic dies; this is a powerful means of physical modeling and, if properly interpreted, the results are relevant. Computer programs based on numerical techniques allow mathematical modeling. The expense involved in die design can be minimized by the adoption of group technology.

Die Pressures and Forces There is no simple yet satisfactory method of calculating impression-die pressures and forces, partly because the strain rate varies tremendously in various parts of the workpiece. A very approximate estimate may be obtained by analogy to forging simple shapes, by dividing the forging into parts (cylinders, slabs, etc.) that can be separately analyzed. Alternatively, the entire forging is considered as a simplified shape (Fig. 4-30):

Step 1: Calculate the *average height* from the volume V and the total projected area A_t of the workpiece (complete with the flash-land area)

$$h_{ave} = \frac{V}{A_t} = \frac{V}{Lw} \tag{4-25}$$

Step 3: The *average strain* is found from

$$\varepsilon_{ave} = \ln \frac{h_0}{h_{ave}} \tag{4-26}$$

and the *average strain rate* is

$$\dot{\varepsilon}_{ave} = \frac{v}{h_{ave}} \tag{4-27}$$

Step 4: The relevant flow stress may now be found for cold (Eq. (4-4)) or hot (Eq. (4-7)) working.

Step 5: *Average die pressure* is found by multiplying the flow stress by a factor Q_c that allows for shape complexity. Its value is taken from Table 4-5. For squat forgings, a cross-check should always be made against Eqs. (4-17) and (4-20). (As

TABLE 4-5
MULTIPLYING FACTORS FOR ESTIMATING FORCES (Q_c)
AND ENERGY REQUIREMENTS (Q_{fe}) IN IMPRESSION-DIE FORGING

Forging shape	Q_c	Q_{fe}
Simple, no flash	3–5	2.0–2.5
With flash	5–8	3
Complex (tall ribs, thin webs), with flash	8–12	4

a rule of thumb, die pressures are usually kept to 350 MPa (25 ton/in^2) in forging aluminum alloys and to below 700 MPa (50 ton/in^2) in forging steels).

Step 6: The required forging force is

$$P = \sigma_f Q_c A_t \qquad (4\text{-}28)$$

Step 7: The energy requirement may be estimated with the aid of a multiplying factor Q_{fe} from Table 4-5

$$E = \sigma_f Q_{fe} V \epsilon_{\text{ave}} \qquad (4\text{-}29)$$

More sophisticated calculations require larger computational effort because the optimum die configuration can be determined only by iteration. Local die pressures must be calculated, and the shape must be changed and calculations repeated if die pressures are found to be too high. Programmable calculators simplify the task. Computer programs are available that perform these calculations as well as the modeling of material flow.

One of the most difficult tasks is to determine the relevant flow stress, especially at high strain rates prevailing in hammers. It is found that, because of reduced cooling, hammer forces are only some 25% higher than press-forging forces.

Elastic deflections of the die can reach a significant proportion of tolerance limits in forgings such as the airfoil section of fan or turbine blades. To forge close-tolerance parts, the pressure distribution is calculated and the die cavity is designed to compensate for elastic deflections of the die.

Example 4-16

A small connecting rod is forged of AISI 1020 steel at 2200 °F on a mechanical press which travels at 600 in/min when the die contacts the workpiece. The volume of the connecting rod is calculated at 1.75 in^2, and 20% of the starting material is expected to go into flash. In the finishing die the projected area is 5.4 in^2 exclusive of the area of the flash land. The flash-land width is 0.3 in all around the 12-in circumference, adding 12(0.3) = 3.6 in^2 to the projected area. Thus, A_t = (5.4) + (3.6) = 9 in^2.

Step 1: h_{ave} = (1.75)/(0.8)(9) = 0.24 in.
Step 3: $\dot{\epsilon}_{\text{ave}}$ = 600/(60)(0.24) = 42 s^{-1}.
Step 4: In Table 4-2, data are available only for 1015 steel. However, in the austenitic temperature range the carbon content makes little difference (compare 1015 and 1045 steels). For

2200 °F = 1200 °C, the value of $C = 50$ MPa $= 7$ kpsi; $m = 0.17$. Hence $\sigma_f = 7(42)^{0.17} = 13.2$ kpsi. (Note that in Tables 4-2 and 4-3 C and m are given for $\epsilon = 0.5$; thus, they represent appropriate mean values for a forging such as this.)

Step 5: The ribs and webs were approximately 0.12-in thick, making the part intermediate in complexity (some experience is needed for this judgement), and $Q_c = 8$.

Step 6: $P = 13.2(8)(9) = 950$ klb $= 475$ tons. (The data for this example were taken from the course material on *Basic Principles of Forging Die Design* of the Forging Industry Association, Cleveland, Ohio. The forging force was actually measured at Battelle Columbus Laboratories and was found to be 430 tons.)

Forging Practices The hot-forging sequence shown in Fig. 4-33 is typical of hammer forging. Short contact times and repeated blows in the same cavity allow the forging of parts with thin ribs and webs and intricate details. In press forging the workpiece enters each cavity only once; more preforming cavities may be required and die design is more critical. A typical sequence of hot forging on a special-purpose press (*horizontal upsetter*) is shown in Fig. 4-34.

Cooling—and thus forces—in hot forging are reduced with heated dies. In isothermal forging (with the die at the workpiece temperature) very slow forging speeds are permissible and complex, thin-walled parts can be forged at low

FIGURE 4-34
A typical hot-upsetting sequence showing the development of the workpiece from the bar, and the associated tooling. (*Courtesy National Machinery Co., Tiffin, Ohio*).

Gripper dies   Heading tools (punches)

FIGURE 4-35

A typical cold-upsetting sequence showing the development of the part and the transfer of the part between forming stages. (*Courtesy National Machinery Co., Tiffin, Ohio*).

pressures, into shapes that have very small or zero draft and require little or no machining (*near-net-shape* and *net-shape forging*). The low forging temperature of aluminum alloys allows isothermal forging in steel dies. Titanium alloys require superalloy or TZM dies; superplastic superalloys may be formed in TZM dies. To prevent oxidation of the molybdenum-alloy dies, special presses with evacuated work spaces are built. Lubrication is vital in isothermal and most nonisothermal forging.

Die pressures are high in cold forging and deformation is usually distributed to several cavities (Fig. 4-35). Lubrication is crucial for success, partly to reduce die pressures and partly to prevent die pickup (adhesion) and subsequent scoring of workpieces.

4-4-5 Closed-Die Forging

In true *closed-die forging* the workpiece is completely trapped in the die and no flash is generated. Economy of forging is thus increased, but die design and process variables must be very carefully controlled. At the end of the stroke the cavity is completely filled with an incompressible solid, and die pressures rise very steeply; this becomes a critical factor in setting up the equipment (Sec. 4-6). Forces are calculated as in impression-die forging.

A special case of closed-die forging is *coining*, in which a three-dimensional surface detail is imparted to a preform. The largest application is, of course, to the minting of coins, but coining is useful for improving the dimensional accuracy, surface finish, or detail of other parts too. The forging pressure is at least $p_i = 3\sigma_f$ but filling of fine details calls for pressures of $5\sigma_f$ or even $6\sigma_f$.

4-4-6 Forge Rolling and Rotary Swaging

These are two of the more specialized forging processes.

Forge rolling performs an impression-die forging operation, but this time the die-half contours are machined into the surfaces of two rolls. Reciprocating roll motion is suitable for the rolling of short pieces while unidirectional rotation is used in high-production lines. Forge rolling often replaces open-die forging for preforming but is also suited for finishing more or less flat forgings such as cutlery and scissors.

A special form of hammers is the *rotary swager*. The workpiece is usually stationary, while the hammer itself rotates. The construction resembles that of a roller bearing (Fig. 4-36a): the anvils are free to move in a slot of the rotating shaft and are thus hurled against the rollers, which in turn knock them back. A rapid sequence of blows is obtained and the workpiece, fed axially, is reduced in diameter by a drawing-out process. While, strictly speaking, swaging should be regarded (and sometimes is used) as an open-die forging process, it is capable of producing exceptionally smooth surfaces to close tolerances. The process can be

FIGURE 4-36
A rotary swager reduces (*a*) solid or (*b*) hollow workpieces with rapid blows.

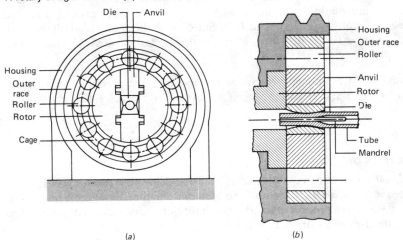

(a)   (b)

employed for pointing, assembling a bar and collar, or shaping the internal contour of a tube on a mandrel (Fig. 4-36b).

In the impression-die forging of shapes with large d/h ratios, high die pressures are reduced by replacing the top die with an *orbiting tool* that makes contact over only part of the surface. Several proprietary designs exploit this principle.

4-4-7 Process Limitations

Impression- and closed-die forging are extremely versatile but are still subject to a number of limitations.

1 Die pressures may become excessive even with the best die design, lubrication, and minimum cooling. The only solution is then to increase wall thicknesses to reduce the effective d/h or L/h ratio. There are a number of handbooks that give minimum rib and web thicknesses attainable in commercial forging; some general guidelines are given in Fig. 12-4.

2 The shape of the part must allow withdrawal of the die. Greater shape complexity is permissible if the die is made in more than two parts. Thus, a *horizontal upsetter* has, in addition to the main ram (similar to the moving crosshead of a press), an auxiliary movement that closes a split die (Fig. 4-34). Thus, shapes that are undercut relative to the ram movement can be forged. Some presses have three or four rams, so that parts such as valve bodies can be forged.

3 Deformation must be as homogeneous as possible to prevent the generation of internal defects. This condition is usually satisfied in impression-die forging but high h/L ratios may develop in swaging; then the center may open up (centerburst defect).

4 Inhomogeneous deformation is intentionally induced in *shot peening*. Many overlapping indentations are made with high-velocity shot, causing localized compressive deformation of the surface. Since the bulk of the workpiece is not affected, compressive residual stresses are set up and fatigue life is increased (Fig. 2-10).

Manufactured components of complex shape are often produced by a combination of forging and extrusion; therefore, extrusion processes will be discussed next.

4-5 EXTRUSION

In *extrusion* the workpiece is pushed against the deforming die while it is being supported in a *container* against uncontrolled deformation. Since the workpiece is in compression, the process offers the possibility of heavy deformations coupled with a wide choice of extruded cross sections.

4-5-1 The Extrusion Process

To initiate extrusion, a cylindrical billet is loaded inside a container and is pushed against a die held in place by a firm support. The press force is applied to the punch and, after the billet has upset to fill ou the container, the product emerges through the die (Fig. 4-37). Initially, deformation is non-steady-state but once the product has emerged, steady-state conditions prevail until close to the end of extrusion when continuous material flow is again disturbed. Two basically different processes are possible:

1 In *direct* or *forward extrusion* the product emerges in the same direction as the movement of the punch (Fig. 4-37a). The really important point is that, for extrusion to take place, the *billet must be moved* against frictional resistance on the container wall.

2 In *indirect* (*reverse* or *back*) *extrusion* the product travels against the movement of the punch (Fig. 4-37c). Most importantly, the billet is at rest in the container; thus, *container friction plays no role*. By definition, piercing in a container (Fig. 4-29a) may be regarded as a case of back extrusion.

Further distinctions may be made according to whether a lubricant is used.

1 The material always seeks a flow pattern that results in minimum energy expenditure. When extrusion is carried out without a lubricant and with a die of flat face (180° die opening), the material cannot follow the very sharp directional changes that would be imposed on it; instead, the corner between the die face and

FIGURE 4-37
Extrusion processes: (a) forward or direct without lubrication (and the associated extrusion-pressure/stroke curve); (b) forward with full lubrication; (c) reverse or indirect or back; (d) reverse can (impact); (e) hydrostatic extrusion.

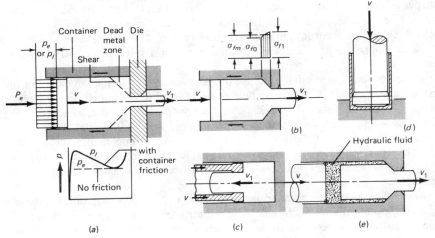

container is filled out by a stationary *dead-metal zone*, and material flow takes place by shearing along the surface of this zone (*unlubricated extrusion*, Fig. 4-37*a*). Thus, the extruded product acquires a completely freshly formed surface.

2 Alternatively, a very effective lubricant is applied to ensure complete sliding on the die face and along the container wall (*lubricated extrusion*). Accordingly, the die is now provided with a conical entrance zone that, ideally, corresponds in shape to the flow pattern of minimum energy (Fig. 4-37*b*).

The movement of the punch must be stopped before the conical die entry is touched or, in unlubricated extrusion, before material from the dead-metal zone is moved, since this would create internal defects. Two basic methods of operation are possible:

1 When the purpose of extrusion is to produce a long bar or tube of uniform cross section (extrusion of semifabricated products), the remnant (*butt*) in the container is scrap which is removed by taking it out with the die. After the butt is cut off, the extrusion can be extricated from the die, and the die is returned for inspection, conditioning, and reuse.

2 When the purpose of extrusion is to produce finished components, with the butt forming an integral head of the component, the extrusion is ejected by pushing it back through the extrusion die and lifting it out from the container. Since ejector actuation can be mechanically synchronized with the punch movement, high production rates are achieved, provided, of course, that the extruded stem is strong enough to take the ejection force.

4-5-2 Hot Extrusion

While hot deformation is often typical of primary processes, the hot extrusion of shapes offers such a wide scope for custom design that this process can justifiably be regarded as a secondary manufacturing technique. Shapes are usually classified into three groups according to their complexity (Fig. 4-38):

Solid shapes are produced by extruding through a suitably shaped stationary die.

Hollow products necessitate the use of a die insert that forms the cavity in the extruded product. This insert may be a *mandrel* fixed to the punch (Fig. 4-39*a*) or moving inside the punch (Fig. 4-39*b*), or a *bridge* (*spider*) section attached to the die (Fig. 4-39*c*). The last method is permissible only if material flow can be divided and then reunited prior to leaving the die, with complete pressure-welding of the separated streams. This is practicable only in the unlubricated hot extrusion of aluminum and lead; even a trace of lubricant would prevent rewelding.

Semihollow products appear to be solid sections, but their shape makes the use of a single-piece die impracticable. The die tongue forming the internal shape is connected to the external contour by such a small cross section that it would

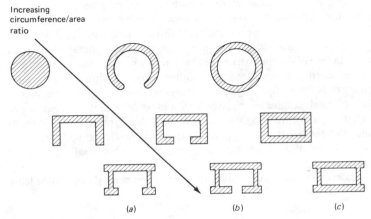

FIGURE 4-38
In the extrusion of (a) solid, (b) semihollow, and (c) hollow configurations, process difficulty increases with increasing circumference-to-area ratio.

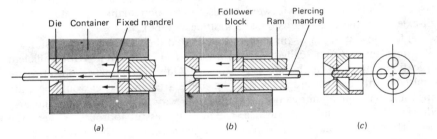

FIGURE 4-39
Hollow products may be extruded with (a) fixed or (b) piercing mandrels or with (c) bridge- or spider-type dies. (*After J. A. Schey, in Techniques of Metals Research, R. F. Bunshah (ed.), vol. 1, pt 3, Interscience, 1968, p. 1494. With permission.*)

break off; therefore, techniques similar to the extrusion of hollow sections must be used.

In designing complex sections it should be kept in mind that the circumscribed circle must be less than the container diameter, and that wall thicknesses should be kept as uniform as possible in order to equalize material flow. Sections with highly varying cross sections are, nevertheless, produced. The rate of extrusion is equalized by retarding flow in the thicker sections, for example, by developing longer frictional surfaces (*bearings*) in the die. Increasing complexity is often expressed as a perimeter/weight or perimeter/cross-sectional area *shape factor*; the higher its value, the more skill required to produce the part. Computer-aided die design is possible, often with expert programs that capture the knowledge of experienced die designers. Location of die opening(s), bearing lengths, and die deflections are readily obtained and die-tryout time and costs are minimized.

Aluminum alloys are extruded isothermally, without a lubricant, and with flat dies made of hot-working die steels. The presence of a dead-metal zone gives all-new, bright surfaces. Wall thicknesses below 1 mm are possible. Copper and brass are extruded mostly unlubricated, nonisothermally. Cooling on the colder container and die limits the complexity and thinness of shapes. This is true also of the hot extrusion of steel, conducted mostly with a glass lubricant that envelopes the billet and melts in a controlled manner to form a die approach of optimum shape; sometimes shorter lengths and thinner sections are produced with graphitic lubricants. The dies are often coated with a ceramic (e.g., partially stabilized zirconia) for protection.

The starting material is often a cast billet; the extrusion ratio should be at least 4:1 to ensure adequate working, but it may rise to 400:1 in the softer alloys.

4-5-3 Cold Extrusion

The purpose of cold extrusion is mostly that of producing a finished part. In most instances, the residue (butt) in the container becomes an integral part of the finished product (e.g., in the forward extrusion of a bolt shank or an automobile half-axle, or in the back extrusion of a toothpaste tube).

The low flow strength of tin and lead facilitated their early cold extrusion for collapsible tubes (often called *impact extrusion*, Fig. 4-37d). With sufficient lubrication aluminum can be similarly extruded. Only smaller extrusion ratios are permissible with copper and brass, and the cold extrusion of steel would be quite impossible without a lubricant that withstands very high pressures while also following the extension of the surface. The most successful approach converts the steel surface into a zinc–iron phosphate (*phosphate coating*); this porous surface, integrally joined to the metal surface, is then impregnated with a suitable lubricant, usually a soap (Table 4-4). Steels of higher carbon content can be extruded after a spheroidizing anneal. Strain hardening offers a valuable increase in strength and, if a workpiece is to be strain hardened uniformly, the butt may be subsequently upset. Cold extrusion has made great inroads in the automotive and general equipment industries, for parts previously made by machining.

Example 4-17

A component previously made by machining is now to be made by plastic deformation. It resembles a bolt, with a 10-mm-diameter, 100-mm-long body and a 20-mm-diameter, 20-mm-high head. It is desired to retain the benefits of cold working, and the part is to have equal strain hardening in the head and body sections. What process should be used?

Equal strain hardening can be obtained only if the head is upset and the body is extruded (or otherwise reduced). A bar of intermediate diameter d_0 must be chosen so that the upsetting strain $\ln(A_{\text{head}}/A_0)$ is equal to the extrusion strain $\ln(A_0/A_{\text{body}})$.

$$A_{\text{head}} = 20^2 \pi/4 = 314 \text{ mm}^2; \qquad A_{\text{body}} = 10^2 \pi/4 = 78.8 \text{ mm}^2$$

for equal strain, $\ln(314/A_0) = \ln(A_0/78.5)$

$$A_0^2 = (314)(78.5) = 24\,650$$

$$A_0 = 157 \text{ mm}^2$$

Initial diameter, from Eq. (4-12b), $d_0 = [4(157)/\pi]^{1/2} = 14.14$ mm.

4-5-4 Hydrostatic Extrusion

In an extensively investigated variant of the process, the billet is extruded by pressurizing a liquid medium inside a closed container (*hydrostatic extrusion*, Fig. 4-37e). This helps to reduce friction on the container wall but does not fundamentally change the stress state inside the deforming workpiece; reduced die friction can even increase the tendency to internal crack formation. However, the absence of container friction permits extrusion of very long billets or even wires, and large reductions can be taken. The process has reached practical application in special cases, including the cold extrusion of copper tubes and the extrusion of composite copper–aluminum billets to produce copper-clad conductor wires and bars.

4-5-5 Extrusion Forces

The routine of calculations is similar to that followed in forging, but there are some significant differences in detail.

Step 1: The dimensions of interest are the cross-sectional areas of the billet A_0 and extrusion A_1. The diameters d_0 and d_1 will also be needed. When the extruded section is not a round bar, an *equivalent diameter* can be calculated from the area A_1 (Eq. (4-12b)).

Step 2: The engineering strain may be calculated as a reduction of area

$$e_e = \frac{A_0 - A_1}{A_0} \tag{4-30}$$

however, at large reductions, a better feel is obtained from the *extrusion ratio*

$$R_e = \frac{A_0}{A_1} \tag{4-31}$$

Step 3: The strain is simply the natural logarithm of R_e

$$\epsilon = \ln R_e = \ln \frac{A_0}{A_1} \tag{4-32}$$

The strain rate is important in hot working and a *mean strain rate* may be calculated from

$$\dot{\epsilon}_m = \frac{6 v d_0^2 \tan \alpha}{d_0^3 - d_1^3} \epsilon \tag{4-33}$$

where v is the ram velocity. The half angle α is the cone angle of the die entry or, in unlubricated extrusion with a dead-metal zone, it may be taken as 45° (unless experiments show it to be different).

Step 4: In cold working the workpiece material strain hardens during its passage through the die, and a *mean flow stress* σ_{fm} must be obtained as was shown in relation to Fig. 4-13b and Eq. (4-10). In hot working, Eq. (4-7) gives a mean flow stress when the mean strain rate (Eq. (4-33)) is used.

Step 5: Deformation is inhomogeneous, and for approximate calculations the *extrusion pressure* p_e may be found from the following formula

$$p_e = \sigma_{fm}Q_e = \sigma_{fm}(0.8 + 1.2\epsilon) \qquad (4\text{-}34)$$

Step 6: The total extrusion force P_e acting on the billet is

$$P_e = p_e A_0 \qquad (4\text{-}35)$$

A word of warning is in order here. We already observed (Sec. 4-5-1) that back extrusion of a can is similar to piercing in a container (Fig. 4-29a). The extrusion force, Eq. (4-35), is based on the pressure p_e acting over the base area A_0; at low reductions the force may really be given by the piercing force. This is obtained by multiplying the punch area $A_p = A_0 - A_1$ by the punch (indentor) pressure p_i which, as discussed in Sec. 4-4-3 under "Piercing," can never be less than $3\sigma_f$ (Eq. (4-22)) and is more likely $4\sigma_f$–$5\sigma_f$. It is advisable, therefore, to calculate the extrusion force from both Eq. (4-35) and from the punch force P_i

$$P_i = p_i A_p = p_i(A_0 - A_1) \qquad (4\text{-}36)$$

and take the *smaller* of the two values. It does not matter whether the indenting punch is solid as in Fig. 4-29a or hollow as in Fig. 4-37c.

In direct extrusion the billet is pushed forward against the frictional resistance developed on the container wall. Correspondingly, the extrusion pressure is higher at the beginning of the stroke when a long length rubs against the container wall (Fig. 4-37a). At high extrusion ratios interface pressures can be very high and the use of a coefficient of friction value could be misleading (Sec. 4-3-2). Therefore, it is better to estimate the shear strength of the interface τ_i and add the corresponding pressure to the calculated extrusion pressure to obtain the ram (punch) pressure at any point in the stroke

$$p_l = p_e + 4\frac{\tau_i l}{d_0} \qquad (4\text{-}37)$$

where l is the length of the billet at the point in the stroke considered, measured from the end of the stroke. Data for τ_i are scarce but an upper limit is given by sticking when $\tau_i = \tau_f$ or $0.5\sigma_f$. With a truly effective lubricant the pressure will drop toward the basic pressure, Eq. (4-34).

Example 4-18

An 1100 Al can (container) of 50-mm OD and 48-mm ID is to be produced by the back extrusion (Fig. 4-37d) of $d_0 = 50$-mm-diameter annealed slugs. Lanolin is used as a lubricant. Calculate the force during steady-state extrusion.

Noting that the extrusion is a hollow tube:

Step 1: $A_0 = 50^2 \pi/4 = 1960$ mm^2, $A_1 = (50^2 \pi/4) - (48^2 \pi/4) = 154$ mm^2.

Step 2: $R_e = A_0/A_1 = 12.7$.

Step 3: $\epsilon = \ln R_e = 2.54$.

Step 4: We are to calculate forces for the steady-state condition; therefore, σ_{fm} is needed. According to Fig. 4-13b, a σ_f versus ϵ curve can be plotted or, from Eq. (4-10) and Table 4-3 ($K = 140$ MPa; $n = 0.25$) σ_{fm} calculated:

$$\sigma_{fm} = (140/2.54)(2.54)^{1.25}/1.25 = 141 \text{ MPa}.$$

Step 5: From Eq. (4-34)

$$p_e = 141[0.8 + (1.2)(2.54)] = 141(3.85) = 540 \text{ MPa}$$

Step 6: This pressure acts on the base of the container

$$P_e = 1960(10^{-6})(540) = 1.12 \text{ MN } (= 125 \text{ tonf})$$

With the given geometry and lubricant, wall friction can be ignored.
Check the punch force, for indentation of a strain-hardening material:

$$p_i = 141(5) = 705 \text{ MPa}$$

$$p_i = (1960 - 154)(10^{-6})(705) = 1.27 \text{ MN}$$

thus the value calculated for extrusion is lower and will be sufficient to perform the operation.

4-5-6 Process Limitations

Even though the material is kept in overall compression and thus the hydrostatic component of stresses is high, the extrusion process is not free of problems.

1 Deformation tends to be inhomogeneous, especially at extrusion ratios below 4. Inhomogeneity (Sec. 4-3-2) is, in general, a function of the h/L (mean height h over compressed length L) ratio. The h/L ratio in Fig. 4-15 refers to a rectangular slab whereas extrusion is usually conducted with axial symmetry. Nevertheless, the same principles apply (Fig 4-40), except that the mean diameter $(d_0 + d_1)/2$ is now substituted for h.

As before, deformation is inhomogeneous when the h/L ratio is large (Fig. 4-40a); in other words, when the extrusion ratio is small and the die half angle α is large. Deformation is now concentrated in the outer zones which are, therefore, directly elongated. The center of the extrusion is not directly deformed; instead, it is dragged along by the surface material. This generates secondary tensile stresses in the core which may ultimately suffer a characteristic *arrowhead fracture* (also

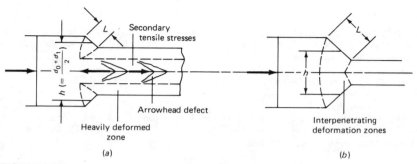

FIGURE 4-40
In the extrusion of materials of limited ductility, (*a*) internal (arrowhead) defects are formed at high h/L ratios but (*b*) a sound product is obtained at low h/L ratios.

described as *centerburst defect*). The danger is greatest at an h/L ratio of 2 and over. The situation can be remedied by lowering the h/L ratio, which implies either a smaller die half angle α or a heavier reduction and thus smaller h and larger L (Fig. 4-40*b*). With a component of fixed geometry, neither of these remedies may be allowable and the only hope is then the use of a more ductile material. Centerburst defects are particularly troublesome when they occur only periodically, affecting the integrity of an unknown number of parts.

2 In very special instances the workpiece material is kept in the compressive stress state throughout the extrusion process, even at critically low extrusion ratios, by extruding the material into a pressurized space, a process usually described as *extrusion against back pressure* (not to be confused with hydrostatic extrusion).

3 In hot extrusion, the heat generated during extrusion may cause the workpiece temperature to rise (Eq. (4-14)) above the solidus temperature of the material. Hot-shortness then leads to the appearance of circumferential surface cracks (*speed cracking*) which can be eliminated by slowing down the press, thus reducing the strain rate and the rate of heat generation (but also losing output).

4 When the extrusion stroke is taken too far, inhomogeneous material flow leads to the generation of a concentric *pipe*.

5 When lubrication breaks down in lubricated extrusion, so that a partial dead-metal zone forms, or some lubricant traces are present in unlubricated extrusion, lubricant trapped at the boundary of the dead-metal zone extrudes into the product to form *subsurface defects*. On subsequent heating, gases cause blistering at these locations.

4-6 FORGING AND EXTRUSION EQUIPMENT

Forging and extrusion are closely related processes. Sometimes they are difficult to distinguish (e.g., piercing in a container versus back extrusion); at other times a distinctly forging-type process is combined with extrusion (e.g., in making a bolt by extruding the shank and then upsetting the head). They also share many types of tooling and equipment.

4-6-1 Tools and Dies

Bulk deformation processes are characterized by high interface pressures coupled with high temperatures in hot working. *Tool and die materials* are selected and manufactured with the greatest care. In general, ductility is sacrificed in cold-working dies but a compromise between hardness and ductility must be struck for hot-working dies that are also exposed to thermal shock (Table 4-6).

In calculating forging and extrusion pressures, the relevant flow stress was σ_f because the workpiece material had to deform. In contrast, interface pressures must be kept low enough not to cause any permanent deformation of the die. Therefore, pressures must not exceed a safe fraction or multiple of the yield strength $\sigma_{0.2}$ of the die material. From the HRC values given in Table 4-6, the tensile strength can be estimated as follows:

HRC	TS, MPa	TS, kpsi
30	960	140
40	1250	185
50	1700	250
60	2400	350

Allowing for some safety, 80% of the above values can be taken as $\sigma_{0.2}$. The allowable stress depends on the relative configurations of the tool and workpiece, and may be calculated now by *regarding the tool as a workpiece*, the deformation of which must be prevented.

1 *Long punch* (Fig. 4-41a). Just as a cylindrical billet will buckle when the h/d ratio is too large, so will a punch. For very long punches, the Euler formula is relevant; for shorter ones—more typical of metalworking—the Johnson formula is suitable

$$p \le \sigma_{0.2}\left[1 - \frac{4\sigma_{0.2}}{\pi^2 E}\left(\frac{L_p}{D_p}\right)^2\right] \qquad (4\text{-}38)$$

where L_p is punch length, D_p is punch diameter, and E is Young's modulus for the punch material (210 GPa for steel, 350 GPa for tungsten carbide).

2 *Short punch* (Fig. 4-41b). This is equivalent to the axial upsetting of a cylinder, and $p = \sigma_{0.2}$. Steel punches are limited to approximately 1200 MPa (180 kpsi) in simple compression. Some cobalt-bonded WC punches operate at pressures up to 3300 MPa (500 kpsi).

3 *Flat platen* (Fig. 4-41c). When a flat die is larger than the workpiece, the workpiece becomes, in effect, a punch. Therefore, by analogy to piercing (Sec.

TABLE 4-6
TYPICAL DIE MATERIALS FOR DEFORMATION PROCESSES*

| Process | Die material† and hardness HRC for working | | | |
	Al, Mg, and Cu alloys		Steels and Ni alloys	
Hot forging	6G	30–40	6G	35–45
	H12	48–50	H12	40–56
Hot extrusion	H12	46–50	H12	43–47
Cold extrusion:				
Die	W1, A2	56–58	A2, D2	58–60
	D2	58–60	WC	
Punch	A2, D2	58–60	A2, M2	64–65
Shape drawing	O1	60–62	M2	62–65
	WC		WC	
Cold rolling	O1	55–65	O1, M2	55–65
Blanking	Zn alloy		As for Al, and	
	W1	62–66	M2	60–66
	O1	57–62	WC	
	A2	57–62		
	D2	58–64		
Deep drawing	W1	60–62	As for Al, and	
	O1	57–62	M2	60–65
	A2	57–62	WC	
	D2	58–64		
Press forming	Epoxy / metal powder		As for Al	
	Zn alloy			
	Mild steel			
	Cast iron			
	O1, A2, D2			

*Compiled from *Metals Handbook*, 9th ed., vol. 3, American Society
for Metals, Metals Park, Ohio, 1980.
†Die materials mentioned first are for lighter duties, shorter runs.
Tool steel compositions, percent (representative members of classes):
      6G (prehardened die steel): 0.5C, 0.8Mn, 0.25Si, 1Cr, 0.45Mo,
        0.1V
      H12 (hot-working die steel): 0.35C, 5Cr, 1.5Mo, 1.5W, 0.4V
      W1 (water-hardening steel): 0.6–1.4C
      O1 (oil-hardening steel): 0.9C, 1Mn, 0.5Cr
      A2 (air-hardening steel): 1C, 5Cr, 1Mo
      D2 (cold-working die steel): 1.5C, 12Cr, 1Mo
      M2 (Mo high-speed steel): 0.85C, 4Cr, 5Mo, 6.25W, 2V
      WC (tungsten carbide)

4-4-3)

when the platen $D > 3d$, $p = 3\sigma_{0.2}$; (Eq. (4-22))
when the platen is smaller, $p = \sigma_{0.2}(D/d)$; (Eq. (4-24)).

4 *Cavity* (Fig. 4-41*d*). This is a much more severe case than the flat die, because the workpiece develops an internal pressure which can burst the die. This is also true of extrusion containers. The design of containers is a specialized

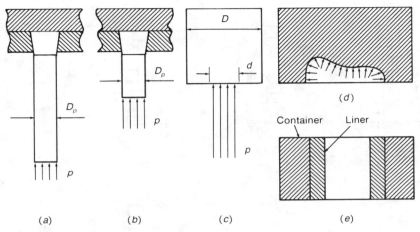

FIGURE 4-41
Tools and dies fail by various mechanisms: (a) long punches by buckling; (b) short punches by upsetting; (c) flat platens by indentation; (d) die or (e) container cavities by internal pressure.

subject; as a very rough guide, $p = \sigma_{0.2}/2$ when $D \geq 3d$, thus a single-piece container made of high-strength die steel can take up to 1000-MPa (150-kpsi) pressure. The inner part of the container (*liner*) may be shrunk (Fig. 4-41e) into a larger outer shrink-ring (*container*) or it may be wrapped with steel band or wire under high tension. Thus, the internal surface of the container is in compression and can stand up to 1700-MPa (250-kpsi) internal pressure. Special constructions permit pressures up to 2700 MPa (400 kpsi).

Dies are finished to a specified surface roughness; this may be a controlled, random roughness for hot working with solid lubricants, and usually a highly polished finish for cold working with liquid or soap-type lubricants. Many dies are now surface-treated for improved wear resistance, by techniques similar to those described for metal-cutting tools (Secs. 8-3-1 and 9-5).

All highly stressed tooling must be surrounded by heavy shielding because a fractured die part becomes a potentially deadly projectile.

Example 4-19

In Example 4-13 we calculated a die pressure of 437 kpsi for flattening a stainless steel pin. Is such a high pressure permissible for a tool-steel die?

The pressure exceeds the yield strength of the best steels. However, if the die is made at least $3(0.8) = 2.4$ in wide, the nonloaded part of the die will give support (Fig. 4-41c) and a tool of HRC 60 will be safe. After forging several thousand parts, an indentation may gradually develop.

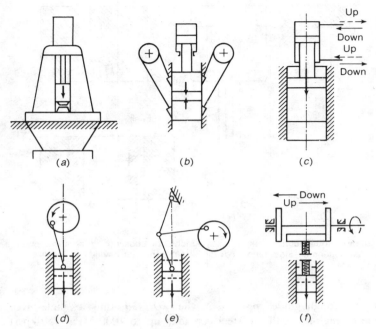

FIGURE 4-42
The deformation force and energy may be delivered by impact devices such as (*a*) hammers or (*b*) counterblow hammers, force-limited devices such as (*c*) hydraulic presses, stroke-limited mechanical presses such as (*d*) crank or (*e*) knuckle-joint presses, or (*f*) screw press. (*After J. A. Schey, as Fig. 4-39, pp. 1474–1477.*)

Example 4-20

In Example 4-14 we found a pressure of 204 kpsi acting on the tool used for pinching the pin. Is this pressure permissible?

If the tool is made in the form of a short punch (or a short extension on a longer but also wider punch), it will be just safe (because it is loaded as a short punch, Fig. 4-41*b*).

4-6-2 Hammers

Hammers are impact devices in which a mass (the *ram*) is accelerated by gravity and/or compressed air, gas, steam, or hydraulic fluid (Fig. 4-42*a*). For a ram mass M and impact velocity v the hammer energy E_h is

$$E_h = \frac{Mv^2}{2} = \frac{Wv^2}{2g} \tag{4-39}$$

where W is the weight of the ram and g is the gravitational acceleration. The

TABLE 4-7 CHARACTERISTICS OF HAMMERS AND PRESSES*

Equipment type	Energy,† kN · m	Ram mass, kg	Force,‡ kN	Speed, m/s	Strokes/ min	Stroke, m	Bed area, m × m	Mechanical efficiency
Hammers								
Mechanical	0.5–40	30–5,000		4–5	350–35	0.1–1.6	0.1 × 0.1 to 0.4 × 0.6	0.2–0.5
Steam and air	20–600	75–17,000 (25,000)		3–8	300–20	0.5–1.2	0.3 × 0.4 to (1.2 × 1.8)	0.05–0.3
Counterblow	5–200 (1250)			3–5	60–7		0.3 × 0.4 to (1.8 × 5)	0.2–0.7
Herf	15–750			8–20	< 2			0.2–0.6
Presses								
Hydraulic, forging			100–80 000 (800 000)	< 0.5	30–5	0.3–1 (3)	0.5 × 0.5 to (3.5 × 8)	0.1–0.6
Hydraulic, sheet m.w.			10–40 000	< 0.5	130–20	0.1–1	0.2 × 0.2 to 2 × 6	0.5–0.7
Hydraulic, extrusion			1000–50 000 (200 000)	< 0.5	< 2	0.8–5	0.06 to 0.6 diam. container	0.5–0.7
Mechanical, forging			10–80 000	< 0.5	130–10	0.1–1	0.2 × 0.2 to 2 × 3	0.2–0.7
Horizontal upsetter			500–30 000 (1–9 in diam.)	< 1	90–15	0.05–0.4	0.2 × 0.2 to 0.8 × 1	0.2–0.7
Mechanical, sheet m.w.			10–20 000	< 1	180–10	0.1–0.8	0.2 × 0.2 to 2 × 6	0.3–0.7
Screw			100–80 000	< 1	35–6	0.2–0.8	0.2 × 0.3 to 0.8 × 1	0.2–0.7

*From a number of sources, chiefly A. Geleji, *Forge Equipment, Rolling Mills and Accessories*, Akademiai Kiado, Budapest, 1967.
†Multiply number in column by 100 to get m · kg, by 0.73 to get 10^3 lbf · ft.
‡Divide number by ~ 10 to get tons. Numbers in parentheses indicate the largest sizes, available in only a few places in the world.

striking velocity v increases with the stroke (drop height) H_d and acceleration ξ

$$v^2 = 2\xi H_d = 2H_d\left(g + \frac{Ap_m}{M}\right) \tag{4-40}$$

where A is the cross-sectional area of the driving piston, p_m is the mean indicated pressure of the pressurized medium, and M is the accelerated mass. Hammers are available in a large range of sizes (Table 4-7) and with increasingly sophisticated controls for metering the energy per blow.

The energy of impact is absorbed mostly by the energy E required for deforming the workpiece (Fig. 4-19 and Eq. (4-29)). Some energy is, however, transmitted to the die, the hammer foundation, the ground, and also the hammer components, setting up shock waves in the ground and air. Ground vibration and noise are objectionable and reduce the efficiency of forging. The total hammer energy E_h to be delivered is

$$E_h = \frac{E}{\eta} \tag{4-41a}$$

where E is the energy required for forging (Fig. 4-19) and the efficiency is*

$$\eta = 0.9\left[1 - \left(\frac{P}{10^3 Mg}\right)^2\right] \qquad (4\text{-}41b)$$

if P is in newtons and M is ram mass in kilograms. Ground shock is avoided in *counterblow hammers* (Fig. 4-42b).

High impact velocities and short contact times minimize cooling; therefore, hammers are used mostly for open-die forging and for impression-die forging of intricate shapes. Except for counterblow and *high-energy-rate forging* (HERF) hammers (counterblow hammers driven by gas pressure), the forging is produced by several blows in any one die cavity; therefore the total energy requirement (Eq. (4-29)) can be delivered by a relatively small hammer. Hammer forging does require, however, considerable operator skill and is less suitable for materials of high strain-rate sensitivity.

4-6-3 Presses

Presses are powered mechanically or hydraulically.

Hydraulic presses (Fig. 4-42c) stall out when their load limit is reached and can be used with dies that make contact (*kiss*) at the end of the stroke. Hydraulic presses are particularly suitable for isothermal forging where very low strain rates are required.

Mechanical presses are of various constructions (two examples are shown in Fig. 4-42d and e). They have a preset stroke and develop an infinite force at the end of the stroke. Therefore, in true closed-die forging the die must allow escape of excess material or the die gap must be set with extreme care. In setting up the press, elastic extension of the frame must be taken into account, as discussed in conjunction with Fig. 2-1. Spring constants of presses seldom exceed 4 MN/mm (12 000 ton/in); thus, a press exposed to a 20-MN (2000-ton) force must be set, in the unloaded condition, 5 mm (0.16 in) closer than the desired final dimension under load. Because of the lower speeds and longer contact times, workpieces must be preformed carefully if complex parts are to be made by hot press forging.

Screw presses (Fig. 4-42f) slow down as the stored energy is exhausted in the blow; hence, they have characteristics between mechanical presses and hammers.

Multiram presses and associated manipulators can be computer-controlled to form the nucleus of a flexible forging system.

As a rule of thumb, a hammer equipped with a 1-tonne ram can do the work of a 1000-ton press, because it delivers the total energy required in several blows. Information on typical equipment is given in Table 4-7.

---

*Private communication, Dr. A. A. Hendrickson, Michigan Technological University.

A special class of presses comprises *horizontal upsetters* for hot and cold working and *cold-headers* for cold working. Both start with straight lengths of bar or wire. For cold-headers the material is fed with indexing pinch rollers, sometimes through a multiroll straightener or even a draw die that delivers bar of tight tolerances. The end of the bar or wire is deformed in successive steps ranging from simple upsetting to the most complex combined forging-extrusion operations. Auxiliary movements are synchronized with the main ram movement and are used to open and close clamping dies, actuate auxiliary punches and shearing dies, and transfer the workpiece from one die cavity to another. The workpiece material is cut off the bar or wire either at the beginning or end of the sequence, and either one workpiece may go through the die sequence at a time or a workpiece may reside in each die during each stroke. An example of a hot-upsetting sequence was given in Fig. 4-34 and one of cold extrusion is shown in Fig. 4-43.

FIGURE 4-43
A typical cold-forging sequence in a seven-station cold former, producing hose connectors by combined forward and back extrusion and forging, at the rate of 60 per minute. (*Courtesy National Machinery Co., Tiffin, Ohio.*)

The construction and mechanization of these machines is often very ingenious and their production rates are difficult to match with other techniques.

Example 4-21

Estimate the size of press or hammer needed for making the part of Example 4-12.

The press size is given by the maximum load; note that this relatively small part requires a 4600-kN (520-ton) press because of the large d/h ratio. Cooling would still further increase the force requirement.

Estimation of the hammer size is more difficult because the speed ranges from a high value at impact to zero at the end of the stroke. However, this particular steel is not very strain-rate sensitive ($m = 0.13$) and the high rate of deformation will actually increase the temperature of the workpiece, so that the calculated energy requirement will not be too far off.

If the striking velocity, Eq. (4-40) is, say, 6 m/s, then the energy available is, from Eq. (4-39)

Ram mass, kg	Energy, N · m
500	9 000
1000	18 000
1500	27 000
2000	36 000
4000	72 000

To make the part in one blow, even a 2000-kg hammer would be just sufficient. Perhaps more economically, a 1000-kg hammer could be used to deliver several blows (see Prob. 4-22).

4-7 DRAWING

Long components of uniform cross section can be produced not only by extrusion but also by *drawing*. Instead of being pushed, the material is now pulled through a stationary die of gradually decreasing cross section. Most wire is of circular cross section, but square, rectangular, and shaped wires (*sections*) are also drawn. In addition to direct applications such as electrical wiring, wire is the starting material for many products including wire-frame structures (ranging from coat hangers to shopping carts), nails, screws and bolts, rivets, wire fencing, etc.

Seamless tubes are made by a variety of hot-working techniques but below a minimum size they must be further reduced cold. One of the options is to draw them, and such cold-drawn tubes perform important functions in hydraulic systems of vehicles, aeroplanes, ships, industrial machinery, water distribution systems, and in such applications as hypodermic needles.

4-7-1 The Drawing Process

The material is deformed in compression, but the deformation force is now supplied by pulling the deformed end of the wire (Fig. 4-44a). Therefore, it is often said that the deformation mode is that of indirect compression.

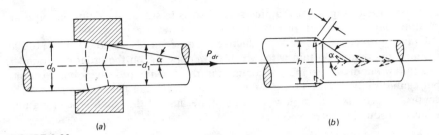

FIGURE 4-44
(*a*) Deformation in wire drawing takes place under the indirect compression developed by a conical die. (*b*) High h/L ratios can lead to centerburst in materials of limited ductility.

The stationary *draw die*, made of tool steel, cemented WC, or diamond, may be replaced with two, three, or four *idling rollers*, all with their axes in a common plane. A *Turk's head* is a tool containing four rollers with adjustable positions.

Seamless tubes are sometimes drawn simply through draw dies, either to reduce their diameter (*sinking*, Fig. 4-45*a*) or to change their shape (say, from round to square). If their wall thickness is to be reduced, an internal die is also needed, which may be of three kinds: a short, conical *plug* held by a long bar

FIGURE 4-45
Seamless tubes are drawn (*a*) by sinking, (*b*) on a plug, (*c*) with a floating plug, or (*d*) on a bar. Half-arrows indicate frictional stresses. (*J. A. Schey, Tribology in Metalworking: Friction, Lubrication and Wear, American Society for Metals, 1983, pp. 353. With permission.*)

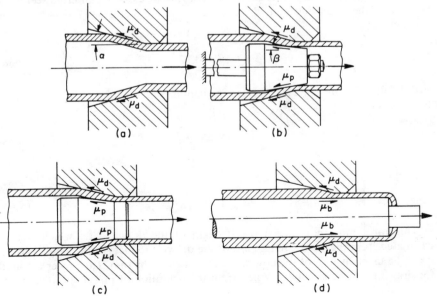

from the far end (Fig. 4-45b); a plug shaped so as to stay in the deformation zone (*floating plug*, Fig. 4-45c); or a full-length *bar* of tool steel (Fig. 4-45d).

Drawing processes are very productive because speeds up to 50 m/s (10 000 ft/min) are possible on thin wire. Much slower speeds, on the order of around 1 m/s (a few hundred ft/min), are common in the drawing of heavier bar. Sections that cannot be bent around a *draw drum* (*bull block*) must be drawn in straight lengths, on *draw benches* at low speeds and, because of the batch-type operation, reduced production rates.

4-7-2 Forces

Most drawing is conducted cold. Initial steps in calculating the draw force follow the routine of extrusion calculations (Sec. 4-5-5). In Step 3, strain is calculated from Eq. (4-32). In Step 4, σ_{fm} (Eq. (4-10)) is again needed because this is a steady-state process.

In Step 5, the effects of die friction and inhomogeneity of deformation must be considered:

$$\sigma_{\text{exit}} = \sigma_{fm} Q_{dr} = \sigma_{fm} (1 + \mu \cot \alpha) \phi \ln \frac{A_0}{A_1} \qquad (4\text{-}42)$$

where μ is the coefficient of friction between workpiece and die, α is the half angle of the draw die (Fig. 4-44a), and ϕ is a factor that takes the inhomogeneity of deformation into account. For reasons discussed in Sect. 4-3-2 and 4-5-6, this factor is a function of the h/L ratio. For drawing wire of circular cross section, h is taken as the mean diameter, L is the length of contact zone (Fig. 4-44b), and the factor is

$$\phi = 0.88 + 0.12 \frac{h}{L} \qquad (4\text{-}43a)$$

For deformation in plane strain (e.g., in shaping a rectangular cross section), the factor is

$$\phi = 0.8 + 0.2 \frac{h}{L} \qquad (4\text{-}43b)$$

In Step 6, the draw force is

$$P_{dr} = \sigma_{\text{exit}} A_1 = \sigma_{fm} Q_{dr} A_1 \qquad (4\text{-}44)$$

This force must not exceed the strength of the drawn wire, which can be calculated from the yield strength $\sigma_{0.2}$ of the drawn product (if this is not known, it can be taken as 80% of the flow stress at exit from the die). The power required for drawing can be simply obtained from the definition of power (power = force × velocity).

Example 4-22

A shaped wire is drawn from annealed, 3-mm-diameter 302 stainless steel wire. The cross-sectional area of the shape is 5.0 mm². A commercial oil-based lubricant is used, the dies have 12° included angle, and drawing speed is 2 m/s. Calculate the draw force and power requirement.

Step 1: $A_0 = 3.0^2\pi/4 = 7.07$ mm².

Step 2: From Eq. (4-30), $e_c = (7.07 - 5.0)/7.07 = 29.3\%$.

Step 3: From Eq. (4-32), $\epsilon = \ln(7.07/5.0) = 0.346$.

Step 4: This is cold working in a steady-state process; thus, use σ_{fm}. From Table 4-2, $K = 1300$ MPa; $n = 0.3$. From Eq. (4-10)

$$\sigma_{fm} = \frac{1300}{0.3}\left[\frac{0.346^{1.3}}{1.3}\right] = 839 \text{ N/mm}^2$$

Step 5: To obtain correction for the inhomogeneity of deformation, the shaped section may be approximated by a circular cross section of equivalent diameter, Eq. (4-12b): $d_1 = 2.52$ mm. From the geometry of a conical die, $L = (d_0 - d_1)/2\sin\alpha = 2.28$ mm. Hence, $h = (3 + 2.52)/2 = 2.76$ mm. From Eq. (4-43a), $\phi = 0.88 + ((0.12)(2.76)/2.28) = 1.145$. From Table 4-4, $\mu = 0.05$ and, from Eq. (4-42)

$$Q_{dr} = (1 + 0.05\cot 6)(1.145)(0.346) = 0.52$$

Step 6: From Eq. (4-44): $P_{dr} = (839)(0.52)(5.0) = 2.2$ kN.

It is always necessary to check whether the draw is feasible. The flow stress of the drawn wire is $\sigma_f = 1300(0.346)^{0.3} = 945$ N/mm²; 80% of this is 756 N/mm². The drawn section of $A_1 = 5$ mm² cross section will support 5(756) = 3.8 kN; thus, the draw is entirely feasible.

Step 7: The net power is simply $P_{dr}v_{dr} = 2200(2.0) = 4400$ W (or 5.9 hp).

4-7-3 Process Limitations

Because the draw force must be less than the strength of the issuing wire, the attainable reduction is typically below 50% (calculated as *reduction of area*, and not as reduction of diameter). Frequent breaking of the wire would severely limit productivity since the end of the wire must be reduced (*pointed*) again so that it can be rethreaded. This is obviously time-consuming, and it is usually more profitable to limit reductions to below 30% per die (usually to 20% per die in multidie drawing). As seen from an inspection of the formula for draw stress (Eq. (4-42)), friction increases the draw stress and limits reduction; therefore good lubricating practices are essential (Table 4-4).

A second limitation arises from possible nonuniformity of deformation. Just as in extrusion (Sec. 4-5-6), the depth of the compression zone may not be sufficient to ensure homogeneous deformation. This is again governed by the h/L ratio: When $h/L > 2$, secondary tensile stresses can lead to the typical arrowhead (centerburst) defect in less-ductile materials (Fig. 4-44b), especially now that the axial stress is tensile.

A further possibility of secondary tensile stresses arises when deformation is limited to one part of a section. This will be discussed in more detail for the rolling of shapes (Sec. 4-8-2). Suffice it to say here that cracking of drawn sections may occur when some part of the cross section is not directly subjected to deformation.

4-8 ROLLING

Of all bulk deformation processes, *rolling* occupies the most important position. Over 90% of all materials that are ever deformed are subjected to rolling (see Table 4-1).

4-8-1 Flat Rolling

The process of reducing the thickness of a slab to produce a thinner and longer but only slightly wider product is commonly referred to as *flat rolling*. It is the most important primary deformation process. It allows a high degree of closed-loop automation and very high speeds, and is thus capable of providing high-quality, close-tolerance starting material for various secondary.sheet-metalworking processes at a low cost.

The cast structure is first destroyed and defects are healed as far as possible by *hot rolling*. The hot-rolled product has a relatively rough surface finish and dimensional tolerances are not very tight; nevertheless, *hot-rolled plate*, over 6 mm (0.25 in) thick and 1800–5000 mm (72–200 in) wide, in weights up to 150 tons, is an important starting material in ship building, boiler making, high-rise and industrial construction, and the manufacture of pipes and miscellaneous welded machine structures. Rolling results in substantial elongation of the cast slab, since the smallest economical weight of a rolled slab, even in the more exotic materials, is around 1 ton. When rolling the long lengths, the thinner sheet issuing from the rolling mill is coiled up. *Hot-rolled sheet* or *band* is typically of 0.8–6-mm (0.032–0.25-in) thickness, up to 2300-mm (90-in) width, in coils weighing up to 30 tons. It is an important starting material for the cold pressing of structural parts of vehicles, heavy equipment, and machinery, and also for making welded tubes.

Thinner gages, better surface finish, and tighter tolerances are obtained by *cold rolling*. After cleaning the surface of the hot-rolled band, it is wound into coils which may be *slit* into narrower widths or *cut* into shorter lengths, or both, depending on the handling facilities of the secondary manufacturing plants.

Standard surface finishes and tolerances are provided at no extra cost; however, exceptionally smooth finish or tight tolerances can also be produced at an often quite slight premium. The cost, of course, goes up as the gage decreases, especially if the thinner gage necessitates an extra pass or passes through a single-stand or multistand (tandem) mill. Large quantities of steel are rolled to around 0.75 mm (0.30 in) for automotive and appliance bodies and down to 0.15 mm (0.006 in) for food and beverage containers (cans). Copper is rolled to various gages for roofing, containers, cooking vessels, and down to 0.07 mm (0.003 in) for radiator fin stock. Aluminum alloy sheet of around 1.0-mm (0.040-in) thickness is extensively used in aircraft fuselages, automotive components, and trailer construction. Aluminum foil of down to 8-μm (0.0003-in) gage is used in large quantities for packaging. Foils of down to 3-μm (0.0001-in) thickness are produced on special mills in all materials.

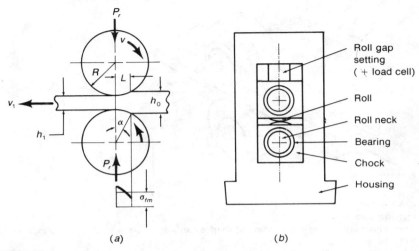

FIGURE 4-46
Rolling is a steady-state process that (*a*) reduces the thickness of the workpiece (*b*) in rolling mills of considerable stiffness.

The process of flat rolling looks deceptively simple (Fig. 4-46*a*). Two driven rolls of cylindrical shape reduce the flat workpiece to a thinner gage. The rolls are supported in housings, and the roll gap can be adjusted by mechanical or hydraulic means (Fig. 4-46*b*). To limit the elastic deflection of rolls, 2–18 support rolls may be incorporated into a mill housing. The finished product must have a uniform thickness in length and width, a flat shape, a controlled and uniform surface finish, and reproducible mechanical properties. Satisfying these requirements taxes the ingenuity of the production engineer, equipment designer, control specialist, and theoretician, and makes the process one of the most complex. More or less complete process models take a long time to run even on the most powerful computers, and studies on the effects of process parameters are run off-line. On-line controls are based on simplified theory and often incorporate empirical, experience-based models.

4-8-2 Shape Rolling

The rolling of shapes has a long history, beginning with the rolling of channels of lead for stained-glass windows (Table 1-1). The largest industrial application is now in the hot-rolling of *structural shapes*, which is a specialized primary deformation process practiced in special-purpose mills. Basically the same techniques can, however, also be applied to the cold rolling of shapes to tight tolerances and excellent surface finish, and these specialized secondary manufacturing processes are gaining popularity as alternatives to drawing and machining.

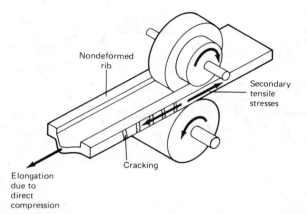

FIGURE 4-47
Nonuniform elongation in the rolling of shapes can lead to cracking due to secondary tensile stresses.

The starting material for *cold shape rolling* is a wire of square, rectangular, or circular cross section, and the finished shape is approached through a number of *passes* (rolling through shaped rolls) that gradually distribute the material in the desired fashion.

The crucial issue is always that of avoiding nonuniform elongation. As seen from the simple example of Fig. 4-47, those parts of the cross section that are directly compressed elongate as required to maintain constant volume, while parts not subject to direct compression elongate only because of their physical attachment to the deforming portion. Elongation in these noncompressed portions generates secondary tensile stresses which, as remarked before, easily lead to crack formation. Therefore, roll pass design aims at equalizing reductions in all portions of the cross section. This aim can be attained by moving the material sideways, especially in the early passes and, if necessary, by the use of vertical rolls that compress the section from the sides. Several rolling stands may be placed in *tandem* (in line) and it is then customary to alternate the axes of rolls from vertical to horizontal.

4-8-3 Ring Rolling

Seamless rings are important construction elements, ranging from the steel tires of railway car wheels to rotating rings of jet engines and races of ball bearings.

The starting material for *ring rolling* is a pierced billet. After making a hole by any suitable technique, the thick-walled ring is rolled out by reducing its thickness and increasing its diameter, as indicated in Fig. 4-10. Larger rings are rolled hot in specialized factories but smaller rings, especially those of small cross-sectional area, are frequently rolled cold. In addition to simple rectangular profiles, rings of a fairly complex cross-sectional profile can be rolled.

4-8-4 Transverse Rolling

When a workpiece is placed between counterrotating rolls with its axis parallel to the roll axes, it suffers plastic deformation (essentially, localized compression) during its rotation between the rolls. The consequences of this deformation depend on the shape and angular alignment of the rolls and, as in all compression (Sec. 4-3-2), on the h/L ratio. The height h is now the workpiece diameter, and L is the length of contact with the roll (equivalent to L of an indenter in plane strain, Fig. 4-15*b*). Several purposes may be accomplished:

1 When $h/L > 1$, deformation is inhomogeneous and the plastic zones penetrating from the point of contact literally try to wedge the workpiece apart; in other words, high secondary tensile stresses are generated in the center of the workpiece. This is the principle of making thick-walled tubes by *rotary tube-piercing methods*. As shown in Fig. 4-10, a mandrel or plug placed against the center of the billet helps in opening up and smoothing out the internal surface. Angular misalignment of the deforming rolls (skewing) forces the billet to progress in a helical path; thus, its whole length is pierced through. Such tube-piercing methods are practiced in specialized plants equipped for hot working.

2 The secondary deformation processes based on the same principle have the roll axes aligned and the workpiece rotates in the same plane (*transverse rolling*,

FIGURE 4-48
Axially symmetric workpieces may be cross-rolled but the h/L ratio must be kept low to avoid opening up the center. (*J. Holub, Machinery (London)* ***102****:131 (1963)*. With permission.)

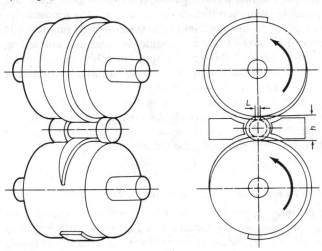

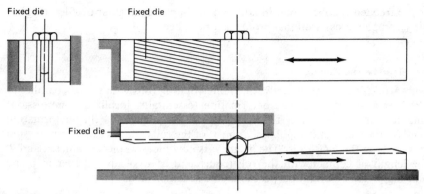

FIGURE 4-49
Strong threads are rolled at high rates in reciprocating flat dies.

Fig. 4-48). The rolls are shaped so as to avoid the generation of large tensile stresses so that a sound workpiece of axial symmetry is formed. For example, a dumbbell shape can serve either as a finished part or as a preform for the further forging of, say, a connecting rod or a double-ended wrench. There are a number of other rotary forging/rolling processes with specialized applications.

3 The rolls may be shaped to roll a thread on the workpiece. Large threads are rolled hot, but most *thread-rolling* operations are conducted cold, most often in thread-rolling machines equipped with so-called *flat dies* (Fig. 4-49). One of the dies is stationary, the other reciprocates; at an appropriate point of the stroke, a workpiece (typically, a cold-headed screw blank) is dropped into the gap, grabbed by the moving die, and rotated against the stationary die, thus the screw-thread profile is gradually developed. Rolled threads have a continuous grain flow and are, therefore, more fatigue-resistant than threads cut on a lathe. The productivity of the process is high. Even large, slow machines roll 60 screws per minute while smaller screws are produced at rates of 500 per minute. In machines containing several die pairs, production rates of 2000 per minute are achieved. The good quality and high productivity of thread rolling has eliminated thread cutting as a competitive process for most mass-production purposes.

Very large internal threads could be made by rolling but, apart from cutting, a more practical way is *cold form tapping*. The tool looks like a screw, except that its diameter changes periodically within the screw envelope, so that the protruding portions displace material from the roots into the threads (Fig. 4-50).

4-8-5 Calculation of Forces and Power Requirements

An acceptable estimate of rolling forces can be obtained if rolling is regarded as a continuous forging (cogging) process. A comparison of Fig. 4-46a with Fig. 4-26 will show that the *projected length of the arc of contact* between roll and workpiece

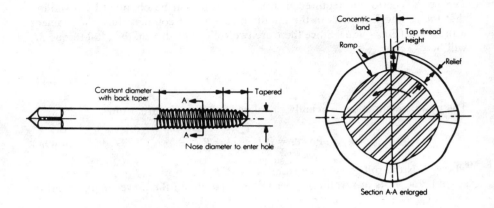

FIGURE 4-50
A cold-form tap forms threads in a hole by displacing rather than removing material. (*National Twist Drill & Tool Division, Lear Siegler, Lexington, S.C.*)

may be regarded as L of the forging tool, because major material flow takes place in the length direction of the slab or strip. The length of contact may be calculated from

$$L = \sqrt{R(h_0 - h_1)} \tag{4-45}$$

where R is the roll radius. Calculations follow those for forging (Sec. 4-4-2), with some differences:

In Step 3, ϵ is again calculated from Eq. (4-5b), but $\dot{\epsilon}$ must now be taken as the average strain rate

$$\dot{\epsilon} = \frac{v}{L} \ln \frac{h_0}{h_1} \tag{4-46}$$

In Step 4, σ_{fm} (Eq. (4-10)) is now needed because rolling is a steady-state process. In hot working, Eq. (4-7) automatically provides a mean flow stress because a mean strain rate was calculated in Eq. (4-46).

In Step 5, to find interface pressure, it is first necessary to check for homogeneity of deformation by calculating the h/L ratio. When $h/L > 1$, the inhomogeneity of deformation predominates and the pressure-multiplying factor Q_i is found from Fig. 4-25. When $h/L < 1$, friction effects are overriding and the pressure-intensification factor Q_p is found from Fig. 4-24 or Eq. (4-19).

Step 6: Roll force is obtained, for plane strain deformation, from

$$P_r = (1.15)\sigma_{fm}Q_i Lw \tag{4-47a}$$

or

$$P_r = (1.15)\sigma_{fm}Q_p Lw \tag{4-47b}$$

where w is the width of the strip.

Step 7: The torque required to rotate the rolls can be obtained by assuming that the rolling force acts in the middle of the arc of contact; thus, the moment arm is $L/2$ (Fig. 4-46a). Since there are two rolls to be driven, the total torque M_r will be

$$M_r = \frac{2P_r L}{2} = P_r L \tag{4-48}$$

The power requirement is readily calculated in units of watts from

$$\text{Power} = P_r L \frac{2\pi N}{60} = P_r L \frac{v}{R} \tag{4-49a}$$

where P_r is the roll force in newtons, L and R are in meters, v is in meters per second, and N is in revolutions per minute. To obtain the power requirement in units of horsepower, take

$$\text{Power} = P_r L \frac{2\pi N}{33000} \tag{4-49b}$$

where L is in feet and P_r is in pounds.

4-8-6 Process Limitations

The rolling process is remarkably forgiving if quality demands are not high, but requires a substantial knowledge and sophisticated control if the product is critical in any respect. There are a number of process limitations:

1 While good lubrication is essential to reduce roll forces at high L/h ratios, some minimum friction is still needed because it is the frictional component of the roll force that pulls the workpiece into the roll gap. The *angle of acceptance* α (Fig. 4-46a) is a function of μ

$$\tan \alpha \leq \mu \tag{4-50a}$$

and, from the geometry of the pass,

$$(h_0 - h_1)_{\text{max}} = \mu^2 R \tag{4-50b}$$

A heavier reduction can be taken by pushing the workpiece into the roll gap.

2 When thin sections are rolled in hard materials, elastic deformation of the rolls may limit the attainable minimum thickness. *Flattening* of the rolls can be minimized with a good lubricant, small roll diameter, and a roll made of a material with a high elastic modulus, such as WC.

3 Under the imposed forces, *roll bending* occurs as with any centrally loaded beam, supported at two ends. This makes the roll gap larger in the middle, thus the workpiece is reduced less and is elongated to a lesser degree in the middle, while the edges elongate more and become wavy. Compensation is possible by *cambering* the rolls (grinding them with a slight barrel shape) and by using back-up rolls. Substantial heat is generated in the rolling process and, if the lubricant/coolant is not fully effective, a *thermal camber* results in a wavy middle.

4 Under the imposed roll force, the entire rolling mill stretches. The spring constant of mills (the *mill elastic constant*) is usually under 5 MN/mm (14 000 ton/in); hence, the roll gap can open up several millimeters (even if a very thin strip is rolled). The rolls must be set closer by the amount required by the roll force, and any variations of roll force during rolling must be compensated for by manual or automatic control.

5 Inhomogeneous deformation, whether from a large h/L ratio or the absence of direct compression, is always harmful. There is one instance, however, when inhomogeneity is purposely induced. In *roller burnishing* the surface of a thick workpiece is superficially rolled. The deformation zone is very shallow and, in the absence of bulk plastic flow, the material of the surface is put in compression (Fig. 2-15*c*), making the part more resistant to fatigue (as in rolling the journal radii on crankshafts or in finish rolling gears).

Example 4-23

An AISI 1015 steel slab of $h_0 = 300$-mm thickness and $w_0 = 1000$-mm width is hot rolled at 1000 °C on a mill with rolls of diameter 600 mm. The presence of scale reduces friction to $\mu = 0.3$. A reduction of 27 mm is taken. Roll speed is 1.2 m/s. Calculate roll force and power requirement.

Check first whether the rolls will pull in the slab. From Eq. (4-50*b*), $\Delta h_{max} = (0.3)^2(300) = 27$ mm; thus, the reduction is just feasible.

Step 1: Thickness after rolling is $h_1 = 273$ mm.
Step 2: Reduction, from Eq. (2-13'): $e_c = 27/300 = 9\%$.
Step 3: Strain, from Eq. (4-5*b*): $\epsilon = \ln(300/273) = 0.094$.

$$h_{ave} = (300 + 273)/2 = 286.5 \text{ mm}$$

From Eq. (4-45): $L = [(300)(27)]^{1/2} = 90$ mm $= 0.09$ m. From Eq. (4-46): $\dot{\epsilon} = (1.2)(0.094)/0.09 = 1.25 \text{ s}^{-1}$.

Step 4: From Table 4-2: $C = 120$ MPa; $m = 0.1$. From Eq. (4-7): $\sigma_f = 120(1.25)^{0.1} = 123$ MPa.
Step 5: Check $h/L = 286.5/90 = 3.2$; thus, deformation is inhomogeneous. From Fig. 4-25, $Q_i = 2$.
Step 6: Roll force from Eq. (4-47): $P_r = (1.15)(123)(2)(0.09)(1) = 25.46$ MN ($= 2860$ tonf).
Step 7: Power from Eq. (4-49*a*): $(25.46)(0.09)(1.2)/0.3 = 9170$ kW.

Example 4-24

After hot rolling, the material of Example 4-23 is cold rolled on a mill of roll diameter 400 mm at a speed of 700 m/min. Calculate the force and power requirement for rolling from 1.0 mm to 0.6 mm, if a lubricant reduces the coefficient of friction to 0.05.

This is cold working, hence σ_{fm} will be needed and strain rate can be ignored.

Step 2: $e_c = 40\%$.
Step 4: $\sigma_{fm} = (620/0.51)(0.51^{1.18}/1.18) = 465$ MPa.
Step 5: $h = (1 + 0.6)/2 = 0.8$ mm; $L = [300(0.4)]^{1/2} = 8.94$ mm (0.00894 m); $h/L = 0.8/8.94 = 0.089$ (thus friction is important); $L/h = 8.94/0.8 = 11.2$, $\mu = 0.05$, and from Fig. 4-24, $Q_p = 1.3$.
Step 6: $P_r = (1.15)(465)(1.3)(0.00894)(1) = 6.2$ MN ($= 698$ tonf).
Step 7: Power $= (6.2)(0.00894)(700)/(0.4)(60) = 1617$ kW.

Note that even though reduction is miniscule compared to Example 4-23, roll force is quite high because of the higher flow stress in cold rolling. Power is high too because of the higher rolling speed.

4-9 SUMMARY

Bulk deformation processes have retained their importance over thousands of years of technological development. They not only provide the starting material for subsequent sheet metalworking, wire and tube bending, and most welding applications, but also ensure the availability of finished components of great structural integrity. The products include hot-forged parts, from turbine blades and gear blanks to garden hoes; cold-forged parts, from nails, screws, and rivets to finished gears; cold-extruded parts, from automotive half-axles and sparkplug bodies to toothpaste tubes; hot-extruded construction sections and valve bodies; and hot- and cold-rolled rings and sections for all purposes. In the design of components and in the control of processes several factors must be considered:

1 Cold working is characterized by strain hardening and offers products of increased strength, good tolerances and surface finish, and thin walls, but usually at the expense of lesser ductility and higher flow stress, die pressure, and deforming forces.

2 Hot working is characterized by dynamic recovery and recrystallization and offers lower (but strain-rate dependent) flow stresses, die pressures, and forces, but at the expense of extra energy consumption for preheating, and poorer tolerances and surface finish of the product.

3 Die pressures are determined by the flow stress of the material, modified by the effects of stress state (as expressed by the yield criterion), friction, and inhomogeneity of deformation. The aim of process control is to minimize pressures and forces by lubrication and the modification of process geometry.

4 The survival of the workpiece material is expressed by the concept of workability which encompasses the effects of the hydrostatic pressure developed in the process, superimposed on the basic ductility of the material. The aim of process development is usually that of increasing the hydrostatic pressure component (except for tube-piercing operations).

5 The pressure developed by the process must be accommodated by tools and dies made of appropriate materials, in configurations designed to give maximum resistance to plastic yielding. Elastic deformation of tooling and machinery must be compensated for if the shape of parts is to be kept within close tolerances.

General characteristics of bulk deformation processes are summarized in Table 12-3.

PROBLEMS

4-1 A 90Cu–10Sn binary alloy is made up in the laboratory under well-controlled melting conditions. On attempting hot rolling, the billet breaks up. (*a*) Review the equilibrium diagram of Cu–Sn alloys and (*b*) identify the possible cause of the problem. (*c*) Suggest a remedy.

4-2 Plot the YS data from Example 4-5 to see if they follow the trends of Eq. (3-5). Check whether hardness is $3 \times$ TS.

4-3 The operator of a riveting machine reports that many rivets crack, and the riveting force also seems to be inadequate as judged by the many incompletely formed rivet heads. The rivets are of 2024 aluminum alloy and are riveted in the solution-treated condition, to attain their full strength by natural (room-temperature) aging after riveting. Determine what could have gone wrong and suggest, step-by-step, what remedial action should be taken.

4-4 A customer specifies an Al-5Mg alloy sheet for its corrosion resistance. Both strength (YS = 300 MPa) and ductility (12% elongation) are required. According to *Metals Handbook* (vol. 2, 9th ed., p. 102), the annealed material has YS = 152 MPa, elongation = 35%; a cold-rolled material (H18 temper) of YS = 407 MPa has only 10% elongation. What material condition would satisfy the requirements and why?

4-5 What method is best for determining the flow stress of (a) bar material, for purposes of cold forging and (b) sheet or plate, for purposes of cold rolling? Identify the precautions to be taken if relevant data are to be generated.

4-6 To gain a feel for the effect of strain hardening on flow stress (Eq. (4-4)), calculate the value of ϵ^n for $\epsilon = 0.1, 0.2, 0.3, 0.5, 0.7,$ and 1.0 and $n = 0, 0.05, 0.1, 0.2, 0.3,$ and 0.5. Plot the family of ϵ^n curves as a function of ϵ.

4-7 A 3000-ton press has a spring constant of 8000 ton/in. If the dies are set to kiss at zero load, what will be the gap between them at the rated load of the press?

4-8 Distinguish between recovery and recrystallization in terms of the operative atomic mechanism and make sketches to describe their effects on mechanical properties and grain structure.

4-9 It is desired to produce sheet with very fine grain size. Suggest a suitable processing sequence.

4-10 To gain a feel for the effect of strain-rate sensitivity on flow stress (Eq. (4-7)), calculate the value of $\dot{\epsilon}^m$ for $\dot{\epsilon} = 0.01, 0.1, 1, 10, 100,$ and 1000 s^{-1} and $m = 0, 0.05, 0.1, 0.2, 0.3,$ and 0.5. Plot the family of $\dot{\epsilon}^m$ curves as a function of $\dot{\epsilon}$.

4-11 Define isothermal hot working. On the basis of data given in Tables 4-2 and 4-3, would you expect great benefits in terms of die pressures in isothermal working (a) 0.15% C steel (b) Ti–6Al–4V alloy?

4-12 List, in a tabular form, the attributes of cold, warm, and hot working (strain-rate sensitivity; flow stress; die pressure; dimensional tolerances; surface finish; lubrication; healing of casting defects).

4-13 A medium-carbon steel (1045) is to be cold extruded. Determine from the equilibrium diagram (a) what phases one should expect, (b) in what proportion in this material. From a consideration of the properties of two-phase structures, suggest (c) the optimum metallurgical condition for this application.

4-14 Calculate how much material is saved by making the part described in Example 4-17 by plastic deformation instead of machining.

4-15 Repeat the calculations for Example 4-18 but assume that extrusion is carried out at 300 C on a fast press with a ram speed of $v = 0.75$ m/s.

4-16 It is proposed that the end of a $d_0 = 0.25$-in, 1015 steel bar be upset over a length of $h_0 = 0.2$ in to form a flat head of $h_1 = 0.03$-in height (thickness). To assess feasibility, calculate the upsetting pressures and forces, (a) first assuming that a good lubricant reduces friction to $\mu = 0.1$, and then (b) for rough, unlubricated dies (as would apply if the lubricant had broken down or the lubricant supply had failed). (c) If the operation is feasible, suggest a suitable die material.

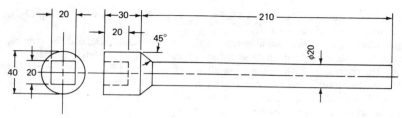

All dimensions in millimeters

4-17 A bolt head is produced by cold-heading (upsetting) the head on an annealed bar. (*a*) Make a sketch of the longitudinal cross section of the bolt, and indicate the grain-size variation one should expect, if the bolt is annealed after cold-heading. (*b*) Point out the weakest cross section and indicate (*c*) a method of production that would avoid this weakness.

4-18 A bolt of the indicated geometry is to be produced for high-temperature service. A material similar to H13 is proposed, and it is planned to forward-extrude the shaft and, in a separate operation, back-extrude the head, both at 1000 °C, in a hydraulic press with a ram speed of $v = 0.5$ m/s. Determine if the proposition is feasible as far as tool loading is concerned, by calculating (*a*) the size of the starting billet, (*b*) maximum forward-extrusion pressure, assuming that a graphitic lubricant of $\tau_i = 70$ MPa shear strength is applied to the container, and (*c*) the back-extrusion pressure (remember to check both container and punch pressures).

4-19 A $d_0 = 1$ in and $h_0 = 2$ in billet of a free-machining (leaded) brass is to be compressed to an $h_1 = 0.4$ in height in a hydraulic press (ram velocity $v = 600$ in/min) at 800 °C between unlubricated anvils. (*a*) For the end of the press stroke, calculate interface pressure and press force. (*b*) What increase in stresses would occur if the workpiece were to cool to 600 °C?

4-20 Take a common (flat head) nail, measure the diameter and the thickness (height) of the head. Calculate (*a*) the head volume and (*b*) the length of wire that had been upset. From these data, (*c*) would you expect buckling during free heading? If the answer to (*c*) is yes, how would the head be formed?

4-21 Returning to the pin forged in Example 4-13, recalculate the average die pressure using m^*. For unlubricated upsetting, $m^* = 1$; for the lubricated case, estimate m^* from μ, as shown in Example 4-11.

4-22 Taking into account the efficiency of a hammer blow (Eq. (4-41*b*)), estimate the size of hammer needed to make the part of Example 4-21. First find the hammer size for a single blow, assuming an efficiency of 0.8, then find the hammer size that will deliver the energy in three blows. (*Hint*: Divide the area under the force-displacement curve into three unequal areas, remembering that the initial, softer blow is more efficient and can thus deliver more energy. You may have to iterate to find a reasonable solution.)

4-23 A component is produced by forward extruding a $d_0 = 20$ mm diameter billet through a $d_1 = 14$ mm die of $\alpha = 45°$ half-angle. Part of the billet remains unextruded to serve as the head. Many components are found to have centerburst defects. What could be done to get out of trouble? (The material cannot be changed.) Make a sketch (to scale) to justify your answer.

4-24 It is proposed to make a tube-shaped part of 0.15% C steel by cold piercing (Fig. 4-29a) a $d_0 = 20$ mm billet with a $D_p = 10$ mm diameter punch. The length of the part (the depth of the hole) is 50 mm; for constructional reasons, the punch is 60 mm long. (a) Calculate the punch pressure. (b) Suggest a suitable punch material. (c) Check the punch for bending and compression. (d) Specify a lubricant suitable for this task.

4-25 A flat, annealed 70/30 brass wire is to be drawn from a 10 mm × 2 mm cross section to a fluted shape of an average thickness $h = 1$ mm and an unchanged width of $w = 10$ mm. The die half angle is $\alpha = 7°$ and the lubricant is an emulsion. Calculate (a) the relevant flow stress and (b) the drawing force. (c) Check if the process is feasible. (d) Suggest a way of achieving the required end result.

4-26 Inspect Fig. 4-45 and note the half-arrows indicating the friction stresses. Consider their effect on drawing stresses, and deduce which of the four processes allows the greatest reduction to be taken.

4-27 Commercial purity (1100 Al) aluminum is routinely rolled, in several passes but without annealing, to a total reduction of over 98%. (a) Find the uniform strain (ϵ), the total elongation (e_f), and the reduction in area (q) for this material. (b) Compare these to the rolling reduction obtainable, and explain the reasons for the difference.

4-28 A copper (99.94% Cu) bar is to be cold rolled into a section which must have a minimum TS of 60 000 psi. If the finished cross section $A_n = 0.05$ in^2, what should be the initial bar diameter? (*Hint:* A look at Figs. 4-3 and 4-4 will show that the true stress σ and σ_{eng} must be very similar in a strain-hardened material.)

4-29 A small, shallow U channel of 5% Sn bronze is cold rolled. The shape is shallow enough to regard it as a $w = 10$ mm wide, $h = 2$ mm thick strip of rectangular cross section. A 30% reduction in height is taken in a single pass, on a mill with 150-mm-diameter rolls, at $v = 0.8$ m/s speed, with a mineral-oil lubricant. Calculate (a) the roll force and (b) the power requirement.

4-30 A screw thread is cold rolled on a 1015 low-carbon steel bolt of 0.25-in diameter. Observation shows that the length of the contact zone between die and screw blank is $L = 0.05$ in. The average strain hardening during rolling corresponds to a strain of $\epsilon = 0.4$. Calculate (a) the applicable flow stress and (b) the average interface pressure in the contact zone. Determine (c) if there is any danger of internal fracture (make a sketch to illustrate the point).

FURTHER READING

A Detailed Process Descriptions

ASM: *Metals Handbook*, 8th ed., vol. 3, *Machining*, 1967, pp. 105–107 (Roller Burnishing), pp. 130–145 (Thread Rolling), pp. 145–146 (Spline Rolling). vol. 4, *Forming*, 1969, pp. 78–88 (Coining), pp. 322–333 (Straightening), pp. 333–346 (Rotary Swaging), pp. 465–496 (Cold Heading and Extrusion); vol. 5, *Forging and Casting*, 1970, American Society for Metals, Metals Park, Ohio.

Avitzur, B.: *Handbook of Metalforming Processes*, Wiley-Interscience, New York, 1983.

Lange, K. (ed.): *Handbook of Metalworking*, McGraw-Hill, New York, 1985.

Wick, C., J. T. Benedict, and R. F. Veilleux (eds.): *Tool and Manufacturing Engineers Handbook*, vol. 2: *Forming*, 4th ed., Society of Manufacturing Engineers, Dearborn, Mich., 1984.

B Textbooks:

Altan, T., S. I. Oh, and H. C. Gegel: *Metal Forming—Fundamentals and Applications*, American Society for Metals, Metals Park, Ohio, 1983.

Avitzur B.: *Metal Forming, the Application of Limit Analysis*, Dekker, New York, 1980.

_____: *Metal Forming: Processes and Analysis*, McGraw-Hill, New York, 1968/Krieger, Huntington, N.Y., 1979.

_____: *Metal-Forming Processes*, Wiley-Interscience, New York, 1981.

Backofen, W. A.: *Deformation Processing*, Addison-Wesley, Reading, Mass., 1972.

Blazynski, T. Z.: *Metal Forming, Tool Profiles and Flow*, Halstead Press, New York, 1976.

Dieter, G. E., Jr.: *Mechanical Metallurgy*, 2d ed., McGraw-Hill, New York, 1974.

Ford, H., and J. M. Alexander: *Advanced Mechanics of Materials*, 2d ed., Halstead Press, New York, 1977.

Harris, J. N.: *Mechanical Working of Metals*, Pergamon, New York, 1983.

Hosford, W. F., and R. M. Caddell: *Metal Forming: Mechanics and Metallurgy*, Prentice-Hall, Englewood Cliffs, N.J., 1983.

Johnson, W., and P. B. Mellor: *Engineering Plasticity*, Van Nostrand, London, 1973.

_____, R. Sowerby, and R. D. Venter: *Plane-Strain Slip Line Fields for Metal Deformation Processes*, Pergamon, New York, 1982.

Rowe, G. W.: *Elements of Metalworking Theory*, Arnold, London, 1979.

_____: *Principles of Industrial Metalworking Processes*, Arnold, London, 1977.

Slater, R. A. C.: *Engineering Plasticity, Theory and Its Application to Metal Forming Processes*, Halstead Press, New York, 1977.

Thomsen, E. G., C. T. Yang, and S. Kobayashi: *Mechanics of Plastic Deformation in Metal Processing*, Macmillan, New York, 1965.

C Specialized Books

Agrawal, S. P. (ed.): *Superplastic Forming*, American Society for Metals, Metals Park, Ohio, 1985.

Alexander, J. M., and B. Lengyel: *Hydrostatic Extrusion*, Mills and Boon, London, 1971.

Altan, T., et al.: *Forging: Equipment, Materials and Practices*, Metals and Ceramics Information Center, Battelle Memorial Institute, Columbus, Ohio, 1973.

ASM: *Source Book on Cold Forming*, American Society for Metals, Metals Park, Ohio, 1975.

Avitzur, B., and C. J. van Tyne (eds.): *Production to Near Net Shape: Source Book*, American Society for Metals, Metals Park, Ohio, 1983.

Burke, J. J., and V. Weiss (eds.): *Advances in Deformation Processing*, Plenum, New York, 1979.

Byrer, T. G. (ed.): *Forging Handbook*, Forging Industry Association, Cleveland/American Society for Metals, Metals Park, Ohio, 1985.

Dieter, G. E. (ed.): *Workability Testing Techniques*, American Society for Metals, Metals Park, Ohio, 1984.

Developments in the Drawing of Metals, Book No. 301, The Metals Society, London, 1983.

Frost, H. J., and M. F. Ashby: *Deformation-Mechanism Maps*, Pergamon, Oxford, 1982.

Geleji, A.: *Forge Equipment, Rolling Mills and Accessories*, Akademiai Kiado, Budapest, 1967.

Hoffmann, E. G. (ed.): *Fundamentals of Tool Design*, 2d ed., Society of Manufacturing Engineers, Dearborn, Mich., 1984.

Kalpakjian, S. (ed.): *Tool and Die Failures*, American Society for Metals, Metals Park, Ohio, 1982.

Krauss, G. (ed.): *Deformation, Processing, and Structure*, American Society for Metals, Metals Park, Ohio, 1984.

Laue, K., and H. Stenger: *Extrusion—Processes, Machinery, Tooling*, American Society for Metals, Metals Park, Ohio, 1981.

Nachtman, E., and S. Kalpakjian: *Lubricants and Lubrication in Metalworking Operations*, Dekker, New York, 1985.

Open Die Forging Institute: *Open Die Forging Manual*, 3d ed., Forging Industry Association, Cleveland, Ohio, 1982.

Pittman, J. F. T., R. D. Wood, J. M. Alexander, and O. C. Zienkiewicz: *Numerical Methods in Industrial Forming Processes*, Pineridge Press, Swansea, 1982.

Pollack, H. W.: *Tool Design*, Prentice-Hall, Reston, Va., 1976.

Poli, C. R., and W. A. Knight: *Design for Forging Handbook*, University of Massachusetts, Amherst, Mass., 1984 (also software version of coding system for Apple II-e).

Proc. ROMP (*Int. Conf. Rotary Metalworking Processes*), IFS (Conferences) Ltd., Bedford, since 1979.

Roberts, W. L.: *Cold Rolling of Steel*, Dekker, New York, 1978.

_____: *Hot Rolling of Steel*, Dekker, New York, 1983.

Schey, J. A.: *Tribology in Metalworking: Friction, Lubrication ana Wear*, American Society for Metals, Metals Park, Ohio, 1983.

Semiatin, S. L., and J. J. Jonas, *Formability and Workability of Metals*, American Society for Metals, Metals Park, Ohio, 1984.

Thomas, A.: *DFRA Forging Handbook: Die Design*, Drop Forging Research Association, Sheffield, England, 1980.

Wang, K. K. (ed.): *CAD/CAM for Tooling and Forging Technology*, Proc. U.S.-Sweden Conference, Society of Manufacturing Engineers, Dearborn, Mich., 1983.

Watkins, M. T.: *Metal Forming I: Forging and Related Processes*, Oxford University Press, Oxford, 1975.

D Journals

Journal of Applied Metalworking
Journal of Mechanical Working Technology
Metal Forming
Metallurgia
Metals Technology
Wire
Wire Industry
Wire Journal International

CHAPTER 5

SHEET-METALWORKING PROCESSES

Because of the low cost of mass-produced sheet of high quality, sheet metalworking has gained an outstanding position among manufacturing processes. Originally, the semifabricated starting material was sheet, rolled and supplied in limited sizes. Since the appearance of continuous tandem rolling mills, sheet has really been produced in coils of wide strip. Coils may be cut up, either in the rolling mill or in service centers, for easier handling in the facilities of the secondary manufacturer; however, there is an increasing trend to ship entire coils (sometimes slit into narrower widths) which are then fed into the presses and press lines of the manufacturer.

5-1 MATERIAL PROPERTIES

All wrought alloys (Sec. 4-2) are suitable for sheet-metalworking applications. The critical properties are, however, somewhat different from those discussed for bulk deformation, partly because deformation now occurs mostly in tension rather than compression, and partly because many sheet parts are large and highly visible, making appearance a major concern.

5-1-1 Formability

We remarked in Sec. 4-3-3 that the survival of a metal in bulk deformation processes can be described by the concept of workability, which encompasses the basic ductility of the material (related to reduction in area in the tension test) and the stress state imposed by the process. The critical failure mode was fracture. Survival in sheet metalworking is linked to *formability*, which is also a complex property, and must now be related to failure definitions relevant to sheet products:

1 The first objection may arise when a stretched sheet becomes *grainy* in appearance (*orange peel*). This is a natural consequence of the polycrystalline structure of metals: Individual grains oriented in different crystallographic directions deform to slightly differing degrees. Roughening of the surface has no bearing on the structural integrity of the part. If the grainy appearance is aesthetically objectionable, a finer-grained material will produce graininess on such a small scale as to be invisible to the naked eye.

2 In some materials *initial yielding* is highly localized and visible surface bands (*Lüders lines, stretcher-strain marks*) form. On continued stretching (Fig. 5-1), families of these lines criss-cross the surface. Stretcher-strain marks (or, as they are called in the shop, *worms*) are harmless but may be objectionable on exposed surfaces. Once the entire surface is covered, they are no longer distinguishable.

3 Appearance suffers and the functional properties of a part may be affected when *localized necking* occurs. Even though the part is not fractured, its load-bearing capacity may be reduced, although in some configurations the part will remain completely functional. In general, materials are chosen to optimize factors that delay the onset of necking (a large uniform elongation, corresponding to a high n value, Sec. 4-1-1) or help to spread out an incipient neck (a high m value, Sec. 4-1-4, or transformations, Sec. 4-2-2).

FIGURE 5-1
The yield-point phenomenon results in the development of visible shear bands, Lüder's lines, on a polished mild-steel strip subjected to tension (*Courtesy S. Kadela, University of Waterloo.*)

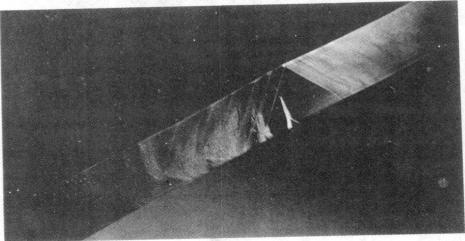

4 Once the neck has localized, further deformation occurs by local thinning until, finally, *fracture* sets in. We saw in Sec. 4-1-4 that post-necking strain is a function of the *m* value. In cold working even a slight increase in *m* (say, from 0 to 0.05) is helpful; in hot working a high *m* allows substantial post-necking deformation while maintaining a reasonably uniform thickness (at *m* = 1, the sheet would thin out completely uniformly). A higher reduction of area allows the sheet thickness to reduce further without fracture, but the load-bearing capacity of the part may be lost if local thinning is too severe.

In summary, a highly formable sheet metal has high uniform elongation (or *n* value) and large post-necking strain (or high *m* value). In industrial practice, a high total elongation in the tension test (Eq. (2-9*b*)) has long been regarded as a desirable attribute; inspection of Fig. 4-2*a* shows that this view translates into a combination of high *n* and high *m*, and is thus fundamentally correct. For a given material, ductility decreases with increasing hardness; therefore, it is common practice to specify hardness rather than elongation. Adequate ductility is a necessary but not sufficient criterion; in addition, a desirable material shows no Lüders bands and has a favorable anisotropy. These topics will be discussed in the following sections.

5-1-2 Yield-Point Phenomena

We mentioned in Sec. 3-1-2 the possibility of forming interstitial solid solutions in which solute atoms, much smaller than the solvent atoms, fit into the spaces existing between atoms in the basic lattice. These solute atoms often seek more comfortable sites where lattice defects have created voids in the structure. Most markedly, this is found with carbon and nitrogen in iron. Their atoms are small enough to fit into the lattice; nevertheless, they tend to migrate to dislocations where distortion of the lattice provides more room (just below the extra row of atoms in Fig. 3-9*c*). In a sense, the solute atoms form a *condensed atmosphere* which completes the lattice and immobilizes, *pins the dislocations*.

In the course of deformation, a larger stress must be applied before dislocations can break away from the condensed atmosphere of carbon or nitrogen atoms. This leads to the appearance of a *yield point* on the stress–strain curve of low-carbon steels (Fig. 5-2*a*). After the dislocations have broken away from the pinning atoms, they multiply and move in large groups in the direction of maximum shear stress (very approximately, at 45° to the applied force). Such localized yielding creates the visible Lüders lines or strain bands (Fig. 5-1). Successive generation of strain bands continues over the whole length of the specimen at a relatively low stress, giving the familiar *yield-point elongation* (Fig. 5-2*a*). Once the strain bands cover the entire surface, normal strain-hardening behavior is evident.

If straining is interrupted and then immediately resumed, the original strain-hardening curve is rejoined. However, if sufficient time is allowed for the interstitial atoms to seek out new dislocation sites (so that the carbon and

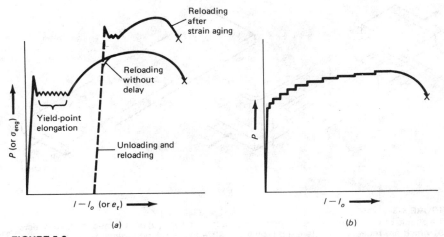

FIGURE 5-2
(a) The yield-point elongation typical of mild steel returns if a steel subject to strain aging is stored after initial deformation. (b) Serrated yielding is typical of solid-solution alloys.

nitrogen atmospheres condense again), the sheet is strengthened and the yield-point phenomenon returns (broken line in Fig. 5-2a). This behavior is described as *strain aging*. It leads to problems in stretching-type sheet-metalworking operations, because ductility is reduced and surface appearance is marred by the Lüders lines.

Abnormal yielding, particularly *stepwise* or *serrated yielding* (Fig. 5-2b) is also observed in other materials and is related to negative strain-rate sensitivity rather than to dislocation pinning. Such serrated yielding is found in some substitutional aluminum alloys and again leads to the development of objectionable stretcher-strain marks.

5-1-3 Textures (Anisotropy)

We saw in Sec. 3-2-1 (Fig. 3-9) that crystals deform by slip on preferred planes. If the crystal shown in Fig. 3-9a is to become longer, the slip planes must rotate into the direction of straining; in compressive deformation the slip planes rotate across the direction of straining. This has important consequences in polycrystalline materials, particularly when only a limited number of slip systems are available. Before deformation, properties will be *isotropic* (the same in all directions), representing the average properties of randomly oriented crystals. However, rotation of slip planes during deformation results in a noticeable alignment (*preferred orientation*) of crystals along common crystallographic orientations. This alignment is often referred to as *texture*. A polycrystalline material possessing a texture will show some of the directional properties typical of single crystals. This *directionality* or *anisotropy* of properties is evident in

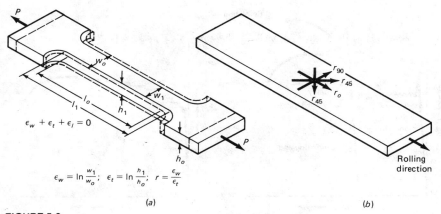

FIGURE 5-3
The effects of anisotropy on the deformation of a material are (a) determined in tension tests, (b) which are repeated in different directions relative to the rolling direction.

variations of the elastic modulus, YS, TS, elongation, and many other properties with the direction of testing. Most importantly for sheet metalworking, the relative magnitudes of strains also change during tensile deformation.

It can be shown that, if the volume of a specimen remains constant, the sum of the three principal *true* strains is equal to zero

$$\epsilon_1 + \epsilon_2 + \epsilon_3 = 0 \tag{5-1}$$

Returning to the definition of true strain as the natural logarithm of new dimension divided by the old dimension (Eqs. (4-3) and (4-5a)), in a tension test the major strain ϵ_1 is positive (tensile) whereas the transverse strains ϵ_2 and ϵ_3 are negative (compressive). For convenience, it is usual to speak of length strain ϵ_l, width strain ϵ_w, and thickness strain ϵ_t (Fig. 5-3a). Then

$$\epsilon_l + \epsilon_w + \epsilon_t = 0 \tag{5-2}$$

This relationship always holds, but ϵ_w and ϵ_t need not be equal in magnitude. By convention, the relative magnitudes of strains are expressed by the *r value*, which is the ratio of width strain to thickness strain

$$r = \frac{\epsilon_w}{\epsilon_t} \tag{5-3}$$

Several possibilities exist:

1 When the material is *isotropic*, $\epsilon_w = \epsilon_t$, and $r = 1$. It does not matter whether the specimen is cut in the rolling direction, across it, or at an intermediate angle (Fig. 5-3b); in an isotropic material

$$r_0 = r_{90} = r_{45} = 1 \tag{5-4a}$$

2 It is conceivable that the r values vary in relation to the rolling direction.

$$r_0 \neq r_{90} \neq r_{45} \qquad (5\text{-}4b)$$

This is denoted as *planar anisotropy* and leads to such problems as earing in deep drawing (Sec. 5-5-2).

3 If the r values measured in the plane of the sheet are identical in all directions but deviate from unity

$$r_0 = r_{90} = r_{45} \lessgtr 1 \qquad (5\text{-}4c)$$

we speak of *normal anisotropy*, because deformation of the test specimen in the thickness direction (normal to the sheet surface) is greater or smaller than in the width direction.

4 It is possible and indeed usual that normal and planar anisotropy occur simultaneously

$$r_0 \neq r_{90} \neq r_{45} \neq 1 \qquad (5\text{-}4d)$$

A measure of normal anisotropy is a mean r, denoted $\bar{r}$ or r_m

$$\bar{r} = \frac{r_0 + r_{90} + 2r_{45}}{4} \qquad (5\text{-}4e)$$

Frequently, the symbol r is used loosely to denote $\bar{r}$ or r_m. A measure of planar anisotropy is Δr

$$\Delta r = \frac{r_0 + r_{90} - 2r_{45}}{4} \qquad (5\text{-}4f)$$

Anisotropy is most evident with hexagonal materials in which the limited number of slip systems leads to the development of a texture after relatively small (20–30%) deformations, with most of the basal planes aligned perpendicular to the application of the rolling force; that is, with basal planes almost parallel to the sheet surface. When a tension test specimen cut from such a sheet is elongated, deformation is highly anisotropic.

1 In hcp materials with a *high c/a ratio* sliding is limited to the basal planes (Fig. 3-2c); thus, the thickness of the sheet is reduced while its width is hardly affected, just as a card pack can be elongated by sliding the cards over each other (Fig. 5-4a). The r value becomes very small, typically, 0.2 for zinc. It may be even smaller for some highly textured metals, such as a cube-textured copper.

2 The deformation of a tensile specimen cut from a hcp material with a *low c/a ratio* shows a dramatically different behavior. Since slip now takes place on prismatic and/or pyramidal planes (Fig. 3-2d), the sheet thickness is hardly reduced at all; instead, most of the deformation takes place by rearrangement of the hexagonal prisms, leading to a marked reduction in the width of the specimen

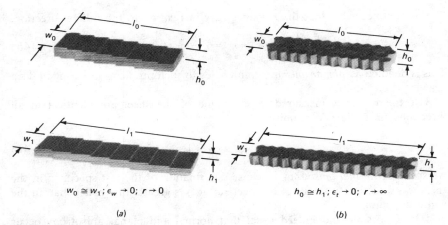

FIGURE 5-4
Deformation of hexagonal metals: (a) A high c/a ratio leads to basal slip and low r value; (b) a low c/a ratio results in prismatic slip and a high r value.

(Fig. 5-4b). The r value could, theoretically, reach infinity, but in practice seldom exceeds 6, the value for titanium.

Metals of fcc structure possess a great many equivalent slip systems (Fig. 3-2a); therefore, only much later—typically after more than 50% reduction—do they develop a texture. A completely randomly oriented polycrystalline fcc material is nearly isotropic ($\bar{r} = 1$). However, after deformation the r value may drop and many aluminum alloys tend to have $0.4 < \bar{r} < 0.8$.

The common slip direction in bcc materials (Fig. 3-2b) can be exploited by appropriate processing to give $\bar{r}$ values ranging from 0.8 to over 2.

It should be noted that annealing does not necessarily restore isotropy; the deformation texture may simply be replaced with an annealing texture.

Anisotropy is of importance in some bulk deformation processes, but its effects are most evident in sheet-metalworking processes, particularly deep drawing.

Example 5-1

A tension test is conducted on a sheet specimen (as in Fig. 2-2b) of $l_0 = 50.0$ mm, $w_0 = 6.0$ mm, and $h_0 = 1.00$ mm. The test is interrupted before the onset of necking; at this time, $l_1 = 60.0$ mm and $w_1 = 5.42$ mm (the thickness h_1 is difficult to measure with sufficient accuracy). Calculate the r value.

We may calculate the average thickness from constancy of volume, or obtain ϵ_t from Eq. (5-1):

$$\epsilon_t = -\epsilon_l - \epsilon_w = -[\ln(60/50)] - [\ln(5.42/6.00]$$

$$= -0.1823 + 0.1017 = -0.0806$$

$$r = (-0.1017)/(-0.0806) = 1.26$$

5-1-4   Metals

We already mentioned that all wrought metals (Tables 4-2 and 4-3) are more or less suitable for sheet metalworking. Some steels, specially developed for sheet metalworking, merit separate discussion.

Low-Carbon Steel   Among carbon steels (Sec. 4-2-1) low-carbon steels with up to 0.15% C (most frequently, 0.06 or 0.08% C) are used in largest quantities. Capped steels are suitable for structural parts and tubing but not for deep drawing. For the latter purpose, rimmed or killed steels are needed.

1 The high ductility and relatively low cost of rimmed steels has made the "commercial quality" (typically, 30% tensile elongation) and "drawing quality" (35% elongation) steels the favorites for appliance and auto-body applications. The very-low-carbon surface is an advantage in enameling. Grain size is controlled by heavy (50–70%) cold rolling followed by annealing (Fig. 4-6). However, the presence of carbon and nitrogen results in yield-point elongation (Fig. 5-5a) and objectionable stretcher-strain marks. Therefore, strip is usually given a *temper pass*, i.e., a very light rolling reduction, on the order of 1% or less (Fig. 5-5b). This is less than the yield-point elongation and thus barely affects ductility, but produces very finely spaced Lüders bands so that on subsequent tensile deformation there is no yield point (Fig. 5-5c) and no visible bands appear. However, if the material is stored prior to drawing, strain aging takes place within a few months or weeks (depending on composition and storage temperature), ductility is reduced, and yield-point elongation returns (Fig. 5-5d). *Roller leveling* (Fig. 5-5e) bends the strip repeatedly and helps to disguise Lüders bands by a mechanism similar to temper rolling, but ductility suffers.

2 Killed steel ("special deep-drawing quality") has uniform properties and high n and r values, and is specified when severe draws (such as the oil pan of an auto engine) are to be made or when storage is unavoidable. Killed steel also

FIGURE 5-5
The undesirable effects of (a) yield-point elongation (b) may be masked by temper rolling. (c) Yield-point elongation is then absent and Lüder's bands are not visible but (d) strain aging leads to their return. (e) Lüder's lines may then be masked again by roller leveling.

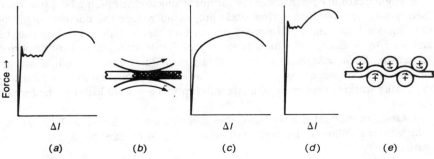

contains carbon and nitrogen but the nitrogen is combined with aluminum into a compound, and only carbon remains in a form that would allow condensation onto dislocations. Annealed sheet again shows yield-point elongation (Fig. 5-5a) but after temper rolling (Fig. 5-5b) the yield point is permanently eliminated (Fig. 5-5c); carbon does not diffuse to dislocation sites unless the steel is heated to 120 °C. These steels can be processed to high (up to 1.8 or even 2.5) r values.

Much steel is used in the hot-rolled condition for automotive wheel rims, axle cases, chassis parts, compressed gas cylinders, etc. However, most steel is used in the cold-rolled form.

High-Strength Steel We already mentioned in Sec. 1-3 the pressures that resulted in reducing the weight of automobiles. The quest for reduced weight in this and other fields prompted the development of stronger materials. All strengthening mechanisms are utilized:

1 Strain hardening is the least expensive mechanism and, if possible, the sheet-metalworking process is directed so that the finished part should have considerable cold work imparted to it. The ductility of the material sets a limit to the attainable tensile strain in the sheet forming operation, and greater strength can be obtained if the starting sheet is already strain hardened by rolling. However, the remaining ductility (Figs. 4-3 and 4-4a) may be too small for all but the lightest deformation.

2 Greater ductility combined with reasonable strength is given by heavy cold rolling followed by recovery anneal (temper letting down, Fig. 4-5).

3 Solid-solution alloys strain harden more steeply than pure metals (Sec. 4-1-2) and often have quite high ductility stemming from their higher n value.

4 A precipitation-hardening alloy, worked prior to aging, will be stronger because the higher dislocation density leads to the formation of finer precipitate particles (Sec. 4-1-5).

5 Cold working strengthens a temper-rolled rimmed steel because of strain aging that follows cold working (essentially, working between the stages represented by Fig. 5-5c and d). Gains from strain aging can be maximized by increasing the nitrogen content (*renitrogenized steels*).

6 Higher carbon contents, needed for quenching and tempering heat treatment (Sec. 3-5-4), are useful in spring steels but would reduce the ductility of deep-drawing steel too much. However, low-carbon steels with 1.4% Mn can be annealed to produce a structure consisting of ferrite strengthened by dispersed martensite. Such *dual-phase steels* have a low yield strength, which is an advantage when springback is objectionable; at the same time, rapid strain hardening during working imparts a high strength (up to TS = 1000 MPa) in the formed product.

7 Grain refinement is a powerful strengthening mechanism (Fig. 3-13). Heavy cold working followed by recrystallization (Fig. 4-6) can be applied to all materials.

8 Further refinement of the grain size is possible in steels by the addition of very small quantities of Ti, V, or Nb. These form carbonitride precipitates that inhibit grain growth in the austenite and thus refine the ferrite formed on cooling from the controlled hot rolling temperature. The combination of grain refinement and precipitation hardening results in high (350–560-MPa) yield strengths. Ductile manganese sulfide inclusions tend to roll out into stringers and the transverse impact properties of the steel suffer. The addition of Zr or Ti reduces the plasticity of inclusions and prevents their spreading, thus removes the harmful effects of inclusions. Such *high-strength low-alloy* (HSLA) steels are increasingly used in vehicles and other structures. One must remember though that their high $\sigma_{0.2}/E$ ratio results in large springback.

Coated Products Much sheet metal is formed with preapplied coatings that improve the service properties or appearance of finished parts.

Tin-coated sheet (*tinplate*) is corrosion resistant as long as the tin layer is free of scratches, and the nontoxicity of Sn makes tinplate suitable for food containers. In zinc-coated sheet (*galvanized sheet*) the steel is protected by the preferred (sacrificial) corrosion of the zinc; thus, even a damaged coating protects. In addition to roofing, galvanized sheet (and more recently, also one-side galvanized sheet) finds increasing application in automotive and appliance construction. Lead-coated sheet (*terne plate*) resists corrosion in some media for which tin or zinc offer no protection but, because of the toxicity of Pb, terne plate is limited to nonfood applications. *Aluminum-coated sheet* is protected from corrosion by hot gases by an aluminum–iron alloy formed at elevated temperatures; thus, the sheet is suitable for heat exchangers, automotive exhaust systems, grill parts, etc.

Prepainted sheet, coated with paints as well as thicker polymeric films (plastics, such as vinyls), offers both protection and a pleasing finish. If formed with care, the coatings remain adhered to the surface. The need for finish painting the part is eliminated, the quality of the coatings is often superior to paint finishes applied after forming, and economies can be realized too.

5-2 SHEARING

Irrespective of the size of the part to be produced, the first step involves cutting the sheet or strip into appropriate shapes by the physical process of *shearing*. It is practiced in several ways:

Cutting a sheet along a straight line is simply called *shearing*. Cutting a long strip into narrower widths between rotary blades is referred to as *slitting*. A contoured part (whether it be circular or more complex in shape) is cut between a punch and die in a press, and the process is called *blanking*. The same process is also used to remove unwanted parts of a sheet, but then one refers to *punching* a hole, of circular or any other shape. Cutting out a part of the sheet edge is called *notching*, and a partially cut hole, with no material removed, is made by *lancing*.

A contoured part may be cut by repeated small cuts in the process of *nibbling*. Drawn products are finished by *trimming* off excess material.

5-2-1   The Shearing Process

The process of separating adjacent parts of a sheet through controlled fracture cannot be described as either purely plastic deformation or as machining. The sheet is placed between two edges of the shearing tools—in the instance of blanking, a punch and a die (Fig. 5-6). The events taking place during the stroke of the press can be followed by recording punch force as a function of stroke (Fig. 5-7) and by inspecting the cut surfaces.

On penetration of the tool edges, the sheet is first pushed into the die, and plastic deformation results in a rounding of the edge of the blank (*roll-over*). Then the blank is pushed into the die by extrusion-like plastic deformation, indicated by the parallel, *burnished zone* on the blank, and characterized by steadily increasing forces. After some critical deformation, cracks are generated at a slight angle to the cutting direction, first usually at the die edge. When these cracks meet, shearing is complete and the cutting force drops (Fig. 5-7) even though the cutting edges had moved only partly through the thickness of the sheet (Fig. 5-6*a*). The *fracture surface* is not perfectly perpendicular to the sheet surface and exhibits some roughness; nevertheless, the finish is acceptable for many

FIGURE 5-6
Sheared parts of acceptable finish are produced (*a*) when blanking with optimum clearance. (*b*) The skirt of torn edge produced with a small clearance and (*c*) the burr produced with excessive clearance are undesirable.

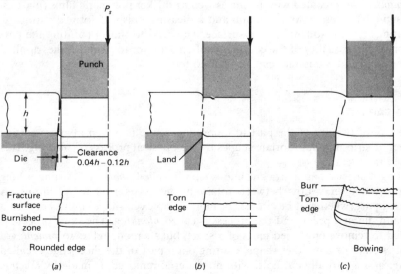

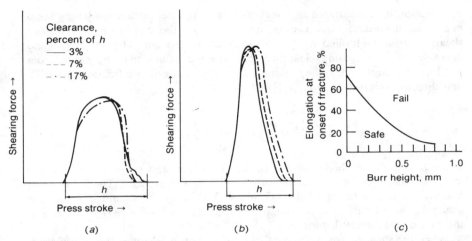

FIGURE 5-7
Total fracture sets in after the cutting edges penetrated (*a*) more than half the sheet thickness when shearing soft materials but (*b*) earlier on hard materials. (*c*) Burr produced in blanking reduces the elongation attainable in subsequent tensile deformation. (*Part* (*c*) *after S. P. Keeler, Machinery 74:101 (1968).*)

applications. The part would hang up in the die and must be pushed by the punch beyond the parallel *die land* (indicated in Fig. 5-6*b*).

The quality of the cut surface is greatly influenced by the *clearance* between the two shearing edges. With a very tight clearance, the cracks—originating from the tool edges—miss each other and the cut is then completed by a secondary tearing process, producing a jagged edge roughly midway in the sheet thickness (Fig. 5-6*b*). Excessive clearance allows extensive plastic deformation, separation is delayed, and a long fin (*burr*) is pulled out at the upper edge (Fig. 5-6*c*).

In the course of shearing thousands of parts, the tool edges wear, become rounded, and burr forms even with an optimum clearance. The jagged edge of the burr with its sharp roots acts as a stress concentrator; the harmful effect may be noted in the reduced elongation measured in the tension test (Fig. 5-7*c*). By reducing ductility, it initiates fracture during subsequent forming or in the service of the part. Therefore, proper choice of the clearance is a vital aspect of the process. A small clearance leads to more rapid tool wear; therefore, greatest economy is obtained when the clearance is chosen as large as permissible for the given application. From experience, the clearance is taken between 4 and 12% of the sheet thickness (the smaller clearance goes with a more ductile material).

5-2-2 Forces

The size of the press required to perform conventional shearing is readily calculated. Since the process involves plastic deformation as well as shear, and deformation is concentrated in a very narrow zone where strain hardening takes place, the maximum force can be obtained from the empirically determined shearing stress multiplied by the cross section to be cut. The shearing stress decreases with increasing clearance, but average values may be taken from handbooks. Alternatively, the shearing stress may be taken as a fraction of the TS and then the *shearing force* P_s is

$$P_s = C_1(\text{TS})hl = C_1 K \left(\frac{n}{e}\right)^n hl \tag{5-5}$$

where h is the sheet thickness, l is the length of cut, and C_1 is 0.85 for ductile materials and 0.65 for less ductile ones (or 0.7 on the average). The TS of most materials is known (as in Tables 4-2 and 4-3). (If only the K and n values are available, the TS may be approximated by substituting $\text{TS} = K(n/e)^n$ where e is the base of the natural logarithm.)

When the shearing edges are parallel, l is the entire length of the contour cut. This can lead to very high forces, which can then be reduced by placing the two shearing edges at an angle to each other (at a *shear* or *rake*); thus, only the instantaneously sheared length l needs to be considered (Fig. 5-8a). In blanking, the *scrap bridge* can be allowed to bend and the rake is on the die (Fig. 5-8b). In punching, the punched-out *scrap* can be bent, and the rake is on the punch (Fig. 5-8c).

The *shearing energy* E_s to be delivered by the press is equal to the area under the force-displacement curve (Fig. 5-7a and b). An approximate value can be obtained from

$$E_s = C_2 P_s h \tag{5-6}$$

where $C_2 = 0.5$ for soft materials (Fig. 5-7a) and 0.35 for hard materials (Fig. 5-7b).

FIGURE 5-8
Shearing forces can be reduced by giving a rake or shear to (a) the blades in a guillotine, (b) the die in blanking, or (c) the punch in piercing.

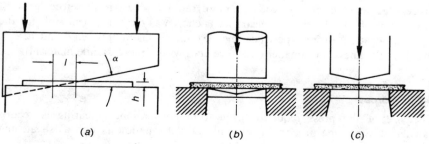

(a) (b) (c)

Example 5-2

Circular blanks of $d_0 = 10$-in diameter are to be cut from $h = 0.125$-in-thick, annealed 5052 aluminum alloy. What press force and energy are needed?

From Table 4-3, TS = 190 MPa = 27 kpsi.

From Eq. (5-5), $P_s = 0.85(27)(0.125)(10\pi) = 90\,000$ lbf.

From Eq. (5-6), $E_s = 0.5(90\,000)(0.125) = 5630$ lbf · in.

Example 5-3

Mild steel plate of 5-mm thickness and 2-m width is cut in the width direction. Estimate the shearing force for cutting (a) with parallel blades and (b) in a guillotine in which the blades are given a 6° shear.

From Table 4-2, for 1015 steel, TS = 450 MPa.

(a) The length to be cut $l = 2$ m; $P_s = 0.85(450)(0.005)(2) = 3.83$ MN.

(b) From the geometry of the operation (Fig. 5-8a), $l = h/\sin\alpha = 5/0.1045 = 48$ mm. Hence $P_s = 0.85(450)(5)(48) = 92$ kN. This value is approximate but shows the large drop in force to be expected.

5-2-3 Finish Blanking and Punching

There is great demand for processes that produce very clean-cut edges, perpendicular to the sheet surface and of a surface finish sufficiently smooth to allow immediate use of the parts, e.g., as gears in lightly loaded machinery and close-tolerance, contacting members in instruments. Several approaches are possible; in most of them, a *counterpunch* cooperates with the main punch and, as an additional benefit, eliminates curvature of the part.

1 We saw that fracture can be delayed by the imposition of a high hydrostatic pressure (Fig. 4-2a). This principle is exploited in *precision blanking* or *fine blanking* (Fig. 5-9a). A specially shaped blankholder (*V-ring, impingement ring*) is pressed into the part just prior to beginning the cut; thus, the deformation zone is kept in compression and the whole thickness is plastically sheared.

FIGURE 5-9
Parts with finished edges can be produced by (a) precision blanking, (b) negative-clearance blanking, (c) counterblanking, or (d) shaving a previously sheared part.

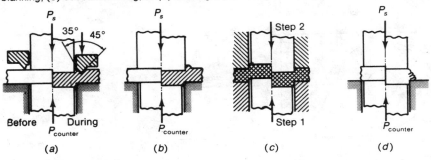

2 A high hydrostatic pressure is also maintained on *shearing with a negative clearance*, and the part is actually pushed (extruded) through the cutting die (Fig. 5-9*b*).

3 In two-sided shearing (*counterblanking*) the sheet is clamped between two dies (Fig. 5-9*c*). The punches penetrate in one direction until cracks are initiated, and then the cut is completed in the other direction.

4 A conventionally blanked part may be *finish-shaved* in a die set with tight clearances (Fig. 5-9*d*). This is equivalent to cutting with a zero-rake-angle tool (Sec. 8-1-1).

5-2-4 Processes and Equipment

Holes of standard sizes and shapes can be cut on general-purpose punch presses. Numerically controlled presses, equipped with an *xy* table and a rotating tool changer, allow rapid and accurate location of the sheet and selection of the punch and die, and thus permit low-cost production of small and medium quantities. Larger holes can be cut by repeated cutting with the same punch or by nibbling.

FIGURE 5-10
A compound die performs all cuts simultaneously.

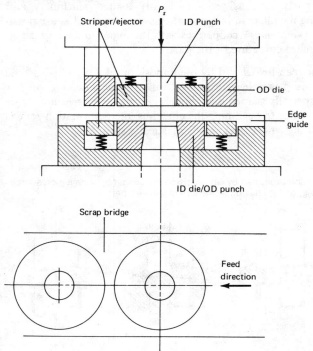

Combined with cutting methods based on welding techniques (laser, electron beam, plasma torch, electric arc, or oxyfuel cutting, Sec. 9-6) for the cutting of curved contours, such *cutting centers* (*CNC punching machines*) become extremely versatile.

In mass production, the punch and die are made of tool steel or sintered WC (Table 4-6). The scrap bridge (*skeleton* or, in punching, the part) would bind onto the punch (or in the die) and must be stripped with fixed, spring-supported, or cam-driven stripper plates (Fig. 5-10) or with a plastic (usually polyurethane foam) pad (Fig. 5-11). Complex geometries can be created in *compound dies* in which several cutting edges work simultaneously (Fig. 5-10). In *progressive dies* several punching and blanking operations are sequentially performed with die elements fastened to common die plates, while the strip is fed in exact increments (*indexed*, Fig. 5-11). Such blanking is a high-productivity process, limited only by the rate of feeding material into the press and by the rate of stroking the press. Multiple punches are used when many parts or holes are to be produced, as in the

FIGURE 5-11
A progressive die performs different cuts in successive stations (die elements for cutting the locating trim are not shown).

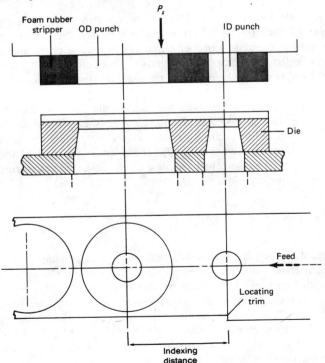

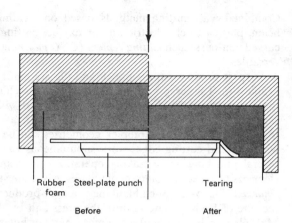

FIGURE 5-12
Low-cost blanking and punching is possible with a rubber foam cushion.

Rubber   Steel-plate punch
foam

Before

Tearing

After

blanking of circles for can making or the punching of holes for perforated metals. *Lamination dies* are used to blank sheet for transformers and motors. Highest production rates are obtained in *roll piercing* with the die and punch located on the surfaces of rolls.

The scrap bridge represents material loss. The minimum width of the bridge is limited by the danger of pulling the bridge material into the die clearance. Material utilization can be optimized by proper *layout and nesting* of parts, an art which is considerably aided by computer programs. Productivity is further increased and material losses cut if several parts are blanked from a wider strip (see Prob. 5-4).

For smaller quantities of, say, a few hundred pieces, the die cost can be lowered if a greater scrap loss is tolerable. In *rubber pad blanking* the die is simply a steel plate cut to size, and the cutting action occurs by pressing the sheet around this die with a rubber cushion (Fig. 5-12). The overhanging part of the sheet is bent down and clamped against the base plate by the cushion, and tearing occurs around the edges of the die plate.

5-3   BENDING

Many parts are further shaped by the relatively simple process of *bending* in one or several places. Characteristic of this process is stretching (tensile elongation) imposed on the outer surface and compression on the inner surface (Fig. 5-13). For a given sheet thickness h, tensile and compressive strains increase with decreasing forming radius R_b (i.e., with decreasing R_b/h ratio). Thus, plastic, irreversible bending differs from elastic, reversible bending (Fig. 2-6) in that the bend radius must be small enough to bring much of the sheet cross section into the state of plastic flow. There is, again, only one line (the *neutral line*) which retains its original length.

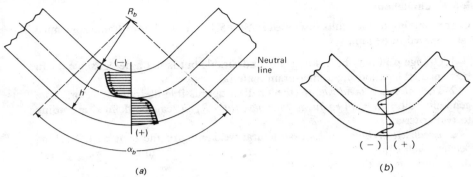

FIGURE 5-13
In the course of bending (*a*) the entire stress–strain curve is traversed; (*b*) elastic stresses result in springback and the retention of a residual stress pattern.

When bending with relatively generous radii, the neutral line is in the center. When bending around tight radii, the neutral line shifts toward the compressive side, the centerline is elongated, and constancy of volume is preserved by thinning of the sheet. The increased length of the centerline is usually taken into account for bends of $R_b < 2h$ by assuming that the neutral line is located at one-third of the sheet thickness. When the sheet is relatively narrow ($w/h < 8$), there is also a contraction in width w.

Example 5-4

The part shown is to be made of 3-mm-thick sheet. Calculate the length of strip.

In bending to $R_b = 10$ mm radius, $R_b/h = 3.3$; hence, the neutral plane will be in the center of the sheet; since the bend is over a 90° angle, $l_4 = 2\pi(R_b + 0.5h)90/360 = 2\pi(10 + 1.5)90/360 = 18.06$ mm.

In bending to $R_b' = 3$ mm, $R_b/h = 1$; hence, the neutral line is at $0.33h$; for a bend of $180 - 45 = 135°$ angle, $l_2 = 2\pi[10 + (0.33)(3)]135/360 = 25.92$ mm.

Thus the total starting length is $l = l_1 + l_2 + l_3 + l_4 + l_5 = 20 + 25.92 + 50 + 18.06 + 20 = 133.98 = 134$ mm. (If we had ignored the shift of the neutral line, l would have been 135.16 mm.)

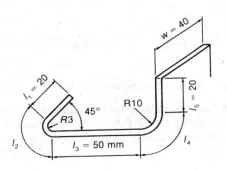

5-3-1 Limitations

Corresponding to the limits discussed in Sec. 5-1-1, a number of limitations must be observed in bending:

1 Orange peel may be aesthetically undesirable but is not a defect since it can be remedied by choosing a finer-grain material.

2 The *minimum bend radius* (the smallest permissible die radius R_b or, more generally, the *minimum radius-to-thickness ratio* R_b/h, can be defined according to two criteria:

a *Localized necking* causes a structural weakening of the bent part. Necking occurs when elongation in the outer fiber, e_t, exceeds the uniform elongation of the material e_u in the tension test

$$e_t = \frac{1}{(2R_b/h) + 1} \le e_u \tag{5-7}$$

For materials that obey the power law of strain hardening, Eq. (4-4), $\epsilon_u = n$ and the engineering uniform strain e_u may be calculated from

$$e_u = (\exp n) - 1 \tag{5-7'}$$

The relationship holds best for steels; for most other materials, the actual e_u measured in the tension test should be used. Because the strain is redistributed to adjacent zones during bending, a somewhat higher strain is usually permissible. A burr acts as a stress raiser and, if on the outer surface, leads to much earlier fracture. Therefore, if at all possible, the burr is oriented toward the punch.

b *Fracture* represents an absolute limit. This is directly related to the reduction in area q measured in the tension test (Eq. (2-10), and Tables 4-2 and 4-3). The minimum permissible bend radius may be estimated for less ductile materials from the following formula

$$R_b = h\left(\frac{1}{2q} - 1\right) \quad \text{for } q < 0.2 \tag{5-8a}$$

and for ductile materials, because of the shift of the neutral radius in tight bends, from

$$R_b = h\frac{(1-q)^2}{2q - q^2} \quad \text{for } q > 0.2 \tag{5-8b}$$

3 *Crushing* on the inside surface may occur when bending to very tight radii.

Anisotropy, of any origin, affects bending. We have seen that mechanical fibering (Sec. 4-1-5) results in greater ductility in the rolling direction, and it is usually more favorable to bend sheet with the bend line oriented across the rolling direction. A textured material of low r value thins down easily (Fig. 5-4a) and thus can be bent around tighter radii than a material of high r value.

Example 5-5

The part of Example 5-4 was originally made of annealed cartridge brass. It is now proposed that, as a weight-saving measure, it should be made of 5052-H34 aluminum alloy. Is there any problem to be expected?

The relevant properties for the two materials, from Table 4-2 and MHDE (p. 633) are:

	Brass	**5052-34**	**5052-0**	**6061-T4**
YS, MPa	100	215	90	145
TS, MPa	310	260	195	240
Elong., %	65	10	25	22
R.A., %	75			

The yield strength is perfectly adequate. The tightest bend is $R_b/h = 3/3 = 1$; from Eq. (5-7), a uniform elongation of $1/(2 + 1) = 33\%$ would be desirable. Since the total elongation of 5052-H34 is only 10%, uniform elongation must be even less and the material will fail in the bend. A soft (5052-0) sheet with 25% total elongation would perhaps survive because of the redistribution of strain, but the YS is slightly low. Aluminum alloy 6061-T4 would do better, although the bend radius may have to be relaxed.

5-3-2 Stresses and Springback

The stress state is extremely complex in bending. The complete tensile and compressive stress-strain curves of the material are traversed on the tensile and compressive sides of the bend, respectively. This means that around the neutral plane the stresses must be elastic. When the forming tool is retracted, the elastic components of the stress cause *springback*, and a residual stress pattern, shown in Fig. 5-13b, develops. Springback increases both the angle and radius of the bent part (Fig. 5-14). The elastic zone is more extensive for a relatively gentle bend (large R_b/h ratio) and for a material with a high ratio of yield strength $\sigma_{0.2}$ to elastic modulus E; therefore springback also increases according to the approximate formula

$$\frac{R_b}{R_f} = 1 - 3\left(\frac{R_b}{h} \frac{\sigma_{0.2}}{E} \right) + 4\left(\frac{R_b}{h} \frac{\sigma_{0.2}}{E} \right)^3 \qquad (5\text{-}9)$$

where R_b is the radius of the bending die and R_f is the radius obtained after the forming pressure is released.

Since the length of the neutral line does not change, the angle after springback, α_f, can be obtained (in radians) from

$$\alpha_f\left(R_f + \frac{h}{2} \right) = \alpha_b\left(R_b + \frac{h}{2} \right) \qquad (5\text{-}10)$$

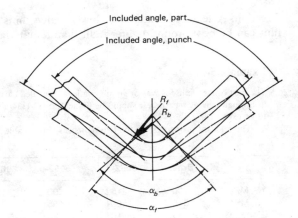

FIGURE 5-14
Dimensions used to characterize springback.

Springback establishes a new force equilibrium with a residual stress distribution typified by a compressive stress on the outer and tensile stress on the inner surface (Fig. 5-13*b*).

Several techniques are used to combat springback:

1 If springback for a given material is known and if the material is of uniform quality and thickness, compensation for springback is possible by *overbending* (Fig. 5-15*a* and *b*).

2 Alternatively, the elastic zone can be eliminated at the end of the stroke by one of two means. First, the two ends of the sheet may be clamped before the punch bottoms out, so that the end of the stroke involves *stretching* of the part, causing tensile yielding in the entire sheet thickness. In the second method the

FIGURE 5-15
Springback may be neutralized or eliminated by: (*a*), (*b*) overbending, (*c*) plastic deformation at the end of the stroke, and (*d*) subjecting the bend zone to compression during bending. (*Part (d*) after V. Cupka, T. Nakagawa, and H. Tyamoto, CIRP **22**:73–74 (*1973*).)

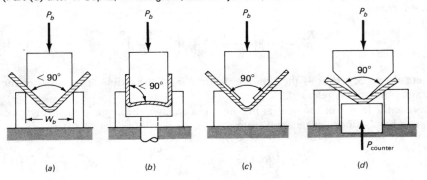

(*a*)   (*b*)   (*c*)   (*d*)

punch nose is shaped to *indent* the sheet, so that plastic compression takes place throughout the thickness (Fig. 5-15c).

3 If a *counterpunch* is used with a controlled pressure, compressive stresses are maintained in the bend zone during the entire process (Fig. 5-15d). Since this also has the effect of imposing a hydrostatic pressure on the bend zone, bending beyond the limits given by Eqs. (5-8a) and (5-8b) is possible.

4 Less-ductile materials may have to be bent at some elevated temperature; because the yield strength is lower, springback is also less.

Bending Force   A very simple estimate of the *bending force* in free bending to 90° may be obtained from

$$P_b = \frac{wh^2(\text{TS})}{W_b} \tag{5-11}$$

where W_b is the width of the die opening (Fig. 5-15a) and w is the width of the strip (the length of the line over which bending takes place).

Example 5-6

How much is the springback in making the 90° bend of the part of Example 5-4, if the workpiece material is (a) annealed cartridge brass or (b) 6061-T4 aluminum alloy. From Table 2-3, $E(\text{brass}) = 140$ GPa. $E(\text{Al}) = 70$ GPa, $R_b/h = 3.3$, and $\sigma_{0.2}$ is taken from Example 5-5.

Thus, for the brass, $R_b/R_f = 1 - 0.0071 + 0.0 = 0.9929$; for the Al alloy, $R_b/R_f = 1 - 0.0205 + 0 = 0.9795$. Springback is negligible with the brass but not with the aluminum alloy.

Example 5-7

Calculate the force required for making the 90° bend in the part of Example 5-4, assuming that the workpiece material is brass.

From Table 4-2, TS = 310 MPa. The minimum die opening must accommodate l_4 plus some straight length (say, twice 10 mm). Thus $W_b = 18 + 20 = 38$ mm. The width of the part is $w = 40$ mm; $h = 3$ mm. From Eq. (5-11), $P_b = (40)(3)^2(310)/38 = 2937$ N. It is usual to allow some 20% more; thus, the force is 3.5 kN.

5-3-3 Bending Methods

The equipment used for bending depends on the size, mostly length, of the bent part. Short lengths can be bent at high rates in mechanical presses in dies (as in Fig. 5-15).

Longer lengths call for special presses with very long beds (*press brakes*). In these, simple tooling suffices for the forming of complex shapes by repeatedly bending a long sheet (Fig. 5-16). The female die may be replaced by a slab of polyurethane foam; thus, tooling costs are reduced. The advantage of press brakes is that a great variety of parts can be produced with a limited number of tools. In conjunction with mechanized sheet feeders, the press lends itself to computer

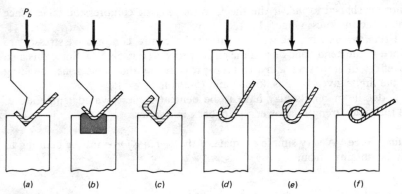

FIGURE 5-16
Press-brake forming of (a) a 90° angle; (b) the same but with a polyurethane female die; (c) a U channel; and (d)–(f) a bead.

control including a back gage for sensing sheet position. More sophisticated control schemes are available which compensate for springback: they derive the essential characteristics of the elastic–plastic stress–strain curve of the material from information obtained by force and displacement transducers.

Bending along a straight line is also possible by a *wiping* motion (Fig. 5-17a). To estimate the bending force, W_b may be taken as $(2R + h)$.

A uniform but adjustable curvature may be imparted to a sheet, plate, or section by passing it through a *three-roll bender*, with the rolls arranged in a pyramidal fashion (Fig. 5-17b). This is an important preparation step for making large welded-plate structures.

Continuous production and very high production rates become possible in *roll forming*. Bending is now done progressively, by passing the strip between con-

FIGURE 5-17
Sheet may also be bent with (a) a wiping die or (b) bending rolls (pyramidal rolls).

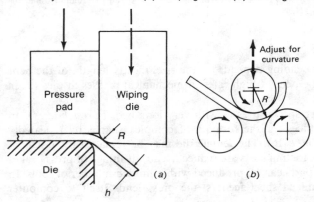

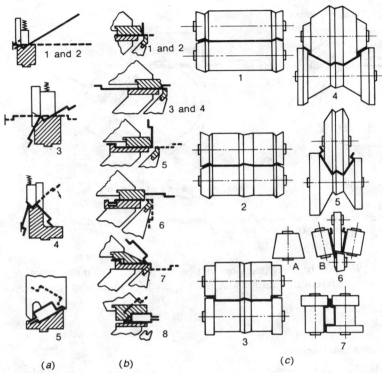

FIGURE 5-18
Complex profiles such as door frames may be formed by a sequence of operations on (*a*) press brakes, (*b*) wiping dies, or (*c*) by profile rolling. (*After G. Oehler, Biegen, Hanser Verlag, München, 1963.*)

toured, driven rolls placed in tandem (in a line). A typical product is corrugated sheet. For many other shapes, idling rollers are used to press the sides of the partially formed shape. Thus, tubes for subsequent welding, sections that replace hot-rolled or extruded sections, as well as complex shapes such as door frames can be formed (Fig. 5-18).

Besides the bending of sheet metal, bending of sections and tubes is an important manufacturing activity. The problem in free bending (e.g., in pyramidal benders) is usually that of distortion and buckling of more complex shapes; better results are obtained when the profile or tube is wrapped around a *form block*. Conformance to the form block is ensured by *winding under tension*, by passing a *wiper roll* or *wiper-block*—hinged at the center of the radius of curvature—around the section or tube (Fig. 5-19*a*), or by a *rotating form block* (Fig. 5-19*b*). To prevent the collapse of tubes when bending over tight radii, the inside can be supported with sand, a low-melting-point metal or, more economically, by a mandrel made up of individual sections (Fig. 5-19*c*), or by a fixed mandrel over which the tube is drawn as it is bent around the shaped, rotating die or form

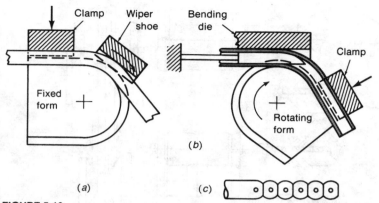

FIGURE 5-19
Tubes and sections may be formed by (*a*) compression bending or (*b*) draw bending, sometimes with the use of a (*c*) linked-ball mandrel.

block (Fig. 5-19*b*). NC bending machines can be programmed to make tubes with several bends in different orientations, as are required for hydraulic systems and exhausts.

5-3-4 Flanging and Necking

Some complex forms of bending are encountered in working the edges of blanks, holes, and tubes:

1 *Flanging of a blank* deforms the outer edge (Fig. 5-20*a*). It is similar to a shallow deep-drawing operation and sets no great demand on ductility.

2 In contrast, *flanging of a hole* (Fig. 5-20*b*) imposes severe tensile strains on the edge of the hole. If burr is present on the cut edge or if the sheet material contains inclusions or other defects, splitting occurs at a much lower strain than would be expected from the tensile elongation measured in the absence of a burr (for the effect of burr on ductility, see Fig. 5-7*c*). Deburring, shaving, and even reaming of the hole may become necessary in critical cases.

3 Severe tensile strain is imposed also in the *expansion* or *flanging the ends* of a tube (Fig. 5-20*c*). In contrast, the necking of a tube (Fig. 5-20*d*) imposes compressive stresses and the reduction that can be taken in a single operation is limited only by the axial collapse of the tube or by the formation of internal wrinkles. Necking is an important step in making cartridge cases and pressurized-gas cylinders.

Examples of flanging a sheet and flanging the end of a tube are encountered in the forming of double seams for sealing food and beverage cans (Fig. 5-21). Tens of billions of such products are made annually in North America alone.

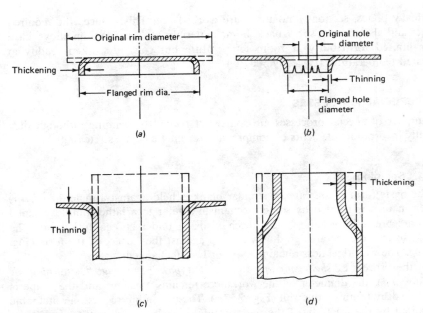

FIGURE 5-20
Deformation is (*a*) compressive when flanging a disk but (*b*) tensile when a hole is flanged (cracks shown are a consequence of excessive tensile strain). Strain is (*c*) tensile in flanging a tube but (*d*) compressive in necking.

FIGURE 5-21
Lids are attached to can bodies by double lock seams formed in two operations.

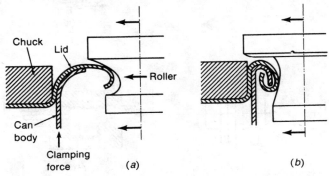

Heavy plates, sections, and tubes are worked hot. This reduces the required forces and also reduces springback. Some materials, such as titanium alloys, have very limited formability at room temperature but can be worked readily at elevated temperatures (typically 500 °C).

5-4 SPECIAL PROCESSES

A number of special processes are somewhat related to bending although they usually incorporate elements of compressive forming and/or stretching.

5-4-1 Spinning

In the basic form of *spinning*, a circular blank is held against a male die (*form*) which in turn is rotated by some mechanism similar to a lathe spindle. Shaped tools are pressed, by hand, tracer mechanism, or under NC control against the blank, so that the metal is gradually laid up against the surface of the form (Fig. 5-22a). The wall thickness remains more or less unchanged.

In the process of *shear spinning* (also called *power spinning, flow turning,* or *spin forging*), the diameter of the workpiece remains constant and the shape is developed by thinning the wall (Fig. 5-22b). The maximum reduction obtainable is limited by the ductility of the material and correlates well with reduction of area in the tension test. Very large thick-walled shapes are spun hot.

FIGURE 5-22
Special techniques include (a) spinning, (b) shear spinning, (c) tube spinning, (d) expansion, (e) expansion applied to the making of bellows, and (f) the production of T fittings.

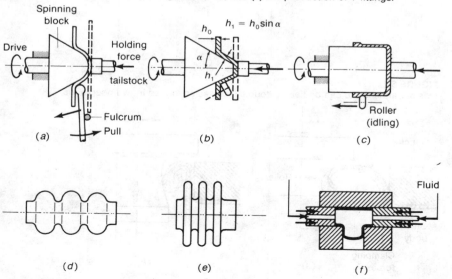

Tube spinning is a form of power spinning in which the wall thickness of a tube or vessel is reduced (Fig. 5-22c).

5-4-2 Bulging

Tensile deformation is typical of the *bulging* of tubes, containers, and similar products, using rubber (polyurethane foam) plugs or hydraulic pressure (Fig. 5-22d). The technique also represents the first step in making metal bellows (Fig. 5-22e); the prebulged tube forms the bellows when axially compressed.

When axial compressive stresses are applied simultaneously with the expanding pressure, very large deformations become possible. A special application is the making of parts such as copper T-fittings (Fig. 5-22f). The tube, constrained in a container, is compressed between two punches while a pressurized fluid is applied internally. Thus a deep bulge, necessary for the T shape, is formed without danger of fracture.

5-4-3 Peen Forming

We saw in Fig. 2-15b that unbalanced internal stresses cause distortion of the part. The principle is exploited in *peen forming* by the judicious shot peening of one of the surfaces. The impacting shot causes localized deformation, expansion of the surface, and the part becomes convex. The technique is used for shaping gently curved surfaces such as aircraft wing skins and also for correcting shape defects in products such as rocket cases.

5-5 SHEET FORMING

Enormous quantities of sheet metal are formed into more or less deep, container-like components of a great variety of shapes. In contrast to bent parts, they are characterized by curvatures in two directions. They can be produced by stretch forming, deep drawing, or their combination.

5-5-1 Stretch Forming

In pure *stretch forming* the sheet is completely clamped on its circumference and the shape is developed entirely at the expense of sheet thickness. Physically this can be achieved in a variety of ways:

1 The sheet may be clamped with a multitude of fixed or swiveling clamps (Fig. 5-23a). The advantage is that only one die (*male die* or *form punch*) is needed, but productivity is low; hence, such stretch forming is most suitable for low-volume production as is typical of the aircraft industry. Very large parts (fuselage skins, wing skins, boat hulls) can be formed. Springback can be substantial when forming very gently curved shapes and then forming at elevated

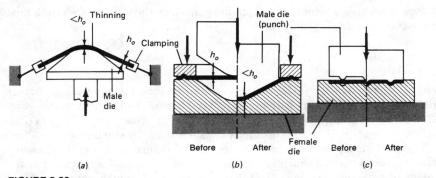

FIGURE 5-23
The shape is developed entirely at the expense of wall thickness in (a) stretch forming, (b) stretch drawing, and (c) embossing.

temperatures (sometimes by allowing creep or superplastic deformation over the die) is helpful. Rolled and extruded sections may also be stretch formed.

2 For mass production, such as is typical of the automotive and appliance industries, the blank is clamped with an independently movable *blankholder* which retains the sheet with the aid of *draw beads* (Fig. 5-23b); the punch cooperates with the *female die* to define the shape. One part is finished for each press stroke; thus, productivity is high but die costs are higher too.

3 In the process of *embossing* (Fig. 5-23c) the sheet is restrained by the sheet itself, through the multiple contact points with the die.

Stretch Formability The first limit is reached in stretching when a localized neck becomes visible, and the ultimate limit is given by subsequent fracture. The *formability limit* is a technological property and the limit strain depends on the material, the strain state, and friction on the punch surface.

The influencing factors are clearly shown when a clamped sheet is stretched by a hemispherical punch (Fig. 5-24a). Localized strain variations (the *strain distribution*) can be revealed simply by applying a grid of small (typically, 2–6-mm-diam) circles (or a *square/circle grid*) onto the sheet surface, usually by electrolytic etching or a photoresist technique. In the course of straining, thinning of the material is accompanied by a growth of the circles, as required by constancy of volume (Eq. (2-2)). When deformation is the same in all directions, as it would be on blowing up a balloon (*balanced biaxial strain*), the circle expands into a circle of larger diameter. When deformation is different in different directions, the circle distorts into an ellipse: the major axis gives the major strain and, perpendicular to it, the minor axis gives the minor strain.

When a sheet of a given material is stretched over the hemispherical punch, strain distribution depends on a number of factors:

1 In the total absence of friction (which could in reality be achieved only in bulging with hydraulic pressure), the sheet thins out gradually toward the apex

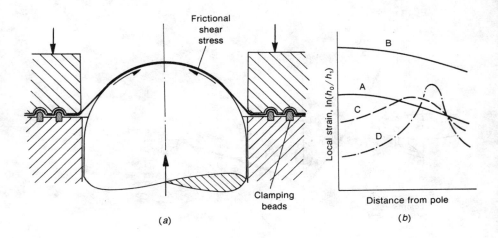

FIGURE 5-24
(a) Friction on the punch surface opposes thinning and leads to (b) changes in strain distribution from pole to flange. (*Adapted from J. A. Schey, Tribology in Metalworking: Friction, Lubrication and Wear, American Society for Metals, 1983, p. 520. With permission.*)

where fracture finally occurs (line A in Fig. 5-24b). Thinning is more uniformly distributed with a material of high n value, and a deeper dome can be obtained before necking becomes localized (line B). It will be recalled that in the tension test localized necking occurs at $\epsilon_u = n$ (Sec. 4-1-1). In balanced biaxial tension the presence of the transverse strain prevents the formation of a localized neck. Straining can continue until a local neck develops at or close to the apex, at some point where there is some inhomogeneity in the material or the sheet was originally thinner. In stretching over a punch, the depth of stretch never reaches that obtained in frictionless, hydraulic bulging but, as long as friction is very low, failure still occurs at the apex.

2 Friction on the punch surface hinders free thinning at the apex, and with increasing friction the position of maximum strain moves toward the die radius (lines C and D in Fig. 5-24b) and strain becomes more localized.

3 Under otherwise identical conditions, a thicker sheet gives a deeper stretch, because bending superimposed on stretching improves ductility.

4 The depth of stretch increases with all material variables that delay necking (high n value, transformations), or increase the post-necking strain (high m value). Indeed, a good empirical correlation is found between total elongation in the tension test and limiting dome height.

Forming Limit Diagram A comparison of various materials is possible with the aid of the *forming limit diagram* (FLD). Gridded sheet metal strips of different widths are tested with a very good lubricant (e.g., oiled polyethylene

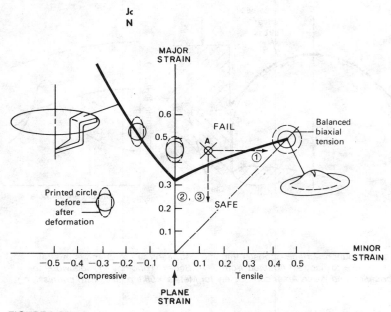

FIGURE 5-25
The forming limit diagram typical of low-carbon steel gives the permissible deformations at various strain ratios. (*Left-hand side typically after G. M. Goodwin, SAE paper 680092, 1968; right-hand side after S. P. Keeler, SAE paper 650535, 1965.*)

film) on the punch. A sheet wide enough to be clamped all around gives the balanced biaxial tension point (Fig. 5-25). As the strip width diminishes, the minor strain decreases too until, at some characteristic strip width, it becomes zero. By definition, this is a condition of plane strain (Fig. 4-12). The FLD is usually constructed for localized necking (another curve could be constructed for fracture). The FLD moves higher for thicker sheet and is, obviously, lower for a material of lower ductility (lower n value).

The FLD is a system characteristic and the FLDs of two materials can be compared only if determined under identical conditions. With decreasing circle size, the FLD moves higher and changes its shape, because more of a small circle falls into the necked zone where strain is high.

The FLD was introduced in the 1960s and quickly became an important tool in diagnosing production problems. When parts are found to split in production, gridded sheets are placed into the production die and are stretched. Distortion of the circles is measured (sometimes with the aid of an instrument called an optical grid analyzer). The circle nearest to the fracture line gives the strain ratio at the critical point and defines, say, point A in Fig. 5-25. Several remedies, some of them not intuitively evident, may then be explored to bring strains within allowable limits:

1 *Increase the minor* strain by clamping more firmly in that direction.

2 If fracture occurred away from the apex, *improve lubrication* to redistribute strains (as in Fig. 5-24b).

3 If all else fails, the part will have to be *redesigned* to reduce the major strain, or some material must be allowed to flow into the die, changing the process into combined stretching-drawing (Sec. 5-5-3).

The FLD of Fig. 5-25 is typical of steel and some aluminum alloys. Some other materials, such as austenitic stainless steels and brass, show no improvement in the forming limit with increasing biaxiality of strains. Nevertheless, some of them have a high *n* value (Tables 4-2 and 4-3) and are then eminently suitable for stretching.

Example 5-8

Complex shapes such as automotive hub caps are often made of mild steel (chromium plated), stainless steel, or aluminum alloy. If the caps are to be made by stretching, which of these alloys allows the deepest stretch?

A first approximation can be obtained by comparing the *n* values. From Tables 4-2 and 4-3, austenitic stainless steels are best ($n = 0.3$), followed by low-carbon steel (0.25), 5052 Al (0.13), and martensitic stainless steel (0.1). If deep details are to be made in aluminum alloy or martensitic stainless steel, draw-in of the metal must be encouraged (see Sec. 5-5-3).

5-5-2 Deep Drawing

The difference between stretching and deep drawing is substantial: In the former, the blank is clamped and depth attained at the expense of sheet thickness; in the latter, the blank is allowed—and even encouraged—to draw into the die, and thickness is nominally unchanged.

In the simplest case of pure *deep drawing* or *cupping*, a circular blank of diameter d_0 is converted into a flat-bottomed cup by drawing it through a draw die with the aid of a punch of diameter D_p (Fig. 5-26). Both the die and punch must have well-rounded edges, otherwise the blank might be sheared. The finished cup is stripped from the punch—for example, by machining a slight recess (a ledge) into the underside of the draw die. After the cup has been pushed through the die, its top edge springs out because of springback, gets caught in the ledge on the return stroke of the punch, and the ledge strips the cup. A central hole is often provided in the punch to prevent the formation of a vacuum and thus aid stripping.

The stress state prevailing in the part *during* drawing is shown in Fig. 5-27a about halfway through the draw. The base is in balanced biaxial tension; the side wall is in plane-strain tension because the punch does not allow circumferential contraction; material in the transition between wall and flange is subjected to bending and rebending (straightening out); and the flange is in circumferential compression, because the circumference of the blank is reduced while it is forced

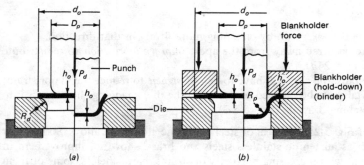

FIGURE 5-26
Containers may be formed by drawing (*a*) without or (*b*) with a blankholder.

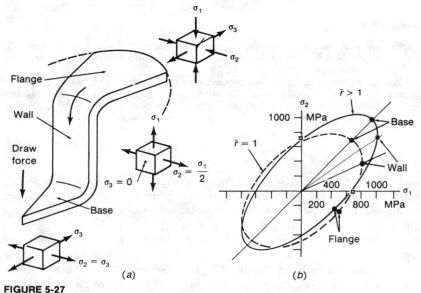

FIGURE 5-27
(*a*) The stress state varies greatly in different parts of a partly drawn cup. (*b*) Yielding is affected by anisotropy of the sheet material, as shown by the experimental data for Ti–4Al sheet. (*After W. A. Backofen, Deformation Processing, Addison-Wesley, 1972, p. 54.*)

to conform to the smaller diameter of the die opening. The force required to perform all this deformation must be borne by the base of the cup. This limits the attainable deformation, expressed as *reduction* $(d_0 - D_p)/d_0$ or as *drawing ratio* d_0/D_p. The maximum diameter of the circle that can be drawn under ideal conditions is expressed as the *limiting draw ratio* (LDR)

$$\text{LDR} = \frac{d_{0(\text{max})}}{D_p} \tag{5-12}$$

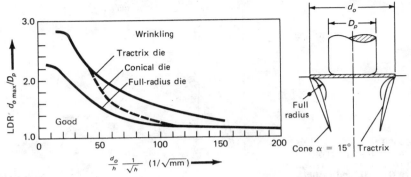

FIGURE 5-28
The limiting draw ratio in cupping low-carbon steel without a blankholder is a function of die geometry and of blank diameter-to-sheet thickness ratio. (*After G. S. A. Shawki, Werkstatts-technik, 53:12– 16 1963. With permission of Springer-Verlag, New York.*)

The circumferential compressive stresses cause the blank to thicken, and the punch-to-die clearance is usually some 10% larger than the sheet thickness to accommodate this *thickening* without the need for reducing (*ironing*) the wall. Compression can also lead to *wrinkling* (equivalent to buckling in upsetting, Fig. 4-21*a*) in the flange. Thus, both forces and wrinkling set limits. In practice, two methods of operation are feasible:

1 In *drawing without a blankholder* (Fig. 5-26*a*), wrinkling can be avoided only when the sheet is sufficiently stiff. This is always the case for very shallow draws, when the drawing ratio $d_0/D_p < 1.2$. Relatively thick blanks give higher drawing ratios (Fig. 5-28); wrinkling depends also on the die profile, which determines the rate of circumferential compression. Most favorable is the tractrix die; it will be noted from Fig. 5-28 that exceptionally high LDRs can be obtained with it.

2 When the blank is relatively thin, and the draw ratio is beyond the limits indicated in Fig. 5-28, deformation must be conducted by *drawing with a*

FIGURE 5-29
Deep drawing of low-carbon steel cups from (*a*) a round blank, with (*b*) insufficient, (*c*) optimum, and (*d*) excessive blankholder pressure. Note in (*c*) the typical earing due to planar anisotropy. (*From J. A. Schey, as Fig. 5-24, p. 527.*)

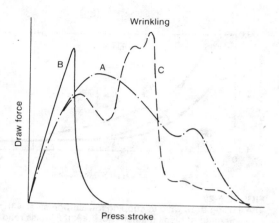

FIGURE 5-30
Draw force curves typical of
drawing with optimum (line A),
excessive (line B), and insufficient (line C) blankholder pressure.

hold-down or *blankholder* (Fig. 5-26b). The blankholder must exert sufficient
pressure to prevent wrinkling (Fig. 5-29b), but excessive *blankholder pressure*
would restrict free movement of the material in the draw ring and thus cause
excessive thinning and finally fracture in the partly formed cup wall (Fig. 5-29d).
To produce a sound cup (Fig. 5-29c), the blankholder pressure may be taken, as a
first approximation, as 1.5% of the yield strength $\sigma_{0.2}$ of the material.

When the optimum blankholder pressure is applied, draw force rises as the
part-drawn flange strain hardens; as the flange diameter decreases, force drops
until the thickened edge of the blank is ironed (Fig. 5-30, line A). Excessive
pressure causes early fracture (line B). Too low pressure allows wrinkling (line C)
and, if the wrinkles cannot be ironed out, the cup fails near the end of the draw
(Fig. 5-29b).

A very approximate estimate of the *drawing force* may be obtained from the
formula

$$P_d = \pi D_p h (\text{TS}) \left(\frac{d_0}{D_p} - 0.7 \right) \tag{5-13}$$

Example 5-9

A low-carbon steel container of 4.125-in height and 2.375-in internal diameter is to be made of
0.067-in-thick strip material. The bottom radius is 0.375 in. Assuming that the average wall
thickness of the container is equal to the sheet thickness, (a) calculate the starting blank diameter,
(b) determine the draw sequence, assuming that the first draw is made with a blankholder, and (c)
estimate the press force for the first draw.

(a) The volume consists of the volumes of the (1) side wall = $H(\text{OD}^2 - \text{ID}^2)\pi/4 = 1.893$ in³;
(2) one-quarter of a hollow toroid (from the theorem of Pappus–Guldin, this is equal to the area
$ABCD$ in the illustration multiplied by the circumference of the circular path described by its
centroid = $2\pi r(\text{area}) = 0.29$ in³); (3) volume of disk = $h(d^2\pi/4) = 0.139$ in³. Thus total volume is
$1.893 + 0.29 + 0.139 = 2.322$ in³. This is equal to $h(d_0^2\pi/4)$. Thus $d_0 = 6.643$ in.

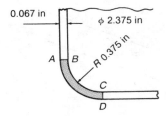

(*b*) From Fig. 5-33, LDR = 2.4; hence, in the first draw $D_p = 6.643/2.4 = 2.77$ in. This leaves a $(2.77 - 2.375)/2.77 = 14\%$ reduction for the redrawing operation. In practice it would be prefer-able to make the first draw less critical, say DR = 2.2. Then $D_p = 3.0$ in and the reduction in redrawing is 20.8%, which is still acceptable.

(*c*) From Table 4-2, for 1008 steel, TS = 320 MPa = 46.5 kpsi. From Eq. (5-13) $P_d = \pi(3)(0.067)(46.5)[(6.643/3) - 0.7] = 44.4$ klbf = 22 ton.

Limiting Draw Ratio When the LDR is reached, the draw force just exceeds the force that the cup wall can support. We have seen that the draw force is composed of the forces required to: compress the sheet in the flange circumferen-tially; overcome friction between blank and blankholder and die surfaces; bend and unbend the sheet around the draw radius; and overcome friction around the draw radius. Therefore, the LDR is not simply a material constant but depends on all variables that affect the draw force and the strength of the cup wall.

1 A high *n* strengthens the cup wall but also increases the draw force; hence, it is fairly neutral. A slight improvement in LDR is often found with higher *n* because of a later development of the force maximum.

2 A high *m* strengthens an incipient neck in the wall while barely affecting the draw force; thus, it is slightly positive in its effect.

3 The most powerful material variable is the *r* value. We saw in Fig. 5-4*b* that a material of high *r* value resists thinning while voluntarily reducing its width. This helps the blank to conform to the reduced diameter of the cup and is thus a positive factor. Furthermore, a high *r* value causes the yield ellipse (Fig. 4-11) to change (an example of experimentally determined yield loci is given in Fig. 5-27*b*). The partly drawn cup wall is in plane-strain tension, in which a high *r*-value material is stronger, whereas the flange is subjected to combined tension and compression, in which it is slightly weaker than an isotropic one. The combined result is that the LDR increases with increasing *r* (or more precisely, $\bar{r}$) value (Fig. 5-31). The effect is more powerful than it appears from Fig. 5-31, because an LDR of 2.0 gives a cup of approximately $0.8D_p$ depth, whereas an LDR of 3.0 gives a depth over $2.0D_p$.

4 Tight punch and die radii impose severe bending strain and thus increase the draw force without affecting the strength of the wall; therefore they decrease the LDR. However, very large radii would leave much of the blank unsupported and *puckering* (wrinkling between punch and die) could occur. Hence radii are

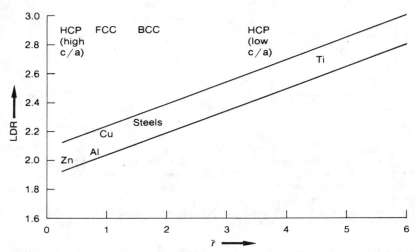

FIGURE 5-31
High normal anisotropy is a powerful factor in increasing the limiting draw ratio.

optimized, usually within the limits of $R > 4h$ for thick (> 5-mm) and $R > 8h$ for thin (< 1-mm) sheet.

5 Friction between blankholder, die, and flange surfaces adds to the draw force and is thus harmful. Contact pressures are below σ_f and, therefore, Eq. (2-18) holds. The friction stress can be reduced by reducing the normal stress (the blankholder pressure), but this is limited by wrinkling. Therefore, a good lubricant must be applied that reduces μ and thus the friction force.

6 In drawing relatively thin sheet, of d_0/h ratios over 50, the frictional force becomes a larger part of the total drawing force; hence, the LDR drops with increasing d_0/h ratio.

7 Friction on the punch is helpful because it transfers the draw force from the cup to the punch. Thus a rough punch, or a blank that is lubricated only over the flange area, gives a higher LDR.

There is still no international standard for LDR determination, and only data obtained under identical conditions are comparable.

The LDR does not necessarily give the usable cup depth. A material with planar anisotropy (Sec. 5-1-3) shows different properties in the rolling, transverse, and 45° directions ($r_0 \neq r_{90} \neq r_{45}$). This leads to *earing*, a periodic variation of the cup height (Fig. 5-29c); the ears reflect the crystal symmetry and come in pairs (4, 6, or 8).

Further Drawing Cups of a depth greater than permitted by the LDR are made by further forming after initial cupping.

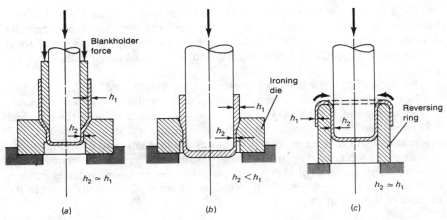

FIGURE 5-32
Cups are further deformed by (a) redrawing, (b) ironing, or (c) reverse redrawing.

1 *Redrawing* (Fig. 5-32a) leaves the wall thickness essentially unchanged.

2 *Ironing* (Fig. 5-32b) leaves the inner diameter virtually unchanged and achieves greater depth by reducing the wall thickness. It will be recognized that ironing is similar to drawing a tube on a bar (Sec. 4-7-1).

3 A basic phenomenon, not mentioned hitherto, is that a cold-worked material exhibits greater ductility when the deformation direction is reversed in successive operations (*strain softening*); this is exploited in the *reverse redrawing* of cups (Fig. 5-32c).

Redrawing is extensively used for food containers, fountain-pen caps, oil-filter housings, shock-absorber pistons, etc. Ironing is used in the mass production of drawn-and-ironed beverage cans and ammunition cartridges.

There is, of course, wide opportunity, but often combined with greater difficulty, to change the basic shape of the drawn part. In drawing square or rectangular containers the degree of difficulty increases with increasing part depth-to-corner radius ratio; earing in the corners is helpful. A punch with a curved or hemispherical end imposes a different, combined deformation state, to be discussed next.

5-5-3 Combined Stretching-Drawing

In many practical applications, most notably in the production of automotive body and chassis parts, the drawing process is neither pure stretching nor pure drawing. The sheet is not entirely clamped (therefore it is not pure stretching), neither is it allowed to draw in entirely freely (thus it is not pure drawing). Instead, the complex shapes are developed by controlling the draw-in of the sheet, retarding it where necessary with *draw beads* inserted into the die and blank-holder surfaces (Figs. 5-23*b* and 5-24*a*). To prevent die pickup and regulate draw-in, a lubricant is applied and the roughness and directionality of sheet surface finish is specified (sheet finish-rolled with shot-blasted rolls is widely used).

The shape of the part is often represented by a "sculptured" surface, one that can be described only with cubic patches or point-by-point in spatial coordinates. The application of CAD/CAM to such shapes has greatly reduced the time and effort involved in the design and analysis of parts and in programming NC machine tools for making the dies. Curvatures can be gentle and nonsymmetrical, resulting in problems of springback and distortion after release from the die, especially with materials of high $\sigma_{0.2}/E$ ratio. In other instances, forming is taken close to the limits allowed by the material, and fracture could easily occur in the absence of tight controls.

Forming Limits In the last few years, there has been a remarkably swift acceptance of formability concepts for production control purposes. The forming limit diagram is useful for analyzing the causes of failures; gridded sheets reveal the local strain distribution and allow the identification of corrective measures, as discussed in connection with Fig. 5-25. Ellipses next to the fracture location give the critical position on the FLD. Then various corrective measures can be taken: The minor strain may be increased (line 1 in Fig. 5-25) by increasing the restraint of the sheet in that direction (by inserting a draw bead or increasing the number of draw beads); the major strain may be reduced (vertical arrow in Fig. 5-25) by reducing the depth of stretch or by allowing more material to draw in (by reducing the number of or completely eliminating draw beads); localized thinning in a deep part of the drawing can be reduced by increasing friction on that part of the male tool.

Shape Analysis The overall severity of the operation is better judged by *shape analysis* which takes the contributions of stretching and drawing into account. For this, combined stretch-draw charts (*forming lines*) are determined in the laboratory. For one endpoint, the LDR is determined. The other endpoint is found in pure stretching, by pressing a steel ball into a clamped sheet until a localized neck is observed. The ball is of 20-mm diameter for sheet of 1.5-mm and lesser thickness, and of 50-mm diameter for up to 3.5-mm sheet thickness ($\frac{7}{8}$-in ball for $< \frac{1}{16}$-in sheet or 2-in ball for up to $\frac{1}{8}$-in sheet). The *stretching limit* SL is the height of stretch h_s divided by the diameter of the die D_s. The stretch-draw limit is obtained by connecting these two endpoints (Fig. 5-33).

When a part is found to fail during drawing (as in Fig. 5-34*a*), a gridded sheet is pressed. The recommended pattern consists of 2.5-mm-diam circles inside 6.4-mm squares (0.1-in diameter inside 0.25-in squares). Analysis begins by drawing an *analysis line* (a vertical cut) through the fracture zone (Fig. 5-34*b*).

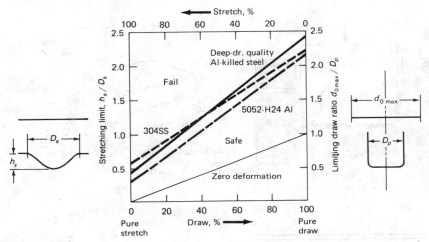

FIGURE 5-33
Combined stretch-draw limit diagram used for judging the severity of combined operations. (*After A. S. Kasper, Metal Progress,* **99**:57–60 (1971). *Copyright 1971, American Society for Metals.*)

The analysis is based on separating the contributions of stretching and drawing to total deformation. To this end, the following steps are taken:

1 The line separating the stretched from the drawn portion of the workpiece is found: This is the line where the die touches the blank at the beginning of draw and is usually visible (the *die impact line*, IL).

2 The entire base of the part is not necessarily stretched; measuring the circles along the analysis line reveals the *inner terminal* (IT) where no deformation had taken place.

3 Similarly, it may be found that the circles are not deformed toward the edge of the remaining flange; thus, the *outer terminal* (OT) is defined.

4 With a flexible ruler, the stretched length L'_s (from IL to IT) and drawn length L'_d (from IL to OT) are measured. The sum of these is the deformed length $L' = L'_s + L'_d$.

5 The original lengths L_s and L_d (Fig. 5-34c) are obtained by counting the number of circles (or squares). The sum of the two gives the starting length of the analysis line: $L_0 = L_s + L_d$.

6 The total elongation of the line is $L' - L_0$. The contribution of the stretched portion is

$$\lambda = \frac{L'_s - L_s}{L' - L_0} \qquad (5\text{-}14a)$$

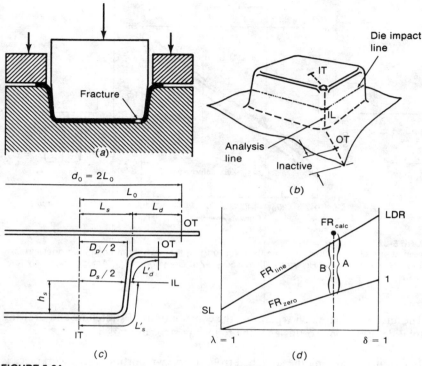

FIGURE 5-34
The relative contributions of stretching and drawing are obtained by shape analysis. (*a*) Cross section through die, (*b*) drawn part, (*c*) characteristic dimensions, and (*d*) severity analysis.

and the contribution of the drawn portion is

$$\delta = \frac{L'_d - L_d}{L' - L_0} \tag{5-14b}$$

(note that $\lambda + \delta = 1$).

7 The stretched portion is now regarded as a case of pure stretch. The horizontal distance between IT and IL is the equivalent of a half die diameter $D_s/2$; the vertical distance is the equivalent of the height of stretch h_s. The stretch forming ratio FR_s is then

$$FR_s = \frac{h_s}{D_s}\lambda \tag{5-15a}$$

8 The drawn portion is regarded as a case of pure draw. The equivalent blank diameter is $d_0 = 2L_0$; the equivalent punch diameter D_p is twice the horizontal distance between IT and the die radius (Fig. 5-34c). Then the draw ratio FR_d

$$FR_d = \frac{2L_0}{D_p}\delta \qquad (5\text{-}15b)$$

9 Since Eq. (5-14) defines the contributions of stretching and drawing, the part can be located on the abscissa of the stretch-draw chart (Fig. 5-34d). The sum of Eqs. (5-15a) and (5-15b) gives the calculated forming ratio

$$FR_{calc} = FR_s + FR_d \qquad (5\text{-}15c)$$

which can be plotted (Fig. 5-34d). Evidently the part falls into the "fail" zone. A "safe" condition can be established by allowing more draw-in, either by removing draw-bead restraint, or by eliminating excess flange material (in Fig. 8-34, clipping the corner of the blank). If this does not suffice, the part configuration has to be changed to reduce FR_{calc} or a different material, of higher FR_{line}, must be used.

The severity of draw Sev may be quantified by expressing the calculated forming ratio as a fraction of the forming ratio of the material. It must be remembered that LDR = 1 signifies no deformation; hence, only the forming ratio above the FR_{zero} line is of significance

$$Sev = \frac{A}{B} = \frac{FR_{calc} - FR_{zero}}{FR_{line} - FR_{zero}} \qquad (5\text{-}16a)$$

since $FR_{zero} = \delta$,

$$Sev = \frac{FR_{calc} - \delta}{\delta(LDR - SL) + SL - \delta} \qquad (5\text{-}16b)$$

The really important application is, of course, to the prediction of success or failure before the expensive tools are built, so that modifications can be made in time. It is still difficult to predict from purely theoretical considerations the relative contributions of draw and stretch, but analyses of new part designs can be made by reference to similar parts for which experience exists. In the most advanced applications, the data base established by CAD is used for preliminary analysis. This is a rapidly developing field which can already boast some successes and should, ultimately, allow fitting of the process to the material (or vice versa) before a production die is finalized. The magnitude of challenge may be sensed from Fig. 5-35, showing a part of complex shape.

5-5-4 Press Forming

In industrial usage, the term *press forming* serves to describe all sheet-metal-working operations performed on power presses with the use of mostly permanent

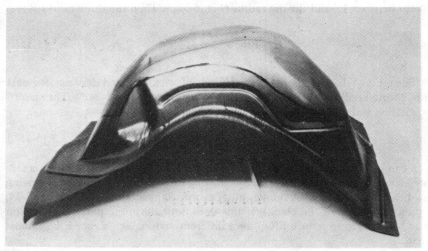

FIGURE 5-35
A severely formed part (an automotive wheel housing) made in a single operation by a combination of stretching and drawing. Note the etched circle grid, the draw beads, and the line marking the boundary between the stretched nose and the drawn-in sides. (*Courtesy A. S. Kasper, Chrysler Corporation, Detroit.*)

(steel) dies. It incorporates all steps required to complete a part of any complexity from the sheet, whether it be blanking, punching, bending, drawing, stretching, ironing, redrawing, embossing, flanging, trimming, and so forth. The die sets used depend on production quantities, required production rates, and the number of operations necessary to complete the part.

1 Minimum complexity is gained if each operation is performed separately, in individual, *single-operation* dies and presses. Die costs still add up and labor and handling costs can be high. Nevertheless, this is the only option when total production quantities are insufficient to justify more complex dies, or when the part is very large. This latter situation prevails in the production of automotive body parts. For increased productivity, presses are lined up behind each other, and the part is moved from press to press with mechanical arms. In-process inventory is reduced and greater flexibility secured if *quick-die-change schemes* are adopted and mechanical arms are replaced with programmable robots.

2 *Compound dies* (Fig. 5-10) perform two or more operations in a single stage and assure the greatest accuracy of the product, but are limited to relatively simple processes such as blanking, punching, and flanging, perhaps combined with bending or a single draw. Special dies are made for multiple draws.

3 Many parts are of a geometry that cannot be directly formed, either because the depth-to-diameter ratio is too large or because the shape has steps, conical portions, etc., requiring several successive draws for which a compound die is often inadequate. The part can still be made in a single press with the aid of *multistation dies* that contain, within one die set, all the die elements needed to complete the part, so that one finished part is obtained for each press stroke. Coil stock is fed at preset increments, and parts are transferred by one of two techniques:

a *Progressive dies* are fed with strip; the blank is only partially cut so as to remain attached with connecting tabs to the remnant of the strip, and this skeleton is used to move the part through the forming stages, with the final separation reserved for the last stage (Fig. 5-36).

b *Transfer dies* are constructed on the same principle, but the blank is cut out first and the scrap bridge is chopped up and disposed of. The blank is moved through successive stages of the die with indexing transfer mechanisms, usually in a straight line, but sometimes along a circular path.

Presses for both progressive and transfer dies have to be large enough to accommodate all die stages on the press bed and to provide the force for all simultaneous operations. Very high die costs are counterbalanced in mass production by low labor costs and high production rates.

FIGURE 5-36
A typical example of progressive die work: forming of two seat-frame parts at a time, by a sequence of blanking, flanging, piercing, flattening of flange and, in the final stage, cutting off and bending. (*Courtesy General Seating Products Division, Lear-Siegler Industries Ltd., Kitchener, Ont.*)

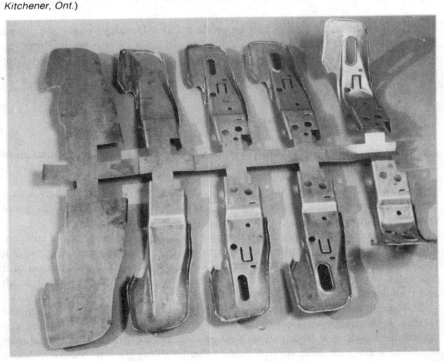

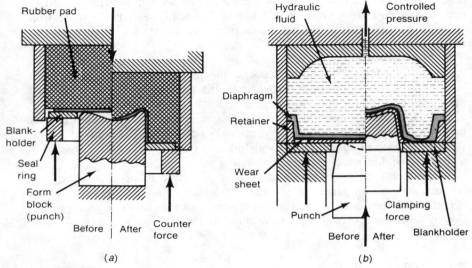

FIGURE 5-37
Deep parts may be drawn with relatively inexpensive tooling using (a) rubber pad or (b) hydraulic forming methods.

4 A great variety of shapes can be produced at high rates in special-purpose machines, some of which are the so-called *four-slide machines*, originally developed for complex wire-bending operations, but now increasingly also used for sheet metalworking.

5-5-5 Special Operations

There are a great number of specialized drawing processes designed to give greater depth of draw, more complex shapes, lower die costs, or a combination of any of these features.

1 Drawing over a male punch (often made of a resin or a zinc alloy) with a rubber cushion (*rubber forming*, Fig. 5-37a) eliminates the need for the more expensive, mating steel dies.

2 Better control of the process is ensured when, instead of the rubber cushion, a liquid—contained by a rubber diaphragm—is used (*hydroforming*, Fig. 5-37b). Hydraulic pressure is programmed throughout the stroke, often with the aid of a computer, to press the sheet onto the punch and thus obtain parts of great depth and complexity.

3 A single die (male or female) is used and the press dispensed with in the various *high-energy-rate forming* (HERF) processes. The energy required for deformation is derived from various sources such as an explosive mat placed over the sheet, a magnetic field applied by a coil surrounding the part, or the pressure shock created in water by the sudden evaporation of a wire. In all these processes the pressure application is very sudden but the rate at which the material deforms is usually not much higher than in a fast mechanical press. Of the many possible applications, drawing-in of necks and internal expansion of tubular and container-like parts is frequently encountered. The latter serves as an alternative to expansion with a rubber plug or hydraulic fluid, and can be used for field repair of condensors and similar tube/header structures.

5-6 SHEET-METALWORKING DIES AND EQUIPMENT

Tool materials are chosen mostly on the basis of the expected size of the production run. Blanking tools are subjected to severe wear and are made from the various cold-working die steels (Table 4-6). Bending and drawing dies are made of similar materials, although cast iron and even hard zinc alloys or plastics are suitable for short production runs or softer workpiece materials.

In contrast to bulk deformation, die pressures seldom limit sheet-metalworking processes. The problem is, more likely, that of finding an economical die material and die-making method. Surface coating of tools exposed to severe wear is gaining in popularity, and lubricants are always chosen to give control of the process as well as reduce die wear.

Dies—their design, manufacture, maintenance, and modification—represent a substantial part of production costs. CAD/CAM techniques minimize the design and tryout effort and allow faster response at lower cost, especially for the design of progressive dies and dies of complex (sculptured) configurations.

Apart from special-purpose equipment, most press forming makes use of mechanically driven and, less frequently, hydraulic presses. Suitable clutches permit operation of mechanical presses in single strokes (initiated by the operator) or continuously, at rates of 30–600 strokes per minute. The principle of construction is similar to presses used in bulk deformation (Fig. 4-42 and Table 4-7) but special features and, for the same rated tonnage, much larger beds make them more adaptable to the working of sheet metal.

Smaller presses often have inclinable press frames which facilitate removal of the stamped part by gravity. Larger presses may have two or even three independently movable rams, one moving inside the other. Such double- and triple-acting presses provide built-in facilities for blank holding or clamping and for ejection, and allow more complex operations. Spring-, air-, or hydraulic-powered cushions provide blankholder pressure on single-acting presses and add flexibility to the operation. Mechanical feeding and part removal speeds up production. Die change and alignment are time-consuming but can be greatly speeded up by quick-die-change techniques, moving prealigned dies in and out of the press through side or front openings in the press frame.

5-7 SUMMARY

Well over one-half of the total metal production ends up in the form of sheet-metal parts. The variety of products is immense, from aircraft skins and automobile bodies to appliance shells, from construction girders and truck frames to furniture legs, from supertankers and bathtubs to beer cans, and from wheel rims and fan blades to watch gears. Combined with joining processes, the scope of sheet metalworking is very broad, but some fundamental limitations must be understood. Some of these limitations may be circumvented, but usually at extra expense.

Because most deformation is the result of an imposed tensile stress, and because many parts are highly visible in service, tensile ductility, yield-point phenomena, and anisotropy of plastic deformation become important. For lower-cost production, the design of the part and of the process must recognize that:

1 Shearing (blanking, punching) does not result in a perfectly smooth and perpendicular cut, but acceptable quality can be obtained with the proper die clearance, provided that bending of the sheet is prevented; this sets a limit on the closeness of two neighboring cuts. Economical production is made possible by NC machines for small production lots but fixed dies remain the choice for mass production.

2 An increased hydrostatic pressure changes the shearing process to resemble extrusion with a resulting smooth "cut" edge.

3 Bending, hole flanging, spinning, and stretch forming are limited by the onset of necking or by fracture; the former is related to uniform elongation (and thus the n and m values); the latter to resistance to triaxial tension (and thus to reduction in area q in the tension test). Localization of the neck may be delayed by changing from uniaxial to biaxial tension and by increasing friction on the punch; nevertheless, minor (and functionally insignificant) adjustments to the shape of the part (typically, more generous radii) often provide the most economical relief from production problems. If a complex shape is genuinely required, it can be produced in a sequence of operations. The suitability of materials for a given task may be judged from limit diagrams.

4 Deep drawing (cupping) is limited primarily by the r value, although the LDR is also a function of process variables such as friction and process geometry. Subsequent redrawing or ironing permit the production of parts of large depth-to-diameter ratio, thin wall combined with thick bottom, zero corner radius, and tapered or stepped shape.

5 Combination of several processes is possible, and the variety of shapes produced by combined stretching and drawing is almost unlimited. Prediction of success is possible by shape analysis and by the application of CAD/CAM techniques. Parts may have varying cross sections (e.g., a neck or bulge on a cup-shaped part) and transverse features (e.g., holes pierced into the side of vessels).

6 The scope of processing can be further expanded if conventional limitations are relaxed. A good example is the multicompartment dinner tray in which wrinkling and folding are not only permitted but even encouraged. This provides the necessary stiffness while also facilitating deep draws that would far exceed the stretchability of the aluminum alloy sheet.

General characteristics of sheet metalworking processes are summarized in Table 12-4.

PROBLEMS

5-1 The edges of shearing punches and dies gradually become rounded in service because of wear. Explain, with the aid of a sketch, the changes one should expect in the quality of cut.

5-2 For a given clearance, which material is likely to give more burr: a soft or a hard one? On this basis, would you use a larger clearance for annealed brass than for hard temper brass?

5-3 Calculate the maximum shearing force in punching a hole of 50-mm diameter with a shear equal to three times the sheet thickness $h = 2$ mm. Use geometric construction to estimate the sheared length. Also calculate the force for zero shear to see the magnitude of force reduction with shear.

5-4 Circular blanks (slugs) of 1100 Al, diameter $d = 25$ mm and thickness $h = 3$ mm, are to be mass-produced as the starting material for toothpaste-tube extrusion. The available presses are of 500-kN capacity and can take maximum 220-mm-wide strip. Economy of material utilization increases with increasing numbers of rows cut from one strip width. Calculate (a) the force required for blanking a single slug and (b) the maximum number of slugs that can be blanked simultaneously with the available press. (c) Suggest the optimum layout for the slugs if the web (remaining material between cuts and at edges) is approximately h, but seldom less than 1 mm (0.40 in).

5-5 Derive Eqs. (5-7) and (5-7′).

5-6 The surface of a bent part shows orange peel. Explain the source of the effect. Is orange peel a structural defect?

5-7 On the basis of data given in Example 4-5, what sheet is best for (a) bending without orange peel; (b) bending to zero radius; (c) greatest resistance to permanent deformation in service. Justify the choices.

5-8 What material properties determine springback?

5-9 Suggest at least three methods of making bends with exactly 90° angle.

5-10 A sheet of 1-mm thickness is bent around radii of 2, 10, and 50 mm; state which bend will give greater springback and why.

5-11 A sheet-metal part used to be made by bending 1015 steel of 5-mm thickness on a sharp edge (zero radius). The part needs to be strengthened and 1045 steel is to be used. Calculate (a) the die radius necessary to prevent necking and the radius required to prevent fracture and (b) the die angle to give a 90° bend.

5-12 From simple elastic bending theory, derive the force necessary for wipe bending a sheet of h thickness and l length. To allow for the effects of plastic deformation, double the end result.

5-13 Automotive body pressings such as trunk lids are traditionally made of low-carbon steel. Should one expect greater or lesser springback when changing to (a) an aluminum alloy such as 5052-H24 or (b) HSLA steel?

5-14 Draw a 5-mm by 20-mm cross section representing the (magnified) cross section of a tension test specimen. Draw the cross section as it would appear at fracture with 50% reduction in area, for materials of $r = 0.2$, 1.0, and 6. Which of these materials would be better for (a) bending around a tight radius; (b) deep drawing.

5-15 A part fails in the course of deep drawing. Fracture occurs toward the end of draw (as in Fig. 5-29d). Suggest possible remedies.

5-16 A flat-bottomed cup has been successfully produced of deep-drawing quality low-carbon steel sheet. For greater corrosion resistance, the part is now to be made of pure titanium. Give an explanation of the changes to be expected in drawing behavior.

5-17 Blanks of annealed 5052 Al alloy, of diameter $d_0 = 10$ in, $h = 0.125$ in are to be drawn, with punches of diameter $D_p = 5$ in, into flat-bottomed cups. Assuming that the drawability of annealed 5052 is at least as good as that of the H24 temper (strain hardened and then partially annealed to give a strength typical of half-hard material), determine from Fig. 5-33 if this is feasible. If the answer is yes, calculate (a) the blankholder force, (b) the drawing force, and (c) assuming that the average wall thickness of the part remains unchanged and the punch nose radius is $R_p = 0.5$ in, the height of the cup to be expected.

5-18 A dish of diameter $d = 1000$ mm and depth $d = 250$ mm is to be produced of 5052 Al, in the shape of a spherical segment. Because of the danger of puckering in deep drawing and because of the small quantity required, it is proposed that the dish be made by pure stretch forming. Determine whether this is feasible. Also, suggest a possible alternative fabrication method.

5-19 Deep-drawing quality, aluminum-killed steel of $r = 1.7$ has an LDR of 2.4 (Fig. 5-33). Calculate, for a sheet of $h = 2$-mm thickness and a punch of diameter $D_p = 100$ mm and nose radius $R_p = 5$ mm, (a) the maximum blank diameter $d_{0\,max}$, (b) the depth of cup, assuming a constant wall thickness of 2 mm, and (c) the height-to-diameter ratio. Repeat the calculation for LDR = 2; note the effect of LDR on height-to-diameter ratio.

5-20 A cylindrical container (a cooking pot) of 200-mm OD, 160-mm depth, 2-mm wall, and 5-mm bottom thickness is to be produced from 5052-H24 Al alloy. (a) Assuming that the wall thickness of the container in the first draw is the same as the starting sheet thickness, calculate the diameter of the starting blank and the blanking force. (b) Check whether the first-draw container can be made in a single draw. If it cannot be, make sketches of the suggested process sequence. (c) Select the punch diameter for the first draw and calculate the drawing force. (d) Suggest two methods of making the finished pots, illustrating with sketches.

5-21 Obtain samples of containers (beverage cans, food cans, sardine tins, etc.). With tin snips, cut them open, measure the wall, base, and lid thicknesses, and make informed judgments on the likely methods of manufacture. (*Caution*: Thin sheet has razor-sharp edges and must be handled with utmost caution, using protective gloves.)

5-22 Many dry-battery cases are made of zinc. A typical height-to-diameter ratio is 4:1. If the base and wall are to be of the same thickness, what process or process sequence could be employed? Answer by drawing sketches of the key features of the dies used. Consider also a competitive bulk deformation process.

5-23 Show that, for a material obeying the power law of strain hardening, TS $= K(n/e)^n$.

FURTHER READING (SEE ALSO CHAP. 4)

A Detailed Process Descriptions

ASM: *Metals Handbook*, 8th ed., vol. 4, *Forming*, American Society for Metals, Metals Park, Ohio, 1969.

Newby, J. R. (ed.): *Source Book on Forming of Steel Sheet*, American Society for Metals, Metals Park, Ohio, 1976.

B Textbooks

Eary, D. F., and E. A. Read: *Techniques of Pressworking Sheet Metal*, 2d ed., Prentice-Hall, Englewood Cliffs, N.J., 1974.
Watkins, M. T.: *Metal Forming II: Pressing and Related Processes*, Oxford University Press, Oxford, 1975.

C Specialized Books

Benjamin, W. P.: *Plastic Tooling*, Macmillan, New York, 1972.
Carlson, H.: *Spring Manufacturing Handbook*, Dekker, New York, 1982.
Dinda, S., K. F. James, S. P. Keeler, and P. A. Stine: *How to Use Circle Grid Analysis for Die Tryout*, American Society for Metals, Metals Park, Ohio, 1981.
Ezra, A. A.: *Principles and Practice of Explosive Metalworking*, Industrial Newspapers, London, 1973.
Halmos, G. T. (ed.): *High-Production Roll Forming*, Society of Manufacturing Engineers, Dearborn, Mich., 1983.
Hoffmann, E. G.: *Fundamentals of Tool Design*, 2d ed., Society of Manufacturing Engineers, Dearborn, Mich., 1984.
Keyes, K. A. (ed.): *Innovations in Die Design*, Society of Manufacturing Engineers, Dearborn, Mich., 1982.
Koistinen, D. P., and M. M. Wang (eds.): *Mechanics of Sheet Metal Forming*, Plenum, New York, 1978.
Morgan, E.: *Tinplate and Modern Canmaking Technology*, Pergamon, Oxford, 1985.
Proceedings of International Deep Drawing Research Group (IDDR) Congress, biennial, since 1958.
Semiatin, S. L., and J. J. Jonas: *Formability and Workability of Metals: Plastic Instability and Flow Localization*, American Society for Metals, Metals Park, Ohio, 1984.
Shingo, S.: *A Revolution in Manufacturing: The Single Minute Exchange of Die (SMED) System*, Society of Manufacturing Engineers, Dearborn, Mich., 1985.
Strasser, F.: *Functional Design of Metal Stampings*, Society of Manufacturing Engineers, Dearborn, Mich., 1971.
_____ : *Metal Stamping Plant Productivity Handbook*, Industrial Press, New York, 1983.

D Journals

Sheet Metal Industries

CHAPTER 6

PROCESSING OF PARTICULATE METALS AND CERAMICS AND OF GLASSES

As indicated in Fig. 2-42, manufactured components or articles of consumption may be directly produced by bringing a powder of the starting material into the desired end shape. The material may be metal, ceramic, or, indeed, a polymer such as PTFE (Sec. 7-7). The term powder is rather restrictive, because many particulate materials of fairly large particle size and of irregular shape may also be processed. The essential feature is that the bond between particles is produced without total melting, although in some instances localized melting may occur. Glasses are the products of melt processing and thus should really be regarded as polymers; however, by composition they are ceramics and will be discussed here.

6-1 PROCESSING STEPS FOR PARTICULATES

Irrespective of the nature of the material, the processing steps are the same. First, particles of the desired composition are produced, brought to the required size and, if several constituents are used, they are blended to ensure their even distribution. Second, particles are compacted to bring them into close proximity while imparting the desired part configuration. Third, compacts are sintered to establish permanent, strong bonds between adjacent particles.

Some of the principles are common to both metallic and ceramic materials, and these will be discussed first.

6-1-1 Preparation of Powder

There are many methods of producing metallic and ceramic powders. Once the powder is available, it is subjected to a sequence of preliminary processing steps.

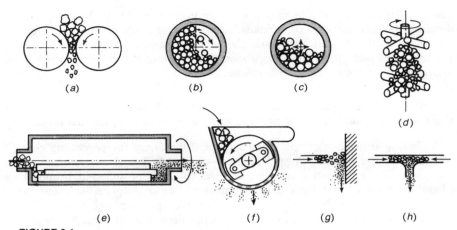

FIGURE 6-1
Particulate material may be comminuted (reduced to a smaller size) by many techniques, some of which are (a) roll crushing, (b) ball milling, (c) vibratory ball milling, (d) attrition milling, (e) rod milling, (f) hammer milling, (g) impact milling, and (h) fluid energy milling.

Comminution (Milling) If the powder is coarser than required, it is reduced in size (*comminuted*) by a number of mechanisms:

1 Brittle materials are crushed in jaw, gyratory, cone, or roll (Fig. 6-1a) *crushers*, especially in earlier stages of preparation.

2 For finer powder, particles are fractured by impacting them between two hard balls, made of a metal or ceramic chosen so as to avoid contamination. The impact energy may be provided by balls dropping in a partially filled, horizontal-axis rotating drum (*ball mill*, Fig. 6-1b). Higher energy is imparted to the balls and milling is accelerated by vibration (*vibratory mill*, Fig. 6-1c) or by the action of horizontal arms attached to a vertical, rotating shaft inside a vertical-axis drum (*attrition mill*, Fig. 6-1d). A similar action is exerted by rods placed in a horizontal-axis rotating drum (*rod mill*, Fig. 6-1e) and by rotating blades called hammers (*hammer mill*, Fig. 6-1f). All these processes are most effective on brittle materials; ductile materials are first substantially deformed, strain hardened, and subsequently broken up into smaller, flake-like particles.

3 Finer particles are obtained when particles are hurled against a hard, stationary surface, either by an air or other gas jet or in a slurry, usually an aqueous suspension (*impact milling*, Fig. 6-1g). Alternatively, particles entrained in two opposing fluid streams are comminuted by impact on each other (*fluid-energy milling*, Fig. 6-1h).

When particle size has diminished to a certain value, secondary bonding forces (van der Waals forces) lead to *agglomeration*, i.e., the formation of larger clusters of particles. In metals, cold welding may also take place. Uncontrolled agglomeration is undesirable and can be prevented by appropriate additives. In dry milling these may be lubricant-type compounds (such as stearic acid); in wet milling they are chemicals that impart an electric charge to the surface of particles so that particles repel each other. Thus *flocculation* (formation of loosely bonded, woolly masses) is prevented by these *deflocculants*.

Sizing Particle shape, size, and size distribution are important variables that affect both the consolidation of powders and the properties of the final product.

1 *Particle shape* is an important factor in determining the processing characteristics of the powder. Formal methods of morphological analysis may be used; in a less quantitative sense, it is usual to speak of spheroidal, nodular (slightly elongated, roundish), irregular, angular, lamellar (plate-like), acicular (needle-like), and dendritic particles. Thin lamellar particles agglomerate into flaky powders; long, thin needles into fibrous powders; and spherical or irregular powders into granular powders. Some powders are porous, while others are more or less complete hollow spheres or other shapes.

2 *Particle size* should be neither too large nor too small. Too large particles may not display the desired structure which is often the reason for choosing the powder route, and they may not allow the development of high densities. Too small particles may be difficult to handle and may tend to agglomerate; furthermore, their large surface area-to-volume ratio may introduce large quantities of undesirable adsorbed substances and, on metals, also oxides.

3 *Particle size distribution* is analyzed by passing the powder through a series of sieves of gradually diminishing hole size (increasing number of holes per unit area). The fraction of particles passing a certain sieve is given in percentages. The sieve size is quoted as mesh number (for mesh number 50 and higher, the particle diameter, in millimeters, is 15 divided by mesh number). Sieve analysis is usually conducted dry, but vacuum or wet sieving is necessary for powders below 325 mesh (45 μm). Techniques based on light scattering, light blockage, electrical pulses, or sedimentation are suitable for analyzing wide size distributions. Optical and scanning electron microscopy can be used for both size and shape analyses.

Classification is the process of separation into fractions by particle size. Excessively large particles are removed and, if required, size fractions are separated by *screening* the entire production lot. Fine particles may be separated by screening a slurry. Settling from a liquid solution (*elutriation*) or classification of dry powders in an air stream within a *cyclone* are also useful for separating fine powders. Superfine particles can be removed by *electrostatic separation*.

Cleaning and Calcining *Cleaning* aims at removing adsorbed films from the surface of the powder, usually by heating in vacuum or in an appropriate atmosphere.

Calcining is a high-temperature heat treatment applied to ceramic powders.

Physical Properties The powder possesses a number of properties that are of importance to further processing:

1 *Surface area* (area/unit mass) indicates the surface available for bonding, and also the area on which adsorbed films or contaminants may be present.

2 *True density* (for powder metals, also called *theoretical density*) is the mass per unit volume of solid, and is a material property. The *apparent density* or *weight per unit volume* (g/cm^3) is a very important value because it defines the actual volume filled out by the loose powder. It is often expressed as a percentage of the fully dense material (as a percentage of true density). *Tap density* is obtained by tapping or vibrating the receptacle, and is a measure of compaction achievable without pressure. Both apparent and tap densities depend on particle shape and distribution as well as interparticle friction.

Technological measures of flow properties are given by the *flow rate* (the time required for a measured quantity of powder to flow out of a standard funnel) and *angle of repose* (the base angle of a cone of powder resting on a circular plate).

Compressibility as a term describes the change in green density with increasing compacting pressure. It is usually given as the density at some specified pressure or, in a graphical or tabular form, at several pressures.

Blending A single powder may not fulfill all requirements of production or service properties, and powders are then *blended*. Blending may serve several purposes: uniformity of size distribution is ensured in a large lot; response to imposed stresses (*rheology*) is controlled for improved handling; density of the compacted body is adjusted; and composition or service properties are changed.

1 Blending a coarser fraction with a finer fraction (Fig. 6-2a) ensures that interstices between large particles will be filled out. Thus, tap densities over 65% of theoretical density can be obtained with powder of favorable (spherical or nodular) shape.

2 Metallic alloys, complex ceramics, and metal–nonmetal composites are made by blending different powders.

3 Friction between particles and between particles and mold wall is reduced by *lubricant* additives.

4 *Binders* are added to powder that would otherwise fail to develop adequate green strength.

5 *Sintering* aids are added to accelerate densification on heating.

6 When the powder is to be made into a slurry, deflocculants and other chemicals are added to ensure favorable rheological characteristics.

Blending must be thorough, with each particle uniformly coated and with the various constituents uniformly dispersed. Ball milling is often employed. *Spray drying* ensures uniform distribution of constituents in fine powders and allows forming free-flowing powder from fines that otherwise would not flow. The ingredients are made—with water or an organic liquid—into a slurry, which is

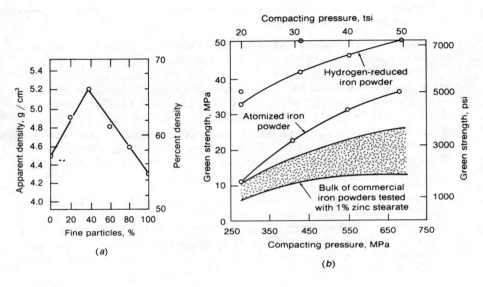

FIGURE 6-2
In compacting powders, (*a*) optimum fill density is obtained by a mixture of fine and coarse powders. (*b*) Green strength increases with compacting pressure and is greater for more irregularly shaped powder. ((*a*) *After H. H. Hausner, in Handbook of Metal Powders, A. R. Poster (ed.), Reinhold, 1966, (b) From Metals Handbook, 9th ed., vol. 7, American Society for Metals, 1984, p. 289. With permission.*)

then *atomized* (broken into small droplets) by air or a gas. The liquid is evaporated from the droplets in their flight and the coarser, roughly spherical powder thus formed is collected. Less favorable shapes are obtained when the powder is *granulated*: a damp mix is forced through orifices in a plate or holes in a screen.

6-1-2 Consolidation

The method chosen for *consolidation*, i.e., for bringing the particulate material into the required shape, depends on the particulate and on the intended density of the product.

Pressing of Dry Particulate Dry powders, which may be coated with a lubricant or dry binder, are *compacted* in molds by the application of pressure to form a so-called *green body* (a body without permanent bonding). For a given particle-size distribution, the density of the compact increases with applied pressure (Fig. 6-2*b*). Density is also a function of particle shape: A spherical powder compacts to a higher density than an irregularly shaped one. In the course of pressing, air is expelled from between particles, and particles slide against each other and against the mold wall. At higher pressures, the applied force is concentrated at contact points between particles; the high local pressures cause local deformation or fracture, and the compact acquires some *green strength*, usually expressed as rupture strength, measured on transverse bend specimens (Fig. 2-6).

There are several sources of green strength. Sliding combined with pressure promotes adhesion (and even cold welding with some powders); therefore, strength increases with increasing pressure (Fig. 6-2*b*). Another source of green strength is mechanical interlocking, especially with particles of irregular shape; therefore, green strength is less for more spherical powders even though they pack more densely. If neither mechanism is available, bonding agents are added which evaporate in the course of sintering. Of course, green strength is lower when the powder is coated with a lubricant. Various compacting techniques are available:

1 Most compacts are made by *pressing* in dies. If the part shape is fairly simple and a die can be made of steel, high applied pressures are permissible and, if the particulate can deform plastically, densities in excess of 90% of theoretical density can be achieved.

a The effectiveness of pressing with a *single-acting punch* is limited because particulate material does not transmit pressures as a continuous solid would and wall friction also opposes compaction. The pressure tapers off rapidly, giving compacts of higher density close to the punch and of diminishing density farther away (Fig. 6-3*a*), limiting the attainable depth-to-diameter ratio.

FIGURE 6-3
The density of a green body depends on the method of compaction: (*a*) it is higher under the punch when compacting with a single punch in a fixed container; better uniformity is obtained with (*b*) a single punch and floating container or (*c*) with two counteracting punches.

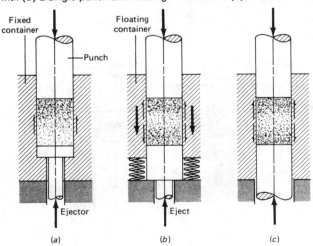

Fixed container

Floating container

Punch

Ejector

Eject

(*a*)　　　　(*b*)　　　　(*c*)

b The situation improves with a *floating container* (Fig. 6-3*b*) which is moved against the stationary punch by the friction force between powder and container.

c Good results are obtained in special presses with two *counteracting punches* advancing from the two ends of the die cavity (Fig. 6-3*c*).

When the thickness of the part changes considerably from point to point, green density is equalized by constructing dies with *multiple punches* guided within each other (Fig. 6-4), so that the same degree of compaction can be applied everywhere. Clearances between moving parts must be kept very small (below 25 μm) to prevent entry of particles. Dies are usually built of high-strength tool steel or, for larger production runs and severe abrasive conditions, of cemented

FIGURE 6-4
Uniform fill density can be ensured with the use of multipunch dies (dimensions are given in inches). (*From Metals Handbook, 8th ed., vol. 4, American Society for Metals, 1969, p. 461. With permission.*)

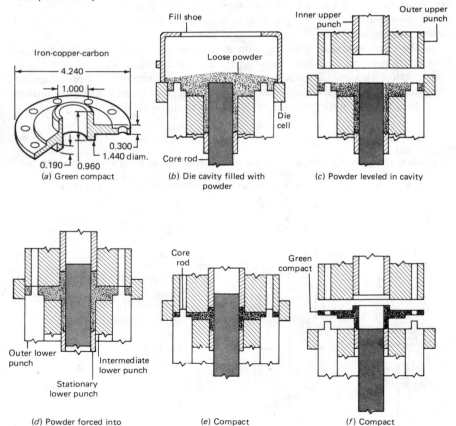

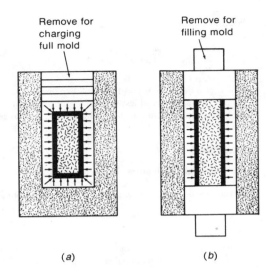

Remove for
charging
full mold

Remove for
filling mold

FIGURE 6-5
Omnidirectional pressure helps
to produce parts with more uniform density distribution in isostatic pressing by the (a) wet-
bag method and, somewhat less
so, by the (b) dry-bag method.

(a)

(b)

tungsten carbide. Allowable die pressures are as in closed-die forging (Sec. 4-4-5), although moderate pressures (of 100 MPa (15 kpsi)) are usual, rising to 500 and even 900 MPa (80–120 kpsi) in some cases. Dies can be quickly and automatically filled by gravity, with the excess powder simply wiped away.

2 More uniform compaction is obtained in *cold isostatic pressing*, in which a deformable (reusable rubber) mold contains the particulate. Hydrostatic (omnidirectional) pressure is applied by means of a hydraulic fluid inside a pressure vessel (Fig. 6-5). The fluid is usually water. Shape limitations are few in the wet-bag method (Fig. 6-5a). Shorter cycle times are achieved in the dry-bag method (Fig. 6-5b) because fluid is introduced only into the space between the fixed die and the elastomeric mold, but shapes are somewhat more limited. Neither lubricants nor binders are needed in many instances. Pressures of 300 MPa (45 kpsi) are usual and up to 550 MPa (80 kpsi) can be achieved.

3 A mold may be filled under *gravity*, perhaps assisted by *vibration*, giving a compact of low density and little strength. Only with careful handling can it be converted into a porous sintered product; more likely, the mold itself will be heated to at least initiate sintering or to set the dry binder. A process known as *loose powder sintering* uses a sugar as the binder for metal powders.

Example 6.1

Perfectly spherical nickel powder of 0.1-mm particle diameter is compacted by vibration. (*a*) What percentage of the theoretical density can be achieved? (*b*) Will this increase or decrease if the particle diameter is uniformly increased to 0.2 mm?

    a In the ideal case, closest packing—equivalent to a fcc structure—will be achieved. From Fig. 3-2a, there are (6 half) + (8 one-eighth) = 4 spheres of R radius in each unit cell of a side. The body diagonal is $4R = a\sqrt{2}$. The packing factor is

$$\frac{\text{Volume of spheres}}{\text{Volume of cube}} = \frac{4(4\pi/3)\,R^3}{a^3} = \frac{4(4\pi/3)\,R^3}{\left(4R/\sqrt{2}\right)^3} = \frac{\pi}{3\sqrt{2}} = 0.74$$

This is also the ratio of powder to solid volume; thus, 74% of the theoretical density is obtained. (In practice, a density of less than 67% of theoretical density is more likely, see Fig. 6-2a.)

b The particle radius R dropped out; thus, no change in fill density will occur.

Example 6-2

Die-wall friction (half arrows in Fig. 6-3) limits the depth to which a compact can be densified. An estimate of the decay of pressure can be obtained if we assume that the nickel powder behaves like an elastic solid, and that the radial stress p_r can be obtained from the axial stress p_a by

$$p_r = \left(\frac{\nu}{1-\nu}\right)p_a = kp_a$$

where ν is Poisson's ratio and is 0.31 for nickel. Calculate the depth at which the axial pressure drops to one-third of its initial value p_{a0} if the die wall is unlubricated ($\mu = 0.5$).

Axial pressure drops to zero at the depth h where the axial force $p_a(d^2)\pi/4$ equals the frictional force $\mu hd\pi p_r$. Since p_r can be expressed from p_a, we may write

$$p_a(d^2)\frac{\pi}{4} = \mu h d\pi\left(\frac{\nu}{1-\nu}\right)p_a$$

$$h = d\left(\frac{1-\nu}{4\mu\nu}\right) = d\left[\frac{0.69}{0.4(0.5)(0.31)}\right] = 1.11d$$

More accurately, the force balance may be written

$$\frac{\pi d^2}{4}\delta p_a = -\pi d(\delta h)\mu\left(\frac{\nu}{1-\nu}\right)p_a = -\pi d(\delta h)\mu kp_a$$

Integrate

$$\int_{p_{a0}}^{p_{ah}}\frac{\delta p_a}{p_a} = -\int_0^h \frac{4\mu k}{d}\delta h$$

$$p_{ah} = p_{a0}\exp\left(-\frac{4\mu kh}{d}\right)$$

$$\frac{p_{ah}}{p_{a0}} = \frac{1}{3} = \exp\left(\frac{-4\mu kh}{d}\right)$$

hence, $h = 1.22d$.

Friction between powder particles further reduces the depth to which uniform density can be obtained; hence, lubricants are used when at all permissible.

Example 6-3

The green compact shown in Fig. 6-4 has a hub of 0.960-in height and a flange of 0.300-in thickness. Calculate the position of the lower punches for filling the die cavity (Fig. 6-4b) if the density of loose powder is 38% and the density of the green compact is 78% of theoretical density.

The mass of loose powder (after leveling, Fig. 6-4c) must be equal to the mass of green compact. Powder cannot be counted upon to flow from hub to flange or vice versa; therefore, the same mass of powder remains in the flange (and in the hub) before and after pressing. Since mass is (volume) × (relative density), we may write

$$V_l \text{ (rel. dens. loose)} = V_g \text{ (rel. dens. green)}$$

where V_l is the loose volume and V_g the volume of the green compact. Since no powder flows from flange to hub and vice versa, we may write

$$V_l = Ah_l \quad \text{and} \quad V_g = Ah_g$$

where h_l is the height of loose powder column and h_g is the thickness of the green compact. Consequently,

$$h_l = h_g \text{ (rel. dens. green)}/\text{(rel. dens. loose)}$$

For the hub, $h_l = 0.960\,(0.78/0.38) = 1.97$ in; for the flange, $h_l = 0.300\,(0.78/0.38) = 0.645$ in.

Plastic Forming When the proportion of the binder or other liquid is high enough to allow the relative displacement of individual particles within a liquid matrix, the mixture acquires rheological properties suitable for processing by *plastic forming* techniques. The mixtures usually behave as Bingham solids or non-Newtonian (pseudoplastic) bodies (Fig. 3-20b) and can be processed by techniques familiar from metalworking and used also in polymer processing (Sec. 7-7).

1 *Wet pressing* is related to forging; *compression molding* and *transfer molding* are related to the similar polymer processing techniques (see Sec. 7-7-3). The liquid phase may be water or, especially with metals and fine ceramics, a polymer.

2 *Injection molding* of finer (< 10 μm) powder is feasible for fairly thin-walled (0.5–5-mm) parts. The powder is combined with 25–45% thermoplastic polymer, injected at pressures of 140 MPa (20 kpsi) into molds held at 135–200 °C (275–400 F) on standard injection-molding machines. Shrinkage on sintering is large, but complex shapes can be produced.

3 *Extrusion* of preshaped billets is practiced mostly for metal powder. Most extrusion of ceramics is conducted by the screw-extrusion technique (see Sec. 7-7-3) which ensures excellent mixing of the constituents and allows continuous operation. Solid bars, hollow tubes, and multihole filters may be extruded and cut up as required.

4 The potter uses clay and refractories in a plastic condition. Hollow shapes, plates, etc. may be formed by *jiggering* (a mechanized form of throwing on the wheel): The plastic mass is pressed against a vertical-axis, rotating mold with appropriately shaped, profiled templates or rollers. In this respect, the process resembles spinning (Sec. 5-4).

Casting *Slurry casting* is based on the non-Newtonian viscous flow of small (< 20 μm) particles suspended in a liquid phase. The suspension is called a *slip* or *slurry*. Slips made of very fine (< 1-μm) powder tend to dilatancy (Fig. 3-20, line B) and are more difficult to handle.

Techniques familiar from casting (Sec. 3-8) may be applied. The difference is that the mold is usually porous, so that fluid from the slurry is absorbed by capillary action, leaving behind a powder compact. To make the mold, water is added to plaster of paris ($CaSO_4 \cdot \frac{1}{2}H_2O$); a hydration reaction takes place which results in the precipitation of gypsum ($CaSO_4 \cdot 2H_2O$) in the form of acicular crystals, arranged in a randomly oriented network. Excess (entrapped) water is driven off by drying the mold. The mold is now a solid structure which has pores below 1 μm in diameter. Mold-release agents are usually applied to the mold surface, and the slurry is then poured.

The most frequently used variant is *drain casting*, also referred to simply as slip casting. This is a relative of slush-casting (Sec. 3-8-5): The mold is filled with slip and, after sufficient time has elapsed to form a dewatered particulate shell, the mold is tipped to pour out the excess slip, leaving behind a hollow product.

To produce a solid casting, the mold cavity must be topped up with slip until it is filled with a powder body. Vacuum aids mold filling and speeds up the process of liquid extraction. Centrifuging (Fig. 3-39) also helps filling.

An important variant of slurry casting involves the forming of thin (< 1.5-mm-thick) tapes, usually of ceramic powders in an organic carrier. Several techniques exist:

The slurry may be cast onto a thin, moving plastic film while controlling the tape thickness with a blade (*doctor-blade process*). The slurry may be cast onto a paper carrier which is subsequently burnt off (*paper-tape process*). A similar end result is obtained by *rolling* the slurry. The green tape obtained by any of these processes is flexible enough to be rolled up; it can be blanked, mechanically scored, or marked with a laser beam prior to firing, so that it can be made into small components such as substrates for integrated circuits (see Sec. 10-3-8).

Green Machining Even though the strength of green compacts is low, they can be machined, usually in vertical-axis lathes, if held in appropriate fixtures. Since drying shrinkage is allowed to take place prior to machining, fairly complex shapes can be created to close tolerances, as in the manufacture of high-tension electric insulators made of porcelain.

6-1-3 Sintering

The green compact is heated to attain the required final properties. In the course of heating, several changes take place.

1 At low temperatures, liquid constituents are driven off (*drying*). The required residence time increases with increasing wall thickness; fast heating would cause sudden vaporization and could result in the disintegration of the compact. Vacuum accelerates drying. If organic binders are to be burnt off, sufficient oxygen must be available for their combustion.

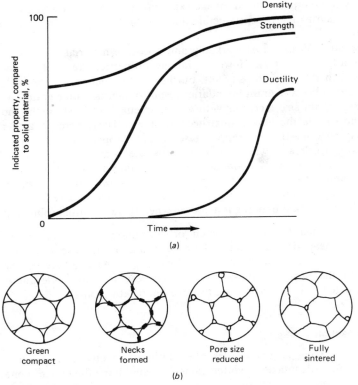

FIGURE 6-6
In the course of sintering (*a*) the compact acquires permanent strength while the volume shrinks (density increases) as a result of (*b*) elimination of most pores between particles.

2 At higher temperatures (in metals, above $0.5T_m$ but, more typically, at hot-working temperatures, about $0.7–0.9T_m$), *sintering* takes place. The compact contains particles of the material in close proximity to each other. The energy of the system will decrease by reducing the total surface area—in other words, the driving force for sintering is the surface energy which is reduced by joining adjacent particles (Fig. 6-6). Several mechanisms come into play, of which evaporation and condensation are usually much less important than solid diffusion. To begin with, interatomic bonds are established between adjacent surfaces, and necks grow by the movement of atoms from the surface and bulk of particles toward the necks. Plastic or viscous flow may also take place and this, together with more massive diffusion, reduces the size of pores (Fig. 6-6*b*). Thus, the volume shrinks and density increases (Fig. 6-6*a*). To achieve the same sintered density, shrinkage is greater for lower green densities. If phase changes take place during heating, shrinkage may be negligible or even growth may occur. It is usually necessary to determine the shrinkage experimentally: If process steps are closely controlled, shrinkage is reproducible and finished parts can be held to close tolerances.

At this stage, strength increases markedly and a ductile material exhibits an increase in ductility (Fig. 6-6a). However, fatigue properties are likely to be inferior as long as there is any porosity left. Further heating does not necessarily improve the situation because grain boundaries begin to migrate, some grains are consumed, average grain size increases with a consequent drop in strength, and pores left behind (inside the new grains) become stable. These pores, together with any porosity resulting from entrapped gases, impair properties.

3 The process is accelerated when one of the constituents melts and envelopes the higher-melting constituent (*liquid-phase sintering*). A liquid that wets the solid particles exerts capillary pressure which physically moves and presses particles together for better densification. The phases formed can be determined from the relevant phase diagrams; best bonding is achieved when there is mutual solubility.

Sintering furnaces may be of the batch or continuous type. Continuous furnaces have a preheat (drying or burn-off) zone, a high-heat (sintering) zone, and a cooling zone. At high temperatures, reactions may take place between the porous body and the furnace atmosphere. If such reactions are undesirable, sintering is conducted in vacuum or a nonreactive (inert) atmosphere.

6-1-4 Hot Compaction

All particulate processing sequences described hitherto involve consolidation followed by sintering. Advantages are often gained by combining the two into one simultaneous operation. Sufficient pressure is applied, at the sintering temperature, to bring the particles together and thus accelerate sintering. If possible, individual grains are deformed to ensure greater compliance; most importantly, grains are moved relative to each other, so that surface films that would slow down or prevent diffusion are broken through. Under such conditions porosity may be completely eliminated. Since a particulate body has no strength whatsoever prior to pressing, the applied pressures must all be compressive—and preferably omnidirectional—so that the generation of secondary tensile stresses is avoided.

Hot pressing in heated graphite or ceramic dies is feasible but suffers from the difficulty of transmitting the pressure uniformly to all parts of the compact (Fig. 6-3). Therefore, *hot isostatic pressing* (HIP) has gained wide acceptance. In the basic form of HIP, the powder is encased in a deformable metal can or shroud which is evacuated and then placed inside a furnace, which in turn is enclosed in a high-pressure chamber. The chamber is pressurized with an inert gas up to 300 MPa (45 kpsi) or, more typically, to 100 MPa (15 kpsi). Furnace temperatures range from 480–2000 °C (900–3500 °F); more typically to 1200 °C (2200 °F).

Hot rolling, extrusion, and *forging* of particulate materials is also feasible, although provision must be made to prevent undesirable reactions with the surrounding atmosphere. Therefore, metallic particulates are sometimes encased in a can made of a metal that resists the high temperatures, or the process is conducted in a protective atmosphere, or carbon is added to produce a reducing environment.

6-2 POWDER-METALLURGY PRODUCTS

In principle, all metals and alloys can be processed in the particulate form (Fig. 2-42). Powder metallurgy may be chosen because of its cost effectiveness in producing certain parts; at other times, it is the only technically feasible process. Some examples of the latter are: dispersion-strengthened alloys in which a high hot strength is secured by the presence of finely distributed, stable oxides which prevent grain-boundary migration and slip (e.g., jet engine components of thoria-dispersed (TD) nickel with 2% ThO_2, or copper welding electrodes with dispersed alumina (Al_2O_3)); ceramics bonded with metals (*cermets*); heavy-duty metallic friction materials such as Cu–Sn or Cu–Zn alloys with embedded SiO_2, Al_2O_3, etc.; nuclear fuel elements with UO_2 dispersed in aluminum or stainless steel; diamond bonded with metal (typically copper and 15–20% tin); mechanically alloyed superalloys; and metal–nonmetal composites such as electrical contacts and brushes that contain graphite.

Processing of metal powders follows the general outline given in Sec. 6-1.

6-2-1 The Powder

Powder Production Powders may be made essentially by five techniques:

1 *Winning* a metal from its compound by: reduction of an oxide (Fe, Cu, Mo, W); thermal decomposition of a compound (Ni, Fe); reduction of a salt in the molten state (Ti); electrolysis (Fe, Cu, Be); or precipitation (cementation) from a chemical solution (Cu, Ni).

2 *Deposition* of an extremely fine powder *from the vapor phase* (Zn).

3 *Atomization*, i.e., the production of small droplets by breaking up a melt stream.

a Largest quantities are produced by *water atomization* in which the melt emerging from a nozzle is broken up with water jets (Fig. 6-7a). The size and shape of particles is readily changed by controlling the process parameters (the median particle size is inversely proportional to jet pressure), but the powder is always oxidized. If this is objectionable, an inert gas (Fig. 6-7a) is used in *gas atomization*; the powder will generally be spherical.

b In *centrifugal atomization* the melt stream is directed at a chilled, rotating disk (Fig. 6-7b) and the powder is hurled off. In *rotating-electrode processes* the alloy to be atomized is in the form of a fast-rotating (15 000-rev/min) electrode which is gradually melted by an electric arc or helium plasma arc (Fig. 6-7c). Melted material is immediately hurled off and the powder solidifies without touching any surface, thus remaining very clean. The particles are spherical.

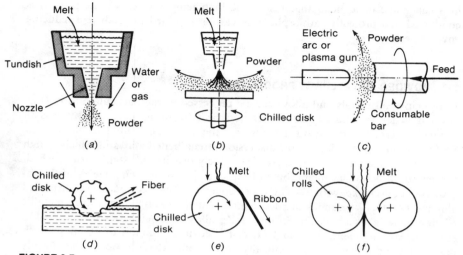

FIGURE 6-7
Metals and alloys may be brought into the particulate form by various processes, including
(*a*) water or gas atomization, (*b*) centrifugal atomization, (*c*) rotating electrode process, (*d*)
melt extraction, (*e*) melt spinning, or (*f*) roller quenching.

In one version of atomizing, the particles formed are immediately deposited
into the form of billets, slabs, or shaped components (*spray forming*).

4 Fiber production: In *melt extraction* a water-cooled, notched rotating disk is
brought in contact with the melt surface (Fig. 6-7*d*). In *melt spinning* the melt is
directed onto a cooled, fast-rotating disk (Fig. 6-7*e*) to form a ribbon of
20–100-μm thickness. A ribbon is also formed in *roller quenching* (Fig. 6-7*f*)
between two chilled rolls.

In all atomizing and fiber (ribbon) processes the cooling rate is much higher
than in conventional solidification, on the order of 100 °C/s in gas atomization,
1000 °C/s in water atomization, and $> 10^6$ °C/s in centrifugal and rotating-electrode atomization (hence the term *rapid solidification technology*, RST). Consequently, the equilibrium conditions described in Sec. 3-1 are never achieved and
unusual and, in many respects, highly desirable structures are formed. At cooling
rates of 1000 °C/s, secondary dendrite arm spacing is very small (on the order of
1 μm) and intermetallic particles are finely distributed. At ultrahigh ($> 10^6$-°C/s)
cooling rates, supersaturated solid solutions are retained, some crystalline phases
are suppressed, and, in the limit, there is no time even for the arrangement of
atoms into a lattice, and the solid (especially the eutectics) remains amorphous.
Amorphous metals (*metallic glasses*) are made by melt spinning of approximately
40-μm-thick ribbons. Some metallic glasses devitrify on heating to 350–500 °C
and become brittle, but others can be transformed into a microcrystalline

structure of high strength (up to 1250 MPa TS) and reasonable ductility. They have already found applications as magnetic materials, reinforcing elements in ceramics, and brazing alloys.

5 Some metal powders (notably beryllium) are produced by *machining* a cast, coarse-grained billet and comminuting the swarf by ball milling and impact grinding. Some powders of ductile titanium alloys are made from castings: hydrogen gas is introduced to form brittle hydrides which can be milled into a powder; ductility is then restored by driving off the hydrogen.

Powder Treatment Some powders are used in the as-atomized form but others are subjected to the preparatory steps indicated in Sec. 6-1-1. The shape of the powder, the distribution of particle sizes, and surface conditions have powerful effects on subsequent consolidation and sintering. All powder is screened to remove large particles. Milling is sometimes necessary to break up agglomerates, flatten (flake) the particles, or modify their properties by strain hardening. Alternatively, the powder may be annealed for improved compressibility.

Some metals, such as iron, are likely to be oxidized, but the oxide is readily reduced by a suitable atmosphere during sintering. Others, such as titanium, dissolve their own oxide and are thus reasonably suitable for powder processing. Still others are covered with a thin but very tenacious and persistent oxide film that greatly impairs the properties of the finished part, and these materials (typically those containing chromium and, in general, the high-temperature superalloys) must be treated by special techniques to keep oxygen content very low. Some powders, such as water-atomized high-speed steel powders, are annealed and deoxidized in a single operation, making them more compressible and easier to sinter. Contaminants that segregate on the surface are bound to create not only consolidation and sintering problems but will also greatly detract from the service properties of the material. Any remnants of a surface film at grain boundaries may act as crack initiators (Sec. 3-2-4).

Alloys can be prepared not only by atomizing the melt of correct composition but also by mixing elemental powders. The alloy is formed in the course of sintering; the driving force is the chemical potential gradient due to concentration differences. Mechanical alloys are prepared by milling the constituent powders to promote cold welding.

Uniform blending is critical. A small amount of lubricant, usually a stearate, is often added to aid densification. Spray drying is used to prepare powder mixes for oxide-dispersion strengthened alloys and cemented carbides.

Finely distributed metal powders can be hazardous. Some (such as beryllium and lead) are *toxic*; others (such as zirconium, magnesium, aluminum) present danger of explosion; many others are *pyrophoric* (ignite spontaneously on contact with air) below some critical particle size. A *thermite reaction*, in which an oxide (such as iron oxide) is reduced by another, more reactive metal (such as aluminum) is also possible, and proceeds at high temperatures.

6-2-2   Consolidation of Metal Powders

Pressing in rigid, multiaction dies is the most widespread, with pressures ranging up to 800 MPa (120 kpsi). Slip casting (with deflocculants added to the water), powder rolling, and hydrostatic extrusion are also employed. All techniques of hot consolidation are used. In a variant of HIP, the powder is placed in a glass mold; in yet another variant, into a ceramic shell mold (Sec. 3-8-4). The high-pressure gas can be replaced with a ceramic powder or a softer metal, and then hot pressing on a conventional press yields results similar to HIP.

6-2-3   Sintering and Finishing

Except for powders of high fill density, covered with a protective oxide (such as aluminum), all sintering is done in an atmosphere chosen to provide a neutral to nonoxidizing or reducing environment. Among the extensively used gases, nitrogen is neutral. Sintering in a vacuum furnace also provides a neutral environment but at high temperatures it favors the deoxidation of many metals. Hydrogen is a very effective reducing agent but must be handled with caution to avoid explosions. Nitrogen with 10% hydrogen plus methane, dissociated ammonia, and partially combusted hydrocarbon gas (exothermic or endothermic) are frequently used. In the sintering of steel the carbon content is also controlled and, in some instances, steels are carburized in a CO-containing atmosphere.

The porosity of a fully sintered part is still significant (4–15%) depending on powder characteristics, compacting pressure, and sintering temperature and time. Density is often kept intentionally low to preserve interconnected porosity for bearings, filters, acoustic barriers, battery electrodes, or when components are to be infiltrated. Powder metallurgy offers unique opportunities for tailoring properties to needs: By pressing different sections of a part to different densities, strength and porosity can be locally adjusted. The residual porosity makes sintered compacts rougher than the compacting die. Impact and fatigue properties are low.

Cold restriking (coining or sizing) of the sintered compact increases its density and improves dimensional tolerances. Further densification and strength improvement can be achieved by resintering the repressed compact (Fig. 6-8).

Instead of the traditional pressing, sintering, repressing, and resintering sequence, the green compact may be heated to the forging temperature and directly hot forged, to close tolerances, at full theoretical density. Such *hot-forged powder-preform* parts, as well as hot-rolled bars, can possess the same properties, including toughness, as conventionally forged pieces, and can be given shapes otherwise too complex to attain. For example, connecting rods can be forged to finish dimensions.

Compaction by *rolling*, followed by sintering and perhaps rerolling, is used both for the manufacture of sheet and for cladding a solid base metal.

Impact compaction on fast hammers or with the aid of explosive charges yields parts of high density.

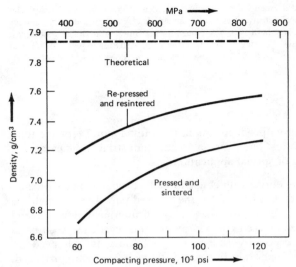

FIGURE 6-8
The density of an electrolytic powder-iron compact can be increased by re-pressing at 700 MPa (100 kpsi) and resintering at 1120 °C (2050 °F) for 1 hour. (*From Metals Handbook, 8th ed., Vol. 4, American Society for Metals, 1969, p. 455. With permission.*)

Spark sintering is a variant of hot pressing, with the application of a high alternating current to generate electric discharges during the early phase of consolidation, thus activating the surface of particles.

Because of the harmful effects of remaining imperfections, parts destined for critical service are nondestructively tested by magnetic particle, eddy current, or ultrasonic inspection.

A sintered compact of interconnected porosity may be *impregnated* with a liquid such as an oil. On immersion into heated oil, capillary action distributes the oil; application of vacuum aids the process. Impregnation with a metal is called *infiltration*, which is carried out by immersion in the molten metal or by placing the infiltrant metal in the form of a sheet above or below the compact in a furnace; again, capillary action fills the pores.

Example 6-4

The green compact shown in Fig. 6-4a is sintered to 96% theoretical density. The composition is Fe-8Cu-2C. Calculate (a) the theoretical density and (b) shrinkage of the 0.960-in dimension.

(a) The densities of constituent elements, from MHDE, p. 1.44, are Fe: 7.87; Cu: 8.96; C(graphite): 2.25 g/cm^3. The volume of 100-g powder is

$$(90/7.87) + (8/8.96) + (2/2.25) = 13.21 \text{ cm}^3$$

Theoretical density is $100/13.21 = 7.566$ g/cm^3

(*b*) In Example 6-3 we had a green compact density equal to 0.78 theoretical density. The sintered density is 0.96 theoretical density. A cube of 1-cm side length and 0.78 density contains material of 0.78 cm^3 volume. Since the mass is unchanged during sintering, it will now occupy a volume of $0.78/0.96 = 0.8125$ cm^3. The side of this shrunken cube is $(0.8125)^{1/3} = 0.933$ cm, and linear shrinkage is $1 - 0.933 = 0.067$ or 6.7%. Thus, the 0.960-in dimension shrinks to $0.960(0.933) = 0.896$ in.

6-2-4   Applications

Shipments of powder metallurgy products amount to only about 2% of the total weight in most metal categories, but their value and industrial significance are much greater, partly because of special applications.

1 Structural parts are competitive with conventionally produced parts because only as much material is used as is needed for the finished part. Even though the starting material may be more expensive, the savings in intermediate processing steps and scrap losses often more than compensate for this, particularly on parts of complex shape. This is especially true when *net shapes* (which need no machining at all) or *near-net shapes* (needing only little machining) are produced.

The largest quantities are made of iron powder, often mixed with 4–6% copper and 1% graphite for greater strength; of porous iron infiltrated with copper; and, increasingly, of atomized steel. They find applications in the automotive, off-road equipment, appliance, and business-machine industries as gears, transmission and pump parts, bearings, and fasteners. Smaller but increasing quantities are made of copper and especially aluminum alloys which are preferred in business machines because of their light weight.

Of great importance are structural components such as near-net-shape or net-shape parts for aircraft, jet-engine, and rocket-motor applications, such as superalloy turbine disks, titanium-alloy bulkheads and fuselage components, and rocket nozzles made of tungsten or molybdenum infiltrated with copper. In superalloys, the powder metallurgy approach avoids the problems of alloy segregation, carbide clustering, and residual cast structures.

2 *Bearings* can be made that combine the load bearing and wear resistance of one component with the lubricating function of another. Examples are oil-impregnated iron or bronze "permanently lubricated" bearings, plastic-filled bearings, lead-filled iron bearings, and bearings pressed with graphite (strictly speaking, all these should be regarded as composites).

3 A small but important application is for *surgical implants* (as in Fig. 1-5). A long-established application is filling of teeth with dental amalgams. They represent room-temperature transient liquid-phase sintering, in which an Ag–Sn alloy is amalgamated with Hg; the mercury is used up in the reaction.

4 Some metals can be produced only by the powder metallurgy route. Beryllium is hot vacuum pressed. Tungsten is sintered and hot forged in preparation for wire drawing into incandescent-lamp filaments; doping with small quantities of alloying elements (e.g., 0.5% Ni) accelerates sintering.

5 In the electrical industry, *contacts* must be good conductors while also resisting wear. Tungsten or molybdenum with 25–50% silver or copper, or tungsten carbide with 35–55% silver fulfill the requirements. *Brushes* consist of graphite bonded with 20–97% copper or silver.

6 Magnetic applications include *magnetically soft materials* such as Fe, Fe–3Si, Fe–50Ni, which, because of their mechanical softness, are difficult to machine but are easily formed into final shape by powder metallurgy. Among *permanent magnets*, Alnico (Fe–Al–Ni–Co) magnets can be sintered instead of cast to final shape. Only powder metallurgy is suitable for the extremely powerful, elongated single-domain magnets which consist of $R \cdot Co_5$ (where R is a rare earth such as Sm) particles. Compaction takes place while the powder is oriented in a magnetic field.

7 Increasing quantities of *tool steels* are made by powder metallurgy. Such high-speed steel tools have a much finer carbide distribution and the carbide content can be raised beyond the limits encountered in conventionally made steels; hence, tool life increases too.

8 Tool, die, and wear-resistant materials of great importance are the *cemented carbides*. Tungsten carbide (WC) powders are milled with cobalt, so that each particle is coated with the metal. After pressing, liquid-phase sintering establishes full density. Sometimes the final shape is given by grinding a presintered compact, which is then finish sintered. Powders of 1–5-μm particle size are sintered for tools and dies. With the cobalt content increasing from 3 to 15%, hardness decreases but ductility increases; die components subjected to bending stresses often contain up to 30% cobalt. Further improvements, at least for steel-cutting purposes, are obtained by replacing some of the WC with TiC. For wear components the particle size is < 10 μm.

9 Cemented carbides belong to the broader class of *cermets* (ceramic–metal composites). Newer members of the family are TiC bonded with a Ni–Mo alloy or with 50–60% tool-steel binder. The latter has the advantage that the part can be machined after sintering; strength is imparted by final heat treatment. Some typical applications are noted in Table 4-6 and in Sec. 8-3-1.

6-3 CERAMICS

In the narrower definition, *ceramics* are compounds of metallic and nonmetallic elements. This leaves out such materials as diamond, SiC, and Si_3N_4, and a broader definition regards as ceramics everything that is not a metal or organic material. In addition to a vast variety of naturally occurring silicates and oxides, the definition includes manufactured materials, sometimes of similar composition but of greater purity, and at other times carbides, nitrides, and other compounds not found in nature. By definition, glass is a ceramic and will be discussed in this chapter, even though it is processed in the melt stage and, strictly speaking, should not be included in a treatment of particulate processing.

6-3-1 Bonding and Structure

Ceramic powders have characteristics that are in many ways different from metals and polymers, differences that stem from the different bonding typical of them.

Nature of Ceramic Bonds   All forms of bonding, except metallic bonds, may play a role:

1 *Covalent bonds* are formed by electrons shared between adjacent atoms. These are very strong bonds. The directionality of bonds often leads to the formation of a spatial framework in which atoms are not necessarily closely packed. The high bond strength reflects in a high melting point, strength, and hardness coupled with brittleness; thermal expansion is often low, electrical resistance high. Carbon in the form of diamond is a purely covalently bonded material (Fig. 6-9a).

2 *Ionic bonds* form when one atom gives up one or more electrons to complete the outer electron shell of another atom or atoms. Electrical charge balance is maintained, but the electron donor is now deficient in electrons and thus becomes a positive charge center (*positive ion* or, because it is attracted to the negative terminal, also called *cation*), whereas the recipient atom has an excess negative charge (*negative ion* or *anion*). Attraction between the opposing charges (coulombic attraction) provides the bond strength.

FIGURE 6-9
The two basic bond types in ceramics are (*a*) covalent bonding, as in diamond in which each carbon atom shares electrons with four adjacent atoms and (*b*) ionic bonding, as in NaCl in which sodium and chlorine ions alternate regularly. (*From L. H. Van Vlack, Materials Science for Engineers, © 1970, Addison-Wesley, Reading, Mass., p. 44 and 60. Reprinted with permission.*)

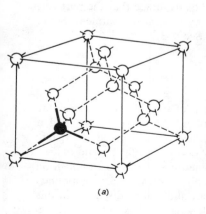

(*a*)

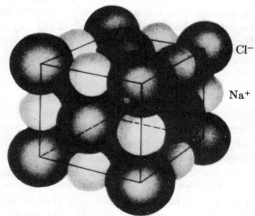

Cl⁻

Na⁺

(*b*)

An example is common salt, NaCl, in which the charge of each Na^+ ion is balanced by the charge of the Cl^- ion. A spatial network is again formed, but this time it is closely packed. For example, the NaCl lattice can be visualized as interpenetrating cubic Na and Cl lattices (Fig. 6-9b). Electrical conductivity is low at low temperatures but increases at higher temperatures when ions move and carry the electric charge (*ionic conductivity*). In such a regular structure dislocations can propagate somewhat like they do in metals; hence, even though the structures are brittle at low temperature, some ductility is evident at high temperatures and when hydrostatic pressure is applied. Bond strength and melting point increase with increasing charge. Thus, NaCl is relatively soft, MgO (double charge) is harder, Al_2O_3 is harder yet, and SiC (in which there are four bonds) is hardest.

3 Many ceramics possess a dual bond character because electrons tend to concentrate toward atomic centers, giving the bonds some degree of covalent character even in ionic compounds. The bond becomes more covalent in character with a decreasing difference in *electronegativity* between the atomic species (electronegativity is a measure of the affinity of atoms to attract electrons). The degree of covalent character can be estimated for different ceramics: MgO = 0.25; $SiO_2 = 0.5$; $Si_3N_4 = 0.7$; SiC > 0.9; C = 1. More complex ceramics may incorporate both bond types: in gypsum ($CaSO_4$) the S is bonded covalently to O but the SO_4^{2-} group is ionically bonded to Ca^{2+}.

Yet other ceramics are made up of two or more simple ceramics such as oxides and can be regarded as the ceramic counterparts of metal alloys. The resulting phases are shown in equilibrium diagrams which reveal features similar to those found in the phase diagrams of metals (Sec. 3-3-2).

4 *Secondary bonds* are extremely important when ceramics form layered structures of platelets. Within the platelets the strong, predominantly covalent bonds ensure great strength; between platelets only secondary bonds act and the strength of these bonds can be influenced by the addition of molecules of a gas or liquid. Such molecules adsorb on the surfaces of platelets, allowing easy movement and thus facilitating processing.

Crystallinity In the stable, lowest-energy form, most ceramics are *crystalline*. Atoms occupy defined lattice sites over long distances. As the temperature or pressure changes, different crystalline structures, *polymorphs*, may become more stable. Because there is a change in crystal structure, polymorphic transformations are accompanied by a volume change. The magnitude of this change is usually larger than that accompanying allotropic transformations in metals, and can lead to crazing, fracture, or total destruction of the part. In a *displacive transformation* the bonds are retained but distorted to allow a new crystal structure to form, somewhat like in the austenite–martensite transformation in steel (Sec. 3-5-4). In a *reconstructive transformation* the new structure is formed by breaking existing bonds; this takes more driving energy and the transformation may be suppressed by rapid cooling.

Some ceramics are noncrystalline (*amorphous*) and are then called *glasses*. A glass forms when a normally crystalline ceramic is heated above its melting point and then cooled so fast that crystallization is suppressed. The bonds are the same as in a crystalline ceramic, but the long-range lattice arrangement is missing in this *glassy state* (also called, from the Latin, *vitreous state*). Whereas extremely fast cooling rates are needed to make metallic glass, ceramic glasses are formed at industrially practical cooling rates. If such a glass is held at elevated temperatures for a long period of time, the more stable crystalline form is again obtained (the glass crystallizes or, as it is often said, it is *devitrified*). A more detailed discussion of glasses will be given in Sec. 6-5. Noncrystalline solids can also be formed by chemical reaction; the resulting *gels* are colloidal structures related to everyday jelly but of substantial strength.

In general, the properties of amorphous ceramics and some ceramics crystallizing in the cubic form are isotropic, whereas the properties of those crystallizing in more complex forms can be highly anisotropic.

6-3-2 Properties of Ceramics

Ceramics are used as engineering materials because of their high strength, hardness, hot strength, corrosion resistance, and desirable electrical, magnetic, and optical properties.

Mechanical Properties   Mechanical properties are measured by techniques described in Sec. 2-1.

1 Ceramics are not only brittle but often have microcracks in them. Therefore, tension testing is difficult, and it is more usual to conduct bending tests to determine the modulus of rupture (Sec. 2-1-3). The four-point bending test is favored because the more uniform stress distribution makes the discovery of cracks more likely (see Example 2-5).

2 Because of their sensitivity to cracks, ceramics are often used in compressive loading, and compressive strength, hardness, and hot hardness are then specified. The high hardness of ceramics also makes them indispensable in many applications where wear resistance is important (see Sec. 8-3-1).

3 Cracks or defects in ceramics can have sharp radii which cause large stress concentrations (Eq. (2-15b)), reduce fracture strength, and make ceramics vulnerable to fatigue failure. Ceramic components are increasingly designed by the fracture-mechanics approach (Sec. 2-1-5).

Tensile properties and toughness improve with decreasing particle size because flaws are generally of the size of the constituent grains; therefore, ceramics of submicron particle size are often used in high-technology applications. Further improvements in fracture toughness are gained by incorporating into the structure features that retard the propagation of large cracks; such features are more effective if they are smaller and more closely spaced. Three approaches are possible:

a Incorporate particles that suffer phase transformation. A prime example is *partially stabilized zirconia* (PSZ). Zirconia has a tetragonal lattice at high temperatures, and on cooling transforms to a monoclinic form with a disastrous (3.25%) volume change. However, by adding small amounts of Y_2O_3, MgO, or CaO, it can be fired to have a predominantly cubic structure. Within this stable cubic matrix, small islands of the metastable tetragonal phase are included. The pressure exerted by a crack propagating through the material causes transformation of these islands into the monoclinic form; some of the energy is absorbed in the transformation and, most importantly, expansion of the transformation product puts the matrix in compression.

b Incorporate fine, < 50-μm-diam fibers (such as graphite or silicon carbide); the crack follows the weak interface and is arrested.

c Create intentionally weak interfaces that produce a multitude of crack-retarding microcracks ahead of the major crack. Propagation of cracks may also be arrested by incorporating particles of differing thermal expansion, so that very fine cracking is induced on cooling; even though these cracks reduce static strength, they prevent the propagation of large cracks.

4 The strength and hardness of crystalline ceramics remains high close to the melting point. They are, however, subject to creep. Creep in single crystals can occur only by dislocation movement, which can be blocked by precipitate particles. In polycrystalline ceramics creep involves diffusion and grain-boundary sliding, both of which are made easier when porosity is present. Hence, minimization of porosity is one of the aims of manufacturing processes. However, if densification were aided by the formation of a glass at particle boundaries, viscous creep of this glass would allow sliding of grains and accelerate the creep of the entire body.

5 Strength properties can be improved by imparting compressive residual stresses to the surface of the component (Fig. 2-15c). This can be achieved by a number of techniques: quenching (Sec. 6-5-2); replacing some surface ions with larger ions (ion exchange or ion stuffing, Sec. 6-5-2); forming a low-expansion surface layer at high temperatures which is then put into compression on cooling (as in coating Al_2O_3 parts with an Al_2O_3–Cr_2O_3 coating); formulating the ceramic so that a displacive polymorphic transformation results in the expansion of a surface layer; and grinding conducted under conditions that lead to surface deformation.

Example 6-5

We saw in Example 2-5 that Si_3N_4 gave a rupture strength of 930 MPa in a three-point bending test but only 725 MPa in a four-point bending test. The uniform stress distribution between the loading points in the four-point test (Fig. 2-6b) brought to light imperfections that were missed in the three-point test. A tension test on the same material gave a maximum load of 11.3 kN on a specimen of 3.2-mm × 6.4-mm cross section. From Eq. (2-8), $TS = 11300/(3.2)(6.4) = 552$ MPa. Thus, the tension test is the most critical and is most likely to detect defects. However, results are greatly affected by bending that may be inadvertently introduced by the slightest misalignment in the test apparatus; therefore, the four-point bending test is preferred.

Physical Properties   Since the covalently bonded ceramics are not closely packed, they can accommodate increasing atomic vibrational amplitudes without a change in macrodimensions; thus, their thermal expansion is lower than that of metals. Some polycrystalline ceramics such as lithium aluminum silicate ($LiAlSi_2O_6$) have zero expansion and can be heated or cooled rapidly without damage.

The electrical properties (Sec. 2-2-2) of ceramics range from good conductors through semiconductors and insulators. Many ceramics also have a high dielectric strength; thus, they withstand high electrical fields without breaking down, and have allowed the miniaturization of capacitors. Some ceramics exhibit *piezoelectricity*. Thus, a crystal subjected to mechanical loading generates a potential difference and can be used as a force transducer; in the reverse mode, a potential difference applied to the crystal causes a dimensional change which can be exploited in ultrasonic transducers and force generators. Ceramics also exhibit the complete range of magnetic properties (Sec. 2-2-3). Without ceramics, the solid-state electronic revolution would have been impossible.

Ceramics can be formulated to provide the full range of *optical properties*. Single crystals of ionically bonded ceramics are usually transparent whereas covalently bonded ceramics may range from transparent to opaque. Grain boundaries and defects such as pores and cracks that create internal reflecting surfaces reduce transparency; only isotropic ceramics are transparent in the polycrystalline form. By appropriate additions, a selective wavelength of the visual spectrum can be absorbed, giving ceramics the widest range of colors. The index of refraction can also be controlled; this affects the change in direction when a light beam enters a solid (at point B in Fig. 2-39a) and is most important in applications such as lenses or decorative "crystal" glass.

Light tubes, video-display terminals, and color television rely on *phosphorescence*: Ceramic phosphors emit light of a characteristic wavelength when stimulated by an electric discharge or electron beam. Of rapidly increasing industrial importance are lasers (Sec. 9-3-7), some of which utilize a single-crystal rod made of a ceramic.

Chemical Properties   A great advantage of ceramics is that they are often resistant to chemical attack by gases, liquids, and even high-temperature melts. Combined with their remarkable high-temperature strength, this makes them suitable for such applications as temperature-resistant furnace linings (*refractories*), insulators, and even mechanical components such as turbine disks, turbine blades, and various components of internal combustion engines.

6-3-3   Preparation of Powders

We already mentioned that some ceramics occur in nature and others are manufactured. Natural raw materials have been dominant for thousands of years and are the starting material for what are now described as *traditional ceramics*. The high demand placed by engineering applications led to the development of modern ceramics: Some natural ceramics are replaced by pure, well-controlled manufactured versions of the same material, and ceramics are made that do not occur in nature at all. Advanced forms of these ceramics are usually called *high-technology ceramics*, or, in Japan, *fine ceramics*.

Natural Ceramics Natural ceramics are mined, in open-pit mines whenever possible, and comminuted in crushers and hammer and ball mills. Undesirable components are removed by screening, magnetic separation, filtering, or flotation. In *flotation* the particulate mass is suspended in water, and a frothing agent is added which preferentially attaches itself to one or the other of the mineral species, causing it to rise to the surface. Thus, either the desirable mineral or the unwanted species (*gangue*) can be separated economically. The most frequently used natural ceramics are:

1 *Silica* (SiO_2) is abundant in nature. It forms a high-viscosity melt at 1726 °C. On cooling it crystallizes and undergoes several polymorphic transformations. Since Si is tetravalent, it forms a tetrahedron with four oxygen atoms (Fig. 6-10). The tetrahedrons then join into a spatial network, with each O atom attached to two Si atoms, resulting in the ratio SiO_2. The hexagonal form is called quartz. The large single crystals found in nature or grown in manufacturing plants are valuable because they exhibit piezoelectricity and, ground to exact thicknesses, are used to control the frequency of oscillators.

2 *Silicates* are obtained when other atoms or oxides are introduced into the silica framework. A tremendous variety exists. Some silicates form chains or fibrous crystals (*asbestos* family). In others the SiO_4 tetrahedra join into sheets, with the negative charge of the top oxygen atoms remaining available for bonding with other cations, giving rise to the vast number of *layer silicates*, including talcs,

FIGURE 6-10
Tetravalent silicon (*a*) forms a SiO_4 tetrahedron with oxygen; (*b*) the remaining valences are available for the formation of a spatial silica network or compounds. (*L. H. Van Vlack, Elements of Materials Science and Engineering, 4th ed., ©1980, Addison-Wesley, Reading, Mass., p. 296 and 297. Reprinted with permission.*)

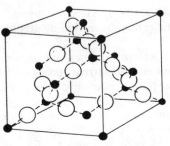

Si

O

(a)

(b)

Ionic
radius,
nm

O^{2-}: 0.13

Si^{4+}: 0.04

Al^{3+}: 0.06

OH^-: 0.13

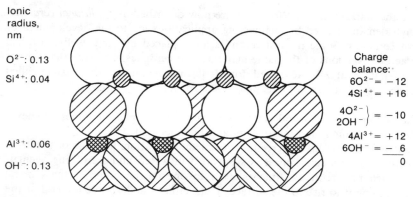

Charge
balance:
$6O^{2-} = -12$
$4Si^{4+} = +16$

$\left.\begin{array}{r}4O^{2-} \\ 2OH^-\end{array}\right\} = -10$

$4Al^{3+} = +12$
$6OH^- = -\ \ 6$
_____
0

FIGURE 6-11
One constituent of clay is kaolinite, the chemical formula of which $2(OH)_4Al_2Si_2O_5$ does not reveal that layers of O^{2-}, OH^-, Al^{3+}, and Si^{4+} combine to form a layer structure which is in electrical charge balance.

micas, and clays. In some sheet minerals such as mica cleavage occurs over long distances on the same plane; prior to the discovery of manufactured dielectrics, mica sheet was used extensively. Now mica is more often comminuted and then bonded with glass to make high-precision insulators. Further ions or oxides may also be introduced into the spatial network of SiO_4. For example, stuffing the framework with Na or Ca ions leads to *feldspars*. *Glasses* are three-dimensional networks in which crystallinity is lost.

3 The most important natural ceramics fall into the family of *clay minerals*. They can generally be described as hydrated aluminosilicates of a layer structure. Each layer crystal consists of several sheets, as shown on the example of the unit cell of kaolinite (Fig. 6-11). Each unit cell is in electrical charge balance, and the layer crystals are held together only by relatively weak van der Waals forces between the surface sheets of O^{--} and OH^- ions. Therefore dry clay is brittle and crumbly. The weak polarization of the surface is sufficient to adsorb water (physically adsorbed water), which facilitates sliding of the thin (approximately 50-nm-thick) plates relative to each other, making the clay plastic.

Manufactured Ceramics Here we discuss only the most important starting materials; specific ceramics—including graphite and diamond—will be discussed in Sec. 6-3-6.

Among manufactured ceramics, *silicon carbide* (SiC) is made in large quantities by passing an electric current through a long mound of coke surrounded by sand (SiO_2); the purest reaction product, found in the core of the mound, is suitable for electronic applications such as high-temperature heating elements; the adjacent, less pure layer is suitable for abrasives.

Alumina (aluminum oxide, Al_2O_3) occurs in nature as corundum or as the single-crystal gems ruby (colored by Cr ions) and sapphire. Most industrial alumina powder is produced by thermal reduction of aluminum hydroxide. By controlling the process and cooling rates, crystallization and properties of the alumina can be adjusted to yield products ranging from relatively soft to hard. *Magnesia* (magnesium oxide, MgO) also occurs naturally but for industrial use it is made from the carbonate or hydroxide.

A ceramic of increasing importance, *silicon nitride* (Si_3N_4) does not occur in nature and may be made by high-temperature reaction of silicon metal with nitrogen gas. Many varieties of oxides, carbides, nitrides, borides, and more complex ceramics are also made.

Hydraulic cements contain calcium silicates and calcium aluminates. The cement is finely powdered and mixed with water, whereupon hydration takes place. The water reacts to form, over a period of days, a partially crystalline ceramic of substantial strength. The main use of cements is in *concrete*, a composite of *aggregate* (sand and gravel) and cement. An even more complex composite is formed when concrete is made more resistant to tensile stresses by the incorporation of steel reinforcing bars and even metal, polymer, or ceramic fibers.

Most man-made ceramics are made at high temperatures and the resulting mass is comminuted to controlled sizes and size distributions. However, the very fine powders used in high-technology ceramics are normally produced by reaction in the vapor phase or from chemical precursors. The processes of spray drying and granulation are often used. A related process is *freeze drying* in which a water solution of salt is atomized, the droplets rapidly frozen, the water *sublimated* (evaporated without melting), and the dry powder thus obtained is calcined to decompose the crystalline salts and obtain the ceramic in the form of dry, pure powder. Because of the short diffusion paths in the fine droplets, such particles are homogeneous and particle size can be well controlled.

Blending Many ceramics are blended with other ceramics or with lubricants and binders. Water may be present or added during comminution, fine screening, or classification. By controlling the proportions of various constituents, the right consistency is obtained for subsequent forming.

Pressing properties and shrinkage can also be controlled by adding comminuted presintered ceramics (*grogs*) to the particulate body.

6-3-4 Consolidation of Ceramic Powders

All techniques described in Sec. 6-1-2 are used.

1 *Dry pressing* requires high pressures and fairly expensive dies but allows mass production of parts to close (typically $\pm 1\%$) tolerances. Thus, tens of millions of spark plug insulators, ceramic capacitor dielectrics, circuit substrates, and enclosures are pressed, sometimes to thicknesses below 1 mm. Lubricants and binders are used as required. Binders may be organic (polymers, waxes, gums, starches, etc.) or inorganic (clays, silicates, phosphates, etc.).

2 *Wet pressing* (deformation in the plastic state) is most frequently employed for clay-type ceramics but finds application for modern ceramics too. Pressing on hydraulic or mechanical presses into steel dies, cold isostatic pressing, extrusion (usually in screw extruders), jiggering, and injection molding are all practiced. The large water or organic carrier content results in a larger shrinkage and less tight (typically, ±2%) tolerances.

3 *Casting* is widespread: Slip casting into plaster of paris molds is the dominant method. Deflocculants and dispersants, combined with acidity (pH) control are used to prevent flocculation of fine (typically < 5-μm) particles. It is usually found that slips of given solid concentration have a minimum viscosity at some specific pH value. The high liquid content makes for large shrinkage and tolerances are fairly wide, except when the part is of simple shape, as in tape casting ceramic substrates for electronic circuits.

4 *Injection molding* is gaining in importance for high-technology ceramics.

6-3-5 Sintering and Finishing

Before ceramics are sintered (*fired*), free water is often reduced by holding at room temperature.

1 The initial, low-temperature heating (*drying*) stage is most important when physically adsorbed water (or, in the case of an organic carrier, organic vapor) must travel long distances to reach the surface. Moisture trapped in the center would blow up the compact. Organic binders must be driven off and burnt to prevent discoloration of the ceramic. At higher temperatures, a number of events may take place.

2 Water contained in the form of *water of crystallization* is removed at relatively low temperatures (between 350 and 600 °C). In clay-type ceramics *dehydroxylation* (breakdown of hydroxy groups) also takes place. In some instances, salts are not calcined prior to compaction and conversion to oxides must take place during heating to sintering temperatures. For all these reasons, the rate of heating is slow and temperature may have to be held for some considerable period of time.

3 At yet higher temperatures, *sintering* begins:

a In single-component ceramics (such as oxides, borides, and carbides) diffusional growth of necks (Fig. 6-6) dominates. On extended heating, grain growth occurs just as in metals. Polymorphic transformations may take place, but the dimensional change is accommodated in the partially sintered ceramic.

b In the many ceramics formulated of more than one component, reactions between adjacent particles takes place. Solid-phase diffusion can lead to the formation of solid solutions and other phases, as dictated by the phase diagram. Various transformations may also take place.

c Most importantly, liquid phases form at higher temperatures, aiding densification but also increasing the danger of grain growth. The relative quantities of liquids can be estimated from phase diagrams. Just as minor elements that form low-melting eutectics in metals lead to hot-shortness, minor contaminants that form low-melting phases in ceramics impair the hot strength.

d In some systems a liquid is present but is then used up in further reactions; such *reactive liquid sintering* often yields products of very good high-temperature properties because the glassy phase is absent.

e The very fine particle size of high-technology ceramics makes for short diffusion paths and allows sintering at significantly lower temperatures, thus avoiding grain growth that would reduce their strength.

4 After sintering is completed, the ceramic is *cooled* to room temperature at closely controlled rates. In ceramics containing a glassy phase, cooling rates determine the degree of crystallization. Ceramics of high thermal expansion could fracture on sudden cooling. Polymorphic transformations may also occur and the accompanying volume change results in microcracking.

The need for slow heating and cooling makes for long cycle times (days or even weeks), even though sintering itself is quite rapid.

Hot pressing of fine ($< 0.1\text{-}\mu m$) powder is possible on presses (using graphite dies supported by ceramics), by HIP, or by pressing the part in a molten glass or metal envelope (*rapid omnidirectional compaction*). Temperatures can be kept lower than in static sintering because the simultaneous application of pressure aids densification. This, together with the short high-temperature exposure, allows the production of fine-grained (microcrystalline) ceramics of high strength. Thus, these techniques find wide application for making high-technology ceramic structural components.

Most ceramics are sintered (fired) to finish dimensions. However, in some critical applications the surface of the sintered body is ground with a yet harder ceramic to improve surface finish, dimensional tolerances, or impart a more complex shape.

Example 6-6

A block of graphite (density: 1.9 g/cm^3) has a length of 36.34 mm, width of 24.68 mm and height of 12.70 mm. Its dry weight is 18.878 g. When weighed suspended in diethyl phthalate (density: 1.120 g/cm^3), its weight is 6.919 g. It is then again weighed (saturated with the fluid), and is found to weigh 19.235 g. Calculate the (*a*) true volume, (*b*) total (bulk) volume, (*c*) open, closed, and total porosity, and (*d*) bulk and apparent density.

(*a*) True volume = dry weight/density = $18.878/1.9 = 9.936 \text{ cm}^3$.

(*b*) Total volume, from geometry: $(36.34)(24.68)(12.7) = 11.390 \text{ cm}^3$. Often the part is of irregular shape and then the Archimedes principle may be used:

Fluid displaced by the fluid-saturated block = $19.235 - 6.919 = 12.316$ g.

Since the density of the fluid is 1.120 g/cm^3, the block displaced $12.316/1.120 = 10.996 \text{ cm}^3$ fluid. This is the total (bulk) volume, including closed porosity.

(*c*) Total porosity = (total volume − true volume)/total volume = $(11.390 - 9.936)/11.390 = 0.12766$ or 12.766%. Open (or apparent) porosity = volume of fluid absorbed/total volume = (saturated weight − dry weight)/(fluid density)(total volume) = $(19.235 - 18.878)/(1.120)(11.390) = 0.027985$ or 2.80%. Closed porosity = total porosity − open porosity = $0.12766 - 0.02795 = 0.099675$ or 9.968%.

(*d*) Bulk density = mass/total volume = 18.878 g/11.390 cm^3 = 1.657 g/cm^3 (since both the graphite and fluid are weighed in the same gravitational field, we may take weight as a direct measure of mass).

Apparent density = mass/(total volume − open pore volume) = (18.878 g)/(11.390 cm^3) (1 − 0.027985) = 1.705 g/cm^3.

Check the calculation:

Density = mass/(total volume − open pore vol. − closed pore vol.) = 18.878/(11.390) (1 − 0.027985 − 0.099675) = 1.9 g/cm^3.

Check the accuracy of measurements:

Bulk volume = true volume + closed porosity = 9.936 + (11.390)(0.099675) = 11.0713 cm^3, in good agreement with the 10.996 cm^3 found in (*b*).

6-3-6 Applications

Ceramics are used in all phases of our life. The emphasis here will be on technical applications.

Clay-Based Ceramics Some products are made from natural clays. Clays of high (60–80%) SiO$_2$ and low (5–20%) Al$_2$O$_3$ content are pressed into bricks and tiles and fired at 900–1000 °C to a porous but reasonably strong condition.

All other clay-based ceramics are made from mixes of controlled composition. Because three components, quartz (flint), clay, and feldspar (aluminosilicates of K, Na, and Ca) are used, one speaks of triaxial bodies. Feldspars reduce the firing temperature by increasing the ratio of eutectic melt. The proportions depend on the field of application (Fig. 6-12). Earthenware is fired only once, at 1150–1280 °C, to a slightly porous body. Stoneware is fired at 1200–1300 °C to dense bodies. Vitrified whiteware for bathroom fixtures is usually slip-cast and coated with a glaze prior to firing at 1260 °C. Dry-pressed floor tiles and lathe-turned

FIGURE 6-12
Traditional ceramics are often triaxial bodies. A: wall tile; B: semivitreous white ware; C: hard porcelain; D: vitreous white ware; E: electrical porcelain; F: floor tile; G: dental porcelain.

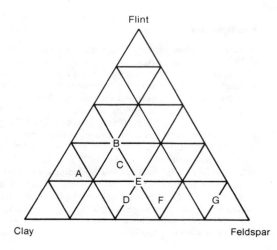

electrical porcelains are also glazed and then fired at 1290 °C. Most other whiteware, such as semivitrified tableware and hard porcelain, is fired twice. The first or *biscuit firing* liquid-sinters the body, making it translucent but at the expense of lower strength and electrical resistivity; after cooling down, the glaze is applied and is then melted in the second or *glost-firing*.

Refractories Like refractory metals, refractory ceramics are noted for their resistance to high temperatures. They are formulated to resist specific melt and atmospheric conditions. Thus, they may be acid (based on SiO_2), neutral (Al_2O_3; mullite $3Al_2O_3 \cdot 2SiO_2$; chromite $FeO \cdot Cr_2O_3$), or basic (magnesite MgO; dolomite Ca–Mg–O). Larger grog particles are usually embedded in a finer ceramic matrix to make bricks. Mortar and furnace hearths produced in situ are made of refractory granules bonded with a cement. Some refractories are melted and cast into shape.

Refractory powders and porous blocks serve as thermal insulation in high-temperature applications.

Oxide Ceramics Fine-grained ceramics of a single oxide can have high strength. Most widespread is *alumina*, Al_2O_3 (melting point 2054 °C) which is sintered into cutting tool bits, sparkplug insulators, high-temperature tubes, melting crucibles, wear components, and substrates for electronic circuits and resistors. Small additions of MgO remain concentrated on grain boundaries, facilitating densification and preventing grain growth. Finest-grain material is obtained by hot pressing. *Zirconia*, ZrO_2 (melting point 2710 °C) is more heat-resistant but, as already indicated, it suffers a polymorphic transformation with a catastrophic volume change. A stable cubic solid solution is obtained by adding 5–15% Y_2O_3 or CaO, and such *stabilized zirconia* is useful to 2400 °C as a furnace lining and, above 1000 °C, as a heating element. Partially stabilized zirconia is used as a hot-extrusion die.

Complex Oxide Ceramics Many of the most important man-made ceramics consist of carefully controlled combinations of several oxides.

1 In the $MgO–Al_2O_3–SiO_2$ system there are several compositions suitable for electrical and electronic applications. For example, *steatites* are used as insulators in high-frequency circuits.

2 *Ferrites* are generally composed of a metal oxide MeO (where Me can be any bivalent metal) and Fe_2O_3. To avoid confusion with ferrite in irons and steels (Sec. 3-1-6), the term ferrospinel is also used. They fall into two major groups:

a The structure of $MeFe_2O_4$ ferrites (where Me is Ni, Mn, Mg, Zn, Cu, or Co) is cubic, the same as that of the mineral spinel ($MgAl_2O_4$ or $MgO \cdot Al_2O_3$). They have a low magnetic hysteresis combined with high electrical resistance; hence, losses due to eddy currents are low. They make excellent cores for high-frequency applications in radios, television, and recording heads. In powder form, they can be deposited on an insulating (plastic) substrate to provide magnetic recording mediums. Rare-earth garnet ferrites, deposited on a nonmagnetic substrate, serve as bubble memory.

b The more complex ferrites, especially those of Ba, Sr, and Pb, are hexagonal in structure. They combine high resistivity with high coercive force and are thus excellent low-cost magnets for loudspeakers, small motors (as used also in automobiles), and, added to polymers, as magnetic elastomeric seals (as used on refrigerator doors).

3 *Titanates* contain TiO_2 as one of the constituents. Most significant is $BeTiO_3$ which has a high dielectric constant, making it suitable for capacitors. It also exhibits *ferroelectricity* (spontaneous alignment of electric dipoles) and, because of anisotropy of properties, piezoelectricity.

Carbides, Nitrides, Borides, and Silicides These ceramics are noted for their high hardness (see Table 8-3). *Carbides* have the highest melting point of all substances; an 80TaC–20HfC ceramic has a melting point of 4050 °C. Silicon carbide (SiC) is difficult to sinter but solid SiC bodies such as high-temperature resistance-heating elements, rocket nozzles, and sandblast nozzles can be obtained by pressure sintering or reactive sintering. Melting crucibles are made with a clay bond. The powder is one of the most important abrasives for grinding (see Sec. 8-8-2).

The extremely hard B_4C is used as a grinding grit and, in a sintered form, for wear-resistant parts and body armor. Other carbides are important as coatings (see Sec. 9-5) and in cemented carbides (Sec. 6-2-4).

Nitrides have only slightly lower melting points than carbides. One form of boron nitride (BN) is hexagonal (also called white graphite). It can be used as a high-temperature lubricant; it is also a good insulator and can be processed into large bodies. The cubic form (CBN) has a structure similar to that of diamond and is, after diamond, the hardest material, suitable for metal-cutting tools.

Silicon nitride (Si_3N_4) has good thermal conductivity, low expansion, and high hot strength, making it the prime candidate for ceramic engine components, turbine disks, and rocket nozzles. It can be processed by hot pressing, reaction bonding, vapor deposition, and injection molding. The *oxynitrides* (trade name Sialon, from Si–Al–O–N) have better oxidation resistance and are used as cutting tools and welding pins.

Borides $(TiB_2, ZrB_2, CrB, and CrB_2)$ have high melting points, strength, and oxidation resistance, and are used as turbine blades, rocket nozzles, and combustion chamber liners.

Molybdenum disilicide $(MoSi_2)$ has high oxidation resistance and serves as a heating element.

Carbon *Carbon* can be amorphous (lampblack), but the industrially most important forms are crystalline.

Graphite, of hexagonal structure, occurs in nature. Adsorption of volatile fluids or gases reduces the bond strength in the c direction (Fig. 3-2c), allowing slip along the basal plane; therefore, graphite is a good solid lubricant up to 1000 °C (although it begins to oxidize at 500 °C). Mixed with clay, it forms the "lead" of pencils. The technically important solid bodies are made of coke, formed into final shape with a pitch or resin binder, and converted to graphite above 2500 °C. It is a good electrical conductor, has low heat expansion, and resists high temperatures; hence, it is used for heating elements, electrodes, EDM electrodes, compacting dies, and crucibles. High-purity graphite is used for moderators and reflectors in nuclear power plants.

Graphite fibers are produced by the conversion of a polymer fiber, such as polyacrylonitrile. Holding the fiber under tension at 2500 °C, approximately 10-μm-diam fibers of oriented graphite are obtained with strengths between 2.0 and 3.5 GPa (the elastic modulus of the weaker fiber is higher, 400 GPa, compared to the 200-GPa modulus of the stronger fiber).

Diamond is totally covalently bonded in a cubic structure (Fig. 6-9a), is an insulator, and is the hardest known material. Natural diamonds are used in wear-resistant applications such as wire-drawing dies, cutting tools, and grinding wheels. Small but relatively defect-free crystals of diamond can be made from carbon at high pressures and temperatures; man-made diamond outperforms natural diamond in many applications. It can be sintered to give polycrystalline bodies ("megadiamonds") or 0.5–1.5-mm-thick layers on a tougher substrate such as cemented carbides.

6-4 PROCESS LIMITATIONS AND DESIGN ASPECTS

Particulate processing technology is subject to limitations that can be readily deduced from comparisons with other techniques previously discussed. The limitations have essentially two main sources: First, the particulate material must be able to fill the mold or die cavity, and second, the completed compact must be of a shape that can be released from the mold or die.

The slip-casting process is the most versatile, and any shapes (including hollow and undercut ones) can be formed provided that they can be released from the mold. It is possible to join several separately molded pieces and assure the virtual disappearance of the joint during sintering. One needs to think only of the complex yet low-cost mass-produced figurines made in porcelain and other ceramics to realize the potential of the process. However, diffusion must be ensured if this technique is to be used for powder metals.

The next-greatest freedom is afforded by flexible isostatic compacting molds which permit undercuts or reverse tapers, but not transverse holes.

The limitations posed by rigid dies are best understood if the die action is contemplated. A single punch cannot assure uniform density if the part is of varying thickness (actually, varying axial height); therefore, steps are limited to one-quarter of height (Fig. 6-13a). Much larger steps are allowable with a multiple-sleeve die; however, it must be remembered that a very thin sleeve is impracticable (Fig. 6-13b), and that the sleeve should be radiused to prevent excessive wear. Knife-edge punches wear excessively and should be changed to present a flat face (Fig. 6-13c). On withdrawal, a deeply penetrating punch would

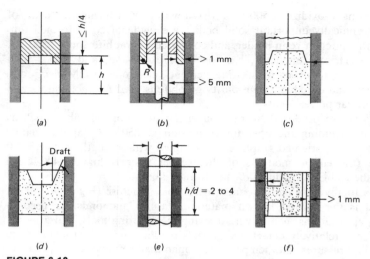

FIGURE 6-13
Some limitations of powder-metallurgy parts: (a) stepped end, (b) sleeve thickness, (c) sharp-nosed punch, (d) draft on punch, (e) length-to-diameter ratio, and (f) minimum wall thickness.

damage the compact and should be of minimum 4–5-mm (0.2-in) diameter to prevent premature core rod failure. The maximum depth-to-diameter ratio is practically limited to 2–4 (Fig. 6-13e). Even under pressure, the powder cannot fill very thin sections (Fig. 6-13f). Despite these limitations, hard tooling is the most suitable for mass production, especially of metal powder parts. Dimensions are well controllable, shape complexity in the plan view can be substantial (as in gears), and production rates are high.

6-5 GLASSES

We already observed that glasses are, by definition, ceramics, which differ from other ceramics in that they are produced by the melt processing route. Accordingly, the starting materials are typical of ceramics, whereas the processing techniques are closer to those of thermoplastic polymers (Sec. 7-7).

6-5-1 Structure and Properties of Glasses

The basic building block of glass is the silica (SiO_4) tetrahedron (Fig. 6-10). In the noncrystalline, glassy form (*fused silica*) the free valences join in a rather loose, three-dimensional, covalently bonded network (Fig. 6-14a). Since each oxygen is shared by two silicon ions, silica glass is really a $(SiO_2)_n$ polymer. A few other oxides (such as B_2O_3 and P_2O_5) are also *network formers* on their own; yet others (such as Al_2O_3) enter into the SiO_2 network. Another group of oxides, the

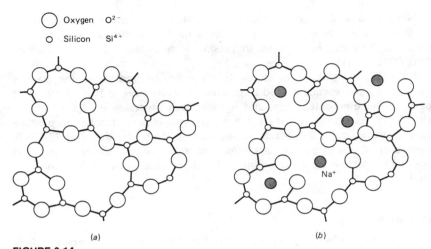

○ Oxygen O²⁻
○ Silicon Si⁴⁺

(a)

(b)

FIGURE 6-14
Simplified two-dimensional representation of (a) fully polymerized silica network and (b) partially depolymerized glass (fourth bonds are outside the plane of the illustration).

network modifiers, depolymerize the network by breaking up the O–Si–O bonds; their oxygen attaches itself to a free Si bond, while the metal cation is distributed randomly, its charge balanced by the negative charge of the dangling oxygen ions (Fig. 6-14b), maintaining overall charge balance. *Intermediate oxides* (MgO, BeO, TiO_2) may enter into the network or depolymerize it. Substantial fundamental knowledge exists that allows the formulation of glasses for special purposes. In general, Al_2O_3 increases hardness and reduces thermal expansion, whereas PbO reduces hardness and increases the refractive index. It should be noted that some elements (S, Se, Te), nonoxide compounds (e.g., As_2S_3), and organic compounds (e.g., abietic acid) also form glasses.

On cooling from the melt state, the spatial network of Fig. 6-14 forms without a sudden change in specific volume (as shown in Fig. 7-5a for amorphous polymers) and the glass can be regarded as a *metastable, undercooled liquid*. Because of the gradually reducing amplitude of thermal vibrations, the specific volume decreases gradually with dropping temperature until, at some *critical temperature* T_g, the glass assumes properties typical of a solid; it is now regarded as being in a nonequilibrium, glassy (vitreous) state. The glass-transition temperature T_g depends on cooling rate and is higher for fast cooling. Hence, as in polymer technology, one speaks of a transformation temperature range.

Mechanical Properties The distinction is important from a mechanical point of view:

1 Above T_g, the undercooled liquid exhibits Newtonian viscous flow behavior (Fig. 3-20b, line A). Therefore, glass cannot be used as a structural material above

T_g; a Newtonian fluid deforms under even the slightest stress (Eq. (3-9)). At the same time, viscous flow is extremely desirable for manufacturing processes because the fluid can be exposed to tension without the danger of localized necking (Sec. 4-1-4).

2 Below T_g, the glass is an elastic–brittle solid. By agreement, the glass is considered to have reached a solid form when viscosity rises to $10^{13.5}$ P (the unit of poise is still used extensively, and is sometimes expressed in the SI unit of dPa $\cdot$ s). The strength calculated from bond strength is never reached in practice because scratches and cracks act as stress raisers in the brittle state (Eq. (2-15b)). The presence of cracks makes glass subject to *static fatigue*: under an imposed tensile load, fracture may occur suddenly after some considerable time has elapsed.

Example 6-7

Prove that a fiber of a Newtonian fluid such as glass thins down in proportion to the applied force.

$$\sigma = P/A = k\eta\dot{\epsilon}$$

where k is a proportionality constant.

By definition, $\epsilon = dl/l = -dA/A$ (integration of this leads to Eq. (4-3)). Strain rate, by definition, $\dot{\epsilon} = d\epsilon/dt = (-dA/A)/dt = -\dot{A}/A$.

Then $P = k\eta\dot{\epsilon}A = -k\eta\dot{A}$ or $\dot{A} = -P/k\eta$. Thus, for a given viscosity, the fiber will thin down more rapidly if a larger force is applied.

Chemical and Physical Properties Glasses are often chosen for their resistance to corrosion by liquids or gases. This does not, however, mean that all glasses are corrosion-resistant even under mild conditions. Indeed, the main source of surface cracks is atmospheric corrosion. Water vapor present in air attaches itself to the glass surface; hydrogen ions replace monovalent cations (chiefly Na^-) by a stress-corrosion mechanism, creating cracks of very small tip radii. The attack is rapid; hence, freshly drawn glass fiber quickly looses the very high strength typical of defect-free glass. Corrosion can be reduced by replacing monovalent alkali metals by calcium. High strength is retained if freshly drawn (or fire-polished) fiber is coated with a polymer that is impervious to water. Corrosion by water is exploited in water glass, a fully depolymerized $Na_2O \cdot xSiO_2$ glass which is dissolved in water; when treated with CO_2, a strong gel forms which serves as a binder for sand molds (Sec. 3-8-3) and grinding wheels (Sec. 8-8-3).

Glasses are electric insulators at low temperature but become ionic conductors in the melt regime, allowing electric heating of melts.

Optical properties are most important; amorphous glass is transparent and may be colored by appropriate oxide or metal additions. Photosensitive eyeglasses are made from glass that contains AgCl. When this glass is energized by ultraviolet rays, Ag^+ ions form and impart a deeper color to the glass.

Glass Ceramics We mentioned that the glassy state is metastable. Therefore, all glasses can be converted, upon heating for a prolonged period of time, into a crystalline form. For some applications, glasses are formulated to avoid unwanted crystallization. In contrast, other glasses are formulated for controlled conversion into the crystalline form. Such *glass ceramics* are usually based on the $Li_2O-Al_2O_3-SiO_2$ system. Nucleating agents (metals such as Cu, Ag, Au, Pt, Pd, or oxides such as TiO_2) are added to promote the formation of many small crystals. Very low thermal expansion combined with high strength makes these glass ceramics suitable for cookware (the Pyroceram of Corning Glass Works) and many industrial applications.

Glass–ceramic construction materials are made from inexpensive natural and waste materials which crystallize spontaneously (e.g., fused basalt and blast-furnace slag for floor tiles, building cladding, and paving blocks).

6-5-2 Manufacturing Processes

Complete melting of a ceramic charge is a slow process; therefore, components of the charge are finely comminuted, blended, and then spread on top of a molten bath held in a batch or, more frequently, continuous melting furnace. Furnaces are usually heated by gas, although auxiliary electric heating is used because it creates intensive convection.

A typical charge for a window glass would be made up of quartz (sand, of 0.1–0.6-mm particle size), limestone ($CaCO_3$), and soda ash (Na_2CO_3). During melting the carbonates decompose and react with SiO_2. Gas evolution helps to homogenize the melt, but bubbles would remain. In the final or *fining* (refining) stage bubbles are allowed to rise, a process which is accelerated by additions such as As_2O_3. Further additions are made to control the color. For example, the green-blue of FeO can be changed to the much lighter yellow of Fe_2O_3, which then can be further masked by adding oxides that give a complementary color. Molten glass is highly corrosive to refractory furnace linings; therefore, linings are chosen for resistance to attack and so as not to introduce harmful components into the glass. The temperature of the bath is controlled, often in a *forehearth* (an extension of the melting furnace), to impart the optimum viscosity for the subsequent forming process. Here the glass is kept agitated, by electric current or mechanical stirrers, to maintain uniformity.

Glass compositions are chosen for specific applications (Table 6-1). The composition also determines the viscosity–temperature relationship (Fig. 6-15). Viscosity η decreases with increasing temperature T according to

$$\log_{10}\eta = C + \frac{B}{T} \qquad (6\text{-}1)$$

where C and B are constants, and T is absolute temperature. The lowest temperature at which forming is still practical is signified by the Littleton softening point at which a standard fiber extends under its own weight.

Example 6-8

The 0080 glass of Table 6-1 is occasionally used as a lubricant for hot extrusion (Sec. 4-5-2). What is its viscosity at 1200 °C?

TABLE 6-1   MANUFACTURING PROPERTIES OF SOME GLASSES*

Property	Corning Glass Works code number and type						
	7940 fused silica†	E-glass	7740 boro-silicate	1720 alumino-silicate	0080 soda-lime–silica	8871 potash–lead	8830 soda-borosilicate
Composition, weight %							
SiO_2	99.9	54	81	62	73	42	65
B_2O_3		10	13	5			23
Al_2O_3		14	2	17	1		5
Na_2O			4	1	17	2	7
K_2O						6	
Li_2O						1	
CaO		17.5		8	5		
MgO		4.5		7	4		
PbO						49	
Viscosity, P‡ at °C							
$10^{14.5}$ (strain point)	956	507	510	667	473	350	460
10^{13} (annealing point)	1084	657	560	712	514	385	501
$10^{7.6}$ (softening point)	1580	846	821	915	695	525	708
10^4 (working point)			1252	1202	1005	785	1042
Coefficient of linear expansion $\times 10^{-7}$/°C	5.5	60	33	42	92	102	49.5
Typical uses	High temperature, aerospace windows	Fiber	Chemical, baking ware	Ignition tube	Container, sheet, plate	Art glass, optics, capacitors	Sealing glass for Kovar (Fe–Ni–Co)

*Data compiled from D. C. Boyd and D. A. Thompson, *Glass*, in *Kirk–Othmer Encyclopedia of Chemical Technology*, 3d ed., vol. 11, Wiley, New York, 1980, pp. 807–880.
†Produced by vapor deposition.
‡Multiply poise by 0.1 to get $N \cdot s/m^2$.

From Table 6-1, viscosity is $10^{7.6}$ P at 695 °C and 10^4 P at 1005 °C. From Eq. (6-1):

$$7.6 = C + \frac{B}{695 + 273}$$

$$4.0 = C + \frac{B}{1005 + 273}$$

Solving the two equations simultaneously we obtain $B = 14366$ and $C = -7.24$. Thus, at 1200 °C,

$$\log_{10}\eta = -7.24 + \frac{14366}{1200 + 273} = 2.51$$

Thus, viscosity is $10^{2.51}$ P at 1200 °C, which agrees reasonably with Table 4-4, where a viscosity of 100–300 P at the working temperature is indicated.

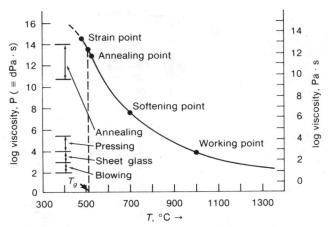

FIGURE 6-15
Glasses soften gradually; their viscosity depends on composition.

Forming Processes

1 *Sheet glass* is formed in large quantities, in a typical thickness of 2.0 mm (but ranging from 0.8–10 mm), by drawing (Fig. 6-16a) or rolling (Fig. 6-16b) from the forehearth. Such sheet is not free of imperfections but is suitable for windows.

2 Heavy *plate glass* used to be cast at higher temperatures and finished by grinding and polishing. Nowadays it is mostly cast onto the surface of a molten tin bath in a controlled atmosphere; the bottom surface of such *float glass* is atomically smooth, and the top surface is smoothed by surface-tension effects.

3 *Glass tube* is made by flowing glass onto a hollow, rotating mandrel through which air is blown (Fig. 6-17a); the gradually stiffening tube is mechanically drawn to thinner dimensions.

FIGURE 6-16
Sheet glass may be (a) drawn or (b) rolled from the forehearth of the melting furnace. (*F. H. Norton, Elements of Ceramics, 2d ed., ©1974, Addison-Wesley, Reading, Mass., p. 189. Reprinted with permission.*)

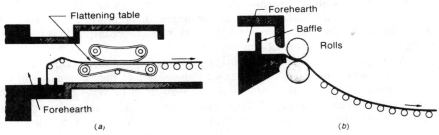

(a)   (b)

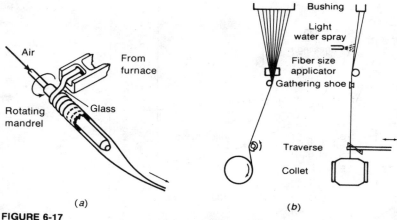

(a)   (b)

FIGURE 6-17
Methods of making (a) tube and (b) fiber by continuous methods [(a) *Adapted from D. C. Boyd and D. A. Thompson, in Kirk-Othmer Encyclopedia of Chemical Technology, 3d ed., Wiley, 1980, vol. 11, p. 864,* (b) *K. L. Loewenstein, The Manufacturing Technology of Continuous Glass Fibers, 2d ed., Elsevier, 1983, p. 29. With permission.*]

4 A similar principle is used in making continuous *glass fiber* of 3–20 μm diameter for insulating fabric and reinforcing fiber for plastics. A glass of high electrical and corrosion resistance (hence called E glass) is melted in (or transferred from the forehearth to) a platinum tundish called *bushing* in which there are 200 to 400 nozzles (Fig. 6-17b). Glass flows out at a rate q determined by nozzle dimensions (radius r and length l), the kinematic viscosity ν (dynamic viscosity η divided by density ρ) of the melt, and the hydrostatic pressure generated by the melt of height h

$$q = \frac{khr^4}{\nu l} \tag{6-2}$$

where k is a constant. The emerging fibers are cooled and subjected to mechanical *attenuation* (stretching) by winding the take-up spool at a higher speed (50–60 m/s). The fiber is coated with an organic size (such as starch in oil) which allows processing with minimum damage. Fibers may be chopped, or short fibers (*staple*) made directly by attenuating the fiber with compressed air or steam which breaks up the fibers while also thinning them down. *Glass wool* of 20–30-μm-diam fibers is spun (ejected) from rotating heads; it is often immediately matted to form insulating blankets.

5 A special class of fibers is used as *light guides* in fiber optics and as fiber-optical wave guides for long-distance transmission of digital signals sent by pulsed lasers or photodiodes. Attenuation (losses) must be very low; hence, all effort is made to ensure unimpeded passage of light while preventing the escape of light from the fiber. The former aim is achieved by melting extremely pure raw materials or by forming the fiber by vapor deposition. The second aim is satisfied by surrounding the core glass with an envelope of lower refractory index, so that reflection takes place at the interface between the two. Optical fiber is a prime candidate for manufacture in space.

6 Individual articles may be made by *pressing* a measured quantity of glass (*gob*) into steel or cast iron molds. The process is related to closed-die forging.

7 Articles with thinner walls and reentrant shapes are often needed. For these, the gob is dropped into the mold (or glass is sucked into the mold by vacuum) and a preform (*parison*) is formed by pressing with a punch. After transfer to a second, split mold, the part is *blow-molded* to final shape (Fig. 6-18). Newtonian viscous flow ensures that the wall thins out uniformly. To maintain a good surface finish, the mold is coated with a mineral oil, an emulsion, or a wax–sawdust mix that is then converted into carbon. Prior to blowing, the carbon layer is slightly dampened; the steam generated during blowing separates the paste-mold from the glass, and gives a smooth finish. Parisons for bottles are often made without a punch, with a puff of air creating the cavity. Light bulbs are produced on rotary, multistage blowing units at the rate of 2000 per minute. Hand blowing is now limited to artistic work.

FIGURE 6-18
A bottle may be made by pressing a gob of glass into a parison which is then blow molded to final shape. (*F. H. Norton, as Fig. 6.16, p. 188.*)

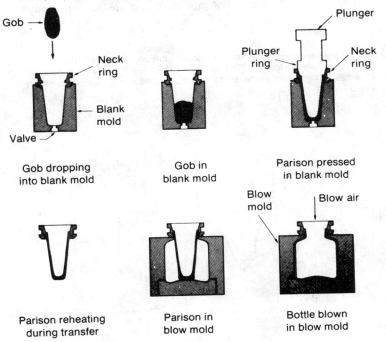

Finishing Operations Some glasses (especially fused silica) have a low thermal expansion and can be cooled rapidly. However, in most glasses rapid cooling sets up residual stresses which could cause explosive disintegration and must be relieved by *annealing* in a furnace called *lehr*. The temperature is chosen to allow stress relief in a reasonable time without loosing the shape of the article. Stress relief occurs (stress is reduced to 2.5 MPa) in 4 h at the lower annealing temperature or *strain point* and in 15 min at the upper annealing temperature or *annealing point* (Fig. 6-15). To avoid reintroduction of stresses, the part must be cooled slowly from the annealing-point to the strain-point temperature.

Glass that has not been in contact with any tool exhibits an extremely smooth surface (*natural fire finish*); therefore, cut edges are *fire polished* by a pencil torch.

The strength of typical glass is around 70 MPa. Resistance to tensile stresses may be increased by inducing compressive stresses in the surface and thus hold cracks in compression. This can be achieved by subjecting the finished glass article (ovenware, containers, eye glasses, etc.) to thermal or chemical treatment.

1 The principle of thermal toughening (*tempering*) is shown in Fig. 6-19 on the example of a plate. The plate is heated above T_g, then its surfaces are quenched with an air blast, causing the surfaces to contract and stiffen. At this point, the center is still soft and follows the contraction of the surface layers. On further cooling, the center cools too and in so doing, contracts; since the surfaces are now stiff, they cannot follow the contraction of the center and are put into compres-

FIGURE 6-19
Glass is toughened by the sequence of operations shown; high compressive surface residual stresses make it resistant to tensile loading. The variation of height across the thickness of the plate is shown schematically for each processing step. The broken line shows the corresponding temperature distribution.

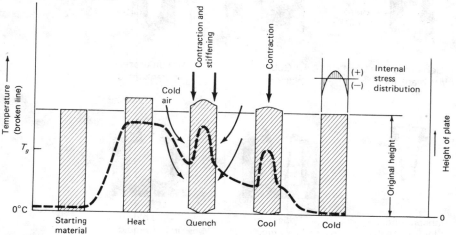

sion. Compressive surfaces stresses on the order of 140 MPa or more are balanced by internal tensile stresses; they are hamless because there are no cracks in the center.

2 Favorable compressive stresses, in a much thinner surface layer, may be developed by the chemical techniques indicated in Sec. 6-3-2. When a soda-lime glass is immersed, below T_g (at about 400 °C), into molten KNO_3, Na ions are replaced by the larger K ions (or Li ions may be replaced by Na ions in a $NaNO_3$ bath). Alternatively, Na or K ions are exchanged for smaller Li ions at temperatures above T_g; the surface has a lower coefficient of expansion and, since it shrinks less, is subjected to compressive stresses by the bulk of the glass.

Example 6-9

A glass container is heated in an oven to 350 °C and is plunged into boiling (100 °C) water. Assuming that the surface stress should not exceed the tensile stress (70 MPa) of the glass, determine which glasses in Table 6-1 can be safely subjected to the above treatment.

In the simplest case, the stress state is uniaxial and, from Eq. (2-5),

$$\sigma = Ee_t = E\alpha \Delta T$$

where E is Young's modulus (typically $E = 70\,000$ MPa) and α is the coefficient of linear expansion (Table 6-1).

Glass	α, 10^{-7} per °C	σ, MPa
Fused silica	5.5	9.6
Borosilicate	33	57.8
Aluminosilicate	42	73.5
Soda–lime	92	161
Potash–lead	102	178.5

Sample calculation for fused silica:

$$\sigma = 7(10^4)(5.5)(10^{-7})(350 - 100) = 9.6 \text{ MPa}$$

The advantage of a low-expansion glass is obvious. A slightly more sophisticated treatment takes into account the fact that the stress state on the surface of a plate or container is usually biaxial (see Prob. 6-14).

6-5-3 Coatings

Ceramic surface layers are deposited on metal and ceramic components for both aesthetic and technical reasons. They thus represent one approach to the manufacture of composite parts.

Glazes and *enamels* are glassy or partially crystalline coatings applied in the form of an aqueous slip (slurry). For optimum control of composition, the glass is often premelted and granulated in water. The resulting *frit* is comminuted and suspended in water with the addition of antiflocculants, etc. To retain the slip on vertical surfaces, its properties are adjusted by the addition of clay or organic binders to impart pseudoplastic or Bingham behavior (Fig. 3-20b).

The composition is chosen so that the glaze or enamel should have a slightly lower thermal expansion coefficient than the part and will thus be put into compression on cooling. Lead reduces firing temperatures but must be avoided in all food-related applications because of its toxicity. Coatings range from transparent to completely opaque, and a wide range of colors and special effects may be obtained.

Vitreous enamels are applied to low-carbon steel, cast iron, or aluminum articles for improved corrosion resistance. Iron carbide reacts with enamels to form CO_2; therefore, better quality and adhesion are obtained on steel of very low carbon content (hence the preference for rimmed steel with its segregation, Fig. 3-18a, and for special decarbonized steel of 0.03% C). Adhesion on cast iron is purely mechanical; on aluminum it is aided by first forming a surface oxide.

Glazes are applied to ceramics to make them impermeable. Some glazes are made up of raw materials, but better control is obtained with frits.

6-6 ELECTROFORMING

The smallest particulate is the atom (or, for a compound, the molecule). Components may be produced through controlled deposition of atoms on a surface; one speaks of *plating* and *coating* when the deposit is to stay in place (as in the chromium plating of a car bumper) and of *forming* when the deposit, stripped from the form (variously called *matrix, mandrel, die,* etc.), serves as a component. Our interest is primarily in the technique of electroforming.

In the process of *electroforming*, a plate or slab of metal (the *anode*) is immersed into an aqueous solution of a salt of the same metal (the *electrolyte*), and is connected to the positive terminal of a low-voltage, high-current dc power supply (Fig. 6-20). An electrically conductive moid (matrix or mandrel) of the desired shape is immersed at some distance from the anode and is connected to the negative terminal (and thus becomes the *cathode*). Metal atoms are removed as positive ions from the anode, transported through the electrolyte toward the cathode (and are, therefore, called *cations*), and deposited on the cathode as neutral atoms.

It takes 96 500 coulomb ($=$ ampere $\cdot$ second) to remove 1 mol of monovalent metal (*Faraday constant*). The metal transfer rate is then

$$W_e = \frac{j}{96\,500} \frac{M}{Z} \eta \qquad (6\text{-}3)$$

where W_e is g/s$\cdot$m^2, j the current density (A/m^2), M the gram atomic weight (g/mol), Z the valence (charge/ion), and η the efficiency (typically around 0.9).

The composition, temperature, and circulation of the electrolyte and the current density need careful control. Once a deposit of sufficient thickness is obtained (and this may take hours or days), it is stripped from the matrix.

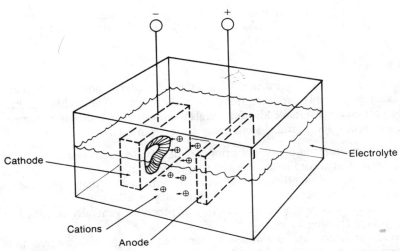

FIGURE 6-20
Complex shapes may be reproduced with great detail and accuracy by electroforming.

Permanent matrixes may be made of metal, or of glass or rigid plastic with a metallized surface (e.g., metallized by a chemical deposition technique). Adhesion is minimized by passivating the metal matrix surface and by the application of a thin coating of a parting compound. A slight taper is allowed to facilitate stripping.

Expendable matrixes are made of a metal (aluminum or zinc) that can be chemically dissolved, or of a low-melting alloy (such as eutectic Sn–Zn alloy), wax, or plastic that can be melted out. Since the finished part is not stripped, great freedom in shape complexity is gained (comparable to investment casting, Sec. 3-8-4).

The atom-by-atom deposit reproduces the matrix surface with the greatest accuracy, and this, together with the attainable shape complexity, defines the economical application range of the process to finished parts (such as waveguides, bellows, venturi tubes, reflectors, seamless screen cylinders for textile printing, filters, and typing wheels) and dies (for stamping of high-fidelity records and for plastic molding in general). Internal stresses can be severe, and there is an art to producing sound parts.

Example 6-10

A highly decorated vase is made by electroforming into a conductive die (matrix). The total surface area is 0.2 m^2; deposition proceeds at 6 V and 80 A from a copper sulfate ($CuSO_4$) solution. How long will it take to attain a wall thickness of 0.5 mm (500 μm)?

Current density $j = 80/0.2 = 400$ A/m^2; $M = 63.54$ g/mol; $Z = 2$. From Eq. (6-3)

$$W_e = (400)(63.54)(0.9)/(96\,500)(2) = 0.1185 \text{ g/s} \cdot \text{m}^2$$

Since the density of copper is 8.96 g/cm^3 = 8.96 Mg/m^3, the thickness of the layer is 0.1185/8.96(10^6) = 0.013 μm/s. Thus, time = 500/0.0132 = 37 880 s = 631 min = 10.5 h.

6-7 SUMMARY

Particulate matter, ranging in size from atoms to coarse powder, has been consolidated into usable products from the earliest times. The technique, first applied to ceramics, is suitable also for metals and, to a limited extent, polymers. Processes comprise some critical steps common to all materials:

1 Powder, whether naturally occurring or man-made, is comminuted when necessary, classified according to size and shape, cleaned, and blended to impart the required composition, fill density, absence of contaminants that would impair properties, and rheological properties that allow easy handling.

2 A green body is produced by a variety of processes generally classified as pressing, plastic forming, or casting.

3 Permanent bonds are established by sintering at high temperatures, thus developing strength while the volume shrinks. Remaining voids or grain-boundary defects impair fatigue and impact properties and one of the aims of advanced techniques is the improvement of the fracture toughness of the finished part.

4 Great improvement in properties is attainable by the application of pressure at the sintering temperature; particles are brought into close proximity and are moved relative to each other, thus promoting adhesion between them.

5 Particulate processing offers the opportunity to produce parts with controlled porosity and composites of unusual properties.

In the application of the technology to metals, powder may be obtained from ores or, by rapid solidification technology, from melts. High cooling rates offer a number of benefits: the size and spacing of features such as dendrite arms and second-phase particles are reduced; supersaturated solid solutions may be retained; in the limit, amorphous (glassy) metals may be produced. In addition to fully densified parts of improved properties, such as jet engine components and metal-cutting tools, parts of great shape complexity and close tolerances, such as gears, may be produced.

Traditional ceramics, even though brittle, have been indispensable in human development as building materials, containers, cooking vessels, and, generally, corrosion-resistant materials. With the development of man-made ceramics, ceramics that also occur in nature have become available in controlled purities, particle sizes, and properties, and have been supplanted by new compounds not known in nature at all. High-technology ceramics of exceptional electrical and magnetic properties have made the microelectronic revolution possible; others have opened up the possibility of producing structural components of hitherto unattainable temperature and wear resistance.

Glasses are ceramics processed by the melt route. Even though fabrication processes are closer to those used for thermoplastics rather than ceramics, the properties of the finished product are more typical of ceramics. Glasses too have entered the high-technology age, tailor-made for specific applications and subjected to special treatments. Crystallization of initially vitreous parts creates glass ceramics.

Processing on the atomic scale makes faithful reproduction of surfaces possible, both for plastics-processing dies and for structural components.

PROBLEMS

6-1 Explain why porosity impairs the mechanical properties of powder-metallurgy parts, and why it impairs tensile and impact properties more than compressive strength or hardness.

6-2 A Cu–5Sn bronze is prone to coring. Show why this should be so and consider whether coring could be eliminated more rapidly in a casting or in a powder-metallurgy part.

6-3 A compacted body of Ni powder is sintered into a 30-mm-diam 50-mm-tall cylinder which, upon weighing, is found to have a mass of 290 g. Calculate (a) the apparent density, (b) the percentage of theoretical density, and (c) the void volume (porosity) in percent.

6-4 Assuming that the cylinder of Prob. 6-3 is sintered until full theoretical density is obtained, and shrinkage is uniform in all directions, calculate the dimensions of the cylinder.

6-5 It is proposed that the part shown in Example 12-6 be made by powder metallurgy (with the contour shown in broken lines). (a) Analyze the part shape and suggest a cold-pressing die configuration. (b) Calculate the press size if the cold-compaction pressure is the maximum indicated in Sec. 6-2-2. (c) If, for a higher density and greater dimensional accuracy, the part is to be repressed and resintered, can the same die be used?

6-6 Is it physically possible to make the part of Prob. 3-16 by powder-metallurgy techniques? If it is, is it likely to be technically and economically attractive?

6-7 A cylinder of $d_0/h_0 = 1$ is compacted to 70% theoretical density by cold pressing an atomized steel powder. Full density and high strength are to be obtained by hot upsetting the cylinder to $\frac{1}{4}$ its original height. (a) What diameter should one expect, approximately, after upsetting? (b) Should one anticipate cracking in upsetting? If yes, where and why? (Illustrate with a sketch.) (c) If cracking is a danger, how could it be prevented?

6-8 Nickel powder is sometimes consolidated by cold rolling into a thin strip. After trimming the edges of the green strip, it is sintered and cold rolled again. During this second cold-rolling operation, edge cracking may occur. (a) What feature of the rolling process is responsible for cracking? (b) What feature of the sintered product contributes to cracking? (c) What could be done to eliminate or reduce cracking?

6-9 An electrical insulator block of $10 \times 20 \times 150$-mm dimensions is slip cast. Immediately after removal from the mold it weighs 48 g. After drying, the weight is 35 g, and the length has shrunk to 130 mm, with proportional shrinkage in the thickness and width directions. Calculate (a) the weight loss, percent; (b) the coefficient of linear shrinkage; (c) the dry dimensions and volume.

6-10 Explain why the mechanical properties of ceramics improve with diminishing crystal size.

6-11 Suggest methods for improving the impact properties of ceramics.

6-12 Suggest appropriate techniques for making a (*a*) common brick; (*b*) face brick with decorative surface configuration; (*c*) alumina tube; (*d*) glass tube; (*e*) toilet bowl; (*f*) field tile; (*g*) porcelain figurine; (*h*) porcelain dinner plate; (*i*) porcelain electrical insulator for high-tension power lines.

6-13 Make a distinction between ceramics, glasses, and glass ceramics.

6-14 Recalculate the stresses generated by the quenching described in Example 6-9. Consider that in balanced biaxial tension $\sigma_1 = \sigma_2$ and $\sigma_3 = 0$. Apply the generalized Hooke's law (ν is Poisson's ratio)

$$e_1 = \frac{1}{E}\left[\sigma_1 - \nu(\sigma_2 - \sigma_3)\right] = \frac{1}{E}\sigma_1(1-\nu)$$

$$\sigma_1 = \frac{eE}{1-\nu} = \frac{E}{1-\nu}\alpha\,\Delta T$$

6-15 Bathtubs are often made of enameled iron sheet. The bathtub is normally at room temperature, but is subjected to sudden heating by hot water. Should the thermal expansion of the enamel be greater or lower than that of iron?

FURTHER READING

A Metals

Ashbrook, R. L. (ed.): *Rapid Solidification Technology: Source Book*, American Society for Metals, Metals Park. Ohio, 1983.

ASM: *Metals Handbook*, 9th ed., vol. 7: *Powder Metallurgy*, American Society for Metals, Metals Park, Ohio, 1984.

Bradbury, S. (ed.): *Source Book on Powder Metallurgy*, American Society for Metals, Metals Park, Ohio, 1979.

Gessinger, G. H.: *Powder Metallurgy of Superalloys*, Butterworths, London, 1984.

Gilman, J. J., and H. L. Leamy (eds.): *Metallic Glasses*, American Society for Metals, Metals Park, Ohio, 1978.

Hanes, H. D., D. A. Seifert, and C. R. Watts: *Hot Isostatic Processi g*, Battelle Press, Columbus, Ohio, 1979.

Hausner, H. H., and M. K. Mal: *Handbook of Powder Metallurgy*, Chemical Publishing Co., New York, 1982.

Hirschorn, J. S.: *Introduction to Powder Metallurgy*, American Powder Metallurgy Institute, New York, 1969.

James, P. J. (ed.): *Isostatic Pressing Technology*, Applied Science Publishers, London, 1983.

Klar, E. (ed.): *Powder Metallurgy: Applications, Advantages, and Limitations*, American Society for Metals, Metals Park, Ohio, 1983.

Lenel, F. V.: *Powder Metallurgy: Principles and Applications*, Metal Powder Industries Federation, Princeton, N.J., 1980.

Luborsky, F. E.: *Amorphous Metallic Alloys*, Butterworths, London, 1983.

B Ceramics

Kingery, W. D., H. K. Bowen, and D. R. Uhlmann: *Introduction to Ceramics*, 2d ed., Wiley, New York, 1976.

McColm, I. J.: *Ceramic Science for Materials Technologists*, Leonard Hill (Chapmann and Hall), Glasgow, 1983.

Norton, F. H.: *Elements of Ceramics*, 2d ed., Addison-Wesley, Reading, Mass., 1974.

Richerson, D. W.: *Modern Ceramic Engineering*, Dekker, New York, 1982.

Samsonov, C. V., and J. M. Vinitsku: *Handbook of Refractory Compounds*, Plenum, New York, 1980.

Schwartz, M. M. (ed.): *Engineering Applications of Ceramic Materials*, American Society for Metals, Metals Park, Ohio, 1985.

Wang, F. F. Y. (ed.): *Ceramic Fabrication Processes*, Academic Press, New York, 1976.

C Glasses

Doremus, R. H.: *Glass Science*, Wiley, New York, 1973.

Hlavac, J.: *The Technology of Glass and Ceramics*, Elsevier, Amsterdam, 1983.

Loewenstein, K. L.: *The Manufacturing Technology of Continuous Glass Fibers*, 2d ed., Elsevier, Amsterdam, 1983.

Vogel, W.: *Chemistry of Glass*, American Ceramic Society, Columbus, Ohio, 1985.

Zschommler, W.: *Precision Optical Glassworking*, Macmillan, New York, 1984.

D Journals

American Ceramic Society Bulletin
Ceramic Industry
International Journal of Powder Metallurgy and Powder Technology
Journal of the American Ceramic Society
Journal of Glass Technology
Powder Metallurgy
Powder Technology

CHAPTER 7

PROCESSING OF POLYMERS

A *polymer* is, as indicated by the Greek roots *poly* (many) and *meros* (part), any substance made up of many (usually, thousands) of repeating units, building blocks, called *mers*. Most polymers are based on a carbon backbone and are thus organic materials. There are many natural polymers and, after concrete, wood is still the most widely used structural material. Our concern here, however, is with synthetic polymers, also called *plastics* (again, from the Greek *plastikos*, derived from *plassein*: to form, to mold) or *resins*.

From relatively recent beginnings (Table 1-1), the growth of the plastics industry has been phenomenal. As shown in Table 7-1, the sale of plastics—on a volume basis—has already outstripped steel production in the United States (which seems to have settled to between 70–100 million tons/year). The growth trend is expected to continue, partly because new uses are still being found and partly because improved polymers can substitute for other materials. Initially, most polymers were used in applications where their low density, high corrosion resistance, electrical insulation, and ease of manufacturing into complex shapes presented advantages and where mechanical strength was of secondary importance. A more recent and most important trend has been the emergence of structural polymers which can be made into load-bearing components and structures, at least for applications in which temperatures are only moderately high, typically, below 150–250 °C.

Many of the principles discussed hitherto apply also to polymers; nevertheless, there are sufficient differences to justify a review of polymer structure and properties, with reference to concepts previously explored for metals.

TABLE 7-1
SALES OF PLASTICS* (UNITED STATES).1000 tonnes.

	1972	1983
ABS	388	460
Acrylic	208	242
Alkyd	161	250
Cellulosics	75	52
Epoxy	78	147
Nylon	67	135
Phenolic	652	1115
Polyacetal	27	44
Polycarbonate	25	110
Polyester, thermoplastic		384
unsaturated	416	450
Polyethylene, low density	2372	3576
high density	1026	2646
Polyphenylene		65
Polypropylene	767	1962
Polystyrene and styrenics	1239	1938
Polyurethane	459	755
PVC and vinyls	2326	3131
Thermoplastic elastomers		200
Urea and melamine formaldehyde	411	602
Other	308	101
Total	11005	18365

*Compiled from *Modern Plastics*, Jan. 1973 and Jan. 1984.

7-1 POLYMERIZATION REACTIONS

Polymers, like metals, are produced from raw materials in specialized plants and are provided to manufacturing facilities in forms suitable for processing into finished articles. Primary manufacturing will be discussed here only to the degree necessary for an understanding of manufacturing properties.

There are many ways of classifying polymers, one of which is the technology used for making them. Details are beyond the scope of this book; it will be sufficient to note that macromolecules may be obtained by one of two processing techniques (Fig. 7-1):

1 *Chain polymerization.* Carbon is a tetravalent element and carbon chains can be formed with single, double, or triple bonds between adjacent carbon atoms. The starting material for chain polymerization is usually a *monomer* in which there is a double bond that can be opened up with the aid of a compound called *initiator* (organic or inorganic substance or catalyst), whereupon polymerization occurs simultaneously in the entire batch in a few seconds. The process is also called *addition polymerization.* The most frequently occurring structures are hydrocarbons, i.e., compounds of carbon and hydrogen (Fig. 7-1) which may

HYDROCARBONS

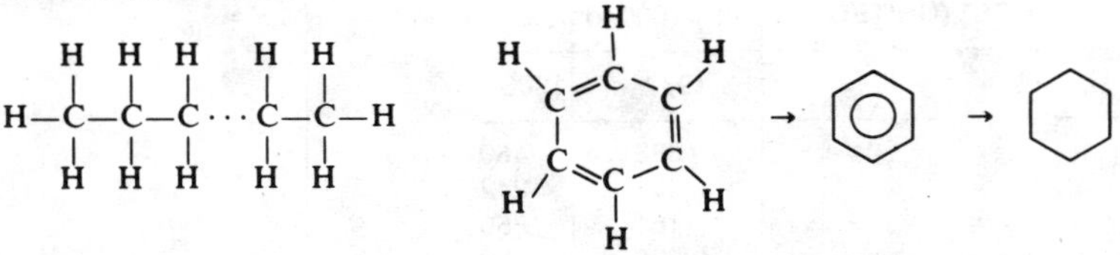

Paraffin (aliphatic)   Benzene (aromatic)   symbolic presentation
(hydrogen not shown)

CHAIN-REACTION POLYMERS

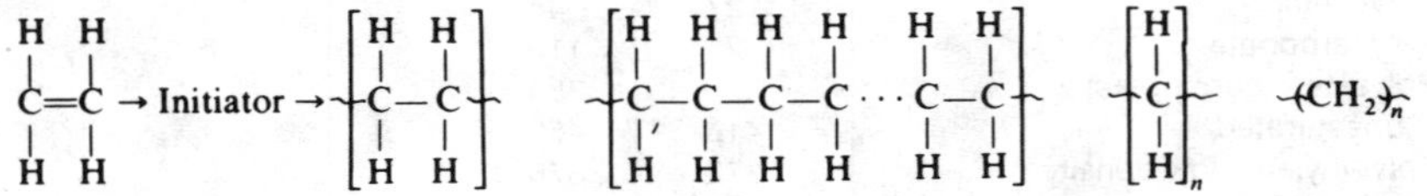

Ethylene   repeat   Polymer: PE (Polyethylene)   alternative
monomer   unit   presentation

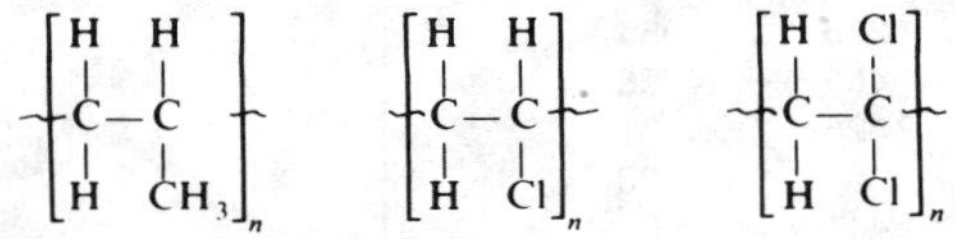

Polypropylene (PP)   Polyvinyl-   Polyvinylidene-
                  chloride (PVC) chloride

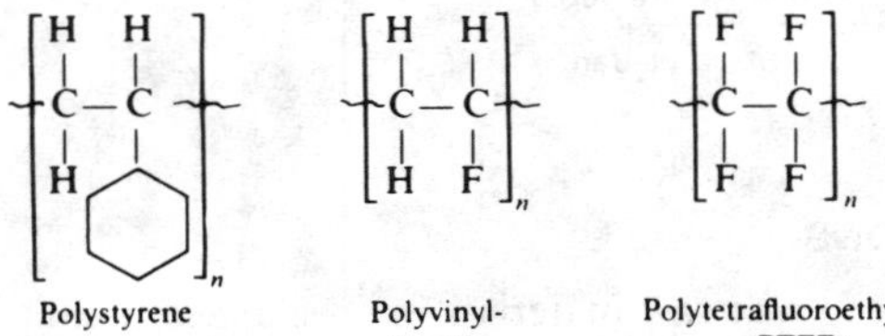

Polystyrene   Polyvinyl-   Polytetrafluoroethylene
          fluoride   (PTFE)

STEP-REACTION POLYMERS

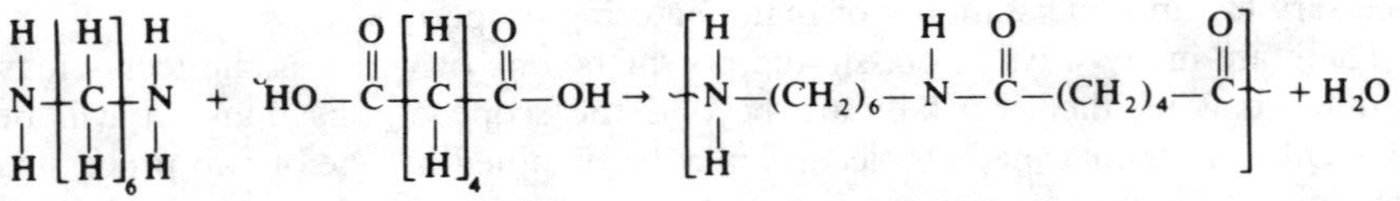

Hexamethylene + Adipic acid   Polyamide (nylon 66)   + Water
  diamine

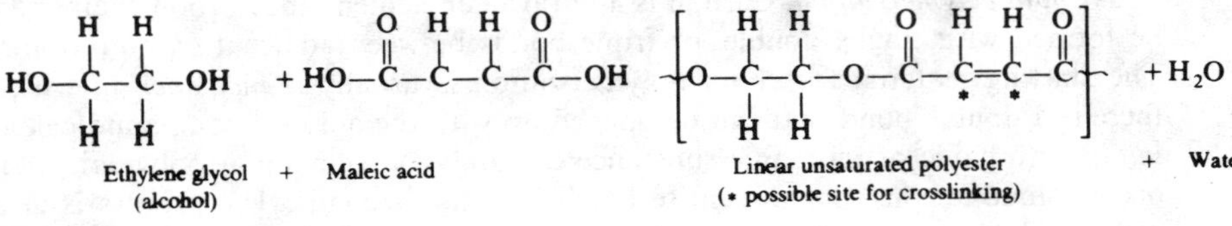

Ethylene glycol + Maleic acid   Linear unsaturated polyester   + Water
(alcohol)   (* possible site for crosslinking)

FIGURE 7-1

Thermoplastic polymers are made up of essentially linear molecules, formed by chain-reaction (addition) or step-reaction (condensation) polymerization. The presence of double bonds in the polymer, as in unsaturated polyesters, makes cross-linking possible.

form straight chains (*aliphatic hydrocarbons*) or benzene rings (*aromatic hydro-carbons*). Other polymer molecules may contain also N, O, S, P, or Si (in the backbone or side chains) while the monovalent Cl, F, or Br may replace H.

2 *Step-reaction polymers.* In the majority of these processes, two dissimilar monomers are joined into short groups, followed by the polymerization and cross-linking of these groups; a byproduct of low molecular weight is often also released (water in the example of nylon-66, Fig. 7-1), and then it is customary to speak of a *condensation reaction.*

Either way, the polymer chemist can control the average length of the mole-cules by terminating the reaction. Thus, the *molecular weight* (the average weight, in grams, of 6.02×10^{23} molecules) or *degree of polymerization* (the number of mers in the average molecule) can be controlled. For example, the length of molecules may range from some 700 repeat units in low-density polyethylene (LDPE) to 170 000 repeat units in ultrahigh-molecular-weight polyethylene (UHMWPE).

7-2 LINEAR POLYMERS

It will be noted that all examples shown in Fig. 7-1 result in the formation of more or less straight chains; hence, these polymers are called *linear polymers.*

7-2-1 Structure of Linear Polymers

In metals the repeating unit was the atom or, at the most, the unit cell of an intermetallic compound, and these units readily conformed to long-range order to give a crystalline structure. The great length of polymer molecules combines with other spatial features to make for a much greater variety of possible structures. Here we review them with emphasis on the significance of structure to manufac-turing and service properties. The molecules of a linear polymer are not simple straight chains for the following reasons:

1 Even the simplest chain, that of polyethylene (PE), is not straight. The C—C bond is at a fixed bond angle of 109.5°. The spacing between C atoms is 1.54 Å along the bond but only 1.26 Å in a straight line. Thus, a PE molecule of 2000 carbon atoms would have a length of 2520 Å (252 nm or 0.25 μm) when fully stretched. However, the single bond between the carbon atoms allows rotation around the bond; thus, the molecule can become randomly coiled and twisted, and the actual average end-to-end distance is typically only 180 Å or 18 nm. Such a molecule will not readily fit into a long-range ordered structure and the polymer will be *amorphous*.

2 The chains of some polymers, such as high-density polyethylene (HDPE), are smooth when stretched out (Fig. 7-2*a*) whereas others, such as polypropylene (PP), have pendant groups (in this case, $-CH_3$) in certain positions. The

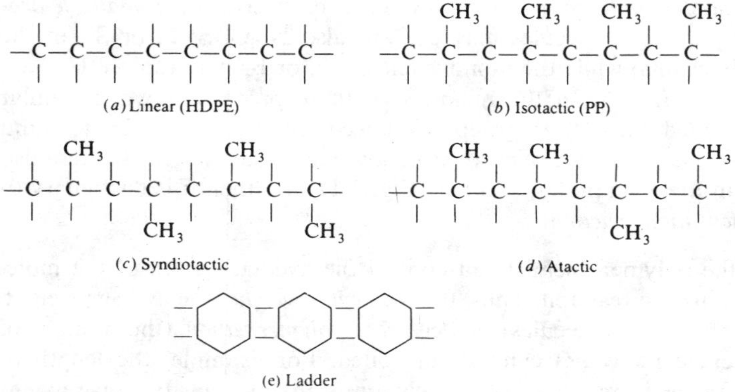

FIGURE 7-2
The type of backbone and the ordering (spatial arrangement) of pendant groups around the backbone determines many properties of linear polymers. See text for significance of various arrangements. (For simplicity of presentation, hydrogen atoms are not shown.)

ordering (in Greek *taktika*) of these groups determines whether the polymer is *isotactic* (with all groups on one side of the backbone, Fig. 7-2b), *syndiotactic* (alternating on the two sides, Fig. 7-2c), or *atactic* (randomly arranged, Fig. 7-2d). The pendant group is still larger in polystyrene (Fig. 7-1). Tight packing of molecules is obviously more difficult if the pendant groups are large and randomly oriented.

3 Even simple molecules such as LDPE may not be truly straight chains but have side branches which further increase the difficulty of close packing and ordering. Only some polymers, such as linear LDPE (LLDPE) or HDPE polymerized in the presence of special catalysts and polytetrafluoroethylene (PTFE), are free of branching.

4 In aromatic polymers the presence of the benzene ring offers the possibility of creating a backbone of a double strand, resembling a ladder (*ladder polymer*, Fig. 7-2d). Because two bonds must be broken before a lower molecular weight product is formed, such structures can be highly temperature-resistant.

5 The examples shown in Fig. 7-1 are made up of one kind of repeating unit and are called *homopolymers*, even if the repeating unit is made up of two precursor molecules.

6 It is possible to polymerize two types of polymers (generally, A and B) to obtain *copolymers* (more exactly, *binary copolymers*) somewhat analogous to solid-solution alloys. In copolymers each repeating unit is capable of forming a polymer on its own, as in an ethylene–propylene copolymer. (A three-component polymer is a *ternary copolymer* or a *terpolymer*, an example of which is ABS, made of acrylonitrile, butadiene, and styrene monomers.)

The repeat units may occur in a *random* (AAABBABAABBA); *alternating* (ABABABAB), or *block* (AAAAABBBBBAAAABBBBBBAA) sequence, or one species may branch off in a *graft polymer*

AAAAAAAAAAAAAAAAAAAAAAAAAA
B	B
B	B
B	B
B	B
B	B

7 A further possibility is to have two incompatible polymers (which do not enter into a joint chain) mixed with one another, with one serving as the matrix. These are called *polymer alloys* or polymer blends and can be regarded as the polymeric counterparts of two-phase metal alloys. Desirable properties of the constituent polymers are combined, as will be seen later in the example of thermoplastic elastomers.

Example 7-1

A UHMWPE has a molecular weight (MW) of 4 million. If this refers to the number average of molecules, calculate the degree of polymerization and the length of the stretched chain.

The building unit is C_2H_4, of a molecular weight of $(2 \times 12) + (4 \times 1) = 28$. The degree of polymerization is $4\,000\,000/28 = 143\,000$ (this is very large; that of PTFE is typically $30\,000$). The length of the molecule would be $(143\,000)(1.26) = 180\,000$ Å $= 18\ \mu$m. Of course, we shall see that it will be folded, as in Fig. 7-4b.

7-2-2 Sources of Strength

An engineering part consists of many macromolecules. Bonding *within* each molecule is provided by the electrons shared between adjacent molecules (*covalent bonding*). Bond energy (the energy required to break 1 mol, i.e., 6.02×10^{23} bonds) is on the order of 350–830 kJ/mol, making the molecule itself very strong. However, this tells us little about the actual strength of a polymeric part; to appreciate the strength or lack of strength of certain polymers, we must look into forces that hold the multitude of molecules together. Entanglement accounts for some of the strength, but the predominant source of strength is the presence of *secondary bonds*. These are of several kinds:

1 At the least, there are always *van der Waals forces* present, even though they are very weak (2–8 kJ/mol) (Fig. 7-3a).

2 When atoms share electrons in covalent bonds, the atom that loses electrons appears to have a positive charge and vice versa; thus, a permanent dipole is set up (the molecule has a polar character). Polar molecules, such as those containing free Cl, F, or O valences, make stronger (6–13 kJ/mol) *dipole bonds* (as in polyvinylchloride, PVC, Fig. 7-3b).

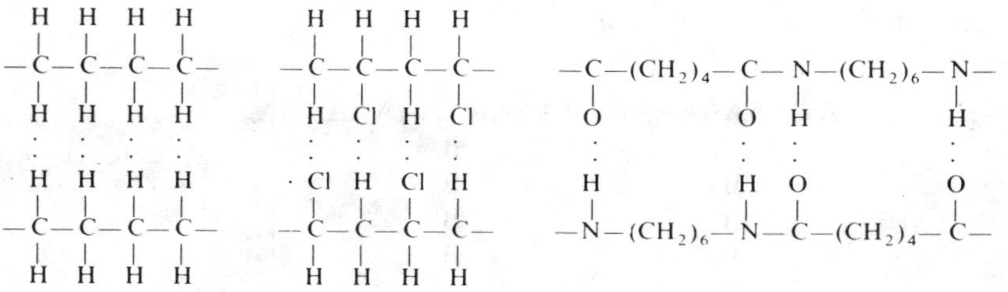

```
 H  H  H  H          H  H  H  H
 |  |  |  |          |  |  |  |
-C--C--C--C-        -C--C--C--C-        -C-(CH₂)₄-C-N-(CH₂)₆-N-
 |  |  |  |          |  |  |  |          ‖              |        |
 H  H  H  H          H  Cl H  Cl         O              H        H
 .  .  .  .          .  .  .  .
 .  .  .  .          .  .  .  .
 H  H  H  H          Cl H  Cl H          H              H  O     O
 |  |  |  |          |  |  |  |          |              |  ‖      ‖
-C--C--C--C-        -C--C--C--C-        -N-(CH₂)₆-N-C-(CH₂)₄-C-
 |  |  |  |          |  |  |  |
 H  H  H  H          H  H  H  H

    (a) PE              (b) PVC                    (c) Nylon 66
```

FIGURE 7-3
Polymers owe their strength to secondary bonds between molecules: (*a*) weak van der Waals forces between nonpolar molecules, (*b*) dipole bonds between polar molecules, and (*c*) the strong hydrogen bond between H and O, N, or F.

3 The *hydrogen bond*, established between hydrogen and O, N, or F, is a special case of dipole bonds. Bond energy is high (13–30 kJ/mol), as in nylon-66 (Fig. 7-3*c*).

The number of secondary bonds increases with chain length, giving increasing strength to the body. Thus, secondary bonds are sources of strength. At the same time, at higher temperatures—where thermal excitation is significant and secondary bonds are easily broken and reformed—they allow molecules to move relative to each other. The ease of movement depends on the number of secondary bonds present; very long-chain plastics, such as UHMWPE and PTFE, may char before ever reaching the moldable state.

7-2-3 Crystalline and Amorphous Polymers

When linear polymers are heated to some high temperature (but not so high as to break primary bonds), one can visualize a mass of polymer molecules as a bowl of spaghetti. Since no long-range order exists, the polymer is amorphous (Fig. 7-4*a*). The comparison to spaghetti is, however, incomplete. Under the influence of the elevated temperature, the molecules are in constant motion and the free volume is large (Fig. 7-5). Upon cooling, one of two events may take place.

Crystallization If the molecule is of relatively simple shape, with chemical regularity along the chain, and conditions are favourable, some long-range order may develop: the polymer becomes crystalline at the melting point T_m. This involves the repeated folding of chains into thin, approximately 100-Å-thick lamellae (Fig. 7-4*b*). When the polymer is allowed to crystallize in a static situation, several lamellae form a *spherulite* (Fig. 7-4*c*). If the polymer is mechanically drawn while it is cooling, an oriented "shish-kebab" structure (Fig. 7-4*d*) may result. Perfect crystallinity—such as is found in metals and ceramics—is never achieved; there is always some amorphous material between the spherulites. Thus, the term *crystalline polymer* refers to a structure in which crystalline regions predominate. During the formation of the crystalline zones,

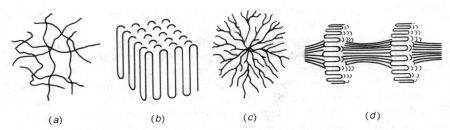

FIGURE 7-4
Linear molecules may be arranged (*a*) randomly (amorphous polymers) or (*b*) in thin, ordered (crystalline) lamellae which form (*c*) spherulites in an amorphous matrix. Crystals that grow during deformation have (*d*) a shish-kebab structure. [*From G. R. Moore and E. Kline, Properties and Processing of Polymers for Engineers, ©1984, p. 31. Reprinted by permission of Prentice-Hall, Inc., Englewood Cliffs, N.J.; (*b*) adapted from P. J. Flory, J. Am. Chem. Soc., 84:2857, 1962. Reprinted with permission of American Chemical Society.*]

FIGURE 7-5
Structural changes are reflected in (*a*) changes in specific volume and (*b*) large changes in viscosity and strain-rate sensitivity.

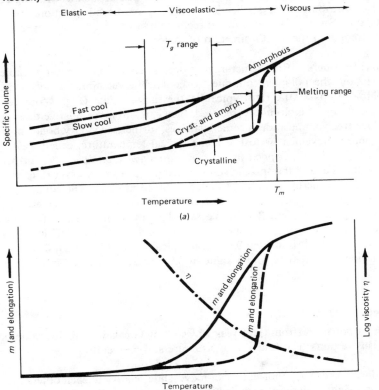

there is a corresponding drop in specific volume, but it is neither as steep nor as complete as in metals (Fig. 7-5).

The degree of crystallinity depends on many factors. It is greater for polymers of regular shape, without side branches or pendant groups. If there are pendant groups (Fig. 7-2), the isotactic form will more readily crystallize than the atactic form. For the same reasons, copolymers in general, and random and graft copolymers in particular, will not crystallize. Since it takes time for molecules to fold, slower cooling promotes crystallization. For rapid crystallization from the melt, the mold temperature is set to $(T_m + T_g)/2$. Above T_g, crystallization may continue for a prolonged period of time. This causes after-shrinkage, as in nylon-66, which has a T_g below room temperature. Heating at elevated temperature results in rapid completion of crystallization.

Cooling rate also affects the size of spherulites. As in metals, the rate of growth reaches its maximum at a higher temperature than the rate of nucleation (Fig. 3-14); hence, spherulites become coarser at slower cooling rates. Nucleation rate can be increased and thus the size of spherulites reduced by seeding, for example, with very fine silica (heterogeneous nucleation).

Mechanical alignment induced by directional deformation such as drawing or extrusion also contributes to crystallization, and the pronounced directionality of structure (texture, Sec. 5-1-3) results in a directionality (anisotropy) of properties, with higher strength in the length direction of molecules.

Amorphous Polymers If the structure and process conditions are unfavorable for crystallization, the polymer continues to cool while remaining amorphous. Specific volume drops at the rate typical of the molten state (Fig. 7-5); even though the excitation of molecules is reduced, they can still move relative to each other to reduce free volume and, in some instances, further secondary bonds may form. Freedom of movement is lost at some typical temperature, called—as in glasses, Sec. 6-5-1—the *fictive point* or *glass-transition temperature*, T_g. No further bonds are established, and the specific volume changes at a much lower rate, as a result of the reduced thermal motion of molecules fixed in space. The situation is somewhat analogous to metals in that T_g is often around $0.5T_m$ (these are homologous temperatures, Sec. 2-1-7). We saw that in metals the onset of hot working can be shifted to higher temperatures by alloying. Similarly, T_g can also be shifted substantially. It is around $0.66T_m$ in typical homopolymers, it can drop to $0.25T_m$ in block copolymers, and a value of $0.9T_m$ may be reached in random copolymers.

7-2-4 Rheology of Linear Polymers

The *rheology* of polymers (from the Greek *rheos* = flow) deals with their response to stresses. This response is a function of structure and temperature.

Viscous Flow Above T_m, molecules can move, slide relative to each other, and low-molecular-weight polymers may exhibit Newtonian viscous flow. This is subject to Eq. (3-9), repeated here because of its importance.

$$\tau = \eta \frac{dv}{dh} = \eta \dot{\gamma} \tag{7-1}$$

Viscosity increases with increasing molecular weight because of the greater number of secondary bonds available along a longer chain (for many high-molecular-weight polymers, viscosity is proportional to the 3.4 power of the weight-average molecular weight).

Viscosity is also a function of molecular structure: entanglement, less open structure (fewer side chains), and lesser ease of molecular segment rotation (chain flexibility) all contribute to higher viscosity, whereas a wide molecular-weight distribution (which signifies the presence of shorter chains) leads to a lower viscosity.

Viscosity increases with decreasing temperature, because of the drop in free volume and the lessened mobility of molecules; as with thermally activated processes in general, viscosity changes exponentially with inverse temperature

$$\eta = A \exp\left(\frac{-E}{RT}\right) \tag{7-2}$$

where A is a material constant, E is activation energy, R the universal gas constant, and T temperature (K).

(The change in viscosity with temperature can be expressed by considering the change in free volume; this leads to the Williams–Landel–Ferry (WLF) equation. With experimentally determined constants, it takes the following form:

$$\log_{10}\left(\frac{\eta_T}{\eta_{T_g}}\right) = \frac{-17.44\left(T - T_g\right)}{51.6 + \left(T - T_g\right)} \tag{7-3}$$

where T is the temperature of interest (K), and η_T and η_{T_g} are the viscosities at T and T_g, respectively. The importance of the glass-transition temperature is obvious.)

Molecules uncoil when subjected to shearing; uncoiling reduces entanglement and hence viscosity. Thus, most polymers exhibit non-Newtonian flow: the shear stress needed for deformation is not constant but is also a function of shear strain rate (Fig. 7-6). Polymers may be pseudoplastic (Fig. 3-20b, line C) and obey a power law

$$\tau = \eta_a \dot{\gamma}^m \tag{7-4}$$

The similarity to Eq. (4-7) (flow stress in hot working a metal) is evident, except that Eq. (7-4) is written in terms of shear stress τ and shear strain rate $\dot{\gamma}$. The strain-rate sensitivity exponent m is high and reaches values close to unity at temperatures above T_m (Fig. 7-5b).

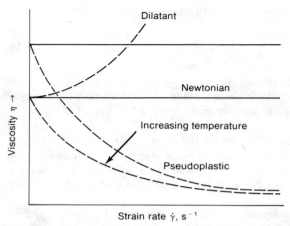

FIGURE 7-6
Many plastics show pseudoplastic flow and are best worked at strain rates where viscosity is almost independent of shear rates.

Viscosity is more appropriately called an *apparent viscosity* η_a, showing that it is not a constant but a function of temperature, pressure, and shear rate. Since viscosity represents resistance to the sliding of molecules,

$$\eta \propto \frac{\text{entanglement}}{\text{free volume}}$$

and since pressure reduces while temperature T (above T_g) increases the free volume, the combined effects of temperature, pressure, molecular weight, and—for a given molecular weight—ordering, can be expressed qualitatively as:

$$\eta \propto \frac{\text{pressure}}{T - T_g} \frac{\text{molecular weight}}{\text{long-chain side branches}}$$

It is generally found that a pressure increase of 100 MPa is equivalent to a temperature drop of 30–50 °C. Many polymers behave like Bingham substances (Fig. 3-20b, line D) and begin to flow in a viscous manner only after a certain initial shear stress has been imposed; this initial stress increases with pressure.

For practical processing purposes, the flow properties of polymers under realistic conditions need be known. Thus, shear stress is determined as a function of temperature and shear rate in *torque rheometers*. For some polymers, a *melt index* is given which is simply the weight of plastic (in grams) extruded in 10 min through a standard orifice (ASTM D1238) at a pressure of 300 kPa (43.2 psi) at a specified temperature (190 °C for PE and 230 °C for PP). Thus, the melt index is inversely related to viscosity.

Upon imposing a load, a viscous material deforms at a rate dictated by its viscosity. When the shear force is removed, the material remains in its deformed

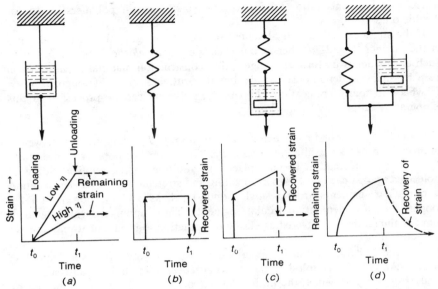

FIGURE 7-7
The deformation of polymers may follow the rules of (a) viscous flow or (b) elastic deformation or viscoelastic flow according to the (c) Maxwell or (d) Voigt (Kelvin) model. (Note: Load is applied at time t_0 and is removed at time t_1.)

shape; hence, it can be modeled by a damper or dashpot in which a piston is displaced against the resistance exerted by the shearing of an oil (Fig. 7-7a). It will be noted from Eq. (7-1) that deformation will occur, no matter how slowly, even under the slightest imposed load. Therefore, the limiting temperature for structural use is below T_m for highly crystalline polymers and well below T_g for amorphous polymers.

Elastic Deformation   When the temperature drops below T_m for a highly crystalline polymer or well below T_g for an amorphous polymer, there is no chain mobility (except on an exceedingly long time scale). Upon imposing a load, the polymer deforms purely elastically, and the original dimensions are regained immediately upon removing the load. The polymer behaves like a spring (Fig. 7-7b). By analogy to Eq. (2-5), the shear stress τ is

$$\tau = G\gamma \tag{7-5}$$

where G is the *shear modulus*. This relates to E according to

$$G = \frac{E}{2(1 + \nu)} \tag{7-6}$$

where v is Poisson's ratio and ranges from 0.25 for stiff polymers to 0.5 for flexible ones.

If the stress exceeds a critical value (the tensile strength), the part breaks in a brittle manner (Fig. 2-3a); hence, one speaks of *glassy elastic* or brittle behavior. Such a polymer is useful as an engineering construction material because it will keep its shape under imposed loads, but its brittleness (lack of toughness) is a drawback in many applications. Obviously, polymers can be processed in this temperature range only by machining.

Viscoelastic Flow From below T_g to above T_m, high-molecular-weight amorphous polymers exhibit *viscoelasticity*. The increased mobility of molecules allows some deformation to take place—in addition to the relative sliding of molecules against each other—by: uncoiling and stretching of molecules; rotation of molecular segments around single bonds (such as a C—C bond); and the cooperative movement of molecular segments. Thus, upon removing the load, some of the strain is recovered as elastic deformation, but part of it may remain as viscous flow.

Viscoelastic behavior can be described by various (and often very complex) models. Among the simplest is the *Maxwell element* (a damper and spring in series, Fig. 7-7c), in which the initial elastic (spring) deformation is followed by viscous flow, and the elastic component is regained immediately upon unloading. Deformation occurs at the rate

$$\frac{d\gamma}{dt} = \dot{\gamma} = \frac{d\tau}{dt}\left(\frac{1}{G}\right) + \frac{1}{\eta}\tau \qquad (7\text{-}7)$$

Alternatively, the polymer may be described by a *Voigt element* (a damper and spring in parallel, Fig. 7-7d), in which elastic deformation is damped by viscous flow in the dashpot and is only gradually recovered after unloading. The active stress is

$$\tau = G\gamma + \eta\frac{d\gamma}{dt} \qquad (7\text{-}8)$$

Most polymers can be modeled by Maxwell and Voigt elements acting in series and/or parallel.

Because of the viscous flow component, polymers can be processed in the viscoelastic temperature regime. Behavior in the region just above T_g is often termed *leathery*, and at higher temperatures (but below T_g) *rubbery*. It is important that these polymers be cooled well below T_g prior to release from the mold, so that the new molecular arrangement is frozen in. This means, however, that the part must not be allowed to heat above T_g in use because distortion, due to recovery of the elastic strain, will follow. Thus, T_g represents the highest service temperature for amorphous and partially crystalline polymers.

From the manufacturing point of view, the viscous and viscoelastic behaviors are of utmost importance. Since m is high, very large total elongation is possible. Even though a neck may appear after only 2–15% uniform elongation, the neck is resistant to fracture because of: strain-rate effects (Sec. 4-1-4); localized heating (which further increases m); and strengthening due to the alignment of molecules. Successive necks may form, and total elongation of several hundreds or even thousands of percent may be attained before fracture occurs.

Example 7-2

Heat-shrinkable sleeves are sold in the expanded state (i.e., they were expanded at some elevated temperature and cooled to room temperature where their shape is stable). Upon heating, they shrink. For maximum shrinkage, is it better to have a behavior characterized by a Maxwell element or Voigt element?

In a Maxwell element (Fig. 7-7c) the deformation corresponding to viscous flow is irreversible. In a Voigt element, the spring is in parallel with the dashpot and, if the viscosity of the fluid in the dashpot is decreased (by heating the polymer), the spring will pull it back.

Thermoplastic Polymers It is evident from the above discussion that the behavior of linear polymers is greatly affected by temperature. When an amorphous polymer is heated above T_g (or a crystalline above T_m), it becomes deformable, and it will retain its shape upon cooling below that temperature. The heating and cooling sequence can, in principle, be repeated; hence, it is customary to speak of *thermoplastic* polymers. It is possible to form such a polymer into a semifabricated shape (pellet, bar, tube, sheet, or film), which is then cooled and shipped to the secondary manufacturer who reheats and forms it into the final shape. Clean scrap can be recycled by adding it to virgin polymer, at least to some limited extent (although some degradation may occur).

7-3 CROSS-LINKED POLYMERS

In the linear polymers discussed hitherto, covalent bonds existed only *within* molecules and molecules were held to each other only by secondary bonds that could be broken and reformed during deformation. When covalent bonds are established *between* molecules, they are firmly bonded to each other, and no permanent deformation is possible because breaking the bond results in failure.

7-3-1 Elastomers

Elastomers are a special class of amorphous linear polymers, used above their T_g. The repeating unit contains a double bond (as in polyisoprene, Fig. 7-8a) which can be opened up to establish *cross-links* (in this case, with the aid of sulfur). The accidental discovery of cross-linking rubber with sulfur by Charles Goodyear dates back to 1839.

When only a few cross-links are formed, say, between every 100 repeat units (as in rubber gloves), the coiled molecules not only unwind, but can also move relative to each other as far as the cross-links permit. Extensions of minimum 200% and often over 500% are obtained, yet no permanent deformation can occur because the cross-links prevent sliding of the molecules (Fig. 7-8b). Upon removing the load, all deformation is regained: The polymer behaves as a spring.

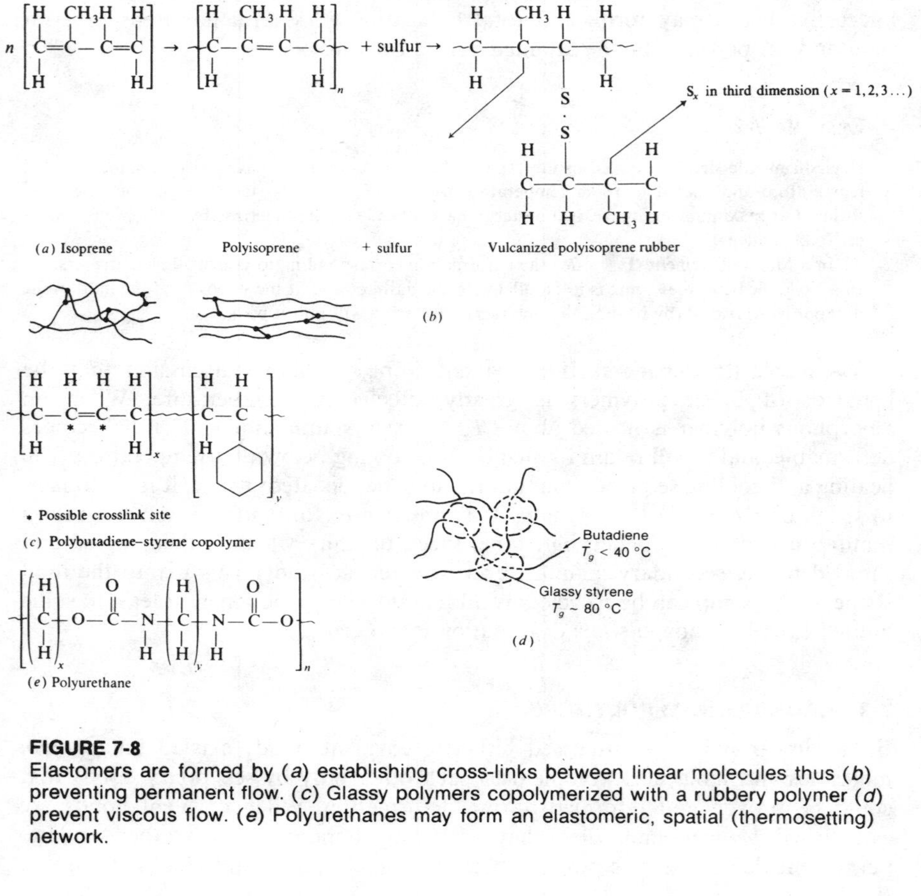

(a) Isoprene   Polyisoprene   + sulfur   Vulcanized polyisoprene rubber

S_x in third dimension ($x = 1, 2, 3 \ldots$)

(b)

• Possible crosslink site

(c) Polybutadiene–styrene copolymer

Butadiene $T_g < 40\ °C$

Glassy styrene $T_g \sim 80\ °C$

(d)

(e) Polyurethane

FIGURE 7-8
Elastomers are formed by (a) establishing cross-links between linear molecules thus (b) preventing permanent flow. (c) Glassy polymers copolymerized with a rubbery polymer (d) prevent viscous flow. (e) Polyurethanes may form an elastomeric, spatial (thermosetting) network.

An increasing degree of cross-linking (say, between every 10 or 20 mers) makes the rubber harder and suitable, for example, for automotive tires.

Once cross-linked, these elastomers cannot be further shaped: The entire structure is a single giant molecule.

A special class of elastomers of much more recent origin owes its spring-like properties not to covalent cross-links but to the presence of *glassy regions in an amorphous matrix*. This is the case with styrene–butadiene block copolymers (Fig. 7-8c). The styrene segments segregate to form polystyrene-like glassy regions which act as cross-links in the polybutadiene-like rubbery regions (Fig. 7-8d). These elastomers are unique in that they are thermoplastic, can be repeatedly heated and cooled, and scrap can be recycled.

Cross-linking of already formed thermoplastic polymers is possible by the application of high-energy radiation such as ultraviolet (UV) light, electron beams, γ-rays, x-rays, and particle beams. Cross-linking by radiation is exploited to increase strength, as of wire and cable insulation, or to endow the polymer with a capacity for large shape changes, as in heat-shrinking sleeves and films. Excessive exposure leads to concurrent chain degradation, oxidation, and gas formation, and then one speaks of *radiation degradation* of the polymer.

Example 7-3

Calculate the amount of sulfur required to establish a cross-link between every 10 mers of polyisoprene. Assume that only one sulfur atom is involved in each cross-link.

Each mer of isoprene (C_5H_8) has a molecular weight of $(5 \times 12) + 8 = 68$ g. Sulfur has an atomic weight of 32 g. Since a link is to be made between every 10 mers, the weight of sulfur is

$$\frac{32}{10(68)} = \frac{4.7 \text{ g sulfur}}{100 \text{ g polyisoprene}}$$

The fraction of sulfur is $4.7/(4.7 + 100) = 0.045$ or 4.5% of the vulcanized rubber. As shown in Fig. 7-8a, sulfur itself forms chains and, for this reason, a higher proportion of sulfur will be needed.

7-3-2 Thermosetting Polymers

A different class of materials is that of the *thermosetting polymers* or *thermosets*, so named because once polymerization is completed (e.g., by the application of heat or a catalyst), no further deformation is possible. In contrast to the cross-linked elastomers (which, by definition, are also thermosets), the cross-link density is so high in thermosetting polymers that no significant elastic deformation is possible. The covalent bonds form a *spatial network*, as in phenol formaldehyde and thermosetting polyester resins (Fig. 7-9), permanently fixing the shape of the part. Prolonged heating of a polymer cured at a relatively low

FIGURE 7-9
Phenol–formaldehyde is the earliest example of a spatial-network, thermosetting polymer.

Crosslinked phenol-formaldehyde resin (Stage C)

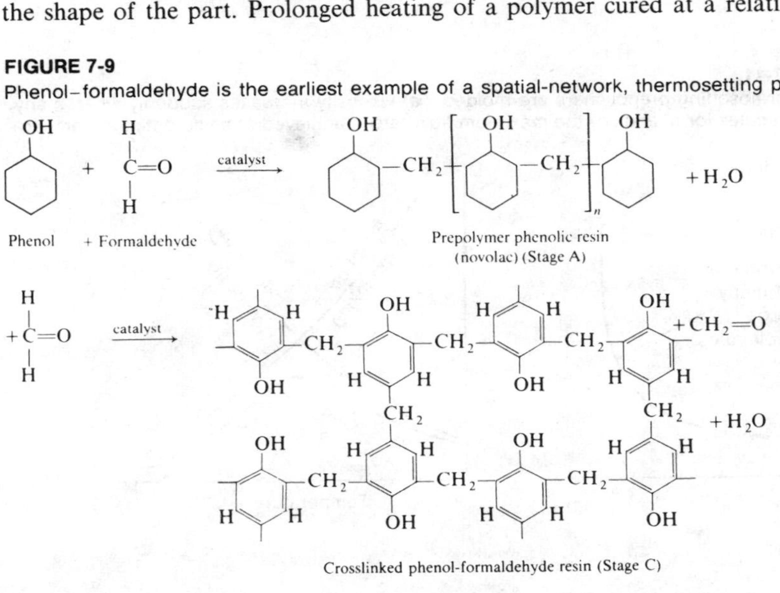

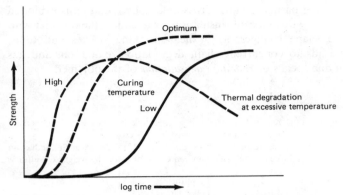

FIGURE 7-10
Cross-linking and polymerization are accelerated by temperature, but excessively high temperatures lead to the destruction of the polymer.

temperature causes further cross-linking; a fully cross-linked polymer is unaffected. Once a thermosetting polymer is cured, the only possible change is destruction of the bonds by overheating (Fig. 7-10). Scrap cannot be recycled.

It should be noted that it is not necessary to complete polymerization and cross-linking in a single step to the final stage. With some thermosets, it is possible to polymerize the precursor materials partially into a linear, so-called *A-stage* resin. After adding fillers, colorants, and some catalysts, the *B-stage* resin may be formed and supplied to the secondary manufacturer. This resin is stable at room temperature and may be thermoplastic for a limited time at elevated

FIGURE 7-11
When thermosetting prepolymers are molded, (*a*) viscosity increases suddenly when many large molecules form, and (*b*) the maximum flow rate is achieved at some optimum temperature.

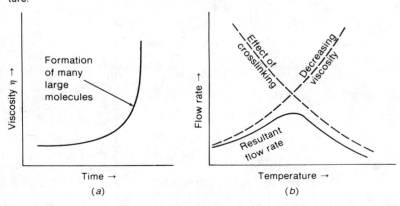

temperatures. Final polymerization and cross-linking (*C-stage*) then occurs after molding into the desired shape, under the influence of heat or catalysts.

The viscosity of monomers or of B-stage resins is generally not unduly high, but it increases suddenly when—under the influence of catalysts or heat—many large molecules form (Fig. 7-11*a*). Because cross-linking and polymerization occur simultaneously with the flow of the material into the mold, the rheology of thermosets becomes complicated, and maximum flow rates are attained at some intermediate temperature (Fig. 7-11*b*).

The essential difference between thermoplastic and thermosetting polymers is, then, that the thermosetting polymer is heated to stabilize its shape, whereas the thermoplastic polymer is heated to make it moldable and must be cooled to fix its shape.

7-4 ADDITIVES AND FILLERS

The wide range of properties available in various polymers can be further modified by the addition of agents which fall into two groups. Those that enter the molecular structure are usually termed *additives*, whereas those that form a clearly defined second phase are termed *fillers*.

7-4-1 Additives

Additives are agents designed to change properties.

1 *Antioxidants* are added particularly to chain-reaction polymers such as PP; they also act as heat and UV light stabilizers. Specific stabilizers must be added to PVC and other chlorine-containing polymers because the release of HCl initiates a destructive chain reaction.

2 *Flame retardants* are important because all carbon-based polymers support combustion. In general, control is exerted for various purposes: increase the temperature at which a flame is supported; slow down the rate of flame propagation; generate an atmosphere that does not support combustion; eliminate or reduce the emission of noxious fumes.

Plastics can never be fireproof in the sense of ceramics or most metals, but they do show some large differences in behavior. Some plastics (e.g., those containing chlorine) are more flame resistant than others. Polymers having lower-strength side groups will not support a flame but are reduced to a solid, carbonaceous body (they carbonize or *char*). In other polymers, breakage of primary bonds results in the formation of liquid or gaseous compounds of lower molecular weight which then ignite.

3 *Plasticizers* make an otherwise rigid thermoplastic polymer flexible. By plasticizing nitrocellulose with camphor, Parkes in England and the Hyatt brothers in the United States introduced, in the 1860s, celluloid, the first moldable plastic. Today, the major application is to PVC. A plasticizer is usually a fluid of high molecular weight (over 300), the molecules of which are interspersed between the polymer molecules, loosening the structure, thus depressing T_g and allowing greater flexibility.

4 *Solvents* are organic substances of short chain length. Their molecules enter between polymer molecules, breaking the secondary bonds. Thus, they are undesirable in service, and polymers are selected to resist specific solvents. Cross-linking prevents dissolution; at the most, the solvent molecules cause swelling. Highly cross-linked polymers are completely resistant to many solvents.

Solvents can, however, serve a useful purpose in manufacturing by allowing processing in the fluid state; subsequently, they must be removed by evaporation. Some solvents, such as methylene chloride, attack amorphous thermoplastic polymers such as polystyrene, PMMA, polycarbonate, and ABS, and this is useful for solvent welding. Crystalline polymers such as PP and PTFE are much more resistant.

5 Many cast or molded plastics are transparent, as are PMMA and polycarbonates (PC); others are translucent (PE, PP). Yet others are opaque or may have a natural color. Organic *dyes* (which are soluble in solvents) and organic or inorganic *pigments* (finely divided insoluble substances such as oxides) are added to impart a desired color. For example, titanium dioxide is an excellent white pigment; iron oxides give yellow, brown, or red color; carbon black is not only a pigment but also a UV light absorbent. Finely divided calcium carbonate dilutes (extends) the color and is used in large quantities as a low-cost filler.

6 *Lubricants* and *flow promoters* are added primarily to PVC, PS, and ABS to facilitate flow in molds and dies. Mold-release agents may be added to the polymer or to the mold surface.

7-4-2 Fillers

Fillers may be incorporated to improve mechanical properties and then they are often called *reinforcing agents*. At other times, their main purpose is to reduce material cost, reduce shrinkage and thermal expansion coefficient, increase the electrical or thermal conductivity of the plastic, or to ease processing, and then they are usually called *extenders* (or simply, fillers).

Since many fillers are structurally or chemically significantly different from the basic polymer, their effect on mechanical properties can be discussed with reference to Fig. 3-11 which deals with the properties of two-phase structures.

1 Fillers that are not wetted and thus produce a weak interface act as stress raisers, reducing the strength and toughness of the polymer (Fig. 3-11, case 4).

2 The effectiveness of fillers wetted by the polymer depends on their shape, size, and distribution.

a More or less equiaxed fillers of relatively large size simply serve to increase the bulk, with little effect on properties (Fig. 3-11, case 3a). However, when finely distributed, they change the character of the polymer because the bond length between particles is reduced, increasing the stiffness of the structure (Fig. 3-11, case 3b). Particulate fillers include fine (between 0.5- and 20-μm) minerals such as calcium carbonate, clay, talc, quartz, diatomaceous earth (siliceous fossils of algae); metal oxides, such as alumina; metallic powders; carbon black; coarser (70–500-μm) natural organic fillers such as wood flour; and synthetic polymers.

b A more powerful effect is obtained when the filler is plate-like (flake), because the short bonds developed over a larger surface area (as in Fig. 3-11, case 2d) are stronger. A typical example is mica. In the course of processing, flakes often become aligned and the structure acquires directional properties.

c The greatest strengthening effect is obtained when the second phase is a stronger fiber, of higher elastic modulus, and of at least 50:1 length-to-diameter ratio. The strength of the plastic is increased in tension because stresses are transferred—through the interfacial bond—to the fiber. Strength is also increased in compression because elastic buckling of the fiber is prevented by the surrounding polymer. Further strengthening is derived from the shorter bond length between fibers. If the fibers become oriented, markedly directional properties can be obtained. Fibrous fillers include cellulosics, asbestos, glass, carbon (graphite), boron, metals, and polymers of higher strength and/or greater toughness. Polymers reinforced with fibers (filaments) will be discussed in Sec. 7-10.

Some fillers such as polymers are immediately wetted by the plastic, whereas other polymers such as PE, PTFE, and silicones need special surface treatments. In many other instances special *coupling agents* are needed; for example, silanes are applied to fiberglass or mineral fibers.

7-5 SERVICE PROPERTIES OF POLYMERS

The vast numbers of available polymers exhibit a wide range of characteristics, making them suitable for a very broad range of applications.

7-5-1 Mechanical Properties

The properties of polymers are determined in tests similar to those described in Sec. 2-1. Prior to testing, the specimens are brought into a standard condition (temperature and moisture) according to ASTM D618.

Strength Standard tension tests are performed on molded or machined specimens (ASTM D638) and films (ASTM D882). The typical stress-strain curve may show purely brittle behavior for thermosets and for thermoplastics well below T_g (Fig. 7-12a). A tougher polymer exhibits post-necking extension before orientation of the molecules raises the load prior to fracture (Figs. 2-12 and 7-12a). For reasons explained in Sec. 4-1-4, post-necking elongation increases with increasing m and thus with temperature above T_g (Fig. 7-5b).

In most engineering applications, permanent deformation is undesirable and the useful loading range is limited to the elastic regime. The initial slope of the force-displacement recording again gives the elastic modulus E (Eq. (2-6)), also called the *tensile modulus*. In polymers the elastic modulus is not well defined,

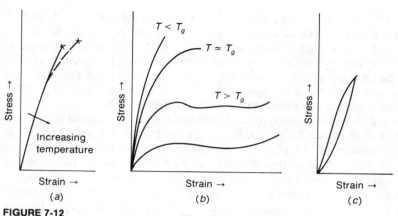

FIGURE 7-12
The mechanical properties of polymers are greatly affected by increasing temperatures: (*a*) The elastic modulus of thermoplastics drops and (*b*) strength decreases while elongation increases. (*c*) A hysteresis loss is often encountered on repeated loading.

because there is a slight deviation from linearity even at very low strains. Furthermore, in most polymers the modulus drops with increasing temperature, and it is some two orders of magnitude lower than in metals (inspect data in Table 2-3). Many thermoplastics show some strain-rate sensitivity even below T_g, and the tensile modulus is then also a function of loading rate. The strength of some plastics, such as nylon-66, is markedly reduced by moisture.

When deformation is taken past the linear deflection regime, deformation may still be largely elastic but a different curve is followed upon unloading (Fig. 7-12c). The area of the *hysteresis loop* represents work which is transformed into heat. On repeated loading this may lead to overheating and fatigue.

In low-impact thermoplastics, such as PMMA and polystyrene, tensile deformation becomes concentrated into localized regions. The difference in light reflection makes them visible as *crazing*.

The stiffness of plastics is often determined in a three-point bending test (Fig. 2-6a). The maximum stress at fracture is the *flexural strength*; for tough plastics, the maximum fiber stress at a strain of 5% is taken. The *flexural modulus* is the initial slope of the bending force-strain curve and is usually somewhat higher than the tensile modulus.

Compression and indentation hardness testing are used mostly for processing checks on specific kinds of plastics. For applications such as gaskets, springs, and force fits, stress-relaxation tests are made to obtain the decay of stress under constant-strain conditions.

Toughness Impact tests provide a quick evaluation of the toughness of polymers. The Charpy test (Fig. 2-8) is suitable, but the *Izod test* (ASTM D256) —in which the notched specimen is clamped at one end—is often preferred. Because of the high strain rates and imposed bending, impact toughness may differ considerably from toughness determined in the tension test, and an *impact tension test* (ASTM D1822) often correlates better with field performance. In thermoplastic polymers, impact strength is meaningful only below T_g. Toughness can be increased with moderate crystallinity, decreasing spherulite size, and, in block copolymers, by incorporating a component which has a T_g much below the service temperature.

For sheat and film, shear strength and puncture strength are often important. For applications involving fatigue, an endurance limit (Fig. 2-10*a*) is determined.

Elevated-Temperature Properties   The great temperature sensitivity of polymers makes creep deformation significant. Creep is a very complex function of temperature, the magnitude and state of stress, and solvents (and for hygroscopic materials, humidity) in the environment. Design is therefore often based on pseudoelastic principles: It is assumed that the maximum temperature and load operate throughout, and creep-stress data are substituted into classical elastic stress-analysis equations.

The upper temperature limit of application is often defined as the *heat-deflection temperature* (ASTM D648). A bar of 12.7-mm width, 127-mm length, and 3–13-mm thickness is loaded in a three-point bending test (Fig. 2-6*a*), on a span of 102 mm. The load is chosen to give a maximum stress (Eq. (2-14*a*)) of 1820 kPa (264 psi) for very rigid polymers and 455 kPa (64 psi) for flexible ones. The bar is heated, under load, at a rate of 2 °C/min; the heat-distortion temperature is reached when the midspan deflection is 0.25 mm.

For highly crystalline thermoplastics, the temperature at which a standard flattened needle penetrates the plastic is taken as the *Vicat softening* point (ASTM D1525).

Gradual changes at elevated temperatures lead to loss of properties and the resistance to *thermal aging* is often specified with reference to Underwriters Laboratories (Northbrook, Ill.) Standard UL 746B.

Residual Stresses   Manufacturing processes often induce residual stresses in plastic components. In thermoplastics, differential cooling rates lead to varying degrees of crystallization and thus shrinkage, and the flow of the plastic may result in varying unwinding of molecules in different sections of a part. Surface and center layers of thicker sections have different thermal histories because of poor heat conductivity, and this can induce further differences in structure. In both thermoplastics and thermosets, the shape of the mold may prevent uniform contraction of all sections, and residual stresses are then again set up.

If creep deformation is possible, residual stresses gradually diminish with consequent distortion (Fig. 2-15*b*). In polymers subject to solvent attack, stresses lead to the development of cracks; a multitude of fine cracks is again referred to as *crazing*. Very fine cracks appear as a clouding of the surface.

Because linear polymer molecules are uncoiled and stretched during manufacture, a part heated to above T_g may return to its original shape. This is exploited in heat-shrinkable plastics but is undesirable in structural parts.

7-5-2 Physical and Chemical Properties

Density is important in many applications. Amorphous hydrocarbons have a density close to that of water (0.86–1.05 g/cm^3; e.g., 0.91 for LDPE and 0.96 for HDPE). Chlorinated plastics are heavier (PVC 1.4, polyvinylidene chloride 1.7) and the crystalline PTFE is heavier yet (2.2).

The tribological properties of polymers are generally good. Many of them give a very low coefficient of friction either because of low adhesion (as do fabric-reinforced phenolic bearings in the presence of aqueous lubricants) or because a

TABLE 7-2
THERMAL CONDUCTIVITY AND COEFFICIENT OF LINEAR THERMAL
EXPANSION OF SOME ENGINEERING MATERIALS

Material	Thermal conductivity, $W/m \cdot K$	Coefficient of thermal expansion at 20 °C, $\mu m/m \cdot K$
Aluminum	240	23.6
Copper	390	16.5
Gold	300	14.2
Iron	74	11.7
Invar (Fe–36Ni)	11	0.6–0.3
304 stainless steel	15	16.5
410 stainless steel	24	10.0
Alumina	17	6.6
Beryllia	218	8.5
E-glass	1.7	6.0
Sealing glass	1.7	4.95
Silicon	1.5	2.0
ABS	0.2–0.33	60–130
Acetal	0.23	100
Cellulose acetate	0.17–0.33	80–180
Nylon 66	0.24	80
glass filled	0.2–0.5	15–20
Polycarbonate	0.2	70
LDPE	0.33	110–220
HDPE	0.48	60–110
Polyimide	0.1	45–55
PMMA	0.17–0.25	50–90
Polypropylene	0.12	80–100
Polystyrene	0.12	50–80
PTFE, glass reinforced	0.3–0.4	77–100
PVC, rigid	0.12–0.2	50–100
flexible	0.12–0.17	70–250
Epoxy	0.17	45–65
glass filled	0.17–0.42	11–50
fused silica filled	0.7	22
silver filled	0.8	53
Phenolic resin	0.17–0.33	30–45
Polyurethane	0.2	100–200

transfer film forms which then ensures low dry friction (nylon, PTFE). Polymers may have intrinsically high wear resistance in particular applications (as does UHMWPE) or their wear resistance may be increased with fillers that increase hardness or impart lubricating properties (as does molybdenum disulfide in nylon).

Polymers are, in general, poor heat conductors, although conductivity can be somewhat improved by filling with metal powders (Table 7-2). Their thermal expansion is large (Table 7-2), and this can create problems when assemblies contain both plastics and metals.

The resistance of polymers to weathering—involving moisture, ozone, UV light, and temperature changes—can be improved by the incorporation of opaque fillers and UV absorbers. Otherwise, darkening, crazing, and embrittlement may occur.

Amorphous polymers tend to be transparent since there are no internal reflecting boundaries in them. Crystalline polymers are translucent or opaque except when crystals are very small or few or the polymer is in the form of a very thin film.

Polymers have many desirable electrical properties: high resistivity, dielectric strength, arc resistance, and dielectric constant, as well as a small dissipation factor (i.e., little heat is generated in them when placed in an electric field). Thus, they enjoy wide application for insulation and, in general, as substrates. If necessary, they can be made conductive by adding metal powder (6–8% if on bead boundaries, 35–40% if randomly distributed). A new class of polymers is intrinsically conductive. Very high static charges may build up in plastics as a result of friction against some other surface. Antistatic additives are designed to dissipate the charge by electric conduction into the polymer or by radiation to the surrounding atmosphere.

7-6 POLYMERIC MATERIALS

There is an ever-growing variety of plastics available and, for detailed data, handbooks such as *Modern Plastics Encyclopedia* (McGraw-Hill, annual) should be consulted. An outstanding attribute of plastics is the relative ease with which they can be manufactured into parts, and "general-purpose" plastics are often employed simply because of their cost effectiveness, for products that do not set very high demands. However, general-purpose plastics also have many engineering applications, and are often reinforced with appropriate fillers for higher strength in load-bearing components. The term "engineering plastics" is loosely applied to describe polymers that are too expensive for general applications but offer high strength, elastic modulus, toughness, temperature and chemical resistance, desirable electrical properties, and good surface finish, so that they can often be used in structural applications, replacing metals. Many of them can be further strengthened by fiber reinforcement or fillers. Therefore, they are finding increasing application in instrument and machinery housings, electrical connectors, gears, cams, impellers, valves, and, in general, structural components.

7-6-1   Thermoplastic Polymers

The number of thermoplastic polymers has increased greatly in the last three decades. Classification is possible from various viewpoints; here, thermoplastics will be discussed in groups defined by their structure. The properties of some frequently used thermoplastics are given in Table 7-3.

1 *Polyolefins*. These are based on straight-chain alkenes (i.e., hydrocarbons with double bonds between the carbon atoms, such as ethylene in Fig. 7-1).

a *Polyethylenes* (PE). These are very versatile plastics with good chemical and electrical properties but need antioxidants and UV stabilizers. The shorter-chain, branched LDPE (density: 0.910–0.925 g/cm^3) finds its main use in packaging, although it is losing ground against linear LDPE (LLDPE). The linear HDPE (0.941–0.965 g/cm^3) is suitable for containers and pipes. UHMWPE (with a molecular weight of minimum 3 million) is a true engineering thermoplastic. It has extremely high abrasion resistance and impact toughness; hence, it is used in many industries for wear surfaces.

b *Polypropylene* (PP, Fig. 7-1) is used in the isotactic form. It has good stiffness, stress-crack resistance, and a higher temperature resistance than PE. Appliance parts and containers are typical uses.

c *Ionomers* contain inorganic compounds which give ionic bonding between chains (cross-linking), while still allowing plastic molding processes at elevated temperatures. They are transparent, very tough, and are used typically for golf-ball coatings, hammer heads, and other high-impact applications.

2 *Vinyls* are based on ethylene monomers in which at least one H is substituted.

a *Polyvinylchloride* (PVC) (Fig. 7-1) is a low-cost, easily processed plastic of good water resistance and strength-to-weight ratio. It needs stabilizers to prevent decomposition at high processing temperatures and in sunlight. Rigid PVC is hard and relatively brittle but finds extensive use in buildings as window frames, gutters, pipe, and also for wire insulation and bottles. Flexible PVC, made by adding some 30–80% plasticizer, finds extremely broad applications, e.g., as garden hose, boots, tubing, packaging, raincoats, etc.

b *Vinylidene chloride* contains two substituted Cl atoms per mer (Fig. 7-1). It is chemically inert, has low permeability to liquids and gases, and is relatively resistant to combustion. It is used for seat covers, pipes, and valves. The copolymer with vinyl chloride (saran) makes a packaging film.

c *Fluorocarbons*. Replacing H with F creates chemically inert polymers (Fig. 7-1). The fully substituted polytetrafluoroethylene (PTFE) is crystalline and has great heat resistance, extreme chemical inertness, and inherently low friction. It does not melt and must be produced by metal technologies. Billets are skived (shaved) to make films.

3 *Styrenes*. These are obtained by substituting a benzene ring for H.

a *Polystyrenes* (Fig. 7-1) are amorphous, transparent, low-cost but brittle polymers. Elastomers may be included to improve impact resistance. Apart from solid forms, polystyrene is also employed in the form of expandable beads. It finds wide use in packaging, trays, and general houseware.

TABLE 7-3 MANUFACTURING PROPERTIES OF SELECTED THERMOPLASTIC POLYMERS*

Type	T_m, °C	T_g, °C	Heat deflection temperature, °C		Compression molding† temp., °C	Injection molding temp., °C	Shrinkage, %	Tensile properties		Flexural modulus, GPa	Izod impact,†† J/25 mm
			1850 kPa	450 kPa				Strength, MPa	Elonga-tion, %		
ABS		110–125	95	100	175–260	190–260	0.4–0.9	30–55	5–25	2.5	4–16
Acetal copolymer	175		110	155	170–205	195–230	2	60	40–75	2.8	2
With 25% glass	175		160	165	170–205	190–250	0.4–1.8	130	3	8	2.5
Acrylic (PMMA)		90–105	80	85	150–220	160–260	0.2–0.8	50–80	2–10	3	0.7
Cellulose acetate	230		55	65	125–220	165–255	0.3–1.0	15–35	6–70	0.8–2	1–8
PTFE	327			120	320–400	330–405	3–6	20	200–400	0.6	4
Fluorinated ethyl-enepropylene	275			70			3–6		300	0.6	No break
PE-TFE copolymer	270		70	104	300–330	300–350	3–4	45	100–400	1.4	No break
With 25% glass	270		210	265	300–330	300–350	0.2–3.0	85	8	6.7	12
Nylon 66	265	50	75	245		270–330	0.8–1.5	85	60–300	1.3	2.5
With 30% glass	265	50	255	260		270–300	0.4–0.6	155	5–7	6	4
Polycarbonate		150	130	135		250–345	0.5–0.7	55–70	100–130	2.4	20
Polyester PBT	250		65	150		225–275	1.5–2.0	55	50–300	2.7	1.5
PET	250	20	40			280–315	2.0–2.5	50–70	50–300	3.0	0.7
Polyethylene LD	110	–120		40	E135–230	150–230	1.5–5.0	8–30	100–650	0.3	No break
HD	135		45	80	E170–275	170–260	1.4–4.0	20–35	10–1200	1.2	0.5–5
UHMW	130			80	200–260		4	40	420–500	1.0	No break
Polyimide		310–365	340	115	330–365	205–290		120	10	3.5	2
Polypropylene	168	–18	55	100	E205–260	175–260	1–2.5	30–40	100–600	1.5	0.5–1
Polystyrene		100	95	80	150–205	175–275	0.4–0.7	35–55	1–2	3.2	0.5
high-impact		95–105	80		E190–260	150–210	0.4–0.7	35–65	20–65	2.0	1.5–4
PVC, rigid		75–105	60	62	140–205	160–195	0.2–0.6	10–25	40–80	3.0	1.5–30
flexible		75–105			140–180	150–220	1.0–5.0	4–20	200–450		Varies
Styrene-butadiene block copolymer			0	0	120–160		0.1–0.5		300–1000	0.02–1	No break

*Compiled from *Modern Plastics Encyclopedia*, McGraw-Hill, New York, 1985. Values given are approximate.

†E = extrusion.

††Izod impact values J/25-mm notch, measured on 3.2 mm-thick specimen (divide by 1.35 to obtain ft·lb/in notch).

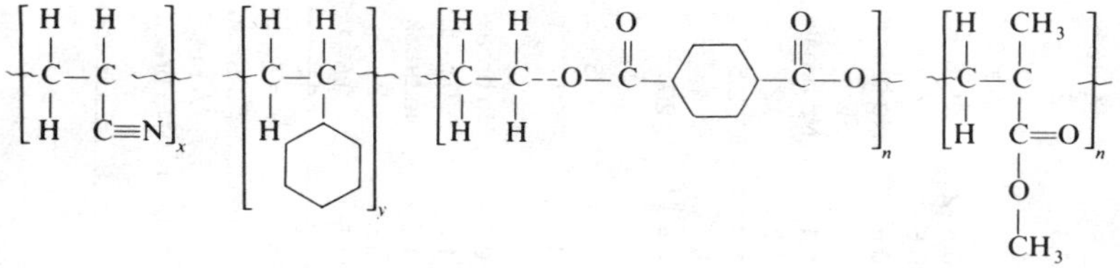

(a) Acrylonitrile-styrene copolymer (SAN)

(b) Poiyethylene terephthalate (PET)

(c) Polymethyl-methacrylate (PMMA)

FIGURE 7-13
The chemical formulas of some important linear thermoplastic polymers.

b *Styrene acrylonitrile* (SAN) random copolymer (Fig. 7-13) is also transparent but has much better stiffness. It is used in appliances and for glazing.

c *ABS* is a copolymer of butadiene rubber grafted with SAN. The very finely distributed rubber-like regions (of $0.1-1.0$-μm diameter) impart low-temperature ductility and toughness while maintaining the high hardness and rigidity of the glassy styrene. It is an engineering plastic, the properties of which can be varied by changing the ratios of the monomers. It needs UV stabilization. It is extensively used for business machines, electrical hand tool housings, camper tops, and pipes.

4 *Polyesters.* Esters are reaction products between an organic acid and an alcohol (Fig. 7-1); thus, ester groups are in the main chain.

a *Thermoplastic polyesters. Polyethylene terephthalate* (PET, Fig. 7-13) is a clear, tough, orientable polymer of higher temperature resistance. It finds wide use in containers for carbonated beverages and liquors. *Polybutylene terephthalate* (PBT) is semicrystalline. After reinforcement, it has good creep resistance, making it suitable for auto and appliance bodies and electrical parts.

b The aromatic ester *polyarylate* is an engineering plastic of high creep resistance.

5 *Acrylics* have side-chain ester groups. PMMA (Fig 7-13) is transparent, brittle, and notch-sensitive. High-impact grades contain a plasticizer or copolymer. It is polymerized in sheet form from a solution of monomer and prepolymer, or the polymer is treated by thermoplastic techniques. It is extensively used for glazing, skylights, fixtures, and lenses (including large Fresnel lenses). Acrylic ester polymers (*polyacrylates*) are less brittle.

6 *Cellulosics.* Cellulose is a natural polymer from cotton fiber or wood pulp (after removal of lignin). It does not melt. Chemically modified (acetate, etc.), it becomes a tough thermoplastic for containers, tool handles, tape cases.

7 *Polyurethanes* can be linear thermoplastic polymers which find use as wire coatings, fascia boards, and other components.

8 *Polyamides* are formed by the condensation of an amine and a carboxylic acid.

a *Nylons* (Fig. 7-1) are tough crystalline polymers. Modified grades are amorphous and transparent. They absorb moisture which plasticizes them, with a

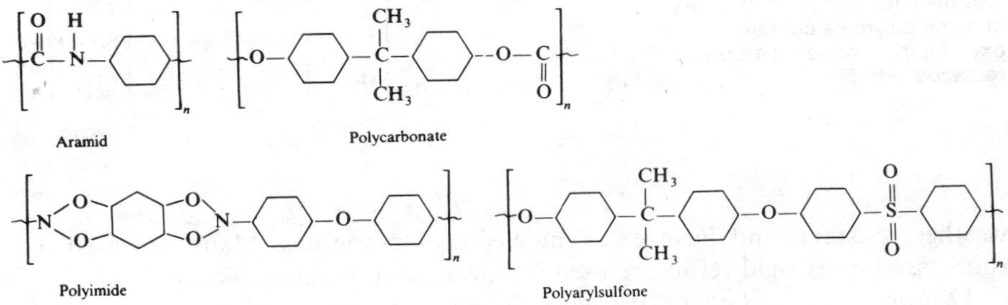

FIGURE 7-14
The benzene ring in aromatic linear polymers makes these polymers stronger and more resistant to temperature.

resultant swelling and loss in strength and electrical properties. They must be processed dry (less than 0.2% moisture). Nylons are self-lubricating. Glass and mineral fiber fillers greatly improve high-temperature properties. Gears, bearings, autobody parts, caps, wheels, and fans are typical products.

b *Aromatic polyamides* (*aramids*, aromatic nylons, Fig. 7-14) have increased thermal stability because of the aromatic backbone. They do not melt. The chain-extending bonds are all parallel, endowing them—after special processing —with exceptionally high tensile strength and modulus. They are used primarily as reinforcing fibers, as in radial tires and in composite materials for aerospace applications, but also on their own in bulletproof vests, as cables, or tough work gloves.

9 *Aromatic thermoplastics.* A great variety of engineering plastics have in common an aromatic backbone which makes for higher temperature resistance.

a *Polycarbonate* (Fig. 7-14) is transparent and impact resistant, and can be used in auto lamps, instrument panels, optical and electrical applications, and returnable milk bottles.

b *Polyimide* (Fig. 7-14) has a semiladder structure, giving it high temperature resistance and toughness. Among the copolymers, polyamide–imide is opaque, whereas polyether–imide is transparent and UV-resistant. It can thus be used, for example, for microwave cookware.

c *Sulfones* include polysulfone, polyarysulfone (Fig. 7-14), and polyether sulfone, all used as engineering plastics.

d *Phenylene-based resins* also find uses in high-temperature applications.

10 *Polyacetals* are made by the polymerization of formaldehyde; thus, they have a $—CH_2O—$ backbone (Fig. 7-15). These are highly crystalline, opaque engineering plastics, often glass filled. Copolymers contain randomly distributed $C—C$ bonds and are highly creep resistant. They are used for plumbing, sinks, fittings, and containers.

11 *Silicones.* The siloxane bond ($Si—O—Si$) in the backbone (Fig. 7-15) makes them usable over a wide temperature range. They are water repellent,

FIGURE 7-15
The backbone of the molecules
of some polymers contain
oxygen; in silicones, carbon is
replaced with Si.

$$\left[\begin{array}{c} H \\ | \\ -C-O- \\ | \\ H \end{array}\right]_n \qquad \left[\begin{array}{c} CH_3 \\ | \\ -Si-O- \\ | \\ CH_3 \end{array}\right]_n$$

Polyacetal
(polyoxymethylene)

Silicone

weather resistant, and have excellent electrical properties. Many are used as rubbers, whereas rigid resins are used for encapsulation and molding.

12 *Copolymers and alloys.* We have already mentioned ABS and polybutene–styrene copolymers. Alloys (polyblends) are made to improve the properties of the matrix; for example, ABS second-phase particles of 1–10-μm diameter may be added to PVC, PS, or PC to improve notch resistance.

7-6-2 Elastomers

We already discussed the styrene–butadiene copolymer thermoplastic and the polyurethane elastomers (Fig. 7-8e). Most other elastomers are, however, cross-linked (thermosets).

Natural rubber is primarily polyisoprene. In its natural form it is a sticky substance because the molecules slide readily. Cross-linking with sulfur (Fig. 7-8a) produces a rubber which, because of its relatively low hysteresis loss, is used in automobile tires.

Most synthetic rubbers are random copolymers. For example, butyl rubber is a copolymer of isobutene and a few isoprene units, with the isoprene providing the sites for cross-linking. The high hysteresis loss of this rubber makes it suitable for vibration-damping engine mounts. There are also styrene–butadiene and ethylene–propylene rubbers.

For all applications, the T_g of the rubber must be below the lowest service temperature. Silicone rubber serves down to -90 °C, most others to -50 to -60 °C.

Polyurethanes can be elastomeric thermosets. They form tough and abrasion-resistant coatings. Large quantities are formed into foams, ranging from soft (seating, bedding), semiflexible (dashboard padding), rigid (insulation), to high-density (structural) foams. They are castable and can be made of a prepolymer or in one-shot processes using the monomers, as in reaction-injection molding (Sec. 7-7-3). Reinforced with glass or other fibers, they can serve as automotive door panels, trunk lids, and fenders.

7-6-3 Thermosetting Polymers

Thermosets offer, in general, greater dimensional stability than thermoplastics, but at the expense of greater brittleness. However, these properties can be modified, and in many applications thermosets and thermoplastics are directly competitive. The properties of some frequently used thermosets are given in Table 7-4.

TABLE 7-4 MANUFACTURING PROPERTIES OF SELECTED THERMOSETTING POLYMERS*

Type	Deflection temperature, °C		Molding temperature, °C		Shrink-age, %	Tensile properties		Flexural modulus, GPa	Izod impact, J/mm
	1850 kPa†	450 kPa†	Compression	Injection		Strength, MPa	Elonga-tion, %		
Epoxy, casting	45–290				0.1–1	30–90	3–6		7–35
Molding, glass-filled		120–260	150–160		0.02	120–180	4	18–28	850–1200
Melamine-formaldehyde (cellulose-filled)	175–200		150–190	90–170	0.5–1.5	35–90	0.5–1	8	7–15
Phenol-formaldehyde, casting	115				1–1.2	36–60	1.5–2		7–15
wood-flour-filled	150–190		145–195	165–205	0.4–0.9	35–65	0.4–0.8	7–8	7–20
Polyester, glass-filled, SMC	190–260		130–175		0.1–0.4	55–175	3	7–15	250–750
BMC	160–175		155–195	150–190	0.05–0.4			7–15	120–450
Polyurethane, casting	Varies		85–120		2	1–70	100–1000	0.07–0.7	850–No break
Urea-formaldehyde, cellulose-filled	125–145		135–175	145–160	0.6–1.4	40–90	0.5–1	9–11	7–15

*Compiled from *Modern Plastics Encyclopedia*, McGraw-Hill, New York, 1984.
†Divide by 7 to get psi.

1 *Phenol-formaldehyde* (Fig. 7-9) is the oldest synthetic resin and is still used extensively (the original trademark "Bakelite" has become a generic term). Dark in color, it is almost invariably filled with wood flour or, for better impact properties, with glass, asbestos, or cotton fibers. Further improvements are obtained by adding butadiene–acrylonitrile copolymer rubber to the A-stage resin. Molding compounds are used for electrical, automotive, and appliance purposes, such as switch gear, ignition parts, handles, etc.

2 *Amino resins* are so named because they contain the amino ($-NH_2$) group. Strictly speaking, only *melamine formaldehyde* belongs here, but *urea formaldehyde*, which contains amide ($-ONH_2$) groups, is often classed with it. They are translucent and can be pigmented. The filler is usually α-cellulose (a high-molecular weight cellulose), obtained from wood pulp or cotton fiber, for applications such as countertops, wall paneling, dinnerware, circuit breakers, and wall plates. Wood flour is also used as a filler for electrical purposes.

3 *Polyesters*, as discussed in Sec. 7-6-1, are thermoplastic when in the linear form (Fig. 7-1). However, these linear esters can be treated as prepolymers to which colorants, fillers (such as ground limestone), or reinforcing agents (chiefly glass fiber) can be added to make a *premix* which then can be molded and cured by heat. The term *alkyd* (*alc*ohol reacted with ac*id*) is, in principle, generic to all polyesters, but it is also used to describe specific types, distinct from the *unsaturated polyester plastics* which are cured by the addition of catalysts (more correctly, initiators), which can be chosen to cause cross-linking polymerization with added styrene at room or elevated temperature. Major applications are pipes, boats, tanks, walls (bathrooms), auto hoods, deck lids, quarter panels, helmets, structural beams, and even oil-well sucker rods.

4 *Epoxy resins* are available as room-temperature-curing, two-component systems (an intermediate resin and a reactant, misnamed a catalyst) or as elevated-temperature-curing, single-component resins. They are widely used for encapsulation (potting) of electrical parts, for fiber-reinforced epoxy structures, or, mixed with sand, glass, or marble, as concrete.

7-7 PROCESSING OF POLYMERS

Many processing techniques for polymers have their counterparts in processes for metals and ceramics; hence, we will discuss them with reference to the principles introduced earlier. In general, plastics can be processed at much lower temperatures than metals, and this removes many processing difficulties and also allows some processes which are impracticable for metals. Many plastics can be processed by several techniques, but some techniques may be more suitable than others, and this will be pointed out when applicable.

7-7-1 Casting Processes

The term *casting* is used mostly to describe *filling a mold by gravity*. In most instances, prepolymers or monomers are cast and polymerization and—for thermosets—cross-linking takes place in the mold.

Epoxy resins and thermosets are prime examples of polymerization casting. When the polymer is used to fix a component in place, one speaks of *potting*; when it surrounds the component entirely, of *encapsulation*.

Among thermoplastics, PMMA sheet is produced by pouring catalyzed MMA between glass plates (*cell casting*) or between endless stainless steel belts (*continuous casting*). Polymerization takes place by heating the liquid for several hours at 90 °C. Processing times are shorter when a partially polymerized "syrup" is cast.

High-molecular-weight, highly crystalline, and, hence, strong nylon parts such as gears and bearings are obtained by melting the monomer, adding the catalyst and activator, and pouring the mix into molds.

In all instances, absence of moisture is critical and gases must either be removed from the liquid by processing in vacuum, or must be kept in solution by the application of pressure during curing. Molds may be made of metal, glass, and rigid or flexible plastics. The latter can be peeled off the castings and thus allow the production of complex shapes. In common with metal casting, shrinkage can present problems, especially with acrylics that shrink greatly during polymerization. Internal voids may result in a collapse of the surface.

Of great importance is the casting of *plastisols*, especially for flexible PVC. A plastisol is a suspension of PVC particles in the plasticizer; it flows as a liquid and can be poured into a heated mold. When heated to 177 °C, the plastic and plasticizer mutually dissolve each other. On cooling the mold below 60° C, a flexible, permanently plasticized product results. Slush casting is used extensively for thin-walled products such as snowboots, gloves, and toys.

Occasionally, solutions of polymers, especially of PVC, are cast onto a traveling stainless steel belt (*solvent casting* of films). Large quantities of solutions are spun from multihole dies (*spinnerettes*) into fibers. Because of the short diffusion path, the solvent is easily removed by heating and can be recirculated.

7-7-2 Principles of Melt Processing

Most plastics are too viscous even at high temperatures to flow under the force of gravity, and the term *melt processing* refers to techniques in which polymers are deformed *with the aid of applied pressure*. It is applicable to both thermoplastics and thermosets.

Melt Processing of Thermoplastics For melt processing of a thermoplastic, the plastic is heated into the regime of viscous flow (above T_m for crystalline polymers and well above T_g for amorphous polymers). The shape is fixed by cooling well below T_g (or, for highly crystalline polymers, below T_m). Some of these techniques, especially those involving the injection of plastic into molds, are direct counterparts of metal casting, even though there are differences in the way the melt is produced and conveyed.

1 The starting material is usually granular, cut up strand, diced sheet, or, in the case of recycled material, chopped up and, sometimes, compacted scrap. For freedom from gas bubbles, the plastic must be free of water. Heating is partially external and partially internal (by transforming the work of viscous shearing into heat). Overheating can cause permanent damage. For example, PMMA depolymerizes and monomer gas bubbles form; PVC needs stabilizers; PE and PS are relatively insensitive; some others (such as polyacetal with PVC) may even form explosive mixtures.

2 The substantial volume change on cooling (Fig. 7-5a) reflects the rearrangement of molecules and establishment of secondary bonds. Since these are time-dependent processes, shrinkage increases with slower cooling (higher melt temperature), decreasing injection pressure, and shortened injection time.

3 Production rates are governed by solidification time which can be shortened by setting low mold temperatures and a small differential between injection temperature and T_g or T_m (corresponding to low superheat in metals). However, fast solidification also means that the orientation of uncoiled molecules will be frozen in. Orientation can be desirable when molecules are aligned in the direction of maximum service stress, but may cause distortion in service. Distortion is minimized if the molecules are given time to recoil before freezing (i.e., with high mold temperatures and fast injection rates). Unfortunately, measures that reduce distortion also increase shrinkage; hence, collapse of the surface is often observed in thicker sections of the part. Because thicker parts cool more slowly, molecules have more time to recoil inside the already solidified shell, setting up stresses. It is often possible to reduce cycle time by removing the part as soon as it cools to a temperature where it holds its shape, so that it can cool to room temperature out of the mold.

4 Molding should be done at temperatures where small changes in shear rates do not result in large changes in viscosity (right-hand end of the pseudoplastic curve in Fig. 7-6), otherwise, complex mold cavities fill unevenly. In general, it is desirable to mold at high shear strain rates (by injecting at high speeds or through small gates) because lower temperatures can then be used (Fig. 7-6) and cycle time can be reduced.

5 Feeding of the mold is critical. The system of runners and gates is similar to that used for metals (Sec. 3-8-2). Gates must not be too large, because melt would flow back when the pressure is released. On the other hand, gates that are too small freeze off prematurely, cutting off the molding pressure before full packing is attained. The location of gates determines the sequence of mold filling and the alignment of molecules (and thus the direction of maximum strength in the finished part). When the polymer is split into two or more streams which then meet to complete mold filling, a *knit line* forms where molecules of the two streams intermingle. Failure to attain complete interpenetration results in a weak line (similar to a cold shut in metal casting). Mold filling is also the critical factor in terms of distortion. Computer programs are available for modeling mold filling, so that filling can be observed on a VDT. This greatly reduces or eliminates the trial-and-error approach otherwise needed for finding the optimum gating scheme.

Melt Processing of Thermosets We saw in Sec. 7-3 that prior to cross-linking (i.e., in the A or B stage) thermosetting polymers are capable of flowing under pressure. They may be granular, and then they can be treated as powders, or they may become thermoplastic on heating. Hence, processing techniques can be similar to those used for thermoplastics. There is, however, a major difference: Whereas thermoplastics are cooled to fix their shape, thermosets must be held in a heated mold for a long enough time for polymerization and cross-linking to occur. Some polymers can be removed from the mold as soon as their shape is fixed and then full cross-linking is obtained during cooling or holding in a separate oven. At other times, cross-linking begins immediately upon heating; then the prepolymer must be introduced into a cold mold, and the mold must be taken through a cycle of heating and cooling for each part, making the cycle time very long.

7-7-3 Melt Processing Techniques

When melt processing takes place in a mold and results in a finished part, it is customary to speak of *molding processes* (equivalent to shape casting or hot forging in metals); when it produces a rod, tube, sheet, or film, one speaks of *extrusion processes* (equivalent, in their end product, to metal extrusion).

Compression Molding   *Compression molding* is the equivalent of closed-die forging (Sec. 4-4-5). A premeasured quantity of polymer is introduced into the mold. When the plastic is completely trapped, one refers to a *positive mold* (Fig. 7-16a); any variation in polymer quantity results in a variation of part thickness. Closer tolerances can be held if a small flash is allowed to extrude—usually along the punch perimeter—in *semipositive molds*. More plastic is lost in *flash molds*, similar to those used in impression-die forging (Fig. 4-30b).

FIGURE 7-16
The processes of (a) compression molding and (b) transfer molding are suitable for both thermoplastics and thermosets. If material in the sprue hardens between shots, the sprue is tapered to break away (c).

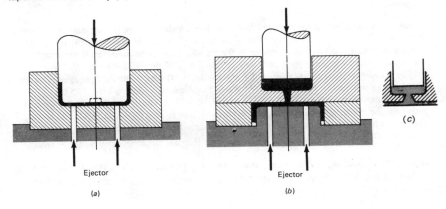

(a)   (b)   (c)

Ejector   Ejector

The molding temperature is chosen from experience (Tables 7-3 and 7-4). The press force is calculated by multiplying the projected area by the empirically determined molding pressure, typically 7–15 MPa (except for thermosets and stiffer thermoplastics; up to 70 MPa for ABS, PMMA, and PS; 140 MPa for PE–TFE and UHMWPE).

Although suitable for thermoplastics, the main application of compression molding is to thermosets. Shortly after closing the mold, it is often opened slightly to vent gases, steam, and air, and then it is closed again until curing takes place (typically, 1–5 min). The advantages of the process are that there is little or no material loss, fillers do not become oriented, and internal stresses are minimum.

Cold molding is a variant in which a powder or fillers (often of refractory materials) are mixed with a binder, compressed in a cold die, and removed for curing in an oven.

Transfer Molding *Transfer molding* utilizes an extrusion-molding principle (Fig. 7-16*b*). An excess quantity of the polymer is loaded into the *transfer pot* from which it is pushed through an *orifice* (sprue) into the mold cavity, by the action of a punch, at approximately 35–100-MPa (5–15-kpsi) pressure, while the die is held closed by another, higher-capacity ram. Pressures up to 300 MPa may be needed for some plastics such as cellulose acetate, rigid PVC, or styrene–butadiene block copolymer. The die is of closed construction and multiple cavities may be used, with appropriate runners and gates to feed them (as in Fig. 3-29). Thus, reasonable production rates are achieved on both thermosetting and thermoplastic compounds.

A great advantage is that the plastic acquires uniform temperature and properties in the transfer pot prior to transfer. The plastic is further heated by shearing through the orifice, viscosity is reduced, and the plastic fills intricate mold details. The technique is favored for making electrical connectors and for the encapsulation of microelectronic devices, because the low-viscosity plastic does not damage delicate wires and inserts.

Example 7-4

A transfer-molding machine is operated from a 3000-psi hydraulic supply. The transfer cylinder is of diameter 5 in and the clamp cylinder of diameter 10 in. If the polymer calls for a transfer pressure of 7 kpsi, calculate the diameter of the transfer punch (the diameter of the transfer pot) and the maximum allowable projected cross-sectional area of the molded parts.

The transfer cylinder exerts a force of $P = pA = 3(5^2\pi/4) = 58.9$ klb. At a required pressure of 7 kpsi, the punch area is $P/p = 58.9/7 = 8.4$ in^2. The diameter is 3.27 in (say, 3.25 in).

The clamp exerts a force of $P = pA = 3(10^2\pi/4) = 235.6$ klb. Since the transfer pressure is acting on the entire projected area, the permissible area is $A = P/p = 235.6/7 = 33.66$ in^2. To prevent the formation of a flash, it is usual to calculate with 1.15 times the transfer pressure. Thus, $A = 235.6/(1.15)7 = 29.3$ in^2.

Injection Molding *Injection molding* is perhaps the most widespread technique, utilized for thermoplastic and, more recently, also thermosetting resins. The process (Fig. 7-17) resembles the hot-chamber die casting of metals.

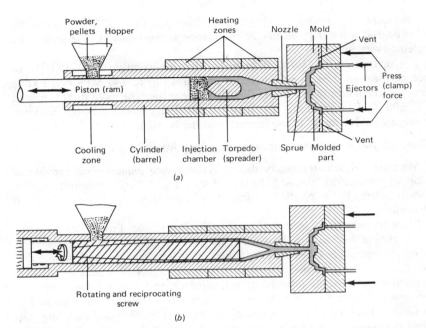

FIGURE 7-17
High productivities can be obtained by injection molding, with controlled amounts of melt injected by a (a) reciprocating plunger or (b) reciprocating rotating screw. (*Adapted from Petrothene ® Polyolefins: A Processing Guide, U.S. Industrial Chemicals Co., New York, 1971.*)

In common with transfer molding, the die is split to allow removal of the product. As in die casting, the die must be kept firmly shut during injection, with the aid of a large hydraulic cylinder, or hydraulically actuated mechanical clamps, or a mechanical clamp combined with a short-stroke hydraulic cylinder. The clamping force is calculated from the projected area of the molding(s) and the recommended injection pressure (which is similar in magnitude to transfer-molding pressures). *Ejectors* are provided for removing the molded component, and fine (0.02–0.08 mm × 5 mm) *vents* ensure that no air remains trapped.

Multiple cavities are readily accommodated, but great care must be taken in designing the running and gating system. Flow rates can be very high, and erosion by hard filler particles may become severe. Temperature control is critical. Very close and rapid control of the temperature of the sprue allows *runnerless molding*: a sudden drop in temperature shuts off the flow, while rapid heat-up prevents freeze-ups.

In its basic form, the equipment is fairly standard. The polymer is fed through a *hopper* to a cylinder or *barrel*, the die-end of which is surrounded with heaters that gradually bring the polymer to the required temperature. Operating temperatures depend on the polymer (Tables 7-3 and 7-4).

1 Thermoplastics are heated above the melting point (170–320 °C) while the mold is held at a lower (typically 90 °C) temperature; typically, 2–6 cycles are completed every minute.

2 For thermosets, the barrel is preheated just sufficiently (to 70–120 °C) to ensure plastication. Injection under high pressures (up to 140 MPa or 20 kpsi) generates enough heat to reach 150–200 °C in the sprue. The mold itself is heated to 170–200 °C. The process is also used for molding glass-filled BMC; however, the charge would hang up in the hopper, and the barrel is stuffed.

There are two basic ways of transporting the polymer:

1 Machines with hydraulically driven *reciprocating plungers* are capable of developing pressures of 70–180 MPa (10–25 kpsi, Fig. 7-17a). Uniformity of flow and shear heating of the plastic is obtained by inserting a *torpedo* (spreader) into the barrel.

2 More frequently, a *rotating screw* (Fig. 7-17b) is used. Proper design of the screw is critical for success. It usually has three sections: the *feed section* of constant root diameter takes in the granules or pellets from the feed hopper and moves them to the *transition section* in which the flights of gradually reducing cross section compress the softened pellets, expel air, and deliver a viscous fluid to the *metering section*. This, like the feed section, has a constant but smaller free cross section (the ratio of the volumes of one flight at the feed and metering ends is called the *compression ratio*, ranging typically from 2 : 1 to 4 : 1). Here the melt is further heated by shearing at a high rate.

The compression ratio and the length (or, more properly, the *length-to-diameter ratio*) of the screw are chosen with due regard for the polymer. Heat-sensitive polymers (such as PVC) are extruded with minimum shear whereas polymers with a sharp melting point (such as nylon) call for short transition and long metering sections. For successful operation, temperatures (heating and cooling), back pressure, screw rpm, injection rates, etc., must be tightly controlled.

To deliver the required amount of molten plastic to the mold, the screw of a *reciprocating screw machine* is supported by a hydraulic ram that is pushed back when the pressure in front of the screw builds up to a preset value (Fig. 7-17b). This results in a pressure drop, which allows the hydraulic ram to push the screw forward and thus inject the plastic into the mold.

Production rates are greatly increased with *multistation, rotary turntable machines* on which loading, injection, and stripping (and, if appropriate, placing of inserts) take place simultaneously.

In some machines the screw is only used for plasticating, i.e., feeding the melt to an injection chamber, from which a separate plunger injects the melt into the mold. Screws are sometimes also used to feed compression- and transfer-molding presses.

Example 7-5

An injection-molding machine of 125-ton capacity (clamp force) is to be used for making a 5-in-wide × 10-in-long × 3-in-deep box. (*a*) What molding pressure can be sustained? (*b*) How will this pressure change if a box of the same width and length but 6-in depth is to be made? (*c*) How will the pressure change if the wall thickness of the boxes is doubled?

(*a*) Only the pressure acting perpendicular to the parting plane (parallel to the clamp force action) has to be resisted by the clamp. Thus, only the projected area is of interest: $A = 5 \times 10 = 50$ in². The sustainable pressure is $125/50 = 2.5$ ton/in² = 5 kpsi (= 35 MPa).

(*b*) The projected area does not change with the depth of the box; hence, the pressure remains unchanged.

(*c*) Any change in wall thickness is immaterial, as long as the projected area remains the same. (However, the quantity of plastic to be delivered doubles.)

Extrusion   *Extrusion* is used mostly for the production of bar, tube, sheet, and film in thermoplastic materials. The equipment is similar to that shown in Fig. 7-17*b*, but an *extrusion die* takes the place of the mold, and the screw does not reciprocate. Because such presses are often devoted to extruding only one kind of polymer, the screw is optimized. The screw is often in two sections, allowing decompression at about the middle of the barrel, so that gases can be vented off, and the pressure is then built up again. To prevent any unmelted polymer or entrapped dirt from entering the extrusion, fine wire screens are placed into the stream of polymer just before it enters the die. Additionally, the screen increases back pressure, thus improving mixing and homogenization.

On leaving the die, the polymer molecules partially recoil, causing the dimension of the extrudate to increase (*die swell*). This must be taken into account in designing the die (Fig. 7-18*a*). In-line after-treatment is often necessary to hold tolerances.

Sheet is extruded, often in great widths, through dies which are designed to promote uniform rates of material flow throughout the entire width (Fig. 7-18*b*). Adjusting screws are used to close the die lips in the center where flow rates would be higher. The emerging sheet is usually guided around a set of three highly polished rolls so that it is cooled while its surface is polished. Thin films are similarly extruded and are guided around a chilled roll prior to further treatment.

FIGURE 7-18
Extrusion dies can have complex shapes to (*a*) compensate for die swell, (*b*) distribute material across the width of a sheet, or (*c*) coat a wire.

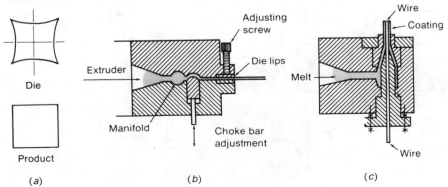

Hollow products are extruded with the aid of spider dies (Fig. 4-39c). The low operating temperatures and pressures allow a crosshead die arrangement (as in Fig. 7-18c). The highest pressure is obtained just in front of the orifice; thus, the separated streams are reunited and a sound product is extruded.

Twin- and multiple-screw extruders are more suitable for heat-sensitive materials such as rigid PVC, because they rely less on shear and drag to move the material: Intermeshing screws provide positive feed with minimum shearing.

The market for extruded products is large. Profiles are used in buildings and automotive applications; tubes, sheet, and film in a great variety of general applications; pipes in drainage, waste, and vent lines. Much film is produced by blowing extruded tube, a process properly discussed in a later section. Wire and cable insulation is applied by feeding the wire through a crosshead die (Fig. 7-18c).

Reaction Injection Molding *Reaction injection molding* (RIM) differs from other processes in that not the polymer but reactants are heated and brought together under high pressure so that they impinge upon each other as they enter the die (Fig. 7-19). Good mixing results, and the polymer is produced directly in the mold. The primary application is to polyurethane (both foams and elastomers), with some applications to nylon and epoxy. Pressures in the mold are low (300–700 kPa (50–100 psi)); thus, the mold-closing forces are low too. Dies can be simple, of low-cost construction. Since polymerization takes place in the mold, internal stresses are minimal, and the process is suitable even for large, complex, filled-plastic parts, such as auto-body and appliance components. To avoid directionality of reinforcement, flake glass is preferred.

Rotational Molding *Rotational molding*, also called *rotomolding*, is suitable for making large, relatively thin-walled, hollow (open or closed) parts. A mea-

FIGURE 7-19
Reaction injection molding (RIM) produces parts by rapid impingement mixing and injection of monomers into the die cavity.

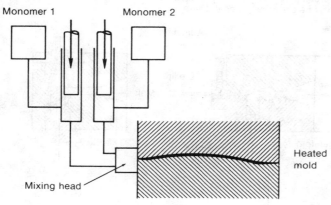

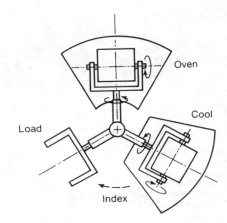

FIGURE 7-20
Rotomolding produces hollow products by rotating the mold around two mutually perpendicular axes; three stations speed up production.

sured quantity of polymer is placed into a thin-walled metal mold, and the mold is heated while rotating around two mutually perpendicular axes (Fig. 7-20). Thermoplastics (such as PE, nylon, or polycarbonate) melt whereas thermosets polymerize. The mold is cooled and the part is removed. To increase production rates, three-arm carousels are often used, with one mold each in the load–unload, heat, and cool positions. Since no pressure is involved, the mold is simple. The part is free of molded-in stresses. Even very large parts (such as 20 000-L containers) can be made. The technique is suitable also for plastisols.

In all molding processes the polymer reproduces the surface finish of the mold, and, once the often very expensive die is made, a smooth surface can be obtained at no further expense. Dimensional variations, however, can be rather large unless sufficient time is allowed for the polymer to become rigid prior to removal from the mold.

Calendering *Calendering* is related to rolling in that the thermoplastic polymer melt is fed to a multiroll calender. The first roll gap (nip) serves as a feeder, the second as a metering device, and the third sets the gage of the gradually cooling polymer which is then wound, with about 25% stretching, onto a drum (Fig. 7-21). As in metal rolling, parallelism of the roll gap must be maintained: temperature distribution is carefully controlled, and the roll camber (crown) is controlled either by skewing the central roll or by roll bending. Calendering is a high-production-rate (typically, 100 m/min) process, mostly for flexible PVC (upholstery, rainwear, shower curtains, tapes, etc.) and rigid PVC (trays, credit cards, laminations). Some ABS is also formed.

Melt Spinning Thermoplastic polymer fibers can be made by *melt-spinning*. The melt is extruded from a spinnerette, typically at 230–315 °C, in an arrangement similar to that of Fig. 6-17b, except that air is used for cooling the emerging

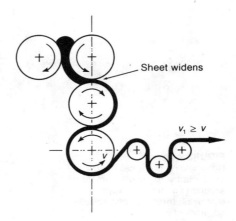

Sheet widens

$v_1 \geq v$

FIGURE 7-21
Wide sheet and foil can be produced by calendering.

fibers. Fiber diameters are small (typically, between 2–40 μm) and are usually expressed in units of *denier* (the weight, in grams, of 9000-m fiber) or the SI unit *tex* (the weight, in grams, of 1000-m fiber). The tensile strength, in units of N/tex, is called *tenacity*.

Particulate Techniques Even though many polymers are processed from particles (powder, pellets, etc.), the techniques described hitherto are not particulate processes in the sense adopted in Chap. 6, since the particles fuse and melt. However, some very-high-molecular-weight polymers such as UHMWPE, PTFE, and polyimide decompose before reaching a temperature where viscous flow is possible. Therefore, they must be processed by techniques typical of metals (Sec. 6-2).

The powder is compacted cold, with pressures up to 350 MPa (50 kpsi), and then sintered at 360–380 °C to form a solid billet or shaped component. Alternatively, the powder is compression molded at a somewhat higher temperature (as in the hot pressing of metal powders), or powder is hot extruded in a ram-type press (as in Fig. 4-37b). The extrusion of solids causes alignment of molecules, and extrusion at a high extrusion ratio is one of the methods of making ultrahigh-strength polymer fibers.

7-7-4 Processing in the Rubbery State

The viscoelastic behavior of thermoplastics above T_g allows the further processing of cast, extruded, or calendered semifabricated products. Processing is based mostly on the high m value which allows substantial stretching without the localization of necks. Molecules are uncoiled and aligned in the direction of stretching; therefore, strength increases in the direction of alignment but strength and, particularly, impact strength, are lower perpendicular to the alignment.

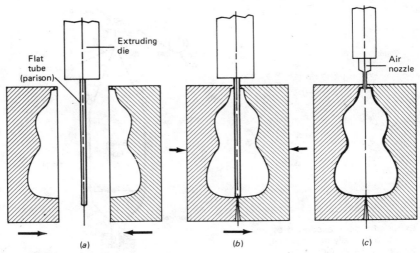

FIGURE 7-22
In extrusion blow molding (*a*) a flat extruded tube is (*b*) pinched off and welded at the bottom by the closing die and (*c*) expanded into the mold by air pressure; the bottle is cooled and removed at third and fourth stations (not shown).

Biaxial stretching increases strength in all directions, and biaxially oriented products show superior performance in many applications.

Blow Molding In *blow molding* an extruded tube or preform emerging from the melt process is expanded by internal pressure (usually by hot air).

1 *Injection blow molding* is the same process as used for glass bottles (Fig. 6-18) except that the parison is injection molded—complete with neck—around a hollow core rod, and is then blown through the core rod. Many PVC, PP, PET, and polycarbonate bottles of smaller (say, up to 1.5 L) sizes are blown on indexing machines.

2 *Extrusion blow molding* differs in that a continuous tube is extruded, pinched off so that a solid-state weld is produced at the pinch line, and the preform thus produced is then blown (Fig. 7-22). This method has the highest productivity. Even very large parts such as 2000-L drums can be molded; to avoid the need for very-high-capacity extruders, melt is gathered in an accumulator from which it can be quickly extruded, with a separate ram, into a tube. Localized thinning of parts can be avoided, or parts with intentional variations of wall thicknesses along the length can be produced by the programmed movement of a tapered mandrel in the extrusion die orifice.

3 *Stretch blow molding* is a term applied to the simultaneous axial and radial expansion of a parison, yielding a biaxially oriented container. The extruded or injection-molded parison is first conditioned, either by controlling its temperature

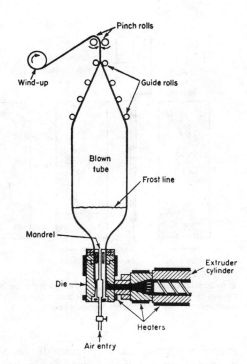

FIGURE 7-23
Extruded tube is expanded by air into a biaxially oriented film; temperatures are closely controlled to keep the frost line (where the film becomes slightly opaque) at a constant position. (*From D. C. Miles and J. H. Briston, Polymer Technology, Chemical Publishing Co., 1979, p. 551. With permission.*)

immediately after extrusion, or by reheating the cold parison. Many PVC, PP, and PET bottles are thus made.

4 *Blown film extrusion* is used to make biaxially oriented thin film by expansion of an extruded tube (Fig. 7-23) in materials such as PE. Bags are obtained by heat sealing or by pinching off, and very wide (over 6 m) flat films are produced by slitting along the length.

Example 7-6

Blown film of 25-μm thickness is extruded for grocery bags. The width of the bags (the lay-flat width of the blown tube) is 50 cm. Optimum properties are obtained by a blow-up ratio (bubble-to-tube diameter ratio) of 2.5 : 1. On emerging from the extrusion die, the tube is pulled at a faster rate; the draw-down ratio (the ratio of die opening to tube thickness) is 15 : 1. Calculate the required extrusion die diameter and opening.

The circumference of the bubble equals twice the lay-flat width. Thus $d\pi = 2(500) = 1000$ mm and $d = 318.3$ mm. The blow-up ratio of 2.5 gives a tube diameter of $318.3/2.5 = 127.3$ mm (a die of 125- or 130-mm diameter is likely to be chosen). The tube thickness, before blowing, is $2.5(25) = 62.5$ μm $= 0.0625$ mm. The draw-down ratio of 15 : 1 gives a die opening of $15(0.0625) = 0.9375$ (an opening of 1 mm will do).

Thermoforming *Thermoforming* is a term applied to forming a sheet into open, container-like, but often very complex shapes.

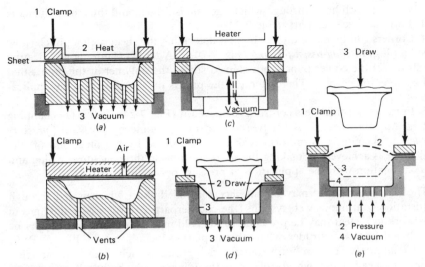

FIGURE 7-24
Thermoforming processes convert a sheet into complex shapes by (a) vacuum, (b) pressure, (c) drape-vacuum, (d) plug-assist, and (e) pressure-bubble plug-assist methods. Numbers refer to the process sequence.

Thermoforming employs a clamp that grips the sheet around its circumference, a heater to bring the polymer above the glass-transition temperature (usually around 55–90 °C (130–190 °F)), and a die which may be male or female. Conformance to the die shape may be achieved by mechanical means or by air pressure. Since the die is cooler, the polymer is chilled (and stiffened) by die contact and portions of the workpiece that first touch the die become stiff while still of a thicker gage. Subsequent deformation is limited to the freely deforming portions of the workpiece, and excessive thinning could lead to fracture. Much of process design is aimed at controlling the wall-thickness distribution by the planned sequence of operations. Numbers in Fig. 7-24 refer to this sequence.

1 In the simplest (straight) techniques all forming is done with vacuum or pressure. In *vacuum forming* (Fig. 7-24a) the sheet is (1) clamped, (2) heated above T_g, and then (3) vacuum is applied to draw the sheet into intricate recesses of the female die.

2 Alternatively, hot *air pressure* may be applied (Fig. 7-24b) to drive the sheet into the female die cavity (provided, of course, that venting holes are furnished at the underside).

The corners of all straight-formed parts will be thin, as will be seen from considering the points of first die contact in Fig. 7-24a and b.

3 The shape can be developed by *drape forming* with a male die (similar to the stretch forming of metals, Fig. 5-23); reentrant shapes can be formed by providing the punch with holes through which vacuum is drawn, and thus the polymer is pulled into the recessed parts (Fig. 7-24c).

4 Corners can be made thicker by first deforming the sheet with a punch (now called a plug) in *plug-assist forming* (Fig. 7-24d). After (1) clamping and heating the sheet, (2) the cooler punch moves in to stiffen the polymer at the points that will become the corners. Thereafter, (3) the part is finish formed by the application of vacuum.

5 Further wall-thickness control is achieved if (Fig. 7-24e), after (1) clamping and heating the sheet, it is (2) free-formed into a dome by pressure (*pressure bubble*) or vacuum; this preform is then (3) deformed by the plug, and (4) the final shape is achieved by making the sheet conform to the plug, die, or plug and die combination with the application of pressure or vacuum.

The starting material may be cut sheet or coiled sheet which is fed through multistation continuous web systems (the counterparts of progressive die lines in metalworking). Heat may be provided to the cold input sheet, or the line may be directly fed from an extruder. Many thermoplastics including ABS, PP, PS, PVC, PMMA, and polyesters are formed into varied products such as aircraft canopies, automobile head liners, wheel-well liners, building panels, bathtubs, refrigerator liners, and vast quantities of packaging products.

Cold Drawing The term is applied to the continuous stretch drawing of filaments and fibers, including fibers emerging from a spinnerette or ribbons slit from a film. The term is, obviously, misapplied, because the plastic must be above T_g which, by definition, classifies the technique as hot deformation. Alignment of linear molecules results in a substantial increase in strength. The process is most important for textile and reinforcing fibers, including PP, polyester, nylon, and aramid fibers.

Solid-State Forming The terminology is somewhat diffuse, but here we use the term *solid-state forming* to describe deformation just above T_g (or, for highly crystalline polymers such as PP, just below T_m). This too is sometimes misnamed cold forming. All metalworking techniques can be used. Forging is applied to a limited extent, e.g., for making gears, but most forming is done by sheet-forming techniques including bending, stretching, and drawing. Many food-packaging tubs and containers are formed of PP essentially by stretching.

Thermoplastic stamping (or *matched-die forming*) is a term used to describe the deformation of filled thermoplastic polymer sheet at melt temperatures, between cold, mating dies, thus giving reduced cycle times. At lower preheating temperatures, elastic behavior becomes more dominant and more springback is to be expected.

7-7-5 Cellular or Foam Plastics

Blowing agents create gas-filled voids, cells in the polymer; thus, one speaks of the production of *cellular* (*foam*) *plastics*. Many techniques are available: Air may be introduced by mechanical whipping; gases can be diffused in at elevated pressures; low-boiling liquids, gases (such as CO_2), or chemical compounds that decompose on heating, may be incorporated; hollow glass or hollow or expandable plastic spheres may be added; or the polymer may be molded with an additive that can be subsequently bleached out.

In general, a highly viscous polymer gives closed cells, whereas a less-viscous polymer results in open (interconnecting) porosity. Porosity is uniform throughout products such as flexible or elastomeric foams used in insulation, packaging, cushions, etc. *Structural foams* have a solid skin. For these, the polymer is poured or injected into a cold mold which suppresses expansion of the polymer, forming a solid skin on the surface; in contrast, the slowly cooling core is foamed. The foam prevents collapse of the surface, and the final product has low density combined with reasonable strength.

Various techniques of manufacture are used:

1 *Expanded polystyrene.* A volatile hydrocarbon (usually, pentane) is added to polymer beads as a blowing agent. When heated with live steam (steam introduced into the polymer), preexpanded beads are obtained. The beads are stored to allow evaporation of moisture, and are then fed into the appropriate mold. Both mold and beads are heated with live steam to cause final expansion and fusing, followed by cooling. Insulating board is composed of large beads at densities of 15–30 kg/m^3; foam cups are made of small beads at 50–65 kg/m^3.

2 *Extruded thermoplastic foams.* Most frequently, the gas (pentane, fluorocarbons) is directly introduced into the molten plastic in the extrusion barrel. Expansion takes place as the extruded sheet or profile leaves the die. To control cell size, a fine, dry powder is mixed in the plastic as a nucleating agent; gas comes out of solution on the powder particles. The main application is for trays and packaging, at densities of 30–150 kg/m^3 in PE and PS. PVC extrusions are used as substitutes for wood moldings in the construction industry.

3 *Structural foams.* All thermoplastics (such as ABS) can be injection molded, at rapid injection rates, to 60–90% of solid density. There are many applications: computer and appliance housings, material-moving pallets and bins, doors and shutters, and automotive instrument panels.

4 *Multicomponent liquid foam processing.* In this, chemical compounds are poured or injected into molds to form thermosetting foams in situ. RIM is also applicable. Most widely used is polyurethane in both the open-cell (flexible) and rigid (up to 500-kg/m^3) forms. Load-bearing structures such as furniture frames and doors are also made. Polyester and silicone rubber foams are also produced by this technique.

Example 7-7

The density of polystyrene is 1.05 g/cm^3. Calculate the volume of air space in the walls of an expanded polystyrene foam cup of 50-kg/m^3 density.

For ease of conversion, consider that 1 $g/cm^3 = 1000$ kg/m^3 (you may wish to check this). Hence the volume of polymer in 1-m^3 bead is 50 kg/1050 kg $m^{-3} = 0.0476$ m^3 or 4.75%. The volume of air is $100 - 4.76 = 95.24\%$.

7-8 PLASTICS-PROCESSING EQUIPMENT

Molds (dies) used in plastics processing can be of relatively light construction if pressures are low, as in thermoforming, rotary molding, and blow molding, and are often made of fully heat-treated 7075 aluminum alloy. Injection molding and extrusion dies are made of heat-treated steel. Surfaces are usually highly polished (unless given a decorative finish) and, for greater wear resistance, may be chromium plated or built up with a wear-resistant surface coating. Wear is of particular concern when polymers containing highly abrasive fillers move at high velocity over die surfaces, as in injection-molding dies. Temperature control of dies is critical and sophisticated heating-cooling systems are needed.

Plastics processing equipment (see Table 7-5) shares many features with metal-processing equipment; however, the greater sensitivity of plastics to temperature, shear rate, and residence time makes process control more critical. Therefore, substantial effort has been directed at the development of sensors (mold pressure and temperature, ram speed and position) that allow data acquisition and closed-loop or adaptive control of all important process variables. The introduction of programmable logic controllers and, for the most sophisticated machines, microcomputers has made processing much more reproducible. For example, adaptive control is used in extruding sheet and film to equalize thickness across large widths: sheet thickness is continuously measured at various points across the width, and the flow of polymer is redistributed by trimming the die shape with the aid of die bolts (Fig. 7-18b). The introduction of CAD/CAM techniques has taken much of the trial-and-error experimentation out of process design. Variable-speed electric drives and servo-controlled hydraulics provide opportunity for CNC control of most processes.

7-9 DESIGN WITH PLASTICS

The availability of a vast variety of plastics makes the designer's task more difficult. The difficulty is further compounded by the different responses of various plastics to service conditions such as static and impact loading at service temperature; the effects of humidity, radiation (including light), and other environmental factors on the long-term deterioration of properties (aging); dimensional stability; wear; the release of undesirable chemical compounds on contact with food or air; and response to electrical fields. In identifying the most suitable plastic, it is often necessary to follow the process of elimination.

As in all design, the interaction of the process with the material must be considered early on, with due regard to the quantities to be produced. It is often found that the initial choice of material must be changed in light of constraints imposed by the economics of processing. Hence, the final material choice also fixes the process, and design then proceeds with allowance for the limitations of that process. Many of the principles discussed for casting (Secs. 3-8 and 3-10), forging (Secs. 4-4-4 and 4-4-7), extrusion (Sec. 4-5-6), rolling (Sec. 4-8-6), sheet metalworking (Secs. 5-3 and 5-5), and particulate processing (Sec. 6-4) are also

TABLE 7-5 CHARACTERISTICS OF PLASTICS-PROCESSING EQUIPMENT*

Equipment	Extruder			Press (clamp)			No. of parisons	No. of stations (shuttle or rotary table)
	Capacity, kg/h	Barrel diam., m	Shot size, kg	Force, kN	Bed, m	Daylight,† m		
Extruders								
Single screw	1–3000	0.01–0.6						
Twin screw	20–1000	0.01–0.9						
Injection molders			0.01–40	20–45000	0.07 × 0.15– 2 × 2.5	0.15–1.8		(1–14) (1–6)
Blow molders								
Continuous extrusion	4–400			10–300	0.15 × 0.25– 0.6 × 0.6	0.15–0.75	1–4	(1–18)
Accumulator	30–1600		0.1–150	30–3000	0.3 × 0.4– 3 × 3	0.15–3.5	1–4	
Injection			0.1–25	100–3000	0.25 × 0.3– 0.6 × 1	0.1–0.75		1–2
Reciprocating screw	30–350		0.06–2.5	50–600	0.25 × 0.25– 1 × 1.5	0.25–1	1–10	1–10
Compression molders				50–35 000	0.15 × 0.15– 2.5 × 3.6	0.75–3.5		
Transfer molders								
Clamp				50–6500	0.01–2.3‡	0.1–1		
Transfer				10–2000				

*Data extracted from *Modern Plastics Encyclopedia*, McGraw-Hill, New York, 1984.
†Press stroke is usually 60–80% of daylight (maximum open height).
‡Molding area, m².

427

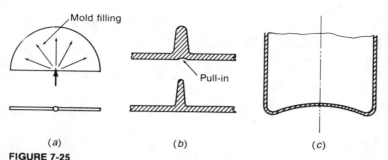

FIGURE 7-25
In the design of plastic parts, care must be taken to (*a*) allow orderly filling of the mold cavity, (*b*) minimize changes in cross section that could cause collapse of the surface, and (*c*) stiffen the part, for example, by doming.

applicable to the processing of plastics, provided that the special characteristics of plastics are taken into account.

1 *Tolerances.* While molds and dies can be made to close tolerances, the great sensitivity of dimensions to processing conditions, post-processing changes (polymerization, crystallization, loss of plasticizer, aging, relief of residual stresses) dictates that tolerances be as wide as permissible for the given application.

2 *Parting line.* As in die casting and forging, the parting line must be chosen to minimize the complexity of the mold, avoid unnecessary undercuts that would necessitate complex movable inserts and cores, and minimize the cost of removing flash, for example, by allowing flash removal by tumbling. Distortion is minimized when the gate is placed so as to give symmetrical mold filling (Fig. 7-25*a*).

3 *Wall thickness.* The removal of gases (produced by reactions or entrapped during the compaction of particulate starting materials) must be allowed and encouraged. This sets a practical limit of 100–200 mm to the maximum thickness attainable without gross porosity. The low heat conductivity of plastics limits heating rates and thus also the economical thickness of thermoformed parts, typically to below 6 mm (0.25 in). The minimum attainable wall thickness (Sec. 12-2-2) is limited in molding by the difficulty of removing very thin parts from the mold, and also by the high pressures required to fill at a high width-to-thickness ratio. Large wall-thickness variations are just as undesirable as in metal casting (Fig 3-40).

4 *Ribs.* Distortion is often minimized by ribbing larger surfaces. The width of ribs is kept small to prevent the creation of large hot spots (Fig. 7-25*b*). Doming is an attractive alternative, especially in cylindrical parts (Fig. 7-25*c*).

5 *Drafts and radii.* Release from the mold requires a draft of 0.5–2°, and even larger drafts on ribs and bosses. Tight corners can be filled, but generous radii increase die life and prevent stress concentrations in service. Minimum radii of 1–1.5 mm are recommended.

6 *Holes.* Through-holes are limited only by the strength of the core pin and are usually held below a length-to-diameter ratio of 8. Freely extending core pins are needed for blind holes; therefore, such holes are limited to a depth-to-diameter ratio of 4 for $d > 1.5$ mm (0.06 in) and to a ratio of 1 for smaller holes. Threaded holes of 5-mm (0.20-in) diameter and over can be molded directly, preferably with a coarse thread. Smaller holes are best drilled.

7 *Inserts*. The use of molded-in metal inserts greatly expands the scope of application for plastics and very often eliminates problems in subsequent assembly, although at some expense. Threaded inserts, binding posts, electric terminals, anchor plates, nuts, and other metallic components are molded into plastics by the millions. Some precautions are necessary, however. The shape of the metal part must ensure mechanical interlocking with the plastic, for example, by heavy knurling, since there is no adhesion between metals and plastics—at least not without special surface preparation. The thermal expansion of plastics is much larger than that of metals (Table 7-2); this helps to shrink the plastic onto the insert, but could also cause cracking of a brittle plastic. The wall thickness around the insert must therefore be made large enough to sustain the secondary tensile stresses.

General characteristics of polymer processes are summarized in Table 12-5.

7-10 COMPOSITES

We have already come across many techniques that are used to make composites which combine the desirable properties of two materials. In a sense, two-phase materials are also composites, but in the general use of the word, *composites* refer to structures which are made of two distinct starting materials, the identities of which are maintained even after the component is fully formed. The starting materials may be metals, ceramics, or plastics.

7-10-1 Types of Composites

Depending on the distribution of the two materials relative to each other, the following groups are distinguished:

1 *Coatings and laminations*. In Chap. 5, we saw examples of tinplate, galvanized sheet, and terneplate, all with a mild-steel base. In Chap. 9, we will discuss roll cladding and explosive cladding of one metal upon another. In Chap. 6, enamel was an example of a ceramic coating on metal, and glaze on tiles an example of a glassy ceramic on a crystalline ceramic base. Tools coated with TiC, TiN, or Al_2O_3 are examples of ceramics on steel. Polymers are applied in large quantities to paper (milk cartons), textiles (seat covers, carpets), metals (wire insulation, beverage cans), and to polymers (multilayer films and bottles, combining the impermeability of one polymer with the mechanical and processing advantages of another, usually lower-cost plastic).

2 *Particulate composites*. We saw several examples in Chap. 6, including metal–metal (Cu-infiltrated iron), ceramic-reinforced metal (dispersion-hardened metals), metal-bonded ceramics (cermets, including cemented carbides), and metal–polymer structures (metal bearings infiltrated with PTFE or nylon).

The largest application is, however, for filled polymers, discussed in Secs. 7-4-2 and 7-7 (holes in cellular polymers can be regarded as fillers of zero strength). If a more or less equiaxed filler is perfectly wetted by the matrix so that the bond is stronger than either the matrix or the filler, many properties may be calculated from the *rule of mixtures*: Each component contributes to the properties of the composite in proportion to its volume fraction. Thus, the yield strength of the composite σ_c is

$$\sigma_c = f\sigma_f + (1 - f)\sigma_m \qquad (7-9)$$

where f is the volume fraction of the filler, σ_f and σ_m are the yield strengths of the filler and matrix, respectively. Similarly, the composite elastic modulus E_c is

$$E_c = fE_f + (1 - f)E_m \qquad (7-10)$$

Thus, the elastic modulus (and flexural modulus) can be increased by incorporating into the polymer a material of higher modulus, typically a metal or ceramic. Tensile strength may be little affected since fracture occurs in the matrix.

3 *Fiber reinforcement.* When the stronger or higher-modulus filler is in the form of thin fibers, strongly bonded to the matrix, properties depend on the strength of fibers (Table 7-6), on the proportion of the fibers, and on their orientation relative to load application (Fig. 7-26):

a When the fibers are parallel to load application and the fibers are still intact when the matrix fails, strength follows the rule of mixtures. A minimum (critical)

TABLE 7-6
PROPERTIES OF SOME REINFORCING FIBERS

	Young's modulus, GPa	Tensile strength, GPa	Density, kg/m³
High-strength steel	210	2.1	7830
Tungsten	350	4.2	19300
Beryllium	300	1.3	1840
E glass	73	3.5	2480
S glass	85	4.6	2540
Alumina	175	2.1	3150
Graphite			
high modulus	390	2.1	1900
high strength	240	2.5	1900
Boron	420	4.0	2600
Nylon	5	1.0	1140
Dacron*	14	1.1	1380
Kevlar 29†	62	2.8	1440
Kevlar 49†	117	2.8	1440

*Trademark for DuPont polyester.
†Trademark for DuPont aramids.

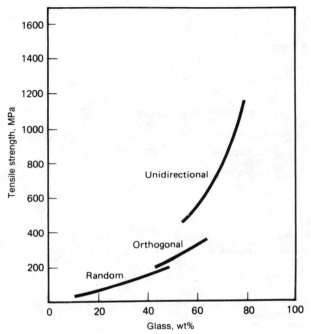

FIGURE 7-26
Orientation of fibers is an important factor in determining the properties obtained in polyester–glass fiber composites [*After R. M. Ogorkiewicz, The Engineering Properties of Plastics (Engineering Design Guide No. 17), The Desing Council, London, 1977. With permission.*]

volume fraction of fiber must be present, and the fiber must be continuous or, if chopped, long enough for stresses to be transferred from the matrix to the fiber. The critical length is given by the condition that it should take more force to shear the matrix at the fiber boundary (to pull out the fiber) than to break the fiber.

b When fibers are oriented perpendicular to the load application, more fiber is needed for stiffening and the transverse properties of oriented composites are poorer.

c When the component is exposed to biaxial stresses (as a pressure vessel or a tube would be), the best results are obtained with biaxially oriented (cross-ply) or randomly oriented fibers.

The potential for improvement can be gaged from data in Tables 7-3 and 7-4 and from Fig. 7-27, where the specific strength and stiffness are shown for various composites. (In principle, strength and elastic modulus of the material should be divided by its density; to keep units in GPa, strength and modulus are divided here by specific gravity.)

The toughness of reinforced structures is a more complex question and depends greatly on the mechanism by which the matrix and reinforcement fail. A

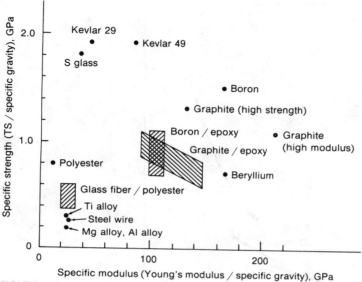

FIGURE 7-27

For a given weight, composite materials with fibers oriented in the direction of loading offer tremendous gains in strength and stiffness, as shown by their specific strength and modulus (shaded areas). For comparison, typical values are shown for bare fibers and for the highest-strength members of alloy groups.

strong fiber, strongly bonded, increases toughness because extra energy is required to break or pull out a fiber. A further advantage is that fracture of one fiber does not impair the total structure; failure is often gradual, and repairs may be possible.

Most advantages of fiber reinforcement are lost when the bond between matrix and fiber is weak. Fibers pull out and delamination may occur at low stresses. Thus, strict process control must be maintained to ensure reproducible quality, and NDT techniques must be used for detecting defects such as improper fiber orientation, lack of bonding, and delamination. Processing of NDT signals by computer imaging and analysis allows the production of such large and critical parts as the carbon-fiber–reinforced wing skin of the Harrier II jumpjet.

Example 7-8

A filament-wound container is to be made of epoxy resin reinforced with E-glass fiber. An elastic modulus of 20 GPa is desired in the structure. Calculate (*a*) the proportion of glass fiber required by volume and (*b*) by weight, (*c*) the density of the composite, and (*d*) how much of the imposed load is carried by the fibers.

(*a*) From Eq. (7-10)

$$E_c = fE_f + (1 - f)E_m$$

The modulus of most glasses is 70 GPa (Sec. 6-5-1). The modulus of an unfilled epoxy is 350 kpsi = 2.4 GPa*

Thus

$$20 = 70f + (1 - f)2.4$$

$$f = 0.26 \quad \text{or} \quad 26\% \text{ by volume}$$

(b) The density of epoxy is approximately 1.2 g/cm^3; that of the glass 2.4 g/cm^3.

In 100-cm^3 composite, there is 26(2.4) = 62.4-g glass and (100 − 26)(1.2) = 88.8-g epoxy.

Total weight is 62.4 + 88.8 = 151.2 g. Of this, 62.4/151.2 = 0.413 or 41.3 wt% is glass and 88.8/151.2 = 0.587 or 58.7 wt% is epoxy.

(c) Density is 151.2 g/100 cm^3 = 1.512 g/cm^3. (This could have been obtained directly from the rule of mixtures:

$$\rho_c = f\rho_f + (1 - f)\rho_m$$

$$\rho_c = 0.26(2.4) + (1 - 0.26)(1.2) = 1.512 \text{ g/cm}^3.$$

(d) If the glass represents 26% of the volume, it also represent 26% of the cross-sectional area. The glass and epoxy are forced to extend together under loading; thus, strain is equal in the two. From Eq. (2-5),

$$e = \frac{\sigma}{E} = \frac{P}{AE}$$

thus

$$\frac{P_f}{P_m} = \frac{A_f E_f}{A_m E_m} = \frac{0.26(70)}{0.74(2.4)} = 10.24$$

In other words, 90% of the load is carried by the 26% fiber.

7-10-2 Reinforcing Fibers

A good reinforcing fiber has a high elastic modulus; high strength; reasonable ductility in both tension and compression; low density; and is wetted by the matrix. Wetting can be improved by the application of coupling agents, coatings, or conversion treatment of the fiber surface.

Fibers are available in *filaments* (or *monofilaments*, i.e., very long, continuous single fibers), *yarn* (twisted bundle of filaments), *roving* (untwisted bundles of gathered filaments), *tows* (bundles of thousands of filaments), *woven fabrics* (made from filaments, yarn, or roving, woven at 90° angles to each other), *mats* (continuous fiber deposited in a swirl pattern or chopped fiber in a random pattern), *combination mats* (one ply of woven roving bonded to a ply of chopped-strand mat), *surface mats* (very thin, monofilament fiber mats for better surface appearance), *chopped fibers or roving* (of 3–50-mm length) and, in brittle materials, *milled fibers* (of 0.5–3-mm length). *Whiskers* are very fine (20–50-nm diam.), short (30-μm long), single-crystal fibers, grown without significant lattice defects, thus possessing very high strength.

Modern Plastics Encyclopedia, McGraw-Hill, New York, 1984–1985, p. 456.

The high modulus of ceramic fibers makes them attractive for the reinforcement of metals (e.g., Al/B, Al/carbon, Mg/carbon). There is also a possibility for increasing the toughness of ceramics by reinforcement. The major application is, however, to polymers which otherwise have low strength, stiffness, and creep resistance. The proportion of reinforcement is seldom below 20% and may go as high as 80% in oriented structures.

For polymer reinforcement, rovings, mats, etc., are sold also as *prepregs* (preimpregnated with the resin). We already indicated in Sec. 7-6-3 that polymers may be premixed with fibers, fillers, colorants, initiators, catalysts, etc. Such premixes of a putty-like consistency are called *bulk molding compounds* (BMC). For large sheet-like parts, *sheet molding compounds* (SMC) are prepared: first, a mixture of resins (chiefly polyester, with some thermoplastics added for improved impact properties) is deposited onto a PE carrier film (with a doctor blade defining the thickness), then chopped fiberglass roving of 6–75-mm length is deposited on top of this, and finally, another layer of resin mix—applied to a carrier film— is used as a cover. The sandwich is then compacted by rollers and is heated to ensure complete wetting of the glass fibers and to begin thickening. This SMC can be rolled up into coils and stored for some limited time. Final fabrication is by pressing between heated dies.

7-10-3 Fabrication Methods

The fabrication method depends largely on the shape and size of the product and on the quantities required.

Open-Mold Processes Only one mold (male or female) is needed and may be made of any material such as wood, reinforced plastic, or plaster.

1 In *hand lay-up* the resin and fiber are placed manually, air is expelled with squeegees and, if necessary, multiple layers can be built up. Hardening is at room temperature but may be speeded up by heating.

2 The unfinished (free) surface is smoother in *vacuum-bag molding*, because a plastic film is placed on that surface and, after sealing the edges, a vacuum is drawn.

3 Similar results are achieved in *pressure-bag molding* in which pressure is applied to the free surface with a tailored rubber bag.

4 Labor costs are lower in *spray-up*, with a spray gun supplying resin in two converging streams into which roving is chopped (Fig. 7-28). Automation with robots results in highly reproducible production.

All these techniques are extensively used for products such as boats, truck bodies, tanks, swimming pools, and ducts.

Filament Winding Spherical, cylindrical, and other shapes are produced by winding a continuous filament in a pre-designed pattern and under constant tension onto a mandrel. CNC winding machines with several degrees of freedom

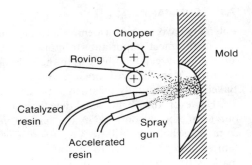

FIGURE 7-28
Spray-up is a low-cost method
of creating medium-strength
composite parts.

are frequently employed. The filament is either precoated with the polymer or is drawn through a polymer bath so that it picks up polymer on its way to the winder. Pipes, tanks, and pressure vessels are made this way because the filament can be oriented in the direction of maximum stress. Carbon-fiber–reinforced rocket motor cases are used for the space shuttle and other rockets (the solid-propellant booster rockets of the space shuttle each have four cases of 3.6-m diameter and 9-m length).

Pultrusion   A bundle of polymer-coated fibers is pulled through a long, heated die where curing takes place. The products are solid rods, profiles, or hollow tubes, similar to those produced by extrusion; hence the name *pultrusion*. Applications are to sporting goods (golf club shafts), vehicle drive shafts (because of their high damping capacity), and structural members for vehicle and aerospace applications.

Matched-Die Molding   For high production rates, close tolerances, and best surface finish, the part is made between two heated *matched dies.*

1 The resin and reinforcement are placed in the lower die in so-called *wet molding.*

2 BMCs are molded essentially by compression molding although, more and more, transfer and injection molding and RIM are also practiced. If the directionality resulting from injection is objectionable, glass flake is used for reinforcement. The GM Fiero was the first mass-produced automobile with a plastic skin attached to a welded steel, precision-machined, drivable space frame. The side panels of the body have been made of RIM, glass-flake–reinforced polyurethane.

3 SMC sheets, cut to size, are formed by techniques similar to metal pressing, except that several molds must be kept in circulation for reasonable production rates. Curing and removal of the part takes place outside the press. SMC has been used for the horizontal surfaces (roof, hood, trunk lid) of the Fiero body.

7-11 SUMMARY

Polymers have been indispensable to humans for millenia. Wood was and still is one of the most important structural materials. Wood and wood products and animal and vegetable fibers were the only organic polymers available until fairly recently. Manufactured organic polymers, beginning with celluloid and later Bakelite, were initially regarded with the suspicion accorded to substitute materials. However, the phenomenal development of the polymer industry has changed our technological outlook in the last 50 years. From containers to packaging, from household utensils to toys, from electrical and electronic components to fabrics, and from gears to aircraft structures, polymers have moved in to take markets from traditional materials.

Much of this success must be attributed to basic research that has given a better understanding of properties and processes, and to the translation of this understanding into practice. By the manipulation of the molecular structure and processing sequence, products of exceptional properties can be obtained. From the manufacturing point of view, there are two major, fundamentally different classes of polymers: thermosetting and thermoplastic substances.

1 Thermosetting plastics must be brought into the final shape before polymerization and cross-linking take place under the influence of heat or catalysts. Once polymerization is complete, the part is one single, giant, spatial-network molecule, and the process is irreversible.

2 Thermoplastic polymers are capable of viscous or viscoelastic flow when heated above some critical temperature. This is the melting point in crystalline (and therefore opaque) polymers, and the glass-transition temperature in amorphous (and transparent) polymers. Thus, these plastics can be formed into the desired shape upon heating, and the shape is fixed by cooling. The process can be repeated.

3 Elastomers have a glass-transition temperature well below room temperature. They are related to thermosetting resins in that cross-linking must take place during or immediately after shaping, although thermoplastic varieties are now available.

Fabrication processes are related to casting and plastic deformation processes in metals:

1 The precursors of polymers can be treated by casting techniques to yield thermosetting or thermoplastic products.

2 By far the largest quantities of polymers, including prepolymers of thermosets, are processed in the melt (viscous flow) regime. Processes include compressive techniques such as: compression, transfer, and injection molding; extrusion; and calendering. Because of their high strain-rate sensitivity (high m value), thermoplastics are amenable also to stretch forming (blowing, thermoforming, and so-called cold drawing).

Composite structures greatly expand the scope of application of all engineering materials by combining the desirable properties of two or more materials. Plastics particularly benefit from the improved strength, stiffness, toughness, and creep resistance imparted by fibers and other reinforcing fillers, and components of exceptional strength-to-weight and stiffness-to-weight ratios can be produced.

PROBLEMS

7-1 The behavior of strain-rate sensitive materials is sometimes expressed by the equation

$$\dot{\epsilon} = b\sigma^n \quad \text{or} \quad \dot{\gamma} = B\tau^n$$

where b and B are material constants and n is a strain-rate sensitivity exponent (not to be confused with our n value, the strain-hardening exponent). What is the value of n for (a) a Newtonian fluid and (b) an ideal rigid-plastic, non-strain-rate-sensitive material? (c) What is the relation of this n to m in Eq. (7-4)?

7-2 Thermoplastic polymer pipes are extruded over a water-cooled mandrel; hence, the inner layers cool first. What residual stress distribution is to be expected? Is it likely to persist over a long period of time?

7-3 Returnable PE milk jugs must be ground up and remanufactured after some 12 cycles to the customer because they shrink. Explain why this should be so.

7-4 Which of the polymers listed in Table 7-3 is viscoelastic at room temperature?

7-5 A bottle is blown, as in Fig. 7-22, from an extruded PET tube of 30-mm OD and 28-mm ID. What will the finished wall thickness be at the waist (diameter 70 mm) and at the bulge (diameter 110 mm)? How could the wall thickness be equalized?

7-6 Collect several different plastic bottles and containers (include, if available, a bottle with a handle). Inspect the outer surfaces for evidences of manufacturing techniques. After sectioning, inspect the inner surfaces and gage the wall-thickness variations. From your conclusions, describe the most likely manufacturing process for each container.

7-7 Measure the wall thickness of a soft-margarine or cottage cheese container at the rim, side, and bottom (use a ball-point micrometer and exert very small pressure): If available, check two containers of identical diameters but different depths. Show that the depth was developed from a circular blank with a thickness equal to the thickness of the rim.

7-8 Grocery sacks are often made of 25-μm-thick, blown LDPE (0.920 g/cm^3) film. A typical sack is 60 cm long and 50 cm wide, and the price of resin is $0.37 per pound. It is now suggested that the sack be made of HMW–HDPE (0.950 g/cm^3, $0.42 per pound). Because this material is stronger and stretches less, the gage can be reduced to 18 μm. Calculate whether the proposition is economical.

7-9 A pressure of 150 MPa is applied to the punch of a transfer molding die. The punch diameter is 50 mm, and the projected area of the mold cavities, including runners and gates, is 5800 mm^2. Calculate the press size and the clamp force required.

7-10 Check the calculation of Example 7-7 by measuring the dimensions of an expanded polystyrene drinking cup and by weighing the cup. Explain the causes of possible discrepancies.

7-11 The part shown in Example 12-6 is to be made of a wood-flour–filled phenol–formaldehyde resin. (a) From data in Tables 7-4 and 12-5, choose an appropriate process. (b) Taking average values from Table 7-4 and the associated text, specify the process conditions. (c) Design, in principle, the die, showing the main die elements. (d) Determine the size of equipment needed.

FURTHER READING

A Handbooks

Ash, M., and I. Ash: *Encyclopedia of Plastics, Polymers and Resins* (3 vols.), Chemical Publishing Co., New York, 1980–1981.

Frados, J. (ed.): *Plastics Engineering Handbook*, 4th ed., Van Nostrand Reinhold, New York, 1976.

Harper, Ch. A.: *Handbook of Plastics and Elastomers*, McGraw-Hill, New York, 1975.

Modern Plastics Encyclopedia, McGraw-Hill, New York, annual.

Saechtling, H.: *International Plastics Handbook*, Macmillan, New York, 1983.

Wallace, B. M. (ed.): *Handbook of Thermoplastic Elastomers*, Van Nostrand Reinhold, New York, 1979.

B General Texts

Brydson, J. A.: *Plastics Materials*, 4th ed., Butterworths, London, 1982.

Driver, W. E.: *Plastics Chemistry and Technology*, Van Nostrand Reinhold, New York, 1979.

DuBois, J. H., and F. W. John: *Plastics*, 6th ed., Van Nostrand Reinhold, New York, 1981.

Hall, C.: *Polymer Materials*, Macmillan, New York, 1981.

Kaufmann, H. S., and J. J. Falcetta: *Introduction to Polymer Science and Technology: An SPE textbook*, Wiley, New York, 1977.

Mascia L.: *Thermoplastics*, Applied Science Publishers, London, 1982.

Miles, D. C., and J. H. Briston: *Polymer Technology*, Chemical Publishing Co., New York, 1979.

Moore, G. R., and E. Kline: *Properties and Processing of Polymers for Engineers*, Prentice-Hall, Englewood Cliffs, N.J., 1984.

Nielsen, L. E.: *Polymer Rheology*, Dekker, New York, 1977.

Patton, W. J.: *Plastics Technology: Theory, Design, and Manufacture*, Reston Publishing Co., Reston, Va., 1976.

Rosen, S. L.: *Fundamental Principles of Polymeric Materials*, Wiley, New York, 1981.

Rudin, A.: *The Elements of Polymer Science and Engineering*, Academic Press, New York, 1982.

Seymour, R. B.: *Modern Plastics Technology*, Reston Publishing Co., Reston, Va., 1975.

_____ and C. E. Carraher, Jr.: *Structure-Property Relationships in Polymers*, Plenum, New York, 1984.

Van Krevelen, D. W.: *Properties of Polymers*, Elsevier, Amsterdam, 1976.

Ward, I. M.: *Mechanical Properties of Solid Polymers*, 2d ed., Wiley, New York, 1983.

Young, R. J.: *Introduction to Polymers*, Chapman and Hall, London, 1981.

C Polymer Processing

Astarita, G., and L. Nicolais: *Polymer Processing and Properties*, Plenum, New York, 1985.

Bernhardt, E. C. (ed.): *Computer-Aided Engineering for Injection Molding*, Macmillan, New York, 1983.

Fenner, R. T.: *Principles of Polymer Processing*, Macmillan, New York, 1979.

Freakley, P. K.: *Rubber Processing and Production Organization*, Plenum, New York, 1985.

Gastrow, H.: *Injection Molds: 102 Proven Designs*, Macmillan, New York, 1983. (English edition ed. K. Stoeckhert.)

Janssen, L. P. B. M.: *Twin Screw Extrusion*, Elsevier, Amsterdam, 1978.

Johannaber, F.: *Injection Molding Machines: A User's Guide*, Hanser, München/Macmillan, New York, 1983.

Kresta, J. E. (ed.): *Reaction Injection Molding and Fast Polymerization Reactions*, Plenum, New York, 1982.

Martelli, F. G.: *Twin-Screw Extruders*, Van Nostrand Reinhold, New York, 1983.

Meyer, R. W.: *Handbook of Pultrusion Technology*, Methuen, New York, 1985.

Middleman, S.: *Fundamentals of Polymer Processing*, McGraw-Hill, New York, 1977.

Miles, D. C., and J. H. Briston: *Polymer Technology*, Chemical Publishing Co., New York, 1979.

Pajgrt, O., and B. Reichstädter (eds.): *Processing of Polyester Fibers*, Elsevier, Amsterdam, 1980.

Sors, L., L. Bardocz, and I. Radnoti: *Plastic Molds and Dies*, Van Nostrand Reinhold, New York, 1981.

Stoeckhert, K. (ed.): *Mold Making Handbook*, Macmillan, New York, 1983.

Suh, N. P., and N. H. Sung (eds.): *Science and Technology of Polymer Processing*, MIT Press, Cambridge, Mass., 1979.

Tadmor, Z., and C. G. Gogos: *Principles of Polymer Processing*, Wiley, New York, 1979.

Throne, J. L.: *Plastics Process Engineering*, Dekker, New York, 1979.

Turner, S.: *Mechanical Testing of Plastics*, 2d ed., Longman, New York, 1983.

D Design with Polymers

Beck, R. D.: *Plastic Product Design*, 2d ed., Van Nostrand Reinhold, New York, 1980.

Benjamin, B. S.: *Structural Design with Plastics*, 2d ed., Van Nostrand Reinhold, New York, 1982.

Brown, R. L. E.: *Design and Manufacture of Plastics Parts*, Wiley, New York, 1980.

Kaeble, D. H.: *Computer-Aided Design of Polymers and Composites*, Dekker, New York, 1985.

Levy, S., and J. H. DuBois: *Plastics Product Design Engineering Handbook*, 2d ed., Van Nostrand Reinhold, New York, 1985.

MacDermott, C. P.: *Selecting Thermoplastics for Engineering Applications*, Dekker, New York, 1984.

Miller, E. (ed.): *Plastics Product Design Handbook, Part A: Materials and Components*, Dekker, New York, 1981.

E Composites

Delmonte, J.: *Technology of Carbon and Graphite Fiber Composites*, Van Nostrand Reinhold, New York, 1981.

Grayson M. (ed.): *Encyclopedia of Composite Materials and Components*, Wiley, New York, 1983.

Hull, D.: *An Introduction to Composite Materials*, Cambridge University Press, Cambridge, 1981.

Kelly, A., and S. T. Mileiko (eds.): *Fabrication of Composites*, Elsevier, Amsterdam, 1983.

Kowata, K., and T. Akasaka (eds): *Composite Materials; Mechanical Properties and Fabrication*, Applied Science Publishers, London, 1982.

NASA Langley Research Center: *Tough Composite Materials: Recent Developments*, Noyes, Park Ridge, N.J. 1985.

Schwartz, M. M.: *Composite Materials Handbook*, McGraw-Hill, New York, 1984.

_____ (ed.): *Fabrication of Composite Materials Source Book*, American Society for Metals, Metals Park, Ohio, 1985.

Sheldon, R. P.: *Composite Polymeric Materials*, Applied Science Publishers, London, 1982.

F Journals

Composite Structures
Composites Science and Technology
Modern Plastics
Plastics Engineering
Plastics and Rubber Processing and Applications
Plastics Technology
Polymer Degradation and Stability
Polymer Engineering and Science
Polymer Journal
Polymer–Plastics Technology and Engineering
Polymer Process Engineering

CHAPTER 8

MACHINING

In the processes discussed so far, the shape of the workpiece was obtained by the solidification or plastic deformation of the material. The amount of material lost in scrap was relatively small, and the scrap particles tended to be large enough and relatively easily separated by alloy type, allowing easy and economical recycling. In contrast, machining aims to generate the shape of the workpiece from a solid body, or to improve the tolerances and surface finish of a previously formed workpiece, by removing excess material in the form of chips. Machining is capable of creating geometric configurations, tolerances, and surface finishes often unobtainable by any other technique (Chap. 12). However, machining removes material which has already been paid for, in the form of relatively small particles that are more difficult to recycle and are in greater danger of becoming mixed. Therefore, developments often aim at reducing or—if at all possible—eliminating machining, especially in mass production. For these reasons, machining has lost some important markets, yet, at the same time, it has also been developing and growing and—especially with the application of numerical control—has captured new markets.

Some feel for the importance of machining may be gained from the observation that in 1983 there were about 2 million metal-cutting machine tools in the United States (of which some 5% were numerically controlled) and that labor and overhead costs amounted to $125 billion, or 3% of the GNP (or 15% of the GNP originating in manufacturing).

If absolutely essential, a machining process can be found for any engineering material, even if it may be only grinding or polishing. Nevertheless, economy demands that a workpiece be machinable to a reasonable degree. Before the concept of machinability can be explored, it is necessary to identify a basic process, that of metal cutting. *Machining* is a generic term, applied to all material removal, while *metal cutting* refers to processes in which the excess material is removed by a harder tool, through a process of extensive plastic deformation or controlled fracture.

8-1 THE METAL-CUTTING PROCESS

The variety of metal-cutting processes is very large; nevertheless, it is possible to idealize the process of chip removal.

8-1-1 Ideal Orthogonal Cutting

As indicated by its name, in *orthogonal cutting* the cutting edge of the tool is straight and perpendicular to the direction of motion (Fig. 8-1a). In the simplest case, the workpiece is rectangular and is of large enough width w for width changes to be neglected (plane strain). Cutting is performed with a tool inclined at a *rake angle* α, measured from the normal of the surface to be machined. To prevent excessive rubbing on the machined surface, the tool is relieved at the back or *flank* by the *clearance angle* θ.

In principle, it makes no difference whether the tool or workpiece is moved. We may visualize a stationary workpiece, with the tool moving at a *cutting speed* v. The tool is set to remove a layer of thickness h. To avoid confusion, this is *not* called the depth of cut, but rather the *undeformed chip thickness* h. In the simplest case, deformation takes place by intense shearing in a plane, the *shear plane*, inclined by the *shear angle* ϕ. The chip thus formed has a thickness h_c. The shear angle ϕ determines the *cutting ratio* r_c

$$r_c = \frac{h}{h_c} = \frac{l_c}{l} \tag{8-1}$$

Frequently, the reciprocal value of r_c, called the *chip-compression factor*, is quoted. Both can be obtained from measured chip thickness or, if the chip is ragged and uneven, from the measured length l_c or, if the width of the chip has changed, from the weight of a chip of measured length.

In practice, the chip is always thicker than the undeformed chip thickness and $r_c < 1$. With a decreasing rake angle, ϕ and, thus, r_c also drop and become particularly low when cutting with a tool of negative rake angle (Fig 8-1b). The value of r_c gives valuable clues regarding the efficiency of the process. From the geometry of the process, the shear angle is defined by

$$\tan \phi = \frac{r_c \cos \alpha}{1 - r_c \sin \alpha} \tag{8-2}$$

Because of constancy of volume, the chip-thickness ratio can also be expressed from the chip velocity v_c and cutting velocity v (Fig. 8-1c)

$$r_c = \frac{v_c}{v} = \frac{\sin \phi}{\cos(\phi - \alpha)} \tag{8-3}$$

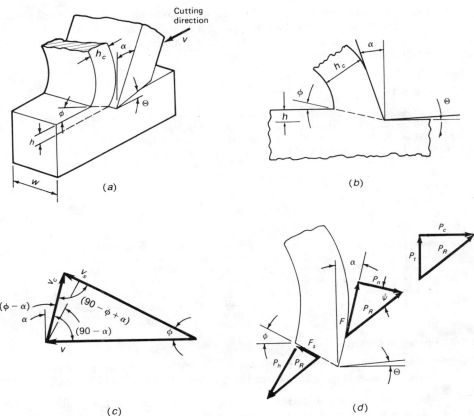

FIGURE 8-1
In orthogonal cutting the cutting edge is perpendicular to the direction of motion. The rake angle may be (a) positive, zero, or (b) negative. (c) Velocities and (d) forces acting on the chip, tool, and tool holder are readily resolved.

thus, with increasing shear angle ϕ, the chip becomes thinner and comes off at a higher speed.

In the ideal case all shear is concentrated in an infinitely thin shear zone. The shear strain γ is (see Fig. 8-2a)

$$\gamma = \frac{AB}{CD} = \frac{AD}{CD} + \frac{DB}{CD} = \tan(\phi - \alpha) + \cot \phi \tag{8-4}$$

and, for an infinitely thin shear plane, the shear strain rate $\dot{\gamma}$ would reach infinity. In reality, the shear plane has some finite thickness Δy, typically 0.03 mm (0.001 in), and the shear strain rate can be calculated from

$$\dot{\gamma} = \frac{v_s}{\Delta y} = \frac{\cos \alpha}{\cos(\phi - \alpha)} \frac{v}{\Delta y} \tag{8-5}$$

The magnitude of the shear angle is of fundamental importance. For any given undeformed chip thickness, a small angle means a long shear plane and, therefore, a high cutting force and energy; also, a small angle results in a high shear strain; hence, the chip will be heavily strain hardened.

Example 8-1

AISI 4340 steel, heat treated to HB 270, is turned with high-speed steel (HSS) tools of $\alpha = 8°$ rake angle. The undeformed chip thickness is 0.3 mm, depth of cut (chip width) is 1.5 mm, and cutting speed is 0.6 m/s. The chip comes off in a continuous helical form; a 1-m-long chip weighs 5.7 g; the chip width is unchanged at 1.5 mm. Calculate the chip-thickness ratio.

Volume of chip = weight/density = $5.7/7.9 = 0.72$ cm^3. Chip thickness $h_c = 0.72/(100)(0.15) = 0.048$ cm $= 0.48$ mm. Chip-thickness ratio, from Eq. (8-1), $r_c = 0.3/0.48 = 0.625$ (chip compression ratio $= 1.6$).

Example 8-2

In the course of cutting the steel of Example 8-1 the chip changes to a straight type. A repeat of calculations shows that r_c is now 0.5. What is the new shear angle and chip velocity?

Shear angles, from Eq. (8-2): for $r_c = 0.625$, $\tan \phi = 0.625(0.9903)/[1 - (0.625)(0.1392)] = 0.678$ and $\phi = 34°$; for $r_c = 0.5$, $\phi = 28°$.

Chip velocity changed, according to Eq. (8-3), from $v_c = 0.625(0.6) = 0.375$ m/s to $v_c = 0.5(0.6) = 0.3$ m/s. Thus the change in chip shape was accompanied by a decreasing chip velocity.

8-1-2 Forces in Cutting

The magnitude of the shear angle ϕ depends on the relative magnitudes of forces acting on the tool face (Fig. 8-1d). Three views may be taken of the situation:

1 With the aid of a dynamometer (a device containing several load cells to resolve forces acting in mutually perpendicular directions), the external forces acting *on the tool holder* can be measured. In orthogonal cutting there are two forces: the *cutting force P_c* is exerted in the direction of cutting, parallel to the surface of the workpiece, while a *thrust force P_t*, acting perpendicular to the workpiece surface, is necessary to keep the tool in the cut. (At high positive rake angles, the thrust force is negative, and the tool is pulled into the material.)

2 From the point of view of forces acting *on the tool*, the *resultant force P_R* may be regarded as being composed of the normal force P_n acting perpendicular to the tool face and the friction force F acting along the face. Their magnitude may be calculated from measured forces and the rake angle

$$P_n = P_c \cos \alpha - P_t \sin \alpha \qquad (8\text{-}6a)$$

$$F = P_c \sin \alpha + P_t \cos \alpha \qquad (8\text{-}6b)$$

A simple way of interpreting the results would be in terms of a model that postulates sliding of the chip along the rake face of the tool. The magnitude of the friction force F would then determine a *friction angle* ψ for which

$$\tan \psi = \frac{F}{P_n} \quad (= \mu) \tag{8-7}$$

Alternatively, from measured forces,

$$\tan \psi = \frac{P_t + P_c \tan \alpha}{P_c - P_t \tan \alpha} \tag{8-8}$$

This presentation can, however, be misleading because with decreasing friction F often drops less rapidly than P_n and μ may actually rise. Therefore, we call μ the *apparent mean coefficient of friction*.

3 From the point of view of forces acting *on the material*, the resultant force P_R can be resolved into a shear force F_s acting in the plane of shear and a compressive force P_h that exerts a hydrostatic pressure on the material being sheared. We saw in Sec. 4-3-3 that hydrostatic pressure does not affect the flow stress of the material but does delay fracture; thus, in a reasonably ductile material, the chip can form in a continuous fashion even though strains are high. The balance of the two forces depends on the shear angle ϕ. Clearly, with increasingly positive rake angle α, ϕ must increase and the chip must thin out. Friction on the rake face has the opposite effect: it resists the free flow of the chip and reduces ϕ. Thus, in contrast to most other metalworking processes, the geometry of the cutting process depends not only on tool/workpiece geometry but also on the process itself. To find a quantitative relationship, the assumption may be made that the material will choose to shear at an angle that minimizes the required energy.* This leads to

$$\phi = 45° - \tfrac{1}{2}(\psi - \alpha) \tag{8-9a}$$

Another approach based on upper-bound analysis† leads to a qualitatively similar result,

$$\phi = 45° - (\psi - \alpha) \tag{8-9b}$$

Thus, the shear angle decreases and the shear force (and with it the work of cutting) increases with a decreasing rake angle (by approximately 1.5% for each degree change) and an increasing friction angle. One may conclude that favorable conditions, in terms of energy consumption, could be secured by using large positive rake angles and by minimizing friction along the tool face.

---

*H. Ernst and M. E. Merchant, in *Surface Treatment of Metals*, American Society of Metals, New York, 1941, pp. 299–378.
†E. H. Lee and B. W. Shaffer, *Trans. ASME J. Appl. Mech.* **73**:405–413, 1951.

8-1-3 Realistic Orthogonal Cutting

While the above conclusions are qualitatively correct, the models are oversimplified in many ways.

The Cutting Zone In the idealized view, metal cutting may be regarded as proceeding by shear in an infinitely narrow zone (Fig. 8-2a). Reality is different (Fig. 8-2b).

1 Most metals strain harden when deformed. The shear plane broadens into a *shear zone* (usually denoted as *primary shear zone*). The thickness of the shear zone is greater for a more heavily strain-hardening material (larger n) and also for a material of high strain-rate sensitivity (high m). The situation is complicated because the energy expended in shear raises the temperature and reduces the flow stress. In general, with a higher n the shear zone becomes wider and longer, and energy consumption increases. Strain hardening extends ahead of and below the shear zone and the newly formed surface is also strain-hardened.

2 Interface pressure on the tool face is high. We saw in Sec. 4-3-2 (Fig. 4-14) that sliding of the workpiece over the tool surface is arrested (sticking friction sets in) when the product of interface pressure and coefficient of friction exceeds the shear flow stress of the deforming material ($\mu p > \tau_f$). Therefore, the coefficient of friction is really meaningless. Since there is no movement on the tool face, the chip must flow up, in the *secondary shear zone*, over the stationary material found next to the rake face. This intense shearing is a second source of heating. Sliding contact is limited to a short distance where the chip begins to curl away (Fig. 8-2b).

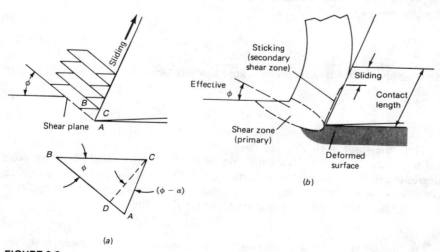

FIGURE 8-2
Chip formation may be visualized as (a) simple shearing but is (b) more complex in reality.

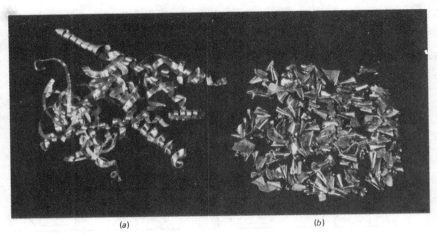

FIGURE 8-3
Chips are (*a*) continuous, straight or helical when cutting ductile materials but (*b*) short, broken up when cutting free-machining metals.

3 Around the tool edge, conditions become even more complex. The workpiece material is upset and *plowed* by the tool edge, and rubbing against the freshly formed metal surface (essentially, a slight ironing) represents a third zone of heat input (sometimes termed the *tertiary shear zone*).

Chip Formation   Realistic metal cutting also differs from ideal cutting in the mode of chip formation.

1 In the ideal case, the shear zone is well defined, primary shear can be assumed to take place on closely placed shear planes (Fig. 8-2*a*), and a *continuous chip* is formed (Fig. 8-3*a*). This situation is approximated under various process conditions:

a At moderately low speeds, in the presence of a cutting fluid which finds access to both the rake and flank faces and acts as a lubricant, the chip slides on the rake face. The newly formed surface is smooth (Fig. 8-4*b*) as is the underside of the chip. The inner side of the chip is jagged, bearing evidence of chip formation by shearing.

b At somewhat higher speeds, heat generation causes a rise in temperatures. Friction increases until sliding at the tool face is arrested, and the system seeks to minimize the energy expenditure by finding an optimum process geometry. It will be recalled that in the processes of indentation (Sec. 4-4-2) and extrusion (Sec. 4-5-1) sticking friction led to the formation of dead-metal zones (Figs. 4-15*a* and 4-37*a*). In metal cutting too, at some intermediate speed, shearing takes place along a nose of stationary material attached to the tool face. This so-called *built-up edge* (BUE) acts like an extension of the tool (Fig. 8-4*c*): Shear takes place along the boundary of the BUE; hence the *effective rake angle* becomes

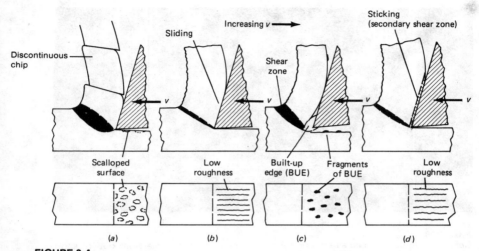

FIGURE 8-4
The process of chip formation changes with cutting speed. When cutting steel, the chip (*a*) is discontinuous at speeds below 2 m/min, (*b*) is continuous and slides on the rake face at 7 m/min, (*c*) forms with a built-up edge at 20 m/min, and (*d*) develops a secondary shear zone at 40 m/min. (*After M. C. Shaw, in Machinability, Spec. Rep. 94, The Iron and Steel Institute, London, 1967, pp. 1–9.*)

quite large and the energy consumption drops. However, a penalty is paid: Dimensional control is lost and, because the BUE becomes periodically unstable, it leaves occasional lumps of metal and damaging cracks behind, and the surface finish is poor. Under certain conditions a small stable BUE may be maintained; this is desirable because it protects the tool without producing an unacceptably poor surface finish.

c With increasing speeds, the material of the BUE heats up and softens, the BUE gradually disappears or, rather, degenerates into the secondary shear zone (Fig. 8-4*d*). The speeds at which the transition to BUE formation and secondary shear zone development occur are indicated in Fig. 8-4 for steel. Similar changes take place when cutting other materials; the critical speeds depend on the temperature reached in the shear zone but are also affected by adhesion between tool and workpiece materials. As in metal deformation processes, temperature effects can be normalized by reference to the homologous temperature scale. Examples of continuous chips are given in Fig. 8-5 with and without BUE formation.

2 Under special conditions, the chip may be continuous yet show a periodic change in thickness. A *wavy chip* (Fig. 8-6*a*) exhibits roughly sinusoidal variations in thickness. Such variations are usually related to *chatter* (vibration) attributable to periodic variations in cutting forces. As in all machines, the imposed forces cause elastic deflections of the workpiece, tool, tool holder, and machine tool. Any variations in forces result in a change of undeformed chip

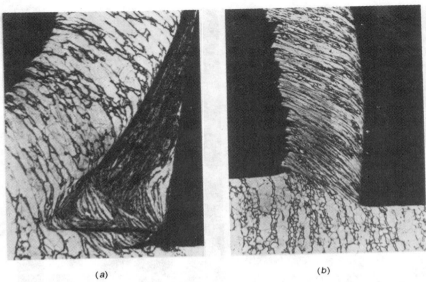

(a)   (b)

FIGURE 8-5
When cutting 60Cu–40Zn brass, chip formation proceeds with (*a*) BUE formation at 30 m/min and (*b*) in a continuous form at 100 m/min. (*Courtesy Dr. P. K. Wright, Carnegie-Mellon University.*)

thickness and hence in a visible and measurable waviness of the machined surface.

a In *regenerative chatter* (*self-excited vibration*) the source of vibration is a change in undeformed chip thickness (from waviness produced in a preceding cut by the presence of a hard spot or other irregularity) or a periodic loss of the BUE. A remedy is usually found in changing the process conditions (speed, feed, workpiece support, tool support).

b Chatter may also originate in *forced vibrations* due to a periodic variation of forces acting within the machine tool (e.g., from a gear box or coupling) or may be transmitted from an external source such as a nearby, vibrating machine tool. Vibration-isolation mountings or moving the offending machine tool eliminate the problem. Interrupted cuts in milling may also set up vibrations and uneven spacing of teeth is then helpful.

3 *Segmented chips* show a sawtooth-like waviness. The thick sections are only slightly deformed and are joined by severely sheared, thinner sections (Fig. 8-6*b*). An extreme form of this is observed in materials of low heat conductivity, such as titanium. The process starts by upsetting ahead of the tool, resulting in a localization of shear. Because heat generated in the shear plane cannot dissipate, the material heats up, weakens, and shears until a chip segment is moved. The process then starts again by upsetting.

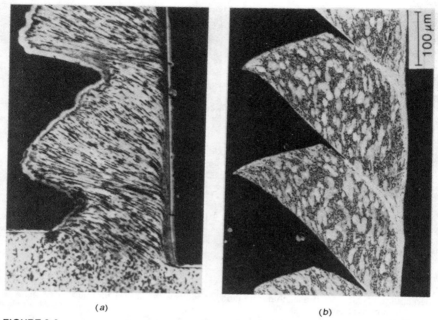

(a)

(b)

FIGURE 8-6
Under some conditions, the chip formed is (a) wavy (AISI 1015 steel, 55 m/min) or (b) segmented (Ti–6Al–4V, 10 m/min) (*Courtesy Dr. R. Komanduri, General Electric Co.*)

4 Under certain conditions *discontinuous chip* forms.

a When ductile materials are cut at very low speeds, severe strain hardening of the material causes upsetting until sufficient strain is accumulated to initiate shear. Elastic members in the system (e.g., the tool holder) allow sudden acceleration and the complete separation of a chip, to be followed by a new upsetting cycle. Cutting forces fluctuate violently; the new surface is torn (Fig. 8-4a) and wavy. High adhesion and low cutting speeds that generate low homologous temperatures favor such chip formation.

b Segmental chips formed at high cutting speeds may also fall apart.

c Discontinuous chip formation is intentionally introduced in some free-machining alloys by the incorporation of inclusions or second-phase particles that serve as stress raisers and cause total separation of tightly spaced chip fragments (Fig. 8-3b). The second-phase particles or inclusions often reduce the shear strength in both the primary and secondary shear zones; thus, cutting forces are low. Because chip separation is facilitated, the surface finish is good and the tendency to chatter is reduced.

The above discussion was based on the assumption of continuous engagement of the cutting edge, as is typical of operations such as turning and drilling. In other processes, most notably milling, cutting is interrupted, with the edge

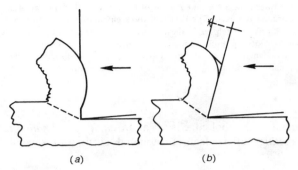

FIGURE 8-7
Chips are forced into tighter curls and break up when (*a*) groove-type or (*b*) obstruction-type chip breakers are used.

emerging from the cut after limited engagement. This has advantages in terms of chip disposal but subjects the tool to impact loading.

Removal of Chips   The short chip produced in cutting free-machining materials (Fig. 8-3*b*) is easily removed from the cutting zone. In contrast, the continuous chip formed when machining ductile materials under stable conditions (Fig. 8-3*a*) is a nuisance: It is difficult to remove; it may clog up the work zone; it may wrap around the workpiece or tool; and it may present danger to the tooling, machine, and operator alike.

A partial remedy is found in *chip breakers*. With some materials, they impart additional strain to the chip, causing it to break into shorter lengths or at least curl up into tight coils that break frequently. At other times, the chip is forced to bend and hit an obstruction such as the tool holder, tool flank, or the workpiece itself. With one end fixed, the chip grows until bending stresses make it snap. Ductile materials always give a continuous chip which must be chopped up or dragged out of the work zone.

A chip breaker may be incorporated into the tool by giving the rake face a curvature, away from the cutting edge (*groove type*, Fig. 8-7*a*) or a separate chip breaker may be attached to the rake face (*obstruction type*, Fig. 8-7*b*). Chip breakers have, in general, little effect on cutting forces but, if moved too close to the cutting edge, they localize heat at the edge and may cause rapid loss of the tool because of overheating. The natural curvature of chips is a function of many process variables; in general, the chip curl radius becomes smaller (the chip is tighter) with increasing h and decreasing speed. Correspondingly, chip breakers work most efficiently at specific undeformed chip thicknesses and speeds.

Example 8-3

Chatter marks are found on the surface of a workpiece being turned. Identify the cause of chatter.
The wavelength of chatter (the distance between consecutive high spots) can be measured. Cutting speed divided by wavelength gives frequency. (*a*) If the frequency agrees with or is a

simple multiple of the rotational frequency of a machine part or of the vibrational frequency of a nearby machine tool, then the vibration is forced. The best remedy is elimination of the source, although changes in cutting conditions are helpful if they increase the cutting forces in the direction of vibration (because heavier loading results in increased stiffness). (*b*) If vibration is not forced, it can often be eliminated by changing the cutting speed, because machine tools have higher stiffness at certain speeds; additionally, conditions may be changed to reduce cutting forces.

8-1-4   Oblique Cutting

In most practical cutting processes, the tool edge is set at an *angle of inclination i* (Fig. 8-8*a*). Such *oblique cutting* differs from orthogonal cutting in several respects:

1 The chip curls into a helical rather than spiral shape and is more readily removed from the work zone. It is normally found that the chip flows, at a velocity v_c, at an angle η_c equal to the angle of inclination *i* (Stabler's rule, Fig. 8-8*b*).

2 The *normal rake angle* α_n is measured in the plane containing the normal to the workpiece surface and the tool velocity *v*. The *effective rake angle* α_e is measured in the plane containing *v* and v_c and is larger than α_n

$$\sin \alpha_e = \sin^2 i + \cos^2 i \sin \alpha_n \qquad (8\text{-}10)$$

Thus, the cutting force is lower than with an orthogonal tool of equal rake angle. In general, for equal effective rake angles, an oblique cutting tool is stronger than an orthogonal one.

FIGURE 8-8
In oblique cutting, the chip flows at an angle over the rake face to form a helix.

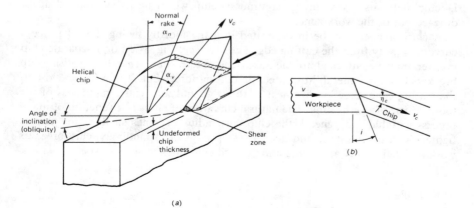

(*a*)

(*b*)

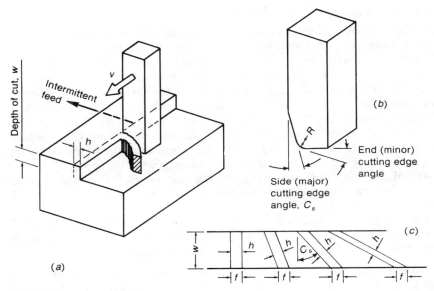

FIGURE 8-9
To machine a large surface, (a) the tool must be given a feed. For easier chip removal (b) the cutting edge is at an angle, which (c) affects the undeformed chip thickness for a given feed.

In many processes the tool edge is not wide enough to take a cut over the whole width of the workpiece and then the surface layer is removed in increments, by feeding the tool across the width of the workpiece. *Feed is the distance between successive engagements of the cutting edge.* In the example shown in Fig. 8-9a, the tool is moved in a straight-line path; a cut is taken during forward movement and the tool is lifted out of contact during the return stroke. The feed f is taken before the next forward movement begins and is equal in magnitude to the undeformed chip thickness h.

The tool geometry shown in Fig. 8-9a is not practical and a number of changes are desirable to allow the chip to flow easily and to prevent damage to the freshly formed surface.

1 The cutting edge is set at an angle in the direction of feed (Fig. 8-9b). An increase in this *side cutting-edge* (*or approach or lead*) *angle* C_S results in a smaller h for the same feed ($h = f \cos C_S$, Fig. 8-9c).

2 The end of the tool is relieved, creating *major* and *minor cutting edges*. The two meet at the *corner* or *nose*; the transition between the cutting edges is *radiused* to obtain a smoother finish.

Further changes may be made, for example to impart a positive or negative rake angle, and the major cutting edge may also be inclined (such a tool will be shown in Fig. 8-25).

8-1-5 Forces and Energy Requirements

There are a number of theories that take a realistic view of the cutting process. Such theories are highly valuable in parametric studies, i.e., in exploring the effects of process variables. For the prediction of forces and energy requirements, there is, however, the problem of determining the relevant flow stress. Large shear strains (in units of natural strain, on the order of 1 to 5) are produced in a narrow shear zone; therefore, even the mean values of strain rates, taken across the zone of intense shearing, reach several thousand per second. The flow stress of most materials is increased by such high strain rates even at cold-working temperatures; however, counteracting this is the large temperature rise which lowers the flow stress. Therefore, only flow stress values determined at the appropriate—and often unknown—temperatures and strain rates are relevant. Even then, predictions of the shear angle and the shear zone width are needed before a reasonable estimate can be made.

For approximate calculations, of sufficient accuracy for all practical purposes, it is usual to calculate forces and power requirements from experimentally determined material constants. Three approaches are usual:

1 The cutting force P_c divided by the cross-sectional area of the undeformed chip gives the nominal cutting stress or *specific cutting pressure* p_c

$$p_c = \frac{P_c}{hw} \quad \left(\frac{N}{m^2}\right) \qquad (8\text{-}11)$$

Note that p_c is not a true stress, even though it has the dimensions of stress.

2 The energy consumed in removing a unit volume of material is called the *specific cutting energy* E_1. Energy (or work) is force P_c multiplied by the distance l over which the force acts. Since the volume of material removed is $V = hwl$, the specific cutting energy may be written as

$$E_1 = \frac{P_c l}{hwl} \quad \left(\frac{J}{m^3} \text{ or } \frac{N}{m^2}\right) \qquad (8\text{-}12)$$

It will be noted that, *when expressed in consistent units*, the numerical values of p_c and E_1 are the same. Since the purpose of calculation is often that of finding the size of the drive motor, E_1 is often given in units of $W \cdot s/m^3$ or equivalent (Table 8-1). For dull tools, E_1 is increased by 25%.

3 The *material removal factor* K_1 is the reciprocal of the specific cutting energy

$$K_1 = \frac{1}{E_1} \quad \left(\frac{m^3}{W \cdot s}\right) \qquad (8\text{-}13)$$

It is convenient because it gives a feel for the amount of material that can be removed in unit time with a drive of unit power.

The above material constants cannot be immediately used for calculations because they are not truly constants but depend also on process parameters such

TABLE 8-1
APPROXIMATE SPECIFIC ENERGY REQUIREMENTS FOR CUTTING*
(Multiply by 1.25 for dull tools.) Undeformed Chip Thickness (feed): 1 mm (0.040 in)

Material	Hardness		Specific energy E_1	
	HB	HRC	hp·min/in³	W·s/mm³
Steels (all)	85–200		0.5	1.4
		35–40	0.6	1.6
		40–50	0.7	1.9
		50–55	0.9	2.4
		55–58	1.5	4.0
Stainless steels	135–275		0.5	1.4
		30–45	0.6	1.6
Cast irons (all)	110–190		0.3	0.8
	190–320		0.6	1.6
Titanium	250–375		0.5	1.4
Superalloys (Ni and Co)	200–360		1.1	3.0
Aluminum alloys	30–150 (500 kg)		0.12	0.35
Magnesium alloys	40–90 (500 kg)		0.08	0.22
Copper		80HRB	0.45	1.2
Copper alloys		10–80HRB	0.3	0.8
		80–100HRB	0.45	1.2
Zinc alloys			0.08	0.22

*Extrapolated from data in *Machining Data Handbook*, 3d ed., Machinability Data Center, Metcut Research Associates, Cincinnati, Ohio, 1980.

as undeformed chip thickness, rake angle, and cutting speed. Undeformed chip thickness is the most powerful factor because the total energy requirement is actually a sum of at least two components:

1 Energy expended in the primary shear zone is proportional to undeformed chip thickness, but so is the amount of material removed. This would make p_c, E_1, and K_1 true material constants.

2 There is, however, additional energy needed to provide the flank friction and plowing forces; since this energy is virtually independent of undeformed chip thickness, it accounts for a larger proportion of total energy when h is small. Thus, the energy required to remove a unit volume of material increases with decreasing undeformed chip thickness. Therefore, material constants such as E_1 must be determined for some agreed-upon h such as $h_{ref} = 1$ mm, and the *adjusted specific cutting energy* E for any other h can then be found from an empirical power law

$$E = E_1\left(\frac{h}{h_{ref}}\right)^{-a} = E_1 h^{-a} \qquad (8\text{-}14)$$

where a ranges from 0.2 to 0.4 and may be taken as 0.3 for most materials. It should be noted that below an undeformed chip thickness of 0.1 mm, the energy requirement increases even more steeply.

The power to be developed by the machine tool can then be estimated if the rate of material removal V_t and the efficiency of the machine tool η (usually around 0.7–0.8) are known:

$$\text{Power (W)} = \frac{EV_t}{\eta} \quad \left(\frac{W \cdot s}{mm^3} \frac{mm^3}{s} \right) \tag{8-15a}$$

or, in conventional units

$$\text{Power (hp)} = \frac{EV_t}{\eta} \quad \left(\frac{hp \cdot min}{in^3} \frac{in^3}{min} \right) \tag{8-15b}$$

The cutting force P_c to be resisted by the tool holder and the machine tool can be calculated by recalling that power divided by speed gives force. If the cutting speed v is in m/s:

$$P_c = \frac{\text{power (W)}}{v} \quad (N) \tag{8-16a}$$

or, if the cutting speed v is in units of ft/min:

$$P_c = \frac{33\,000 \text{ power (hp)}}{v} \quad (lbf) \tag{8-16b}$$

Alternatively, the force can be determined from Eq. (8-11). To a first approximation, the thrust force P_t may be taken as one-half of P_c when cutting with zero or low positive rake tools; with increasingly positive rake angle the thrust force diminishes and, in the extreme, the tool is pulled into the workpiece.

Example 8-4

As discussed in Example 8-1, a 4340 steel bar of HB 270 hardness is cut at a speed of 0.6 m/s (120 ft/min). The undeformed chip thickness is 0.3 mm and the width of the chip is 1.5 mm. Calculate the power requirement and cutting force.

From Table 8-1, $E_1 = 1.6$ W·s/mm³. Hence, the adjusted specific cutting energy is, from Eq. (8-14), $E = 1.6(0.3)^{-0.3} = 2.3$ W·s/mm³. The rate of material removal is simply chip cross section multiplied by cutting speed: $V_t = 0.3(1.5)(600) = 270$ mm³/s. Power, from Eq. (8-15a): $2.3(270)/0.7 = 887$ W or, for dull tools, $887(1.25) = 1110$ W ($= 1.5$ hp). Cutting force, from Eq. (8-16a): $1110/0.6 = 1850$ N ($= 416$ lbf).

8-1-6 Temperatures and Their Control

The energy expended in machining is concentrated in a very small zone. Only a small fraction of it is stored in the workpiece and chip in the form of increased dislocation density, and the vast majority of energy is converted into heat.

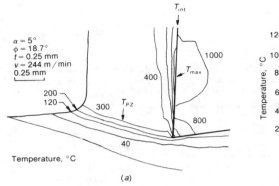

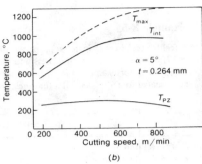

FIGURE 8-10
(a) Calculated temperature distribution in chip and tool and (b) variation of temperature with cutting speed when cutting AISI 1016 steel with a carbide tool. (*After A. O. Tay, M. G. Stevenson, G. DeVahl Davies, and P. L. B. Oxley, Int. J. Mach. Tool Des. Res. 16:335–349 (1976). Reprinted with permission of Pergamon Press, Oxford.*)

Temperatures Because the cutting zone keeps moving into the workpiece, there is little heating ahead of the tool and, at least at high cutting speeds, most of the heat (over 80%) is carried away by the chip. However, the tool is in continuous contact with the chip and, in the absence of an effective heat-insulating layer, the rake face of the tool heats up. Rubbing on the rake face (or deformation in the secondary shear zone) is also a substantial source of heating. Detailed calculations show that the maximum temperature is developed at the rake face some distance away from the tool nose but before the chip lifts away (Fig. 8-10). As would be expected, both the maximum (T_{max}) and average interface (T_{int}) temperatures increase with increasing cutting speed.

A rough estimate of temperatures may be obtained by dimensional analysis, assuming that all energy (Eq. (8-14)) is converted into heat.* Then the mean tool-face temperature T_T is

$$T_T = E\left(\frac{vh}{k\rho c}\right)^{1/2} \tag{8-17}$$

where k is heat conductivity, ρ density, and c specific heat (heat content per unit mass) of the workpiece material. Thus, higher temperatures are to be expected in cutting stronger materials (higher E) at higher speeds, especially if the workpiece material is a poor heat conductor, of low density, and low specific heat. Materials

*M. C. Shaw, *Metal Cutting Principles*, Oxford University Press, Oxford, 1984.

such as titanium and superalloys are difficult to machine, whereas aluminum and magnesium are easy.

Example 8-5

A feel for the effect of material properties on permissible cutting speeds may be obtained from Eq. (8-17) if undeformed chip thickness and temperature rise are kept constant for a given tool (say HSS). Take, for example, the Ni-based superalloy IN-100 of TS = 1000 MPa and the steel 4140 heat treated to TS = 1000 MPa (hardnesses are about equal and around 300 HB).

	IN-100	4140	Source
E_1 (W·s/mm³)	3.0	1.6	Table 8-1
k, at 500° C (W/m·K)	17	37	Metals Handbook,
ρ (g/cm³)	7.75	7.85	9th ed., vol. 1, pp. 148,
c at 500 °C (J/kg·K)	480	520	149; vol. 3, p. 243

Substituting

$$\Delta T = \text{const.} = 3.0\left[\frac{v_{\text{super}}}{(17)(7.75)(480)}\right]^{1/2} = 1.6\left[\frac{v_{\text{steel}}}{(37)(7.85)(520)}\right]^{1/2}$$

$$v_{\text{steel}} = \frac{1\,359\,306}{161\,894}v_{\text{super}} = 8.4\,v_{\text{super}}$$

(From Fig. 8-44, $v_{\text{steel}} = 0.45$ m/s; from Fig. 8-45, $v_{\text{super}} = 0.06$ m/s; the ratio is $0.45/0.06 = 7.5$, in reasonable agreement with our calculation.)

Action of Cutting Fluids Some cutting operations are performed *dry*, i.e., without the application of a *cutting fluid* or, as sometimes called, a *coolant*. In the majority of instances it is, however, essential that a fluid be applied to the cutting zone. The fluid fulfills basically three major functions:

1 *Lubrication.* Access of the fluid to the rake face is difficult, especially at higher cutting speeds. The fluid does, however, enter the sliding zone, and some fluid may seep in from the sides of the chip too. Effects attributable to lubrication can frequently be observed, especially when contact with the cutting tool is intermittent. In low-speed cutting with sliding friction, rake-face friction is reduced; therefore, the shear angle increases, the chip becomes thinner and curls more tightly, and power consumption drops. At speeds where a BUE forms in the absence of a lubricant, the onset of BUE formation is shifted to higher speeds. At higher speeds, where a sticking zone develops, the length of the sticking zone is reduced.

At all speeds, lubricant access to the flank face is possible and rubbing is reduced. The combined effect is that, in general, surface finish also improves.

2 *Cooling.* Because shear is highly concentrated and the shear zone moves extremely rapidly, temperatures in the shear zone are not affected. However, a cutting fluid reduces the temperature of the chip as it leaves the secondary shear zone, and it cools the workpiece. It may also reduce the bulk temperature of the tool. While relationships are by no means straightforward, it is often found that a cutting fluid reduces temperatures sufficiently to allow cutting at higher speeds.

3 *Chip removal.* Cutting fluids fulfill an additional, and sometimes extremely important function: they flush away chips from the cutting zone and prevent clogging or binding of the tool.

Cutting Fluids Cutting fluids fall into two main categories.

1 *Cutting oils* are based on mineral oils with appropriate additives, and are used mostly at lower speeds and with high-speed steel (HSS) tooling.

2 *Water-based (aqueous) fluids* may be emulsions (oils dispersed in water with the aid of surface-active substances), semisynthetic fluids (also called semichemical fluids or chemical emulsions, in which large quantities of surface-active agents are used to reduce oil particle size to the point where the fluid becomes translucent or transparent), or synthetic fluids (also called chemical fluids, which contain no oil, only water-soluble wetting agents, corrosion inhibitors, and salts).

Because of intimate tool-workpiece contact, high temperatures, and the danger of wear, many cutting fluids contain boundary and EP agents. Some typical fluids are listed in Table 8-2.

Application of Cutting Fluids The method of applying cutting fluids is as important as their selection.

Manual Application The application of a fluid from a squirt can or in the form of a paste (for low-speed operations) is commonly practiced even though it is not really acceptable even in job-shop situations.

TABLE 8-2 COMMONLY USED MACHINING FLUIDS*

Process	Tool	Steel (< 275 BHN)	Steel (> 275 BHN)	Stainless steel, nickel alloy	Cast iron	Aluminum alloy	Magnesium alloy	Copper alloy
Turning	HSS	O1, E1 C1	O2, E2, C2	O2, E2, C2	E1, C1	E1, C1, Sp	O1, Sp	E1, C1, Sp
	Carbide	D, E1, C1	D, E1, C1	D, E1, C1	D, E1, C1	D, E1, C1	O1, Sp	E1, C1
Milling	HSS	O1, E1, C1	O2, E2, C2	O2, E2, C2	E1, C1	D, O1, Sp	O1, Sp	E1, C1, Sp
Drilling	Carbide	D, E1, C1	O1, E1, C1	O2, E1, C1	D, E1, C1	D, O1, Sp	O1, Sp	E1, C1
Form turning	HSS	O2, E2, C2	O2, E2, C2	O2, E2, C2	E1, C1	E1, C1, Sp	O1, Sp	E1, C1, Sp
	Carbide	D, E1, C1	E2, C2	O2, E2, C2	D, E1, C1	D, E1, C1	O1, Sp	E1, C1
Gear shaping	HSS	O2, E2, C2	O2, E2, C2	O2, O3	E1, C1	O1, Sp	D, O1, Sp	O1, Sp
Tapping	HSS	O1, E2, C2	O2, E2, C2	O2, O3	E1, C1	D, O1, Sp	D, O1, Sp	O1, Sp
Broaching	HSS	O2, E2, C2	O2, E2, C2	O2, E2, C2	E2, C2	E1, C1, Sp	D, O1, Sp	E1, C1, Sp
	Carbide	O1, E1, C1	O1, E1, C1	O1, E1, C1	D, E1, C1	D, E1, C1	D, O1, Sp	E1, Cl, Sp
Grinding		O1, E1, C1	O2, E1, C1	O2, E2, C2	E1, C1	O1, Sp	O1, Sp	O1, Sp

*From J. A. Schey, *Tribology in Metalworking: Friction, Lubrication and Wear*, American Society for Metals, Metals Park, Ohio, 1983. Code: D—Dry. O1—Mineral oil or synthetic oil. O2—Compounded oil. O3—Heavy-duty compounded oil. E1—Mineral-oil emulsion. E2—Heavy-duty (compounded) emulsion. C1—Chemical fluid or synthetic fluid. C2—Heavy-duty (compounded) chemical or synthetic fluid. Sp—Specially formulated fluid, with boundary and / or E.P. additives.

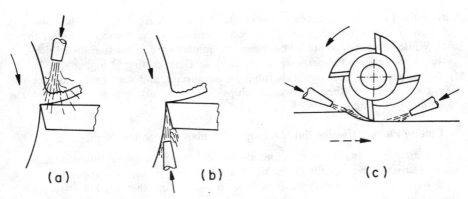

FIGURE 8-11
Cutting fluids are usually applied (*a*) to the chip although (*b*) better cooling is obtained by applying it to the flank face. In milling, (*c*) an additional jet removes the chip. (*From J. A. Schey, Tribology in Metalworking, American Society for Metals, 1983, p. 633. With permission.*)

Flooding   Most machine tools are equipped with a recirculating system that incorporates filters. The fluid is applied at a rate of up to 15 L/min for each simultaneously engaged cutting edge. For convenience, the tool is usually flooded from the chip side (Fig. 8-11*a*), although better cooling is secured by application into the clearance crevice (Fig. 8-11*b*), especially when the fluid is supplied under a pressure of 300 kPa (40 kpsi) or more. A second nozzle may be necessary to clear away the chips in some operations (Fig. 8-11*c*). Flow rates in drilling are typically 5-L/mm drill diameter. However, fluid access to the cutting edges is limited and chip removal is difficult.

Coolant-Fed Tooling   There are drills and other tools available in which holes are provided through the body of the tool so that pressurized fluid can be pumped to the cutting edges, ensuring access of fluid and facilitating chip removal.

Mist Application   Fluid droplets suspended in air provide effective cooling by evaporation of the fluid, although separate flood cooling of the workpiece may be required. Measures must be taken to limit airborne mist, for example, by the use of demisters.

8-1-7   Tool Life

In deformation processes (Chaps. 4 and 5) tool lives are measured in thousands of parts or in weeks or hours of operation, and concern over wear is often overshadowed by considerations of die pressure or material flow. In contrast, tool wear is the dominant concern in metal cutting. This is not surprising since the tool of relatively small mass is exposed to high pressures and temperatures and often also to shock loading. Tool lives on the order of tens of minutes (and, in high-cost machine tools, 5–10 min) are common, and the economy of the process is controlled very largely by *tool life*.

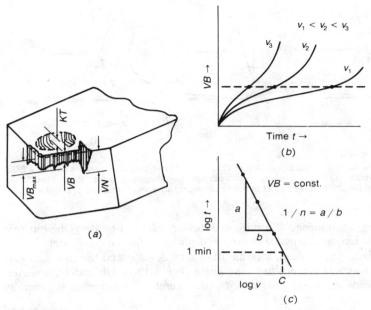

FIGURE 8-12
Flank and crater wear (*a*) may be characterized by the dimensions shown. (*b*) From the progression of flank wear, (*c*) the tool-life constants *C* and *n* may be extracted.

Tool Wear As might be expected, *tool wear* can take several forms (Fig. 8-12) and all wear mechanisms discussed in Sec. 2-2-1 may play a role.

1 *Flank wear*. Intense rubbing of the clearance face of the tool over the freshly formed surface of the workpiece results in the formation of a wear land. The rate of wear can be characterized by interrupting the cut and measuring the average width of the wear land *VB* (Fig. 8-12*a*). After rapid wear during the first few seconds, wear settles down to a steady-state rate only to accelerate again toward the end of tool life (Fig. 8-12*b*). Flank wear is due usually to both abrasive and adhesive mechanisms, and is generally undesirable because dimensional control is lost, surface finish deteriorates, and heat generation increases. It is, nevertheless, the normal wear mode.

2 *Notch wear*. A notch or groove of *VN* depth often forms at the depth-of-cut line where the tool rubs against the shoulder of the workpiece (Fig. 8-12*a*). Abrasion by surface layers is often accelerated by oxidation or other chemical reactions. In the limit, notch wear may lead to total tool failure.

3 *Crater wear*. The high temperatures generated on the rake face (Fig. 8-10*a*) combine with high shear stresses to create a crater some distance away from the tool edge. Wear is usually quantified by measuring the depth *KT* or the cross-sectional area of the crater perpendicular to the cutting edge. Crater wear progresses

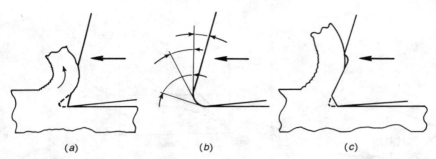

(a)   *(b)*   *(c)*

FIGURE 8-13
The shape of the tool edge has significant effects: (*a*) groove-type chip breakers promote BUE formation, (*b*) rounding by wear results in increasingly negative rake angles toward the flank face; (*c*) a small negative rake ground at the nose encourages the formation of a stable BUE.

linearly under the influence of abrasion, adhesion followed by dragging out tool material, diffusion, or thermal softening and plastic deformation. Crater wear in itself is not damaging; indeed, a stable BUE may develop and the tool then acts as though it had a larger positive rake angle (Fig. 8-13a). Ultimately, however, crater wear leads to catastrophic edge failure; therefore, crater wear is generally avoided.

4 *Edge rounding.* The major cutting edge may become rounded by abrasion. Cutting then proceeds with an increasingly negative rake angle toward the root of the cut (Fig. 8-13b). When the undeformed chip thickness is small, cutting action may cease and all energy may be expended in plastic or elastic deformation. At high cutting speeds (high temperatures) the tool edge may deform plastically; the nose of HSS tools may be entirely lost. Problems with edge rounding may be avoided, at least when hard tools are used, by grinding a double rake (Fig. 8-13c) so that cutting proceeds with a stable BUE.*

5 *Edge chipping.* This may be caused by periodic break-off of the BUE or when a brittle tool is used in interrupted cuts. Surface finish suffers and the tool may finally break.

6 *Edge cracking.* Thermal fatigue may cause cracks to form parallel or perpendicular to the cutting edge of brittle tools (*comb cracks*).

7 *Catastrophic failure.* Tools made of more brittle materials are subject to sudden failures (breakage). This is a problem of all brittle materials such as ceramics and cemented carbides, especially in interrupted cuts. Improved tool manufacturing processes, zero or negative rake, and selection of the proper machining conditions all help.

Cutting fluids are designed to extend tool life, although under certain conditions (chemical reactions, thermal stressing) they may shorten it.

---

*T. Hoshi, in *Cutting Tool Materials*, American Society for Metals, Metals Park, Ohio, 1981, pp. 413–426.

Tool Life Criteria Tool life affects the choice of tool, process conditions, economy of operation, and the possibility of automation and computer control. Unfortunately, no simple definition of tool life is possible: Tool life must be specified with proper regard to the aims of the process. Thus, in finishing operations surface quality and dimensional accuracy are most important; in roughing, greater deterioration of surface quality and dimensional accuracy may be tolerated in exchange for high metal removal rates; an absolute limit is reached when cutting forces increase to high enough values to cause tool fracture.

All these considerations are usually translated into some easily measurable values. Most frequently, flank wear VB or VB_{max} (Fig. 8-12a) is specified as the end of useful tool life:

HSS tools, roughing	$VB_{max} = 1.5$ mm
finishing	$VB = 0.75$ mm
Carbide tools	$VB = 0.4$ mm (or $VB_{max} = 0.7$ mm)
Ceramic tools	$VB_{max} = 0.6$ mm

Other criteria include a specified crater wear, total loss of the tool edge or nose, or total (flank and crater) wear volume.

Tool life is usually given as the time (in minutes) it takes to reach the specific wear criterion under specified process conditions (speed, feed, depth of cut). For tools such as drills and taps a more practical measure is the number of holes drilled or tapped under specified conditions.

Prediction of Tool Life Even though various wear mechanisms come into play, gradual wear is produced by temperature-dependent mechanisms (even abrasive wear is accelerated by temperature because the strength and abrasion resistance of the tool drops at high temperatures). We saw that temperatures are greatly affected by cutting speed (Eq. (8-17)), and it is known that gradual wear is a function of rubbing distance which, for a given cutting speed, is proportional to time. It is to be expected then that, for a given tool life criterion such as flank wear, tool life should drop as a function of speed. It was first observed by Taylor[*] that the relation follows a power law

$$vt^n = C \tag{8-18}$$

where v is the cutting speed (m/min or ft/min), t is tool life (min), and C is the cutting speed for a tool life of 1 min. Strictly speaking, the equation should be

---

[*]F. W. Taylor, *Trans. ASME*, **28**: 31–279, 1907

written as

$$vt^n = Ct_{ref}^n \qquad (8\text{-}18')$$

where $t_{ref} = 1$ min.

Accordingly, there is a straight-line relation when tool life is plotted against speed on log-log paper (Fig. 8-12c). Since metal cutting is a complex system, the constants also depend on a number of variables. Nevertheless, C is basically a constant for a given workpiece material whereas the *Taylor exponent n* (not to be confused with the strain-hardening exponent of Eq. (4-4)) is characteristic of the tool material. Its value is typically 0.1 for HSS, 0.25 for cemented carbides, 0.3 for coated carbides, and 0.4 for ceramic tools.

A better feel for the importance of the Taylor exponent is gained by rearranging the formula to express tool life:

$$t = \frac{K}{v^{1/n}} \qquad (8\text{-}19)$$

It will be noted that for $n = 0.1$, tool life decreases extremely rapidly with the tenth power of speed.

Heat generation is affected by the total heat input (or energy input), which increases with undeformed chip thickness h and chip width (or depth of cut) w. The Taylor formula can be extended to take these into account:

$$t = \frac{K}{v^{1/n_1} f^{1/n_2} w^{1/n_3}} \qquad (8\text{-}20)$$

where, in general, $n_1 < n_2 < n_3$. For example, typical values for HSS are $n_1 = 0.1$, $n_2 = 0.17$, $n_3 = 0.25$. Therefore, for increased material removal rates, it is preferable to increase first the depth of cut, then feed, and only last, speed. Of course, when tool life is limited by catastrophic tool failure, the Taylor equation must be replaced by a statistical life criterion. Even when the Taylor equation is used, the statistical distribution of tool lives must be taken into account, especially, if the equation is used to program tool changes under unattended automatic (computer) control.

An approximate value of C may be obtained by finding the recommended cutting speed v_s for the workpiece material (e.g., from Figs. 8-44 or 8-45) and multiplying it by 1.75 for HSS tools and by 3.5 for carbide tools (see Ex. 8-7 in Sec. 8-7).

8-1-8 Surface Quality

Machining aims to create a part of a given geometry, to specified dimensions and dimensional tolerances. To permit proper function of the part, the surface finish (Sec. 2-4-5) is also specified. Beyond these geometrical considerations, it is also important that the surface produced should be free of defects such as cracks, have no harmful residual stresses, and not be subjected to undesirable metallurgical changes. These are particularly important aspects when the part operates in a hostile environment, is subject to fatigue loading, or when its failure could have catastrophic consequences. With the growth of such critical applications, particu-

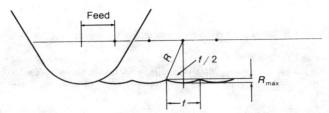

FIGURE 8-14
An ideal roughness value may be calculated from the geometry defined by the nose radius.

larly in the aerospace industries, the term *surface quality* has acquired a complex meaning.

Surface Roughness The surface formed in simple orthogonal (Fig. 8-1a) or oblique (Fig. 8-8a) cutting is, ideally, perfectly smooth (roughness is zero). When a tool of R radius is moved by the feed f between successive cuts (Fig. 8-9), the *ideal transverse roughness* can be calculated approximately by considering the geometry (Fig. 8-14). The peak-to-valley height is

$$R_{max} = \frac{f^2}{8R} \tag{8-21a}$$

The arithmetic average for a triangular roughness is $R_a = R_{max}/4$; hence,

$$R_a \approx \frac{f^2}{32R} \tag{8-21b}$$

The longitudinal roughness will still be zero. Similar relationships can be developed for other processes.

Superimposed on the ideal roughness are features introduced by the chip-forming process itself. This results in a measurable roughness in the longitudinal direction and a modification of surface profile (and hence of roughness values) in the transverse direction. Several features may be observed:

1 In cutting at very low speeds and typically also with all discontinuous chip formation, the surface is scalloped (Fig. 8-4a) and cracks may develop transverse to the cutting direction.

2 In cutting with an unstable BUE, heavily strain-hardened fragments are welded to the surface, covering some 5–10% of it (Fig. 8-4c).

3 When a continuous chip is formed without a BUE, the surface configuration comes close to the ideal one, even though localized wear or chipping of the tool edge gives some roughness increase in the transverse direction (Fig. 8-4b and d).

4 Chatter introduces a periodic variation of surface geometry which is readily visible and shows up as waviness on a recorded longitudinal trace (Fig. 2-25).

5 The surface finish changes in the course of cutting and, in general, deteriorates with the progression of wear. Indeed, tool life is sometimes specified as the time for which an acceptable finish is produced.

Example 8-6

A bar of free-machining steel (HB 200) is to be finish turned to $R_a = 1.6$ μm with a carbide tool. Suggest the appropriate cutting conditions.

We shall see (Fig. 8-44 and Table 8-5) that a suitable feed would be $f = 0.38/2 = 0.19 = 0.2$ mm. From Eq. (8-21b), $R = f^2/32(R_a) = 0.22/32(0.0016) = 0.78$ mm. This is the nose radius for ideal roughness; to allow for some roughening due to chip formation, a tool of minimum 1-mm nose radius should be used.

Surface Integrity The term *surface integrity* has been introduced to indicate the absence of undesirable features on the surface as well as in the subsurface region of the workpiece.

1 Strain hardening of a surface layer is a natural consequence of chip formation (Fig. 8-2b). A residual stress may also be generated which is, most of the time, compressive and is thus beneficial.

2 Cracks formed in low-speed cutting are harmful, as are those sometimes found when cutting with an unstable BUE.

3 Cutting of heat-treatable steels at high speeds can result in heating above the transformation temperature. As the tool leaves the heated zone, the cold mass of the workpiece quenches the surface at a high enough rate for martensite to form. Such transformed surfaces are resistant to attack by common etching agents and are, therefore, referred to as white layers. Since untempered martensite is very hard and brittle, cracks are often formed, if not during machining then in service. The danger is more acute when machining quenched-and-tempered steels. The problem is aggravated when excessive tool wear gives large flank-friction forces.

Some aspects of surface integrity can be evaluated only by destructive techniques (metallography), whereas others can be explored under the microscope, particularly, SEM. On the basis of such tests, cutting conditions that ensure good surface integrity can be specified. For the most critical applications, NDT techniques—including x-ray analysis for residual stresses—are employed.

8-2 WORK MATERIAL

The discussion of the metal cutting process in Sec. 8-1 made it clear that the response of metals must depend on the process itself. Thus, machinability is a system property and no general ranking of materials is possible. Nevertheless, it is customary to speak of *machinability* as a material property, and in the most general sense a material is highly machinable when satisfactory parts can be made from it at low cost, with minimum difficulty.

8-2-1 Machinability

A closer definition of machinability requires that quantitative judgements be made. There are several possibilities.

1 A *machinability index* is often quoted, which is an average rating stated in comparison with a reference material: for steels, a free-machining Bessemer steel B1112, very similar to the present AISI 1212 steel; for copper-based alloys, a leaded free-machining brass; and for aluminum alloys, 7075-T6 aluminum. The system can be misleading because the ranking is different for different processes.

2 A more quantitative measure is *tool life* to total failure by chipping or cracking under specified conditions. Specifications are given as the cutting speed for a given tool life in minutes or seconds, or as the volume of material removed for a given tool life criterion.

3 Another measure is *tool wear*. This can be related to the gradual wear of the flank face or development of the crater. It is given as the change in the dimension of the machined part due to wear, per unit time for a given cutting speed and feed, or as the time required for a standard flank-land wear to develop. In other cases crater depth is specified.

4 Another quantitative measure is *surface finish* produced at standardized cutting speeds and feeds.

Since machinability is a system property, all parts of the system must be well defined if reproducible data are to be obtained. The principles of such testing are laid out in ISO Standard 3685-1977 on "Tool-life testing with single-point turning tools." Evaluation is based on tool wear. Wear is quoted as a function of time when testing at a single speed, and as tool wear–time curves (or Taylor constants) when testing at several speeds. Full evaluation is time-consuming and expensive. Some shortened tests are also available, although they tend to have limited validity.

8-2-2 Machinable Materials

Since machinability is such a many-faced property, it is influenced by a number of material properties. Good machinability may mean one or more of the following: cutting with minimum energy, minimum tool wear, good surface finish. This means that:

1 A material of low ductility is desired, so that chip separation occurs after minimum sliding and the chip breaks up easily. This is exactly the opposite of what one looks for in plastic deformation (Sec. 4-3-3); thus, desirable properties now include a low strain-hardening exponent (n), a low resistance to void formation and thus a low reduction in area (q), and a low fracture toughness.

2 To minimize cutting energy, the shear strength or—what is more practicably measured—the strength (TS) and hardness of the material should be low.

3 A strong metallurgical bond between tool and workpiece, usually expressed as adhesion (Sec. 2-2-1), is undesirable when it also promotes diffusion and weakening of the tool material by depletion of alloying elements. When diffusion does not take place, high adhesion helps to stabilize the secondary shear zone.

4 Very hard compounds (such as some oxides, all carbides, many intermetallic compounds, and elements such as silicon) embedded in the workpiece material act as cutting tools themselves and accelerate tool wear. They are particularly damaging when in the form of platelets with sharp edges.

5 Second-phase particles that are soft or softened at the high temperatures reached in the shear zone are beneficial because they promote localized shear and contribute to chip breaking, making the material free-machining. Because of their low shear strength, such inclusions also reduce the energy expended in the secondary shear zone, and some even act as internal lubricants by smearing on the rake face. Thus, cutting force and energy also decrease.

6 High thermal conductivity is helpful in keeping cutting temperatures low (Sec. 8-1-6).

7 A low melting point of the workpiece material means that cutting temperatures will also remain low, below the temperatures at which the tool softens or reacts with the workpiece.

The above properties must be examined over a range of temperatures. A higher temperature lowers the shear strength of the material, thus making possible the machining of some very difficult materials. Indeed, in special cases metal removal rates can be greatly increased by localized heating of the workpiece just ahead of the cutting zone. To prevent dissipation of heat, the rate of heat input must be high, usually provided by induction heating or with the aid of a plasma torch or laser. High temperatures do, however, have the undesirable side effect of increased adhesion and accelerated diffusion, and tool life can drastically drop. If this is the case, every effort is made to keep the work zone cool with large quantities of cutting fluid.

Some of the requirements are seldom satisfied simultaneously. Some of the most ductile materials favored for plastic deformation are difficult to machine because of their ductility. Even more difficult are the ductile but also high-strength materials. Two-phase materials are often desirable because ductility is impaired by the presence of platelike or, in general, sharp second-phase particles, especially if they are also brittle and of low strength. In many instances it is economical to bring the material into a more machinable condition through metallurgical control (usually by a heat treatment) and then heat treat it again after machining to impart the required service properties.

8-2-3 Ferrous Materials

The full range of machinability is encountered in ferrous materials.

Carbon Steels The term *plain carbon steel* represents a great variety of materials, ranging from very-low-carbon iron to hypereutectoid steel. These steels are commercially available in three different forms (Fig. 8-15):

1 In the fully annealed condition; strength increases while ductility decreases with increasing amounts of carbide present in the lamellar pearlitic form.

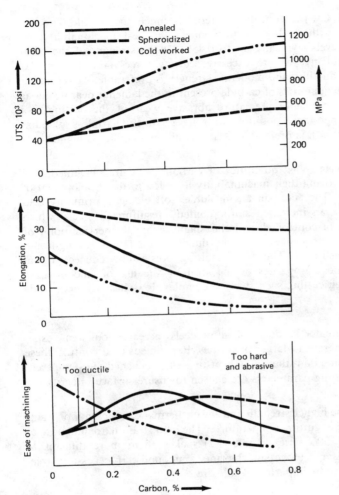

FIGURE 8-15
Carbon steels, like many other materials, are most machinable when brought into a condition
that gives minimum strength combined with minimum ductility.

2 Heat-treated to bring the carbide into a spheroidal form; a spheroidized
steel has lower strength and higher ductility (consider Fig. 3-11).
3 Cold-worked (usually cold drawn); strength is higher and ductility is depressed, while surface finish and tolerances are improved.

On this basis, one can readily choose the optimum treatment that ensures the
best machinability for a given carbon content (Fig. 8-15). At low carbon levels

(typically below 0.2% C), the annealed material is much too ductile, and the cold-worked material with its low ductility offers the best machinability. At intermediate carbon levels (typically up to 0.45% C) the strength of the cold-worked material would give rise to excessive cutting forces and the lamellar pearlite with its lower ductility and moderate strength is preferable. At yet higher carbon levels, the large quantities of carbide present in the lamellar pearlite act as minute cutting tools and cause premature abrasive wear of the cutting tool proper. Thus, the spheroidal condition with its relatively harmless globular carbides and lower strength is preferable even though the ductility is higher.

Free-Machining Steels Vast quantities of carbon steels are machined, and efforts directed at improving their machinability have led to the development of free-machining grades. They contain an insoluble, soft element, primarily lead (leaded steels) or have an increased sulfur content (resulfurized steels) which forms MnS inclusions of controlled, globular shape. From the service point of view, an undesirable consequence is reduced ductility and fatigue strength and slightly reduced tensile strength. The wear of cutting tools can be reduced without impairing the mechanical properties of the steel by the use of calcium as a deoxidizing agent; when cutting such steels, a complex, low-shear-strength oxide forms on the rake face.

Alloy Steels The greater hardness of alloy steels increases tool wear, especially if carbides are present in larger quantities. For dimensional control, these steels are often machined in the fully heat-treated (quenched-and-tempered) condition, and then cutting parameters are chosen to ensure surface integrity.

Stainless Steels The higher strength and lower thermal conductivity of stainless steels results in higher cutting temperatures. The high strain-hardening rate of austenitic steels (AISI 300 series, Table 4-2) makes them more difficult to machine. Cutting fluids must contain chlorine compounds. If necessary, free-machining properties can be imparted by alloying.

Cast Irons The presence of primary cementite makes white cast irons very difficult to machine, and white zones (chill zones) in graphitic cast irons are responsible for much tool wear and breakage.

The machinability of graphitic cast irons is a function of graphite shape and distribution and of the microstructure of the matrix.

1 Gray irons are basically free-machining because the graphite lamellae break up the chip. However, the machined surface is rough because graphite particles break out. Refining the graphite particle size improves the finish without impairing the free-machining properties. Tool life decreases with an increasing proportion of pearlite in the matrix and is lower for finer pearlite. The same factors also contribute to increased hardness, hence machinability decreases with increasing hardness. Gray irons are often cut dry because the fine chips clog filters.

2 Nodular cast iron is more ductile and stronger but, surprisingly, can give a longer tool life.

8-2-4 Nonferrous Materials

In keeping with the convention adopted in Chaps. 3 and 4, nonferrous materials will be discussed in order of increasing melting point.

Low-Melting Materials Only zinc alloys are machined in significant quantities. Their low strength and limited ductility make them highly machinable.

Magnesium Alloys The low ductility imparts free-machining properties, making magnesium a highly machinable material. Finely divided chips ignite spontaneously; therefore, finish cutting with chip thicknesses below 25 μm is always done with an oil-based cutting fluid.

Aluminum Alloys Pure aluminum and its ductile alloys are best machined in the cold-worked condition because their high ductility makes them "draggy" in the annealed condition: Cutting forces are higher than would be expected from their hardness, and the high adhesion leads to a poor surface finish. Precipitation-hardened alloys can be readily machined in the fully heat-treated (solution-treated and aged) condition in which their ductility is low yet their strength is not unduly high. The high thermal conductivity and low melting point allow high machining speeds even with HSS tools, provided that a cutting fluid—which contains boundary-lubricating additives—is applied in a flood. Free-machining properties may be imparted by the addition of lead, bismuth, or tin. Castings that contain elementary silicon give rapid tool wear and must be cut with very hard tools.

Beryllium Beryllium is easily machinable, dry.

Copper-Based Alloys Pure copper, like pure aluminum, is best machined in the cold-worked condition. This applies also to most single-phase alloys which, nevertheless, can often be cut with less energy than pure copper. Chip disposition is difficult. In contrast, $\alpha + \beta$ brasses machine very well. Free-machining additions, usually lead, make all brasses more machinable, and the leaded $\alpha + \beta$ brass serves as a reference base in the machinability scale (Sec. 8-2-1). Free-machining coppers contain lead, sulfur, or tellurium; the chip may still be continuous but cutting force is greatly reduced and surface finish is improved.

Nickel-Based Alloys and Superalloys For lower ductility, it would be desirable to cut these alloys in the cold-worked or fully heat-treated condition. However, their high adhesion and low thermal conductivity is often combined with high strength, and this dictates their cutting in the annealed or overaged condition. Sulfur must be avoided in cutting fluids because it forms a low-melting eutectic with nickel.

Titanium The high reactivity and hence high adhesion of titanium, combined with its low thermal conductivity, make chip formation discontinuous at most speeds and machining is difficult. The best tool protection is given by a stable BUE. For low speeds, HSS tools are used with a highly chlorinated oil or emulsion. At higher speeds (30–60 m/min or 100–200 ft/min), cemented carbides are preferred.

8-3 CUTTING TOOLS

Specific features of cutting tools are varied to suit the process, but some basic characteristics are common to all.

8-3-1 Tool Materials

Because machining can be, in general, regarded as competition for survival between workpiece and tool material, one can expect that the tool material should have properties just opposite to those of the workpiece:

1 The tool should be harder than the hardest component of the workpiece material, not only at room temperature (Table 8-3), but also at operating temperatures. High *hot hardness* ensures that the tool geometry is maintained under the extreme conditions presented by the chip formation process, and it also

TABLE 8-3
HARDNESS OF TYPICAL TOOL MATERIALS OR
THEIR CONSTITUENTS*

Material or constituent	Hardness, HV
Martensitic steel	500–1000
Nitrided steel	950
Cementite (Fe_3C)	850–1100
Hard chromium coating	1200
Alumina	2100–2400
WC (Co-bonded)	1800–2200
WC	2600
W_2C	2200
$(Fe, Cr)_7C_3$	1200–1600
Mo_2C	1500
VC	2800
TiC	3200
TiN	3000
B_4C	3700
SiC	2600
Cubic boron nitride	6500
Polycrystalline diamond/WC	5500–8000
Diamond	8000–12000

*From J. A. Schey, *Tribology in Metalworking: Friction, Lubrication and Wear*, American Society for Metals, 1983.

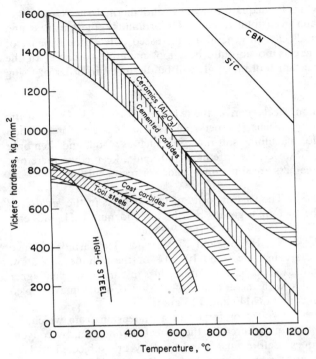

FIGURE 8-16
High temperatures developed in cutting hard materials at high speeds are better resisted by some tool materials. (*From J. A. Schey, Tribology in Metalworking, American Society for Metals, 1983, p. 113. With permission.*)

aids in resisting wear. Some feel for the wide range of hot hardnesses may be gained from Fig. 8-16.

2 Toughness is needed to survive mechanical shocks (impact loading) in interrupted cuts. Shocks occur even in continuous chip formation processes, when the tool encounters a localized hard spot.

3 Thermal shock resistance is needed when rapid heating and cooling take place in interrupted cuts.

4 Low adhesion to the workpiece material helps to avoid localized welding. Paradoxically, high adhesion is desirable when a secondary shear zone is to be stabilized; however, a *diffusion barrier* is then needed.

5 Diffusion of constituents of the tool into the workpiece material results in rapid wear; therefore, solubility of the tool in the workpiece material should be low.

Low hardness and high adhesion are undesirable because they allow distortion of the tool profile, rounding of the tool nose, gradual flank wear, and, combined with diffusion, crater wear. Inadequate toughness and thermal shock resistance lead to edge chipping and even total fracture. Unfortunately, the hardness and heat resistance of materials can, in general, be increased only at the expense of toughness; therefore, there is no absolute best tool material available. In the following, the most important tool materials will be discussed in order of rising temperature resistance.

Carbon Steels   Carbon steels derive their hardness from the martensitic transformation. Martensite softens (tempers) above 250 °C; therefore, carbon steels are suitable only for machining soft materials such as wood, and then only at low production rates. However, they are hard and hold a keen edge; therefore, high-carbon steel hand reamers are sometimes made for metal cutting.

High-Speed Steel (HSS)   The vast majority of tool steels is in the *high-speed steel* (HSS) category. The two main groups are the molybdenum (M1, M2, etc., typically with 0.8% C, 4% Cr, 5–8% Mo, 0–6% W, and 1–2% V) and tungsten (such as T1, with 0.7% C, 4% Cr, 18% W, and 1% V) types. The carbides formed with the alloying elements constitute some 10–20% of the volume and allow repeated heating and cooling to 550 °C without any loss in hardness. Even higher temperatures are permissible with the addition of 5–8% Co, sometimes coupled with an increased carbon content (M40 and T15 grades).

All these steels can be hot rolled or forged to a dimension from which the cutting tool can be readily manufactured, in the annealed condition, by conventional machining techniques. Before final grinding, they are subjected to heat treatment which imparts great strength and high (HRC 63 and over) hardness coupled with reasonable toughness. They can be repeatedly reground. They remain important for the metal-cutting industry, especially for drills, reamers, broaches, and other kinds of form tooling (Fig. 8-17). Improvements in melting and casting techniques have improved their quality; some grades are made by consolidation of prealloyed atomized powder (Sec. 6-2), ensuring more uniform distribution of finer carbides.

Surface coatings are important. Tempering in steam (*blueing*) creates a hard, porous Fe_3O_4 layer which increases tool life. More effective are n triding and TiC and TiN coating (using techniques described in Sec. 9-5) which give two- to sixfold increases in tool life.

Cast Carbides   When the carbides reach very high proportions, the tool material is not hot-workable any more and must be cast to shape. The matrix of *cast carbides* (around 45%) is usually a cobalt alloy into which carbides of Cr and W, formed with 2–3% C, are embedded. Softening is gradual (Fig. 8-16) and higher cutting speeds are permissible, but ductility is much reduced.

Cemented Carbides   *Cemented carbides* produced by powder-metallurgy techniques (Sec. 6-2-4) have achieved a dominant position. The matrix is usually cobalt, 3–6% for greater hardness, from 6–15% for greater toughness. Carbide

FIGURE 8-17
Some high-speed steel (HSS) tools commonly encountered: (a) gear-tooth cutter, (b) shell-end mill, (c) slab mill, (d) side mill, (e) slotting mill, (f) combined drill and countersink, (g) countersink, (h) ball-end mill, (i) square-end mill, (j) single-angle cutter, (k) tap, (l) thread-cutting die, (m) reamer, and (n) angular cutter.

grades are classified according to codes developed in various countries (e.g., the C-system in the U.S.) and by the ISO (R 513). The carbide phase may be made entirely of WC for cutting nonferrous metals and gray cast iron (C1 and C2 grades; ISO group K), but diffusion would lead to rapid cratering in cutting steel (Fig. 8-18). Therefore, 10–40% TiC or TaC (or both), which form a carbide-rich diffusion-resistant interface, are added to grades destined for the machining of steel (C4 to C8; ISO group P). Malleable and spheroidal cast iron present the same diffusion danger and are cut with steel-cutting grades. General-purpose carbides (ISO group M) contain smaller quantities of mixed carbides. Cemented carbides soften only gradually and work best at higher temperatures (over 600 °C).

Cermets Cemented carbides are a subclass of *cermets*, ceramics bonded with a metallic phase. For cutting steel, TiC bonded with nickel and molybdenum has gained acceptance. Better thermal conductivity and higher cutting speeds characterize the mixed TiC–TiN grades.

Coated Carbides Ideally, the tool should possess a very hard, nonreactive surface that also acts as a diffusion barrier, yet it should have a base of sufficient fracture toughness to allow interrupted cuts. *Coated carbides* achieve this aim by

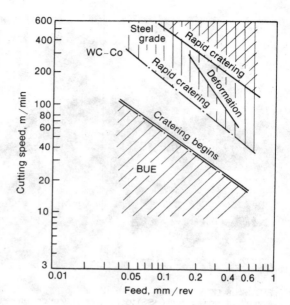

FIGURE 8-18
The dominant wear mechanism is a function of cutting speed and feed in cutting 0.4% C steel of HV 200. (*After E. M. Trent, Inst. Prod. Eng. J.* **38**: *105–130 (1959). With permission of The Institution of Production Engineers, London.*)

combining the virtues of a cemented WC base with those of a thin (typically 5-μm) coating of a ceramic such as TiC, TiN, Al_2O_3, or HfN. Several layers may be deposited—by PVD or CVD (Sec. 9-5)—on top of each other to cater to various functions (e.g., a base layer of TiC, followed by Al_2O_3 and TiN). Some feel for the benefits of coated carbides may be gained from Fig. 8-19. Coated carbides are extensively used in production turning and milling of steels and cast irons. They have taken more than half of the cemented carbide market.

Ceramic Tools   *Ceramics* such as Al_2O_3 may be used other than as coatings; they can be made, by sintering or hot pressing, into solid tool inserts. Since they are self-sintered (with the help of a sintering aid but without a metal binder), they are suitable for very high speeds, but only at light and continuous loads. However, great advances have been made in improving the reliability of these tools and their range of application is growing. Tools of Al_2O_3 reinforced with SiC whiskers, and those made of silicon nitride (Si_3N_4) and Si–Al–O–N ceramics, are tougher and more wear-resistant, and are extensively used in cutting super-alloys and gray cast iron.

Cubic Boron Nitride (CBN)   Made by high-temperature, high-pressure techniques similar to those used for making synthetic diamonds, *cubic boron nitride* (CBN) has a hardness second only to that of diamond (Table 8-3). It can be sintered into a 0.5-mm-thick layer onto a cemented carbide base, or made into

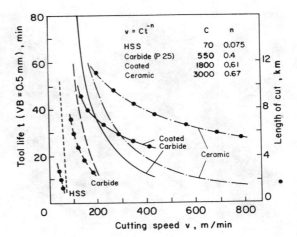

FIGURE 8-19
Tool life is enhanced and the length of chip that can be cut increases greatly on using coated carbide or ceramic tooling for cutting steel. (*R. Abel and V. Gomell, Ind. Anz. **102**: 27–30 (1980). With permission.*)

inserts with or without a ceramic binder. Its great advantage is that it does not suffer diffusive wear in cutting hardened alloy steels.

Diamond The hardest material, *diamond* has long been used in the form of natural single crystals for high-speed finishing of aluminum and other nonferrous materials. Natural diamond suffers from unpredictable early failure, and manufactured single crystals give more reliable performance. More recently, polycrystalline tool tips have become available as self-sintered inserts or as 0.5-mm-thick layers sintered onto a carbide base. Diamond outperforms all other materials on highly abrasive workpieces. However, at high temperatures it changes into graphite which diffuses into iron; therefore, it is not suitable for cutting steel.

8-3-2 Tool Construction

High-speed steels have sufficient toughness to be made into *monolithic* (single-piece) tools. Solid cemented carbide tools can be made and are sometimes used, but the risk of total fracture is great and the cost can become high. Therefore their broadest application is in the form of tool *inserts*, which are either brazed (Fig. 8-20*a*) or clamped (Fig. 8-20*b*) to a tough steel body. Specially constructed cutters (*indexable cutters*, Fig. 8-21) permit moving the insert to compensate for wear, and can thus be used for extended periods of time. Ceramic tools are always made as inserts.

High-speed steel and many cemented carbide tools are reground several times in the machine shop. Some carbide and most ceramic tools are of the throwaway type, and are made so as to have several usable cutting edges.

As discussed in Sec. 8-1-1, a large positive rake angle shortens the shear zone and reduces the energy consumption. This, however, also weakens the tool;

(a) (b) (c)

FIGURE 8-20
Turning tools may have (a) brazed or (b) clamped carbide inserts. Inserts (b) may be fitted with obstruction-type chip breakers or (c) may have preformed chip-breaker grooves.

FIGURE 8-21
This indexable face mill holds peripheral carbide inserts for roughing and face inserts (replaced by dummy inserts during roughing) for finishing. (*Courtesy Ingersoll Cutting Tool Division, Rockford, Ill.*)

Direction of rotation

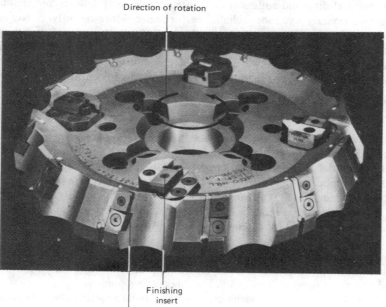

Finishing
insert

Roughing insert

therefore large rake angles are permissible only for cutting lower-strength materials with a tough tool material. Other tool materials, particularly the more brittle varieties, must be made with a small positive, zero, or even negative rake angle. A three-cornered cutting insert can then have six usable cutting edges (as in Fig. 8-20c). Positive-rake inserts in negative-rake tool holders combine advantages of both. In cutting with a negative rake angle, the force pushing the tool out of the workpiece is large and vibrations are easily generated; therefore, an extremely stiff machine tool is needed.

8-4 METHODS OF MACHINING A SHAPE

Irrespective of the machining process employed, the shape of the workpiece may be produced by two basically different techniques: *forming* and *generating*.

8-4-1 Forming

A shape is said to be *formed* when the cutting tool possesses the finished contour of the workpiece. All that is necessary, in addition to the relative movement required to produce the chip (the *primary motion*), is to feed (*plunge*) the tool in depth.

How the primary motion is generated is immaterial. The workpiece can be rotated against a stationary tool (*turning*, Fig. 8-22a), or the workpiece and tool can be moved relative to each other in a linear motion (*shaping* or *planing*, Fig. 8-22b), or the tool can be rotated against a stationary workpiece (*milling* and

FIGURE 8-22
The cutting tool is made to the profile of the part in forming processes such as : (a) form turning, (b) shaping or planing, and (c) drilling.

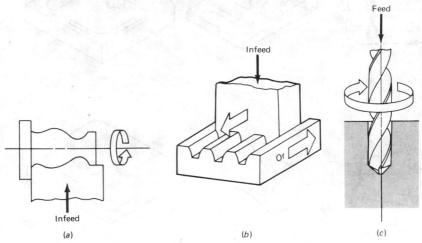

(a) (b) (c)

drilling, Fig. 8-22*c*) or against a rotating workpiece (as in cylindrical grinding). The accuracy of the surface profile depends mostly on the accuracy of the forming tool.

8-4-2 Generating

A surface may be *generated* by combining several motions that not only accomplish the chip-forming process (primary motion) but also move the point of engagement along the surface (described as the *feed motion*, Fig. 8-9). Again, the workpiece may rotate around its axis, as in turning; the tool is set to cut a certain depth and receives a continuous, longitudinal feed motion. When the workpiece axis and feed direction are parallel, a cylinder is generated (Fig. 8-23*a*); when they are at an angle, a cone is generated (Fig. 8-23*b*). If, in addition to the primary and feed motions, the distance of the cutting tool from the workpiece axis is varied in some programmed fashion—e.g., by means of cams, a copying device, or numerical control—a large variety of shapes can be generated.

When the tool (or the workpiece) is fed perpendicular to the primary linear (shaping or planing) movement, a flat surface is generated (Figs. 8-9 and 8-23*c*).

FIGURE 8-23
Programmed tool motion (feed) is necessary in generating a shape: (*a*) turning a cylinder and (*b*) a cone; (*c*) shaping (planing) a flat and (*d*) a hyperboloid; (*e*) milling a pocket; and (*f*) grinding a flat. (Principal motions are marked with hollow arrows, feed motions with solid arrows.)

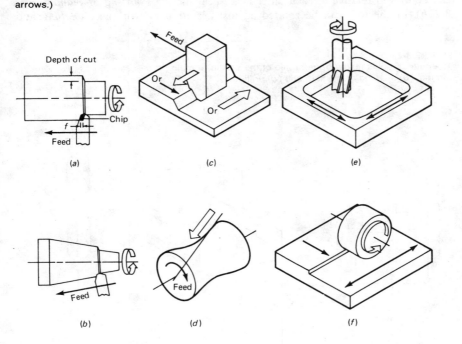

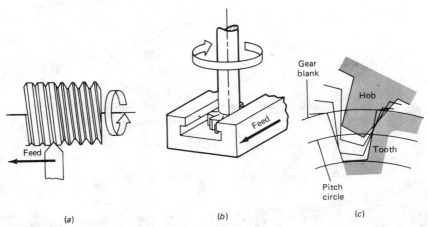

FIGURE 8-24
Forming and generation may be combined: (*a*) thread cutting, (*b*) T-slot milling, and (*c*) gear hobbing.

If the workpiece were given a feed motion by rotating it around its axis parallel to the tool motion, a cylinder could be machined. The workpiece axis could be set at an angle and then a rotational hyperboloid would be generated (Fig. 8-23*d*). In principle, any surface that can be described by a straight generatrix may be produced by this technique. A tool of axial symmetry may rotate while the workpiece is being fed, leading to *milling* (Fig 8-23*e*) or *grinding* (Fig. 8-23*f*).

Frequently, *combined forming and generating* offers advantages. Thus, a thread may be cut with a profiled tool fed axially at the appropriate rate (Fig. 8-24*a*). A slot or dovetail may be milled into a workpiece (Fig. 8-24*b*). A gear may be cut with a *hob* that gradually generates the profile of the gear teeth (Fig. 8-24*c*) while both hob and workpiece rotate.

8-5 SINGLE-POINT MACHINING

It is obvious from the previous discussion that one of the most versatile tools is a *single-point cutting tool* moved in a programmed fashion.

8-5-1 The Tool

The tool must accommodate not only the primary motion (as an orthogonal tool would, Fig. 8-1) but it must also allow for feeding and chip disposal. Therefore the cutting edge is usually inclined (oblique cutting, Fig. 8-8), and the chip is wound into a helix rather than a spiral. The tool is relieved both in the direction of feed and on the surface that touches the newly generated surface, and thus has major and minor flank surfaces (Fig. 8-25). Intersections of these with the rake

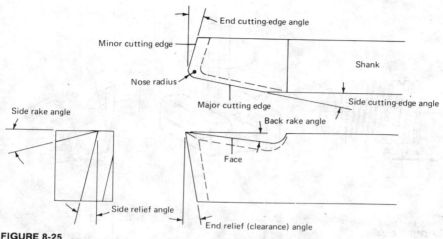

FIGURE 8-25
Nomenclature used for describing the geometry of single-point cutting tools (compare with Fig. 8-20a).

face of the tool constitute the *major and minor cutting edges*, respectively. The nose is rounded with an adequate (typically, 1-mm or $\frac{3}{64}$-in) radius.

The all-important rake angle should really be measured in a plane perpendicular to the major cutting edge, but, for convenience, all angles are measured in a coordinate system that coincides with the major axes of the tool bit (Fig. 8-25). While this systems appears simple, it creates various problems; these are resolved, however, by the ISO recommendation on cutting tools.* In any case, it must be recognized that tool angles have meaning only in relation to the workpiece, after installation in the machine tool.

Some recommendations on cutting-tool angles are contained in Table 8-4. They represent a compromise to give minimum cutting force with maximum tool strength.

The single-point tool may be replaced with a *rotating tool*, which is a disk held at an appropriate angle. The disk may be rotated or it may rotate as a result of its contact with the workpiece; thus, all parts of the circumference are used.

8-5-2 Turning

The most widely used machine tool is the *engine lathe* (*center lathe*, Fig. 8-26), which provides a rotary primary motion while the appropriate feed motions are imparted to the tool.

The workpiece must be firmly held, most frequently in a *chuck* (Fig. 8-27a). Three-jaw chucks with simultaneous jaw adjustment are self-centering. Other

*Draft International Standard ISO/DIS 3002, 1973.

TABLE 8-4 TYPICAL SINGLE-POINT CUTTING TOOL ANGLES*

Workpiece material	BHN	High Speed Steel					Brazed WC		Throwaway		All WC		
		Back rake	Side rake	End relief	Side relief	Edge	Back rake	Side rake	Back rake	Side rake	End relief	Side relief	Edge
Steels	<225	10	12	5	5	15	0	6	−5	−5	5	5	15
	to 325	8	10	5	5	15	0	6	−5	−5	5	5	15
	to 425	0	10	5	5	15	0	6	−5	−5	5	5	15
	>425	0	10	5	5	15	−5	−5	−5	−5	5	5	15
Stainless													
Ferritic		5	8	5	5	15	0	6	0	5	5	5	15
Austenitic		0	10	5	5	15	0	6	5	5	5	5	15
Martensitic		0	10	5	5	15	0	6	−5	−5	5	5	15
Cast iron	<300	5	10	5	5	15	−5	−5	−5	−5	5	5	15
	>300	5	15	5	5	15	−5	−5	−5	−5	5	5	15
Zn alloy	80–100	10	10	12	4	5	5	5	0	5	5	5	15
Al, Mg alloy		20	15	12	10	5	3	15	0	5	5	5	15
Cu alloy		5	10	8	8	5	0	8	0	5	5	5	15
Superalloy	0	10	5	5	15	0	6	0	0	5	5	45	
Ti alloy	0	5	5	5	15	0	6	−5	−5	−5	5	5	
Thermoplastic	0	0	20–30	15–20	10	0	0	0	0	20–30	15–20	10	
Thermosetting	0	0	20–30	15–20	10	0	15	0	15	5	5	15	

*Extracted from *Machining Data Handbook*, 3d ed., Machinability Data Center, Metcut Research Associates, Cincinnati, Ohio, 1980.

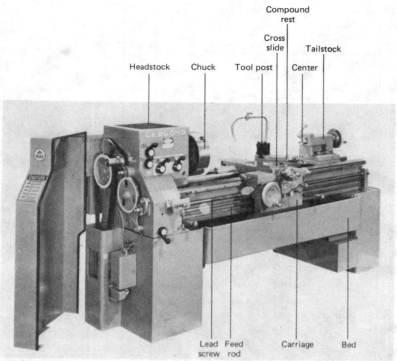

Compound
rest

Cross
slide

Tailstock

Headstock   Chuck   Tool post   Center

Lead   Feed   Carriage   Bed
screw   rod

FIGURE 8-26
A typical engine lathe. Capacity: 380-mm- (15-in-) diameter swing: 1370-mm (54-in) length.
(*Courtesy LeBlond Inc., Cincinnati, Ohio.*)

FIGURE 8-27
Workpieces may be held in a (*a*) chuck, (*b*) collet, or (*c*) face plate. ((*a*) *and* (*b*) *courtesy of DoALL Co.*)

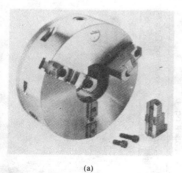

(a)   (b)   (c)

chucks have two, three, or four independently adjustable jaws for holding other than round workpieces. Bars may also be held in *collets*, which consist of a split bushing pushed or pulled against a conical surface (Fig. 8-27*b*). Workpieces of awkward shape are often held by bolts on a *face plate* (Fig. 8-27*c*).

The *headstock* contains the drive mechanism, usually incorporating change gears and/or a variable-speed drive. Long workpieces are supported at their end with a *center* held in the *tailstock*. The tool itself is held in a *tool post* which allows setting the tool at an angle (horizontally and vertically). The tool post is mounted on a *cross slide* which provides radial tool movement. The cross slide is guided in a *carriage*, which in turn receives support from the *ways* machined in the *bed* that ensure rigidity and freedom from vibrations. An overhanging part, the *apron* of the carriage, may be engaged with the *feed rod* to give continuous feed motion, or with a *lead screw* for the cutting of threads. Very long workpieces are secured against excessive deflection by two fingers of a *center rest* or *steady rest* bolted to the lathe bed; a *follow rest* is clamped to the carriage.

Sometimes the tool post sits on a *compound tool rest* which incorporates a slide that can be set at any angle; thus, conical surfaces may be formed by hand feeding the tool. A *four-way tool post* can be rotated about a vertical shaft and allows quick changing of tools in preset positions, thus speeding up successive operations.

8-5-3   Boring

When the internal surface of a hollow part is turned, the operation is referred to as *boring* (Fig. 8-28*a*). For short lengths, the tool may be mounted on a cantilevered bar in the tool post. A long bar is prone to excessive vibration and it is then preferable to have the workpiece secured to the lathe bed while the *boring bar*, clamped in jaws at one end and supported in the tailstock at its other end, is driven. A number of patented solutions exist that aim at reducing or damping out vibrations. Simultaneous cutting with two or three boring inserts equalizes the forces and reduces vibration. A special-purpose machine performing a similar operation, but with more firmly guided boring bars, is the *horizontal boring machine*.

Heavy and large-diameter workpieces that need to be machined on both inside and outside surfaces may be better supported on a lathe turned into a vertical position; called a *vertical turning and boring mill* or *vertical boring machine*, such a lathe can work on several surfaces of a workpiece fastened to the rotating, vertical-axis face plate of the machine (Fig. 8-29).

Holes may be produced in solid workpieces by single-point machining techniques resembling boring. In *gun drilling* the cutting forces are balanced by guide pads placed at angles of 90 and 180° to the cutting edge (Fig. 8-28*b*). To start a hole, a hardened steel guide (*boring bush*) is held against the face of the workpiece. Once the hole has started, the tool guides itself. The tool is usually held stationary while the workpiece, clamped in a chuck and stabilized by steady rests, rotates.

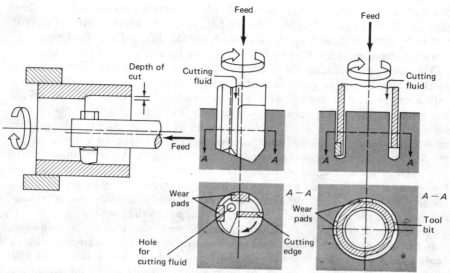

FIGURE 8-28
Machining of holes with single-point tools: (*a*) enlarging (improving the surface finish) by boring, (*b*) gun-drilling, and (*c*) trepanning.

FIGURE 8-29
Large workpieces are often machined on vertical boring machines. The illustration is of a vertical CNC turret lathe. (*Courtesy Bullard Co.*)

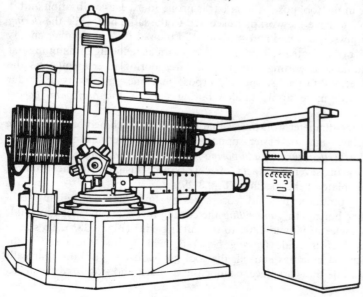

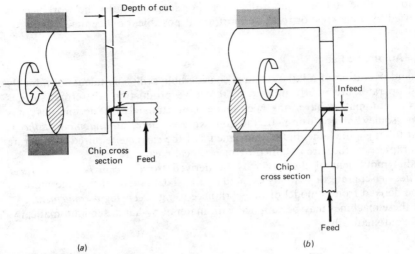

FIGURE 8-30
Flat end faces may be generated by (a) facing and (b) parting off.

Larger holes (diameter of 20 mm ($\frac{3}{4}$ in) and over) can be made by *trepanning*: The cutting tool bit is fastened on the end face of a tube, and the hole is machined by removing an annulus while leaving a center core (Fig. 8-28c). Again, greatly improved patented tool varieties exist.

Both techniques are suitable for making relatively deep holes, of a depth-to-diameter ratio of 5 and over. Force-fed cutting fluid lubricates and helps the removal of the chips, and is vital to success.

8-5-4 Facing

In *facing*, a plane perpendicular to the lathe axis is produced by moving the single-point tool in the carriage so that the feed motion is toward the center of the lathe (Fig. 8-30a). *Parting off* accomplishes the same task but two surfaces are now simultaneously generated (Fig. 8-30b). The cutting speed diminishes as the tool moves toward the center unless the rotational speed is increased in a programmed manner, using a variable-speed drive.

8-5-5 Forming

This method of producing complex rotational shapes (Fig. 8-22a) is fast and efficient, but cutting forces are high and the workpiece could suffer excessive deflection. On cantilevered workpieces the length of the *forming tool* is usually kept to 2.5 times the workpiece diameter. For longer lengths the workpiece is supported by a backrest or roller support, or, if possible, on a center.

8-5-6 Automatic Lathe

The hand operation of a lathe requires considerable skill. The talents of a highly skilled operator are poorly utilized in repetitive production; therefore, various efforts at automation have long been made. Unfortunately, the terminology has become somewhat confusing. In the context used here, an *automatic lathe* is similar to an engine lathe, but all movements of the carriage required to generate the workpiece surface are obtained by mechanical means.

Radial movement of the tool may be derived from a *cam bar* or a *tracer template*, or separate drives may be actuated by NC. Alternatively, the motions may be derived from a model of the workpiece using a *copying arrangement*.

All these machines may be supplied with material by hand, semiautomatically, or fully automatically.

8-5-7 Turret Lathe

When the surface can be generated or formed with relatively simple motions but requires a larger number of tools and operations (such as turning, facing, boring, and drilling) for completion, the requisite number of tools can be accommodated by replacing the tailstock of a lathe with a *turret*. Equipped with a quick-clamp device, a turret brings several (usually six) tools into position very rapidly. All tools are fed in the axial direction, by moving the turret on a slide (*ram-type lathe*) or, for heavier work, on a saddle, which itself moves on the ways (*saddle-type lathe*). Axial feed movement is terminated when a preset stop is reached. Four additional tools are mounted in a square turret on the cross slide and two more tools on a rear tool post. The number of possible operations and the variety of combinations is very large, because several tools may be mounted at any one station for multiple cuts, or simultaneous cuts may be performed at several stations (*combined cut*). Once the machine is set up, it requires relatively little skill to operate.

The *NC lathe* (Fig. 8-31) ensures great flexibility of operation because all motions can be programmed by software. Equipped with a tool changer and linked to material-moving devices, it becomes the center of a flexible manufacturing cell.

8-5-8 Automatic Screw Machines

As the name suggests, these machines were originally developed for making screws at high production rates. Cold heading followed by thread rolling has almost eliminated this market, but machines have been developed to mass-produce more complex shapes.

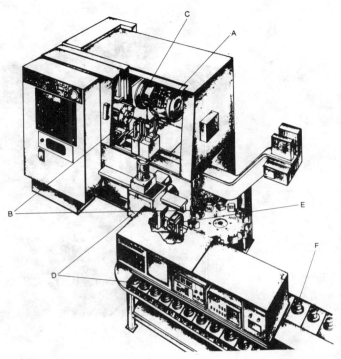

FIGURE 8-31
A CNC turning machine with: (A) 60-tool tool storage and automatic tool changer, (B) automatic chuck jaw changing, (C) touch-type part gaging, (D) machine-mounted robot for delivery and removal of workpieces, (E) visual identification of parts delivered on (F) carousel. Protective guards removed for illustrative purposes. (*Courtesy The Warner & Swasey Co., Cleveland, Ohio.*)

Single-Spindle Automatics *Single-spindle automatics* fall into two basically different groups:

Single-spindle automatic screw machines are based on the principle of the turret lathe, but operator action is replaced by appropriately shaped cams that bring various tools into action at preset times. The stock (a bar drawn to close tolerances) is indexed forward, with cam-operated feed fingers, by the length of one workpiece, at the end of each machining cycle.

Swiss automatics are radically different in that all tools are operated in the same plane, extremely close to the guide bushing through which the rotating bar is continuously fed in a programmed mode. Individual tools are moved radially inward with the aid of cams. Since there is no workpiece overhang, parts of any length may be produced to unsurpassed accuracies and tolerances (down to 2.5 μm (0.0001 in)). More recent machines are numerically controlled (Fig. 8-32).

FIGURE 8-32
Six-slide Swiss-type CNC auto-
matic with three-spindle drilling/
threading attachment.
(*Courtesy Tornos Bechler U.S.
Corporation, Norwalk, Conn.*)

Even though several tools may be set to cut at the same time, the total machining time on single-spindle automatics is the sum of individual or simultaneous operations required to finish the part.

Multispindle automatics Productivity may be substantially increased if all operations are simultaneously performed. In *multispindle automatics* (Fig. 8-33) the head of the lathe is replaced by a *spindle carrier* in which four to eight driven spindles feed and rotate as many bars. The turret is replaced by a tool slide on which the appropriate number of tool holders (sometimes separately driven) are mounted. Additional tools are engaged radially, by means of cross slides; the number of these is often less than the number of spindles, because there may be insufficient room for them. The tool slide with the tool holders moves axially forward, and the cross slides move in radially under cam control, complete their assigned task, withdraw, and the spindle carrier indexes the bars to the next position. Thus, for each engagement of the tools, one part is finished.

Automatic screw machines produce mostly parts of axial symmetry (including threaded parts), but special attachments permit auxiliary operations such as milling or cross-drilling while the rotation of one spindle is arrested. Workpieces of irregular shape can be handled on so-called *chucking machines*.

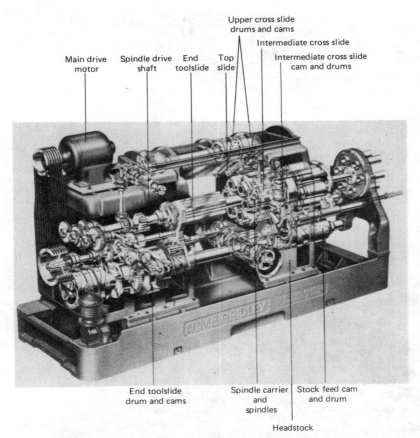

Main drive motor · Spindle drive shaft · End toolslide · Top slide · Upper cross slide drums and cams · Intermediate cross slide · Intermediate cross slide cam and drums · End toolslide drum and cams · Spindle carrier and spindles · Stock feed cam and drum · Headstock

FIGURE 8-33
Six-spindle automatic bar machine, without tooling. (*Courtesy National Acme, Cleveland, Ohio.*)

8-5-9 Shaping and Planing

As indicated in Fig. 8-23c, a surface can be generated with a linear primary motion.

In the process of *shaping*, the primary motion is imparted to the tool and the feed motion to the workpiece (Fig. 8-34a). The tool is moved back and forth by an overhanging ram, the deflection of which limits the length of stroke.

In the process of *planing*, a longer stroke (of practically unlimited length) is obtained by attaching the workpiece to a long, horizontal, reciprocating table while attaching the tool to a sturdy column or arch or, rather, a cross rail with a lead screw that generates the feed movement (Fig. 8-34b).

(a)

FIGURE 8-34
Machine tools with linear motion: (a) 600-mm (24-in) stroke ram shaper and (b) $0.9 \times 0.9 \times 6$ m (36 in $\times$ 36 in $\times$ 20 ft) planer with magnetic chucks mounted on table. (*Courtesy Rockford Machine Tool Co., Rockford, Ill.*)

8-6 MULTIPOINT MACHINING

In multipoint machining at least two cutting edges of the same tool are simultaneously engaged at any one time.

8-6-1 Drilling

Holes are the most frequently encountered machined features. In Sec. 8-5-3 we already discussed two methods of making deep holes, based on single-edge

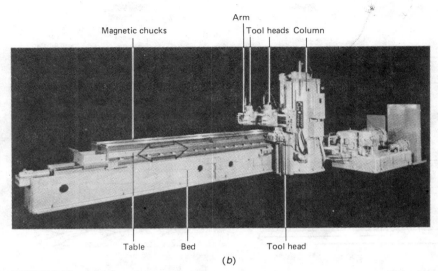

FIGURE 8-34   Continued

techniques. The vast majority of holes, however, are made by the familiar two-edge tool, the *twist drill* (Fig. 8-35).

The twist drill has several advantages: two cutting edges are more efficient; cutting forces are balanced; helical *flutes* allow access of cutting fluid and help to dispose of the chip; and small *margins* left on the cylindrical surface provide guidance.

Nevertheless, the twist drill also has its problems:

1 The two cutting edges must not come together into a point which, because of its small mass, would quickly overheat and lose its strength. A *chisel edge* is usually left and, because of the highly negative rake angle, no real cutting action takes place in the center of the hole. The material is plastically displaced by a process resembling indentation (Fig. 4-29), to be subsequently removed by the cutting edges. The force required for this rotary indentation accounts for much of the total thrust (feed) force in drilling. Modified point geometries allow easier penetration but require more complex grinding. If necessary, a pilot hole, of a diameter equal to the chisel-edge diameter of the larger drill, will greatly reduce the required feed force.

2 When starting a hole, the chisel edge tends to wander. The drill must be kept in place by a *drill bushing*, or an indentation must be created with a *center punch*, or a *center-drill and countersink* is used.

3 The helix angle determines the rake angle at the periphery of the drill; the rake angle becomes smaller on moving along the edge to the center. *High-helix* (*fast-spiral*) *drills* with wide flutes aid in chip removal, and their increased

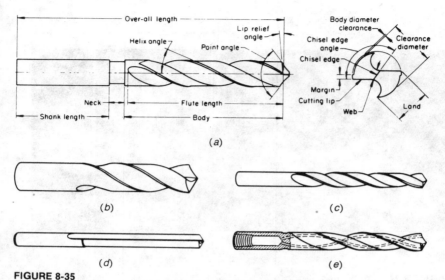

FIGURE 8-35

Twist drills are available in a great variety of shapes, including: (*a*) jobbers drill, (*b*) automatic screw-machine drill, (*c*) low-helix drill, (*d*) straight-flute drill, and (*e*) straight-shank oil-hole drill. (*From Metals Handbook, 8th ed., vol. 3, American Society for Metals, 1967, p. 78. With permission.*)

bearing surface gives better guidance. When drilling thin sheet or materials such as free-machining brass, a reduced helix angle or even a *straight-flute drill* (zero helix-angle drill, Fig. 8-35*d*) is preferable. Drilling in increments (*peckering*) helps chip removal.

4 The surface finish of the hole is not as good as that of a bored hole, and the drill begins to drift at greater depths. Nevertheless, the quality is adequate for a great many purposes, in diameters ranging from 0.15–75 mm (0.006–3 in), at depth-to-diameter ratios of up to 5 (although deeper holes are often drilled).

5 Significant increases in drill life are obtained with coated (particularly, TiN-coated) HSS drills. Yet higher wear resistance is obtained with carbide inserts. Holes of larger diameters may be cut with specially constructed drills equipped with several indexable inserts.

Spade drills (Fig. 8-36), in various configurations, are suitable for drilling holes of all diameters and, when made of carbide, also for hard materials. Straight-shank, monolithic carbide spade drills have a larger cross section than twist drills of the same diameter and are preferred for small (below 0.5-mm-diameter) holes.

Drilling equipment. The simplest *drill press* has a single rotating spindle which is fed axially, at a set rate or under a constant feed force, into a workpiece held rigidly on a table (Fig. 8-37). A *radial arm drill* has a swinging arm which provides much greater freedom. When several holes have to be produced in a large number of workpieces, simultaneous drilling with a *multispindle drill head* or

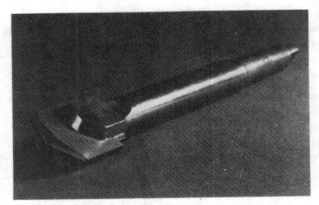

FIGURE 8-36   Spade drill.

FIGURE 8-37
Components of a drill press of 380-mm (15-in) capacity (drill center to column distance).
(*Courtesy Delta International Machinery Corp., Pittsburgh, Penn.*)

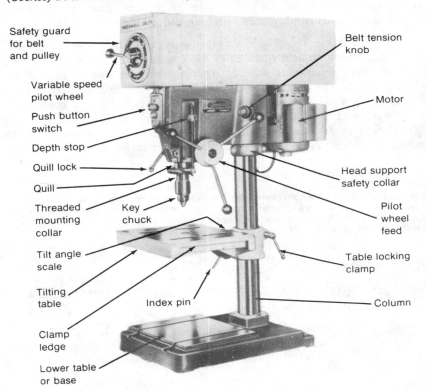

Safety guard
for belt
and pulley

Belt tension
knob

Variable speed
pilot wheel

Push button
switch

Motor

Depth stop

Quill lock

Head support
safety collar

Quill

Threaded
mounting
collar

Key
chuck

Pilot
wheel
feed

Tilt angle
scale

Table locking
clamp

Tilting
table

Index pin

Column

Clamp
ledge

Lower table
or base

drill press assures better accuracy of relative hole location. Exceptional accuracy of hole location is achieved on the *jig borer*, which is really a drill press equipped with a high-precision table movement in two directions. Numerical control is often employed.

Drills can be held in the tailstock of a lathe to machine holes of good concentricity, and drills are important tools for all automatics.

The quality of drilled holes is greatly improved by *reaming*, which could be classified as a milling operation (a reamer is shown in Fig. 8-17*m*). Seats for countersunk screws are prepared by *spot facing*, essentially an end-milling operation in the plunging mode (a countersink is shown in Fig. 8-17*g*).

8-6-2 Milling

Milling is one of the most versatile cutting processes, and it is indispensable for the manufacture of parts of nonrotational symmetry. There are innumerable varieties of milling cutter geometries, but, basically, they can all be classified according to the orientation of tool (or, rather, the orientation of the cutting edges and axis of rotation) relative to the workpiece.

Horizontal mills have the axis of the cutter parallel to the workpiece surface. The cutter axis is usually horizontal, with both ends of the cutter supported.

1 In *plain* or *slab milling* (Fig. 8-38*a*) the cutting edges define the surface of a cylinder and can be straight (parallel to the cylinder axis) or helical (see Fig. 8-17*c*). The milling cutter is wide enough to cover the entire width of the workpiece surface. The primary motion is the rotation of the cutter while feed is imparted to the workpiece. Both the primary and feed motions are continuous. The chips are thickest on the surface of the workpiece and diminish toward the base of the cut. In *down* or *climb milling* (Fig. 8-38*b*), feed motion is given in the direction of cutter rotation; thus, the cut begins at the surface with a well-defined

FIGURE 8-38
In horizontal-axis milling the cutter axis is parallel to the workpiece surface. Direction of feed and cutter rotation determines whether milling is (*a*) up or (*b*) down, resulting in (*d*) different undeformed chip thicknesses. A narrow cutter (*c*) performs slotting. ((*d*) *after M. C. Shaw, Metal Cutting Principles, Oxford University Press, 1984. With permission.*)

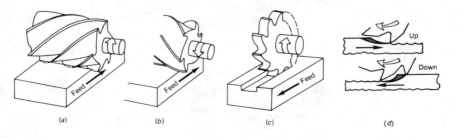

(a)   (b)   (c)   (d)

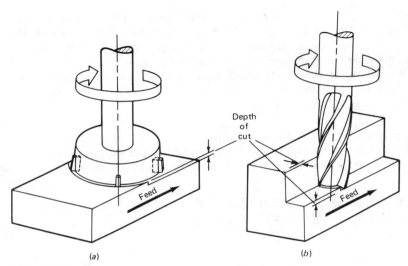

FIGURE 8-39
In vertical-axis milling the cutter axis is perpendicular to the workpiece surface: (*a*) face milling
and (*b*) end milling.

undeformed chip thickness (Fig. 8-38*d*), and surface quality is good. However,
the initial force is high, there is a force reversal in the direction of feed and the
machine must be of sturdy construction and equipped with backlash-free drives.
In *conventional* or *up milling* (Fig. 8-38*a* and *d*) the workpiece is fed in a
direction opposite to the cutter motion; the tooth engages at a minimum depth,
the surface may become smeared and more wavy, but starting forces are lower.
Therefore, this was the preferred method before the arrival of more rigid machine
tools.

2 When a cylindrical cutter is narrower than the workpiece, the cutting edges
must be carried over the end faces of the cylinder (Fig. 8-38*c*; see also Fig. 8-17*d*
and *e*). Because of their action, these mills are called *slotters* or slitting cutters.
When only the side teeth are engaged, one speaks of *side-milling cutters*.

Vertical mills have the axis of the cutter perpendicular to the workpiece surface
(Fig. 8-39). The cutter axis used to be vertical, but in newer CNC machines it is
often horizontal. The cutter is always cantilevered (supported at one end only).

1 When the teeth are attached to the cutter face which is perpendicular to the
axis (see Fig. 8-21), one speaks of *face milling* (Fig. 8-39*a*). In many ways, this is
similar to machining with many single-point tools moving in circles. Since the cut
always starts with a definite chip thickness, face milling uses some 40% less power
than slab milling. A variant frequently employed is *fly-cutting*, that is, cutting
with a single-point tool fixed to the end of an arm protruding from the perpendicular milling shaft.

2 Cutting edges carried over onto the cylindrical surface of the cutter create an *end mill* (Fig. 8-39*b*; see also Fig. 8-17*b*, *h*, *i*, and *j*). End mills are among the most versatile tools because they can be made to follow any path in the plane of and perpendicular (or at an angle) to the workpiece surface. Thus, pockets and contoured surfaces of almost any shape, depth, and size can be machined (Fig. 8-23*e*). Even huge surfaces are sometimes fully machined, as, for example, in making aircraft wing skins.

Milling machines The various feed motions of a milling machine may be controlled by hand although the milling of complex shapes requires considerable skill. For quantity production, the milling machine can be automated to various degrees:

1 *Copy millers* use a model of the finished part to transfer the movement from a copying head to the milling head.

FIGURE 8-40
Machining centers are CNC milling machines, often with horizontal spindle. An automatic tool changer loads tools from a magazine. (*Courtesy Cincinnati-Milacron, Cincinnati, Ohio.*)

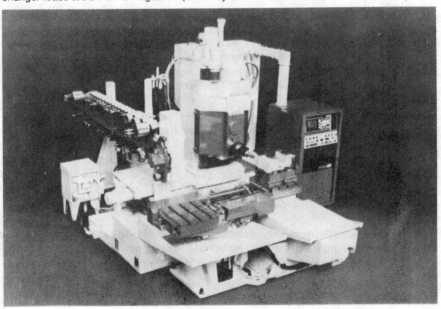

FIGURE 8-41
Typical parts made on a machining center in a single setup. (*Courtesy Cincinnati-Milacron, Cincinnati, Ohio.*)

2 *NC* and *CNC milling machines*, which move some of the skill of the operator into the programming stage (Sec. 1-4-4), have been rapidly developing. If properly utilized, they speed up production by eliminating much of the setup time and the trial-and-error procedures inevitable with manual control.

3 *Machining centers* (Fig. 8-40) are CNC milling machines of extended capability, performing not only a variety of milling but also drilling, boring, tapping, and possibly also turning operations in one setup. More than one machining head may be used, and the *x-y* table may incorporate a rotary table. Some machining centers have a modular design: Tool heads can be changed for optimum production. After a reference surface has been prepared, sometimes on a separate machine, a machining center can work on the workpiece from five sides. The machines often form the core of flexible manufacturing cells (see Sec. 11-2-4).

An example of relatively simple parts produced on a machining center is shown in Fig. 8-41.

8-6-3 Sawing and Filing

A very narrow slitting cutter becomes a *cold saw*. The teeth need not go deep in the radial direction, and are usually made as inserts attached to a larger saw blade. For less-demanding applications the very accurate formation of various tool angles may be relaxed, and the teeth can be formed by bending them into position. The basic cutting action of such *circular saws* is still closely related to milling.

When the teeth are laid out into a straight line, one obtains a *hacksaw* or, if the saw blade is flexible and made into an infinite loop, a *band saw*.

A fine-pitch slab mill laid out into a flat becomes a *file*. The individual cutting edges are broken up into a series of teeth in a *crosscut file*.

All these tools are form tools, and the cut progresses by a positive in-feed or by the pressure exerted on the tool.

8-6-4 Broaching and Thread Cutting

These processes differ from those discussed thus far in that the only motion is the primary motion of the tool. The feed is obtained by placing the teeth progressively deeper within the tool; thus, each tool edge takes off a successive layer of the material. Most of the material is removed by the roughing teeth which are followed by a number of finishing teeth designed to give the best possible surface finish. The shape of the tool determines the shape of the part (pure forming).

1 *Broaching* proceeds with a linear tool motion. A separate broach has to be made up for each shape and size; therefore, broaching is primarily a method of mass production. The workpiece must be rigidly held and the broach firmly guided. Rigidity of the machine tool is particularly important when a surface is broached with a *flat broach*, since the broach would be lifted out of the workpiece by the cutting forces. An *internal broach* is pulled through hollow parts (Fig.

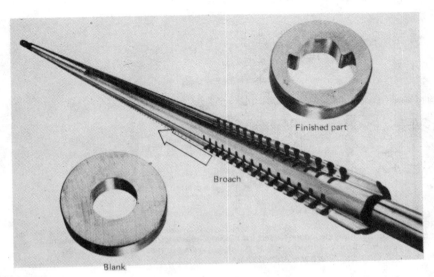

FIGURE 8-42
A broach used for finishing the internal profile of hammers for air-powered impact wrenches.
(*Courtesy Apex Broach and Machine Co., Detroit, Mich.*)

8-42). An *external* (*pot*) *broach* works on the outside surface of the part and the parts are pulled or pushed through. Both techniques are, to some extent, self-guiding. Large broaches may be of segmented construction which allows also some flexibility by using interchangeable broach elements. The machine tool resembles a hydraulic press of long stroke.

2 *Thread cutting* of a hole is an internal operation using a *tap* (Fig. 8-17k), whereas threading of a shaft is an external operation using a *thread-cutting die* (Fig. 8-17l). For both, the primary motion is helical. Threads may also be cut on a lathe (Fig. 8-24a). An alternative to thread cutting is thread forming (Sec. 4-8-4, Fig. 4-50). External thread rolling is, of course, widespread (Fig. 4-49).

8-6-5 Gear Production

Gears are among the most important machine components whose importance has not diminished in recent years.

Gear Making   All processes discussed hitherto are used:

1 Many gears used in more critical applications are still produced by metal-cutting processes.

a *Form cutting* is conducted with a tool that has the profile of the space between two adjacent teeth. A form tool operated in a reciprocal (shaping) motion (Fig. 8-22b) or a form-milling cutter (Fig. 8-17a) installed on a horizontal

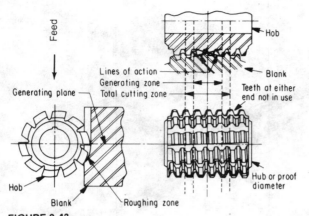

FIGURE 8-43

Hobbing is one of the many processes used for making gears. (*From The Tool and Manufacturing Engineers Handbook, 4th ed., vol. 1, p. 13.48. With permission of the Society of Manufacturing Engineers, Dearborn, Michigan.*)

milling machine may be used. Special gear cutting machines, working on either principle are available. When the axis of the gear blank is set at an angle to the tool movement, helical gears are cut.

b A multiplicity of cutting edges is engaged in *gear hobbing* (Fig. 8-43). The hob looks like a worm gear, the thread of which is interrupted to make several cutting teeth. When cutting spur gears, the hob axis is skewed relative to the gear axis by the helix angle of the hob. When cutting helical gears, the gear angle is added to the helix angle of the hob.

c The cutting of bevel gears with straight or spiral teeth requires a combination of forming and generating and is performed on special-purpose machines that incorporate ingenious mechanisms to develop the required relative motions of tool and gear blank. Some machines use NC or CNC to guide the tool and blank.

d Spur and helical gears may be cut by broaching, with the blanks pushed or pulled through pot broaches.

2 Lower costs and gear teeth of higher fatigue strength are often attainable by plastic deformation processes:

a Spur gears may be produced by cutting up a cold-drawn bar of the appropriate cross section.

b Spur and helical gears may be cold extruded (Sec. 4-5-3).

c Spur gears, bevel gears, and, with the use of more complex, rotating tooling, also spiral bevel gears may be hot forged (Sec. 4-4-5) to near-net shapes.

d All forms of gears can be rolled. Indeed, transverse rolling (Sec. 4-8-4) is the standard method of production for worms of worm-gear drives and linear actuators, as well as for many spur and helical gears.

4 Spur gears are mass produced by various forms of blanking, including fine blanking (Sec. 5-2-3).

5 Many gears are made by powder-metallurgy techniques, with or without restriking and resintering (Sec. 6-2-4).

Finishing Gears produced by the above techniques are often suitable for immediate use in many applications. For smooth, noise-free running at high speeds and also for elimination of surface defects that would reduce fatigue life, many gears used in more critical applications are finished to close tolerances and a specified surface finish.

1 *Gear shaving* is, in some ways, related to broaching. The shaving tool is a meshing gear into which circumferential slots have been cut to make each tooth into a broach. The tool and gear are run in contact, slightly skewed, while imposing an axial oscillating motion, thus removing thin chips. A total of only 25–100 μm is removed. This is still the most frequently used technique.

2 Cold rolling (*burnishing*) between hardened meshing gears imparts a good surface finish and induces a residual compressive stress on the surface. This is a high-productivity process suitable for mass production.

3 Hardened gears are finished by form grinding, using a wheel dressed to the shape of the space between adjacent teeth, or by generating the tooth profile on special machines using straight-sided wheels. Ground gears are accurate but of high cost.

4 Hardened gears may be lapped against cast-iron lapping gears. In some gear assemblies, such as the hypoid rear axles of automobiles, the assembly is lapped together by applying a lapping oil, containing a very fine abrasive, under controlled conditions.

8-7 CHOICE OF PROCESS VARIABLES

The decision to machine a component is usually part of a broader decision-making process extending to the entire manufacturing sequence. Once the decision to machine the component is made, the sequence of machining steps is planned. This planning will take into account the machinability of the workpiece material, the shape, dimension, dimensional tolerances, and surface finish of the finished part, the characteristics of the machining process that is judged to be suitable for the purpose, the availability of machine tools at the plant or at outside vendors, and the economic aspects of production. In the simplest form, the result of these deliberations would be a decision to perform a sequence of operations, such as boring, milling, drilling, etc.

Metal cutting is somewhat unusual in that at this point of planning some important operational decisions have to be made: Even though the process is set, the speeds and feeds appropriate for the workpiece material and tool have to be chosen. When substantial volumes of material are to be removed, production is speeded up by taking one or more *roughing cuts* with large feeds and depths of cut, and then the required surface finish and dimensional tolerances are obtained by taking a *finishing cut* with a small feed and depth of cut.

In a small-shop environment the choice of feeds and speeds may be based on the personal experience of the operator; almost always, such a choice will be conservative. In a competitive production environment the choice is more critical because low speeds and feeds result in low production rates, whereas excessive speeds and feeds reduce tool life to the point where the cost of tool change outweighs the value of increased production and, beyond a certain point, even production rates drop because of time lost in tool change. An initial choice of reasonable speeds and feeds is usually based on collective experience, gathered in many production plants and laboratories. Compilations have been prepared and are continually updated by various organizations, not only within large corporations, but also in specialized organizations such as the Machinability Data Center, Metcut Research Associates, Cincinnati, Ohio; Machining Data Bank, Production Engineering Research Association, Melton Mowbrey, U.K.; Information Center for Cutting Data (INFOS), Technical University of Aachen, West Germany. Collected data are published in handbooks and are also available in computer data banks.

8-7-1 Cutting Speeds and Feeds

We saw in Sec. 8-1-7 that tool life decreases rapidly with increasing temperature which, in turn, depends on the energy expended per unit time and therefore on the cutting speed and shear stress or, more generally, hardness of the workpiece material. Hence, within any one material group, cutting speeds and feeds decrease with increasing hardness. However, large variations in speeds and feeds are noted between different material groups, and the general recommendations given in *Machining Data Handbook* (3d ed.) form the basis of Figs. 8-44 and 8-45. In using these figures, the following should be noted:

1 For some of the nonferrous materials speeds change little with hardness, and in Fig. 8-45 the speed is then simply indicated by the position of the alloy identification. For all materials, the feeds given apply to the hardness ranges indicated by the arrows.

2 The data provide a conservative starting point for rough turning with a maximum 4-mm (0.150-in) depth of cut and a typical tool life of 1–2 h. The tool materials are identified generically. The HSS is typically M2 or, for heavier duties, T15; the symbol WC signifies cemented carbides, uncoated, and of appropriate grade for the workpiece material. Speeds may be increased by 20% for throwaway carbide inserts. If it is found that the actual tool life is much longer than 2 h, the operation may be speeded up. Remembering that heavier cuts are more efficient (Eq. (8-14)), the depth of cut and then feed (which determines the undeformed chip thickness) are increased. Higher cutting speeds generate more heat and should be used only when tool life is still excessive.

3 While the data given are valid primarily for rough turning and boring, they can be used also as a guide for most other processes. The speed v and feed f for any particular process are found by multiplying v_S and f_S from Fig. 8-44 or 8-45 by the factors Z_V and Z_f, respectively (Table 8-5). It will be noted that the feed in shaping and planing is usually quite heavy; the lower feed values should be taken for harder materials.

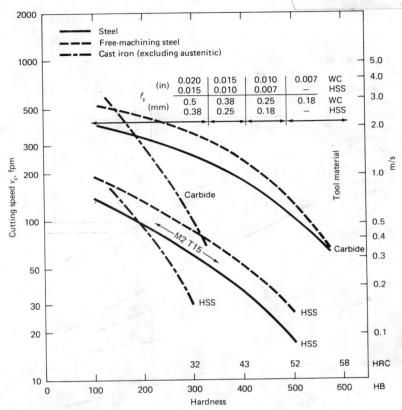

FIGURE 8-44
Typical speeds and feeds for roughing ferrous materials with a 3.8-mm (0.150-in) depth of cut.
Increase speed by 20% for throwaway carbide inserts; reduce speed by 20–30% for austenitic
stainless steels and for tool steels containing over 1% carbon.

4 Special conditions apply to drilling with HSS twist drills. The cutting speed
is $v = 0.7v_S$ for ferrous and $v = 0.5v_S$ for nonferrous materials. Feed is typically a
function of drill diameter D and is $0.02D$ per revolution for free-machining
materials, $0.01D$ per revolution for tougher or harder materials, and $0.005D$ per
revolution for very hard (HB > 420) materials. For deep holes, speeds and feeds
must be reduced:

Hole depth, D	Speed reduction, %	Feed reduction, %
3	10	10
4	20	10
5	30	20
6	35	20
8	40	20

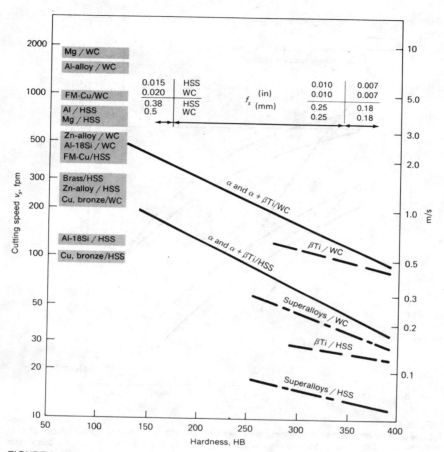

FIGURE 8-45
Typical speeds and feeds for roughing nonferrous materials with a 3.8-mm (0.150-in) depth of cut (FM stands for free-machining metal, WC for carbide tool).

5 For broaching with HSS tools, speeds range from 0.2 m/s (40 ft/min) on free-machining material to 0.025 m/s (5 ft/min) on hard material, while the undeformed chip thickness reduces from 0.12 to 0.05 mm (0.005 to 0.002 in) per tooth.

6 Coated carbides allow higher speeds, and the recommendations of the manufacturer should be taken for first attempts. In general, speed can be increased by 25% with TiN, 25–50% with TiC, and 50–75% with Al_2O_3 coatings.

7 Ceramic tools can be tried at twice the recommended speed for cemented carbides in finishing cuts and then raised to higher speeds (up to five times higher).

TABLE 8-5   SPEEDS AND FEEDS IN VARIOUS METAL-CUTTING OPERATIONS*

Process	Z_v (speed† $v = v_s Z_v$)	Depth of cut in	Depth of cut mm	Z_f (feed† $f = f_s Z_f$)	Other
Rough turning	1	0.15	4	1	
Finish turning	1.2–1.3	0.025	0.65	0.5	
Form tools, cutoff	0.7				In-feed 0.1f–0.2f
Shaping	0.7	0.15	4		Feed: HSS, 1.5–0.5 mm
					WC, 2–1 mm
					(×2 on Cu, Al and Mg)
Planing	0.7	0.15	4		
Face milling	1	0.15	4	0.8–1‡	
Slab milling	1	0.15	4	0.5‡	
Side and slot milling	0.5–0.7	0.15	4	0.5‡	
End mill, peripheral	1	0.05	1.2	0.5–0.25‡	For 1-in-diam cutter
End mill, slotting	1	0.05	1.2	0.2‡	
Threading, tapping	0.5–0.25				Slower for coarser thread

*Approximate values, compiled from *Machining Data Handbook*, 3d ed., Machinability Data Center, Metcut Research Associates, Cincinnati, Ohio, 1980.
†Take v_s and f_s from Fig. 8-44 or 8-45.
‡Feed per tooth.

8 Diamond tools are suitable only for finishing cuts of 0.05–0.2-mm (0.002–0.008-in) depth at 0.02–0.05-mm (0.0008–0.002-in) feed and speeds of 4–15 m/s (800–2800 ft/min) for light nonferrous metals.

The data given here are sufficiently accurate for initial planning purposes. Optimum conditions depend, however, on many factors, including the rigidity of the tooling, workpiece, workpiece holder, and machine tool. It is not uncommon for practical metal removal rates to reach twice the recommended values. In specially constructed machines, very high-speed machining is possible with coated carbide or ceramic tools; speeds of 25 m/s have been exceeded in machining aluminum alloys and speeds of 20 m/s have been reached in machining cast iron.

8-7-2   Cutting Time and Power

Once feeds and speeds are selected, details of the process can be planned and the appropriate equipment chosen. The following steps are usually followed:

1 The volume V to be removed is calculated. Handbooks contain numerous formulas but calculations from simple geometrical considerations are usually just as fast.

2 The chip removal rate V_t is calculated from cutting speed multiplied by chip cross-sectional area.

3 The net machining (cutting time) is simply

$$t_c = \frac{V}{V_t} \qquad (8\text{-}22)$$

(In some operations, such as turning, the total chip length l is easily calculated and then $t_c = l/v$.)

4 The adjusted specific cutting energy is taken from Eq. (8-14), the power of the machine tool from Eq. (8-15), and the cutting force from Eq. (8-16).

Example 8-7

The constant C for Taylor's equation, Eq. (8-18), may be found by multiplying v_S by 1.75 for HSS and by 3.5 for WC. Since v_S in Figs. 8-44 and 8-45 is based on a tool life of 1–2 h or, on the average, $t = 90$ min, what is the implied value of n?

From Eq. (8-18): $v_S\ 90^n = C$; for HSS, $v_S\ 90^n = 1.75 v_s$; $n = 0.125$; for WC, $v_S\ 90^n = 3.5 v_s$; $n = 0.28$.

Example 8-8

The bore of a steel casting conforming to ASTM A27-77, 70-36 is to be machined out using disposable carbide insert tooling. The hole diameter is 130 mm in the as-cast condition; the finished diameter is 138 mm. Suggest cutting speed and feed, and calculate the power and cutting force.

From MHDE (p. 4.47), the minimum TS = 485 MPa. This corresponds to HB = 3(485)/9.8 = 150 kg/mm^2.

Boring with a $w = (138 - 130)/2 = 4$ mm depth of cut is equivalent to rough turning. Hence, from Fig. 8-44, $v_S = 1.8$ m/s. For throwaway insert, increase by 20%: $v_S = (1.8)(1.2) = 2.16$ m/s. From Table 8-5, $Z_v = 1$ and $v = 2.16$ m/s.

From Fig. 8-44, $f_s = 0.5$ mm; from Table 8-5, $Z_f = 1$, and $f = 0.5$ mm/r.

To find the power required, the rate of material removal V_t must be calculated. Chip cross section $A = fw = (0.5)(4) = 2$ mm^2. $V_t = Av = 2(2160) = 4320$ mm^3/s.

From Table 8-1, $E_1 = 1.4$ W · s/mm^3. From Eq. (8-14), $E = E_1 h^{-a} = 1.4(0.5)^{-0.3} = 1.72$ W · s/mm^3. Thus, from Eq. (8-15a), the power, at an efficiency of 0.7 is $1.72(4320)/0.7 = 10.6$ kW. The cutting force, from Eq. (8-16a), $P_c = 10600/2.16 = 4.9$ kN.

Example 8-9

Holes of 10-mm diameter are to be drilled, with a twist drill, into a free-machining steel of HB 180. Determine the recommended cutting speed and feed.

From Fig. 8-44: for HSS, $v_S = 0.75$ m/s. From Table 8-5: $v = 0.7\ v_S = 0.5$ m/s.

Since the circumference of the drill is $10\pi = 31.4$ mm, the rotational speed is $500/31.4 = 16$ r/s.

The feed is $0.02D$ per revolution or 0.2 mm/r, and the feed rate is $0.2(16) = 3.2$ mm/s. Because there are two cutting edges, $f = 0.2/2 = 0.1$ mm/edge.

8-7-3 Choice of Machine Tool

The variety of commercially available equipment is immense, in terms of both size and type of operation. Table 8-6 gives but a general feel for machine tools

TABLE 8-6 CHARACTERISTICS OF CUTTING EQUIPMENT*

Machine tool	Workpiece, max. dimensions†			Main motion, max.‡		Drive, kW	Other
	Width, m	Diameter, m	Length, m	Speed, m/s	r/min		
Lathe, center		0.1–2	0.3–5		3000–100	1–70	
Turret (bar)		0.02–0.3	0.1–1.5		3000–300	1–60	
Automatic (single-spindle)		0.01–0.15	0.05–0.3		9000–500	1–40	
Automatic (multispindle)		0.01–0.15	0.1–0.3		4500–300	5–50	
Automatic (screw)		0.01–0.1	0.03–0.3		10000–1500	2–20	
Automatic (Swiss)		0.005–0.03	0.05–0.3				
Boring machine, horizontal		0.5–1.5	0.4–2		1000–150	2–70	
Vertical		1–6	0.7–2.5		300–30	20–200	
Shaper	0.2–0.8		0.15–1	0.4–1		1–7	
Planer	0.6–2.7		1–10	0.5–1.7		10–100	
Drills					12000–400	<1–10	Drill diam: 0.3–100 mm
Milling machine	0.1–0.4		0.5–2.5		4000–1000	1–20	
Broaching machine			8–24 >	0.2–0.02		1–40	Stroke: 0.5–2 m and up
Grinder, surface	0.1–0.9		0.2–6			<1–30	
Cylindrical		0.02–0.8	0.2–6			<1–20	
Centerless		0.01–0.3				<1–30	

*Commercially available equipment, selected sizes from *Machine Tool Specification Manual*, Maclean-Hunter, London, 1963.
†The range indicates the maximum dimensions taken by equipment of various sizes; smaller workpieces are usually accommodated.
‡Speeds indicated are maximum for equipment of different sizes; variable speed is usually provided down to 1/50 or 1/200 of maximum.

commonly manufactured; special-purpose machines, some of enormous, others of minute size, are in operation. Thus, a lathe (made by Farrell Co.) for turning steam turbine rotors has a swing of 1.9 m (75 in) and a bed 14 m (46 ft) long. A six-gantry NC milling machine (made by Cincinnatti Milacron Inc.) for sculpturing airplane parts has a table 8 m (160 in) wide and 110 m (360 ft) long.

It may turn out, of course, that no machine tool of sufficient power or stiffness is available and then the speed or feed, or both, must be reduced.

Numerical Control and Automation We have seen that the operation and control of machine tools requires considerable skill. This is particularly evident in milling three-dimensional contoured surfaces; it is not surprising that NC was first developed for machining such contoured aircraft components (Sec. 1-4-4). Parallel with the development of software, computer and control hardware developed at a rapid rate, and this was matched by advances in tool materials. Improved tools permitted material removal at higher rates; with the aid of NC, the time during which actual cutting takes place could be increased; and new demands could be set regarding tolerances. These developments also brought changes in machine tools and their operation.

1 To employ NC, the hand-wheels of manually operated machines and the cams, tracer templates, and clutches of automatics had to be dispensed with. The table of the machine tool is fitted, for example, with *backlash-free lead screws*, actuated with stepping motors that move the table in small increments of, say, 2.5 μm (0.0001 in); the number of increments depends on the number of pulses received from the NC control unit. For higher torque, dc or ac motors are employed, or the table is moved by *hydraulic actuators*. Displacement transducers or encoders (Sec. 2-4-6) are installed for closed-loop control.

2 *Spindle drives* suitable for operating at variable speeds had to be installed.

3 The *rigidity* of machine tools had to be increased to minimize tool deflection and reduce the tendency to vibration (some machine tools have reinforced concrete or epoxy-granite beds). Better compensation for distortions due to thermal expansion had to be found.

4 Improved methods of *removing* large quantities of *chips* had to be developed. Partly for this reason, machining centers with horizontal spindles, but using cantilevered tooling (Fig. 8-40) have found acceptance.

5 *Automatic tool changers and magazines* holding 20 to 60 tools had to be installed (for FMS, Sec. 11-2-4, magazines holding as many as 100 tools are needed).

6 To eliminate setup time, tools are *qualified*, i.e., their dimension is premeasured and entered into the NC program. Alternatively, tools are stored in the magazine in special, individual holders which ensure that they are always gripped in the same position relative to the machine tool.

7 The task of *programming* has been greatly simplified. Two general trends have emerged.

First, programming languages—often based on APT—have been simplified for specific tasks such as two-axis contouring (ADAPT), and postprocessors have been added for specific processes such as drilling, milling, and lathe work (EXAPT). The Electronics Industries Association (EIA) Standard RS-494 (Binary Cutter Location Data Exchange Format for NC Machine Tools, or BCL) defines a specific format for cutter location so that no postprocessor is needed for different machine tools.

Second, powerful software has been developed that performs the task of programming for relatively simple configurations. The operator has to enter only the dimensions of the part and the dimensions and material of the tool and workpiece, and the software executes programming.

8 *Unattended machining* has become possible in limited cases with the introduction of automatic monitoring and adaptive control. Tool force is sensed (e.g., from spindle deflection, using four position sensors) and torque and/or power is measured. Process optimization is possible; feed is often the controlled variable, within the constraints set by maximum force, torque, and/or power. Techniques for in-process sensing of tool wear, vibration, and acoustic emission are utilized, and surface roughness is measured.

9 A vital element of automation is *automatic gaging*. Some techniques are suitable for in-process measurement. At other times, a gaging station is set up within the machine tool or at an adjacent station, and the information gained is fed back to the control computer to make appropriate adjustments in cutting the next part.

The growth of NC/CNC has been rapid. Even though NC machine tools still represent a minority of the number of machine tools installed or produced, they account for more than half the value of current machine tool production. Some 40% of the currently manufactured NC machine tools are machining centers and 40% are lathes. In manually controlled machining, much time is wasted in waiting and material movement, and actual cutting takes place only about 20% of the time. Advances in integrating NC machine tools with material-moving devices such as special-purpose robots and pallet changers raised machine-tool utilization to 40% and, in special cases, even 70% of the time, greatly increasing productivity and reducing the number of machines required for a given output.

8-7-4 Optimization of the Cutting Process

The above-described procedure for selecting feeds and speeds is adequate for relatively small-scale production. In a competitive environment, *optimization of the process* becomes necessary. First, a management decision must be made regarding the criterion best applicable to the situation. If the backlog of orders is large and contractual obligations require fast delivery, the criterion will be *maximum production rate*. More usually, the criterion is *maximization of profit*; this requires a rather complex model of the entire production system. A less detailed analysis is sufficient to define production conditions for minimum cost per part, and only optimization according to the *minimum-cost criterion* will be discussed here. It will be assumed that constraints such as machine-tool capacity, surface finish, or surface integrity are not overriding, and that cost factors are

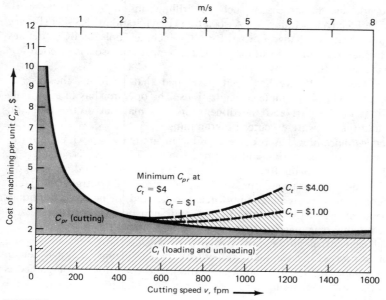

FIGURE 8-46
The optimum cutting speed for lowest-cost production is sensitive to tool cost and life (see Example 8-10).

known. Furthermore, it will be assumed that the constants in Taylor's equation are indeed constant; in reality, they have a certain statistical distribution and some analyses take into account the probabilistic nature of coefficients.

The cost of workpiece material and the loading and unloading time t_l (and hence the associated cost C_l) are independent of cutting speed and do not enter into this optimization (Fig. 8-46). The cutting time per part is t_c (min) and is charged at a total rate R_t (\$/min), which includes operator pay, overhead, and machine-tool charge. Cost must include an allowance for tool costs. If one tool costs C_t, takes time t_{ch} to change, and cuts N_t pieces before it has to be removed, the tool cost per piece C_{tp} is a $1/N_t$ fraction of the total cost

$$C_{tp} = \frac{1}{N_t}(t_{ch}R_t + C_t) \tag{8-23}$$

The total speed-dependent production cost is

$$C_{pr} = t_c R_t + \frac{1}{N_t}(t_{ch}R_t + C_t) \tag{8-24}$$

The cutting time t_c is simply the length of cut l divided by the cutting speed v

$$t_c = \frac{l}{v} \tag{8-25}$$

The number of pieces cut with one tool (N_t) is equal to tool life t divided by the time t_c it takes to cut one part

$$N_t = \frac{t}{t_c} \tag{8-26}$$

Substituting Eqs. (8-25) and (8-26) into Eq. (8-24)

$$C_{pr} = \frac{l}{v} R_t + \frac{t_c}{t} (t_{ch} R_t + C_t) \tag{8-27}$$

From Taylor's equation, Eq. (8.18'), tool life is

$$t = t_{ref} \left(\frac{C}{v} \right)^{1/n} \tag{8-28}$$

Substituting into Eq. (8-27)

$$C_{pr} = \frac{l}{v} R_t + \frac{l}{t_{ref} C^{1/n}} (t_{ch} R_t + C_t) v^{(1-n)/n} \tag{8-29}$$

The production cost C_{pr} is minimum where the first derivative with respect to v is zero

$$\frac{dC_{pr}}{dv} = 0 = -\frac{R_t l}{v^2} + \frac{1-n}{n} \frac{l}{t_{ref} C^{1/n}} (t_{ch} R_t + C_t) v^{(1-2n)/n} \tag{8-30}$$

Rearranging, we obtain the *cutting speed for minimum cost*

$$v_{c\,min} = C \left[\frac{n}{1-n} \left(\frac{t_{ref} R_t}{t_{ch} R_t + C_t} \right) \right]^n \tag{8-31}$$

Even this simple model shows that the most economical speed increases with increasing C and n (as one might expect) and also with R_t, i.e., increasing labor and equipment cost. Conversely, there is not much point in increasing speeds if much time t_{ch} is wasted on tool changing or, as evident from Fig. 8-46, if the nonproductive time t_l of loading and unloading is a large fraction of the total time. All too often, this point has been overlooked and the effort spent on increasing metal removal rates could have been better spent on improving material movement and providing workpiece-locating jigs and fixtures. The importance of these factors is recognized in the application of the computer to machining processes, as was discussed in Sec. 8-7-3.

Example 8-10

Bars of free-machining steel of diameter $d_0 = 4$ in and length $l_0 = 40$ in are rough turned with throwaway cemented carbide inserts. Calculate the variation of production cost with cutting speed if $R_t = \$12.60/h$, $C_t = \$4.00/edge$, $t_l = 8$ min, and $t_{ch} = 10$ min.

First, find cutting speed. From Fig. 8-44, $v_s = 500$ ft/min with 0.150-in depth of cut and feed of 0.020 in/r. From Sec. 8-1-7, prediction of tool life, $n = 0.25$; take $C = 3.5v_s = 1750$ ft/min.

Next calculate the cutting time. The diameter decreases by twice the depth of cut or $2(0.15) = 0.3$ in. The new diameter is $d_1 = 4.00 - 0.3 = 3.7$ in. The volume removed is

$$V = 40(4^2 - 3.7^2)\frac{\pi}{4} = 72 \text{ in}^3$$

On dividing the volume by the cross-sectional area of the chip, the length of cut is obtained:

$$l = \frac{72}{0.15(0.02)} = 24\,000 \text{ in} = 2000 \text{ ft}$$

(More simply, there are $40/0.02 = 2000$ turns, each of $\pi(4 + 3.7)/2 = 12.095$ in = 1-ft length.) Thus, cutting time is

$$t_c = \frac{l}{v} = \frac{2000}{v} \text{ min}$$

Net cost of cutting

$$C_c = t_c R_t = 2000\frac{12.60}{60v} = \frac{420}{v} \text{ \$}$$

The cost of loading and unloading

$$C_l = t_l R_t = \$1.68$$

The total production cost per piece, C_{pr}, is next calculated from Eq. (8-27) and is plotted in a solid line. Recalculation with $C_t = \$1.00$/edge shows (broken line in Fig. 8-46) the magnitude of speed increase one could afford if the cost of tooling were lower. Obviously, the loading time t_l is the most likely target in a production-organization effort.

8-8   ABRASIVE MACHINING

The term *abrasive machining* usually describes processes in which metal is removed by a multitude of hard, angular abrasive particles or grains (also called *grits*) which may or may not be *bonded* to form a tool of some definite geometric form.

From the earliest times, these processes have been of greatest importance because they are capable of producing very well controlled—and if required, smooth—surfaces and, if the process is properly conducted, also very tight tolerances. Furthermore, the hardness of the abrasive grit makes it possible to machine hard materials and, until the development of nontraditional machining processes, abrasive machining was often the only process for the manufacture of parts in certain materials. Grinding is still the dominant process for sharpening cutting tools. If properly conducted, abrasive machining can produce a surface of high quality, with a controlled surface roughness combined with a desirable residual stress distribution and freedom from surface or subsurface damage. Therefore, abrasive machining is often the finishing process for heat-treated steel and other materials.

In contrast to metal-cutting processes, in abrasive machining the individual cutting edges are *randomly distributed* and more or less *randomly oriented*, and the

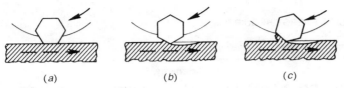

FIGURE 8-47
Depending on depth of engagement and attack angle, contact with an abrasive grit may result in (*a*) elastic deformation, (*b*) plowing, or (*c*) chip formation.

depth of engagement (the undeformed chip thickness) is small and not equal for all abrasive grains that are simultaneously in contact with the workpiece. These factors change the character of the process substantially.

8-8-1 The Process of Abrasive Machining

Abrasive grit is manufactured to present sharp edges; however, the orientation of individual grains is mostly a matter of chance, and a grain may encounter the workpiece surface sometimes with a positive, zero or, in most instances, a negative rake angle. There is thus no guarantee that cutting will actually take place:

1 With a very slight engagement (Fig. 8-47*a*), only elastic deformation of the workpiece, grit, and—if present—binder takes place. No material is removed, but substantial heat is generated by elastic deformation and friction.

2 With a larger depth of engagement, events depend on process conditions. At highly negative rake angles, the grit may simply plow through the workpiece surface, pushing material to the side and ahead of the grit in the form of a prow (Fig. 8-47*b*). Occasionally the prow is removed, but the process is inefficient. Friction on the grit is significant.

3 Only at less negative rake angles, higher speeds, and with less-ductile workpiece materials will typical chip formation take place (Fig. 8-47*c*). The clearance angle is often zero and may even be negative, thus introducing large energy losses in overcoming friction and elastic deformation.

4 Much work is expended in a small space, therefore, temperatures rise substantially. When materials such as steel are machined in air, combustion takes place and sparks are seen to fly.

Because only a fraction of grit–workpiece encounters results in actual material removal and because there are many sources of friction, abrasive machining is inefficient; the energy required for removing a unit volume (Table 8-7) may be up to 10 times higher than in cutting.

Most of the energy is converted into heat, only some of which is removed with the chips; much heat remains in the workpiece with a number of undesirable consequences. Discoloration of the surface is due to oxidation and the color is indicative of temperatures reached. The surface of heat-treated steels may reach high enough temperatures to cause a transformation to austenite. After the grit

TABLE 8-7
APPROXIMATE SPECIFIC ENERGY REQUIREMENTS
FOR SURFACE GRINDING*
Depth of Grinding: 1 mm (0.040 in)

Material	Hardness, HB	Specific energy, E_1	
		$hp \cdot min / in^3$	$W \cdot s / mm^3$
1020 steel	110	4.6	13
Cast iron	215	3.9	11
Titanium alloy	300	4.7	13
Superalloy	340	5.0	14
Tool steel (T15)	67 HRC	5.5	15
Aluminum	150 (500 kg)	2.2	6

*Extracted from *Machining Data Handbook*, 3d ed., Machinability Data
Center, Metcut Research Associates, Cincinnati, Ohio, 1980.

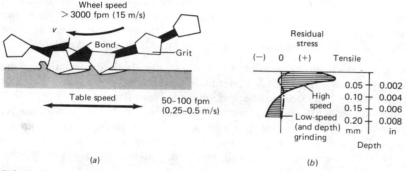

(a)   (b)

FIGURE 8-48
(*a*) A grinding wheel contains abrasive grit in a bond. (*b*) Grinding results in residual surface
stresses that can become highly tensile in abusive grinding.

has passed, sudden chilling causes the formation of martensite and associated
cracks. A deep blue color is indicative of such "burning."

Thermal cycling leads to residual tensile stresses; these are superimposed on
the residual compressive stresses generated by local deformation due to plowing
and cutting with a highly negative rake angle. Hence, a properly controlled
abrasive process can ensure a controlled surface roughness combined with high
surface integrity, whereas an improperly controlled process may result in substantial surface damage (Fig. 8-48*b*).

8-8-2 Abrasives

Abrasive grits must fulfill a number of often contradictory requirements.

1 High hardness at room and elevated temperatures helps to resist abrasion by
hard particles.

2 Controlled toughness or, rather, ease of fracture (*friability*) allows fracture to occur under imposed mechanical and thermal stresses. Thus, new cutting edges are generated on a worn grit, but at the expense of a loss of abrasive material.

3 Low adhesion to the workpiece material controls BUE formation, redeposition of grinding debris on the workpiece, and dislodging of grains from a bonded structure. At the same time, adhesion to the bond ensures the strength of a bonded abrasive.

4 Chemical stability increases wear resistance and resistance to corrosion by oxygen and cutting fluids.

5 The grain must have a shape that presents several sharp cutting edges. Because in-feed rates are low, grain size is several times larger than the depth of engagement. Grit size is specified as for other particulates (Sec. 6-1-1).

Only a few naturally occurring abrasives, primarily various forms of SiO_2 and Al_2O_3, find use and then only for softer workpiece materials or in finishing processes. The vast majority of industrial abrasives are manufactured. *Alumina* (Al_2O_3) is made in grades of varying hardnesses: the hardest grades are friable, suitable for light-duty precision and finishing operations, whereas the less-hard, tough grades (such as sintered alumina and alumina–zirconia abrasives) are suitable for heavy-duty stock removal. *Silicon carbide* (SiC) is harder than alumina (Table 8-3) but wears rapidly on low-carbon steel in which the carbon dissolves. This is the problem also with *diamond*, therefore, CBN is preferred for hardened steels.

8-8-3 Grinding

Grinding is the most widespread of all abrasive machining processes.

Grinding Wheels The abrasive is bonded, with an appropriate bonding agent, into a wheel of axial symmetry, carefully balanced for rotation at high speeds. The *grinding wheel* is a sophisticated tool made in a strictly controlled manufacturing environment.

1 The *strength of the bond*, which is governed by the quantity, kind, and distribution of bonding agent, is chosen so that the grit is held firmly, well supported, yet is still allowed to fracture. At some point, wear would result in unacceptably high forces on the grit, excessive heating, and poor surface quality; the grit must then be released to allow new grit to come into action. A natural consequence of this is that a wheel volume Z_s is lost in unit time. During this time, a volume Z_w is removed from the workpiece. It is usual to express the ratio of the two volumes as the *grinding ratio G*

$$G = \frac{Z_w}{Z_s} \qquad (8\text{-}32)$$

Since grit wears more rapidly when grinding harder materials, a general rule states that a softer (less strongly bonded) wheel should be used for grinding harder materials.

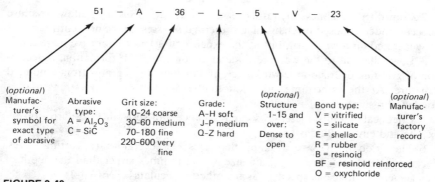

FIGURE 8-49
Grinding wheels are described by standardized nomenclature. (*From ANSI B74.13-1977, American Society of Mechanical Engineers, New York; also ISO 525-1975E.*)

2 Most frequently, the wheel is not solid but has a controlled porosity, a more or less open *structure*. Grains are bonded to each other by *bond posts* (Fig. 8-48a). A more open structure has slimmer posts, accommodates chips until they are washed away by the grinding fluid, and allows grinding fluid to move through the wheel.

The majority of grinding wheels are bonded with glass. Wheels with such *vitrified bonds* are the strongest and hardest, and the composition of the glass can be adjusted over a wide range of strengths. The so-called *silicate wheel*, bonded with water glass, is the softest.

Organic bonding agents are of lower strength but are available in a wide range of properties. *Resinoid wheels* are bonded with thermosetting resins and can be readily reinforced with steel rings, or fiberglass or other fibers, to increase their flexural strength. With the more flexible polymers such as shellac or rubber, very thin *cutoff wheels* can be made.

The size and size distribution of the grit and the openness of the structure all contribute to determining grinding performance; therefore, standard grinding wheel designations refer to all these factors (Fig. 8-49).

Process Variables *Operating (surface) speeds* are usually between 20 and 30 m/s (4000–6000 ft/min). There is usually an optimum speed at which the G ratio is highest. The material removal rate increases with the force imposed on the wheel but, beyond a certain limit, the G ratio drops steeply. Practical conditions represent a compromise, to give a high material removal rate with an economically tolerable grinding ratio. Speeds up to 90 m/s are achieved in high-speed grinding with specially constructed wheels. High rotational speeds impose large stresses on the wheel and all grinding wheels must be inspected for flaws prior to installation in balanced flanges.

Even the best-chosen grinding wheel undergoes gradual changes. It may develop a periodic wear pattern which leads to chatter and the emission of a howling sound. When grinding hard materials, wear causes glazing of the surface. Alternatively, the wheel surface may become clogged (*loaded*) with the workpiece material. For all these reasons, *dressing* of the wheel becomes necessary. This is usually done in the grinder itself so that alignment and wheel balance are not lost. Two basic dressing methods are used: the wheel is dressed by cutting with a *diamond point*, or is *crush-dressed* by pressing a high-strength steel roller against its surface. The latter method is very fast and particularly economical for dressing forming wheels.

An indispensable part of the grinding system is the *grinding fluid* (Table 8-2). It fulfills a triple function: first, it keeps the ground surface cool and may prevent burning and cracking of the surface of hard materials; second, it affects the cutting process and reduces loading and wear of the grinding wheel (and thus the grinding ratio); third, it reduces friction and thus greatly reduces heat generation. Therefore, as in cutting, both cooling and lubrication are important. However, the lubricating function is much more important for grinding fluids than for cutting fluids. Greatest advantages are gained with oils and aqueous fluids that incorporate lubricating additives. It is not unusual to find that specific power requirements drop by a factor of 4 or more; metal removal rates and the volume that can be removed before the onset of chatter increase fivefold or more; grinding ratios may increase by a factor of 10 or more.

Grinding Processes The geometry of grinding can be as varied as that of other machining processes (Fig. 8-50).

1 *Surface grinding* is practiced with the cylindrical surface of a wheel (as in slab milling), but the wheel is normally narrower than the workpiece and must be cross-fed (usually by giving the workpiece a transverse feed motion, Fig. 8-50a).

2 *Cylindrical grinding* (Fig. 8-50b) is similar in its results to turning, except that the fast-rotating grinding wheel now works on the surface of the more slowly rotating part, and individual cuts are short.

3 Cylindrical parts of great accuracy are obtained at high rates in *centerless grinding* (Fig. 8-50c): the workpiece is lightly supported on a workrest while the grinding pressure is taken up by the *regulating wheel* that rotates at about $\frac{1}{20}$ the grinding wheel speed.

4 In *internal grinding* (Fig. 8-50d) a small wheel works on the cavity of the workpiece and individual cuts are longer than in cylindrical (external) grinding.

5 The entire width of a flat workpiece may be ground with the annular end face of a *cup wheel* (Fig. 8-50e); this resembles face milling. Smaller pieces may also be ground on the end face of a cylindrical wheel; this is called *side-grinding*.

6 Apart from generating basic geometric surfaces, grinding is used for finishing many parts of complex shape, including threads and gears. As in other machining, both forming (Fig. 8-50f) and generation of surfaces (Figs. 8-22 to 8-24) are practiced.

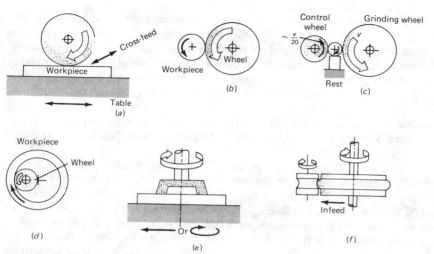

FIGURE 8-50
Various grinding processes: (*a*) surface grinding, horizontal spindle; (*b*) cylindrical grinding; (*c*) centerless grinding; (*d*) internal grinding; (*e*) surface grinding, vertical spindle; and (*f*) form (plunge) grinding.

For a given process geometry, the undeformed chip thickness and length of cut increase with increasing depth of wheel engagement, increasing feed rate, and decreasing wheel speed.

Grinding was originally regarded as a finishing process, however, its scope has been expanded considerably. Processes can be categorized according to undeformed chip thickness:

1 *Precision grinding.* Originally most grinding was performed to improve tolerances and surface finish. Undeformed chip thickness is small and specific energy requirements are high (Table 8-7). The process is sometimes controlled by applying a constant force rather than constant feed.

2 *Coarse grinding.* More recently, grinding has become a material removal process. Wheels are made to release worn grit without dressing, yet without excessive wear. Speeds are increased because higher metal-removal rates are then obtained; chip formation dominates, and specific energy drops to $5\text{--}10 \text{ W} \cdot \text{s}/\text{mm}^3$. Rough grinding (*snagging*) of castings and forgings—to remove in-gates or flash —has long been practiced.

3 *Creep-feed grinding.* The full depth is removed in a single pass but at a low workpiece feed rate. Heat buildup ahead of the wheel accelerates metal removal but without harmful effects, because the heat-damaged material is ground away.

A fine surface finish is no guarantee of good surface quality; the surface integrity of hardened components is ensured by the use of soft wheels, low speeds and in-feeds per pass, high work speeds, frequent dressing, and a copious flow of grinding oil.

Grinding lends itself to NC and CNC control. Adaptive control is possible, and CBN wheels with their long life are particularly suitable for producing precision parts.

Example 8-11

A heat-treated H13 steel block of HRC 55 hardness and 2×4 in surface area is ground on a vertical-spindle, reciprocating-table grinder, using a cup wheel of 6-in diameter (as in Fig. 8-50 e). *Machining Data Handbook* (3d ed., vol. 2, p. 8.45) recommends a wheel speed of 5000 ft/min, a table speed of 100 ft/min, and a downfeed of 0.002 in, using a grade A80HB wheel. A total of 0.015 in is to be removed. Calculate the wheel r/min and the total grinding time, assuming that reversal of the table takes 0.5 s.

It is usual to grind in the length direction. To obtain a uniform surface finish, the wheel must clear the workpiece at the end of the stroke. Thus the total stroke is $6 + 4 = 10$ in. Surface speed is 5000 ft/min $= D\pi N = 6\pi N/12$. Thus, $N = 3183$ r/min $= 3200$ r/min. The number of passes, at 0.002 in/pass, is $0.015/0.002 = 7$ passes.

Time/pass $=$ stroke/table speed $= 10(60)/12(100) = 0.5$ s/pass. Total time for 7 passes $= 7(0.5 + 0.5) = 7$ s.

8-8-4   Other Abrasive Processes

There is a large number of processes that use, for a variety of purposes, abrasive grit in one form or another.

Coated Abrasives   Traditionally, abrasive grains attached to a flexible backing such as paper or cloth have been used for low-speed finishing of surfaces. However, with the development of stronger adhesives and backings, *coated belts* operating at high speeds (typically up to 70 m/s) have become important production tools, capable of high metal-removal rates. Up to 6 mm (0.25 in) of material can be removed in one pass, at rates of 200 cm^3/min · cm width, replacing turning, planing, or milling in mass production (e.g., in machining the gasket surfaces of engine blocks and cylinder heads). *Double-disk grinding* is the fastest method of producing parallel surfaces.

Grains are deposited on a layer of adhesive applied to the backing (*make coat*) and are held in place by a second layer (*size coat*). Bond strength is balanced to prevent stripping of new grains while allowing release of worn grains. Grains with sharp edges and of elongated shapes are electrostatically aligned to give cutting edges of low negative rake angles, spaced some 10 times further apart than in grinding wheels.

Chip formation is the dominant metal-removal mode, and specific energy requirements and surface temperatures are relatively low. Grinding fluids serve to lubricate, cool, flush away the debris, and reduce clogging. A slotted contact wheel or platen is used behind the belt. Therefore, cutting is intermittent; heat

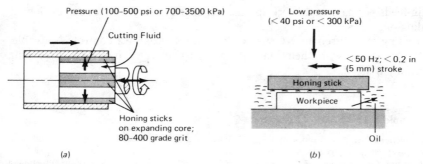

FIGURE 8-51
Schematic illustration of random-motion abrasive machining with bonded abrasives: (*a*) honing and (*b*) superfinishing.

can flow into the workpiece during the noncutting period, allowing high removal rates without burn.

Abrasive wire cutting utilizes CBN or diamond grit bonded to a wire surface.

Honing   In *honing* the abrasive is made into a slab (*stone, stick*). In the most frequent application (Fig. 8-51*a*), an internal cylindrical surface is finished with a number of honing sticks carried in an expanding, axially oscillating head, while the workpiece is rotated. Thus, a cross-hatch pattern is produced. A honing fluid is applied to wash out abrasive particles.

Superfinishing is a variant in which oscillating motion is imparted to a fairly large stone and surface pressure is kept very low. Thus, as the surface becomes flatter, it builds up its own hydrodynamic lubricant film which terminates the action of the abrasive (Fig. 8-51*b*).

Lapping   The process utilizes a form of three-body abrasive wear (Fig. 2-18*c*). The abrasive is introduced as an oil-based slurry between the workpiece and a counterformal surface called the *lap*.

In the best-known form, the lap is a relatively soft, somewhat porous (e.g., cast iron) table, rotated in a horizontal plane. The workpieces are loaded onto the surface (sometimes in cages driven by a gear from the center of the rotating table, Fig. 8-52*a*). The workpiece describes a planetary movement and acquires a very uniformly machined, random finish of excellent flatness. When the lap is made into a three-dimensional shape, curved surfaces (e.g., glass lenses) may be lapped. Lapping is also a very fast process for breaking in mating gears or worms, thus eliminating the need for a running-in period (e.g., for the hypoid gears of an automotive rear axle).

Less frequently, the lap is made of a bonded abrasive and the process is then similar to finish grinding but at a low speed.

Ultrasonic Machining   A piezoelectric transducer is used to generate ultrasonic (about 20 000-Hz) vibrations of small (about 0.04–0.08-mm) amplitude

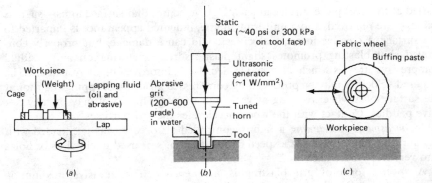

FIGURE 8-52
Machining with loose (unbonded) abrasives; (*a*) lapping, (*b*) ultrasonic machining, and
(*c*) buffing.

which drive the *form tool* made of a reasonably ductile material. Abrasive grit is
supplied in a slurry to the interface, and the workpiece is gradually eroded (Fig.
8-52*b*). The process is particularly suitable for machining less-ductile materials.

Buffing, Polishing, and Burnishing   In most of these processes, the abrasive is
applied to a soft surface, for example, the cylindrical surface of a wheel composed
of felt or other fabric (Fig. 8-52*c*). For *buffing*, the abrasive is in a semisoft
binder. For *polishing*, the abrasive may be used dry or in oil or other
carrier/lubricant. Both buffing and polishing are capable of producing surfaces of
high reflectivity which is attributable not to greater smoothness but to a smearing
(plastic deformation) of surface layers. Localized deformation of the surface
induces compressive residual stresses and improves the fatigue strength of the
components.

Barrel Finishing   A completely random abrasive process, *barrel finishing* or
tumbling is of great value in removing burrs and fins from workpieces and,
generally, in improving their surface appearance. In principle, the workpieces are
placed into a barrel which often has a many-sided cross section which makes
workpieces drop when the barrel is rotated. Mutual impact removes surface
protuberances. Much improved finish is obtained when a *tumbling medium* is
added, either in a liquid carrier (*wet medium*) or by itself (*dry medium*). The
medium is chosen according to the intended purpose, and may range from
metallic or nonmetallic balls to chips, stones, and conventional abrasives. Similar
results are obtained in *vibratory finishing*, with the barrel vibrated by some
mechanical means.

Grit Blasting   All contact with a tool or the other workpiece is eliminated in
grit or *shot blasting*. Particles of abrasives or of some other material (*shot*) are

hurled at the workpiece surface at such high velocities that surface films—such as oxides—are removed, and a uniformly matte, indented appearance is imparted to the surface. Excessive impact velocities could cause damage, but properly controlled shot blasting promotes slight plastic surface deformation and residual compressive stresses which, as noted in Sec. 4-4-7, improve fatigue resistance. The required velocities may be produced by compressed air or by dropping the shot at the surface of a wheel rotating at very high speeds. The wheel may or may not have paddles: Contact with the wheel accelerates the grit particles.

In *abrasive jet machining* a tightly controlled air or CO_2 gas jet, loaded with dry abrasives and flowing at a speed of 150–300 m/s, is used to cut slots or holes into very hard materials.

A special form of grit blasting is *hydrohoning*. The abrasive medium is suspended in a liquid which is then directed onto the surface in the form of a high-pressure jet. At very high (300 MPa (40 kpsi) and over) pressures, abrasives suspended in water can be used for *jet cutting* hard materials. *Abrasive flow machining* is a finishing process in which a semisolid, abrasive-filled medium is forced through a hole or across a surface.

All these processes are extensively used also for deburring. Burrs are formed in many operations, including die casting, forging, blanking, injection molding, and machining, and their removal represents one of the high-cost stages of processing.

8-9 CHEMICAL AND ELECTRICAL MACHINING

All machining processes discussed to this point are characterized by the mechanical removal of material in the form of chips, even though chip formation may be imperfect. There are a number of processes that remove material purely by chemical or electrical action, or both. Some of these processes are not new but have gained wider industrial application primarily because of the demands set by the aerospace and electronics industries; therefore, they are often denoted as *nonconventional* or *nontraditional* processes. As a group they are characterized by an insensitivity to the hardness of the workpiece material; hence, they are suitable for shaping parts from fully heat-treated materials, avoiding the problems of distortion and dimensional change that often accompany heat treatment.

8-9-1 Chemical Machining (CM or CHM)

It has been known for many years that most metals (and also some ceramics) are attacked by specific chemicals, typically, acids or alkalies. The metal is dissolved atom by atom and converted into a soluble compound over the entire exposed surface. In industrial metal-removal applications, only part of the surface is etched away and the remaining parts must be protected by a substance such as wax, paint, or polymer film (the *maskant* or *resist*). Thick films are deposited by dipping or spraying over the entire surface; the pattern to be etched is cut with a knife along a template, and the resist is peeled off (the coating is stripped in steps when parts of varying thickness are to be produced). Greater accuracies are

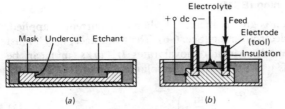

FIGURE 8-53
Material removal by (a) chemical etching is (b) accelerated on electrically conductive materials by electrochemical machining.

obtained by applying the resist through a silk or stainless-steel screen, using a stencil. Highest accuracies (better than 1 μm) can be obtained with photoresists (see Secs. 10-3-5 and 10-4). There are several applications:

1 *Engraving* has been practiced for hundreds of years by the artist and printer, and is now used for nameplates and instrument panels.

2 *Chemical milling* serves to remove pockets of material, as in thinning down integrally stiffened wing skins and other aircraft components.

3 *Chemical blanking* is used to cut through thin sheet. Printed circuit boards and parts made of thin sheet are fabricated by this technique.

The etchant dissolves material in all directions; therefore, it undercuts to approximately the same width as the depth of cut (Fig. 8-53a). Metal removal rates are given in Table 8-8.

TABLE 8-8
CHEMICAL AND ELECTRICAL MACHINING PROCESSES*

	Chemical machining	Electrochemical machining	Electrodischarge machining
Metal removal rate	0.012–0.07 mm/min (0.0005–0.003 in/min)	1.5 cm³/1000 A · min (0.1 in³/1000 A · min)	0.15–400 cm³/h (0.01–25 in³/h)
Surface finish			
μm AA	2 on Al 1.5 on steel 0.6 on Ti	0.1–1.4	0.75 at 0.25 cm³/h 5 at 8 cm³/h 10 at 50 cm³/h
μin AA	90 on Al 60 on steel 25 on Ti	4–50	30 at 0.015 in³/h 200 at 0.5 in³/h 400 at 3.0 in³/h
Electric current			
Volts		4–24	< 300
Amperes		50–40 000	0.1–500
Frequency		dc	500 000–200 Hz

*Extracted from *Machining Data Handbook*, 3d ed., Machinability Data Center, Metcut Research Associates, Cincinnati, Ohio, 1980.

8-9-2 Electrochemical Machining (ECM)

The rate of material dissolution is greatly increased when a dc current is applied. The process is the reverse of electroforming (Sec. 6-6). The workpiece, which must be conductive, is now the anode. Metal removal rates W_c (g/s·m²) can be calculated from Eq. (6-3). It is usual to express the volume removed in unit time

$$V_t = \frac{W_c}{\rho j} \quad \left(\frac{m^3}{A \cdot s} \right)$$ (8-33)

where ρ is density (g/m³) and j is current density (A/m²). The electrode feed rate is then

$$v_e = V_t j$$ (8-34)

On some metals an insulating oxide film may build up and can be broken down by intermittent spark discharges produced by an ac or pulsed dc circuit. Several versions of the process are used:

1 *Electrochemical milling* serves to remove material from large surfaces; the cathode is in the form of a distantly mounted flat plate.

2 *Electrochemical machining* (ECM) uses a cathode which is the negative of the shape to be produced. The cathode is fed into the workpiece at a controlled rate (Fig. 8-53b). The electrolyte is circulated—often through the cathode—to wash out the metal-hydroxide sludge and to ventilate the hydrogen formed in the course of electrolysis. The machine tool is rigidly constructed to prevent vibration and consequent inaccuracies. Rough guidelines on process variables are given in Table 8-8.

3 *Electrochemical grinding* uses a conductive grinding wheel (copper-bonded Al_2O_3 or metal-bonded diamond) as the cathode. Most of the material is removed by electrolysis. In contrast to conventional grinding, the metal remains cool, pressures are low, and the process is suitable for sharpening carbide cutting tools and for grinding delicate parts such as hypodermic needles and honeycomb structures.

Since metal removal occurs in the ionic state, the hardness of the material is of no consequence in either CHM or ECM; surface integrity is excellent; there is no heat damage; residual stresses are minimal or absent; surface finish is nondirectional. Superalloys, fully heat-treated steels, and aluminum alloys are often cut. Alloys susceptible to hydrogen embrittlement should be heated at 200 °C for a few hours after CM.

8-9-3 Electrical Discharge Machining (EDM)

In this process, chemical action is abandoned and metal is removed by the intense heat of electric sparks. The workpiece and the cathode (tool), made of metal or graphite, are submerged in a *dielectric fluid*, commonly a hydrocarbon oil

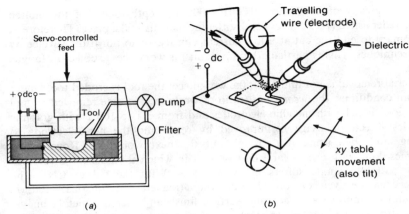

FIGURE 8-54
Electrodischarge machining is practiced with (a) form tools and (b) travelling wire.

(mineral oil). A direct current at a potential of up to 300 V is applied to the system; if a non-solid-state power supply is used, a capacitor is included in parallel with the spark gap (Fig. 8-54a).

At low voltages the fluid acts as an insulator; as the voltage builds up, the fluid suffers dielectric breakdown (large numbers of electrons appear in the conduction band) and a spark passes through the gap. Temperatures increase sufficiently to cause local vaporization of some of the workpiece material. After a discharge of controlled length, the voltage is dropped to a low value for a short time (waiting time) to reestablish the insulating film by deionization of the dielectric. The cycle is repeated at a rate of 200–500 000 Hz. The dielectric is supplied to the tool–workpiece interface to provide cooling and to flush out debris. Discharge always takes place at the closest gap; therefore, the electrode is fed continuously to cut the desired shape. Optimum conditions and spark gaps are maintained by servocontrol. Overall process control is now frequently performed by CNC.

Metal removal rate is a function of current density (Table 8-8). Empirically it is found to decrease for metals of higher melting point

$$V_{EDM} = 4T_C^{-1.23} \quad (cm^3/kA \cdot min) \tag{8-35a}$$

where T_C is the melting point in degree Celsius. In conventional units

$$V_{EDM} = 0.243T_C^{-1.23} \quad (in^3/kA \cdot min) \tag{8-35b}$$

Surface finish is governed by a number of factors and becomes rougher with higher current densities (which give higher discharge energies), more viscous dielectric, and lower frequency. The same factors that contribute to a rougher finish also result in more overcut and in a deeper heat-affected (damaged) zone, which is typically of 2–120-μm (0.0001–0.005-in) depth. Some of the molten metal is redeposited (*recast layer*) and softening may also take place. Therefore, it is customary to end the cut at a low current density, or to finish the surface by other techniques. This is particularly important for workpieces subject to fatigue loading.

Material removal is not limited to the workpiece: the tool is eroded too. Under optimum conditions, the *wear ratio* (ratio of workpiece volume removed to tool volume lost) is 3:1 with metallic electrodes and from 3:1 to 100:1 with graphite electrodes. Electrode usage is improved by restricting the new electrode to finishing, and then using it to rough out the next workpiece. Roughing is sometimes done with a no-wear EDM process in which the polarity is reversed and the graphite anode suffers no weight loss. With the aid of CNC, the workpiece may be given controlled lateral motion (planetary motion of 10–100-μm amplitude) to improve accuracy and surface finish and increase metal removal rates. It also becomes possible to machine complex shapes with simple electrodes moved in a complex path, rather like in three-axis contour milling. Automatic CNC tool changers may also be used.

The process is insensitive to the hardness of the material; therefore, it has found wide application for making forging, extrusion, sheet-metalworking, die-casting, and injection-molding dies into hardened steel die blocks. The copymilling machines are then used only to machine the electrodes. Special abrasive processes have also been developed for making graphite electrodes.

When tungsten wire is used as the electrode in conjunction with an aqueous dielectric, holes of small diameter (between 0.05 and 1 mm) can be made to great depths, as for cooling holes in turbine blades made of superalloys.

An important development is *electrical discharge wire cutting* (EDWC or *wire EDM*). The electrode is now a brass, copper, tungsten, or molybdenum wire of 0.08–0.3-mm diameter (Fig. 8-54b). The wire acts like a band saw, but sparks instead of teeth do the cutting. A slot (*kerf*) somewhat wider (by about 25 μm (0.001 in)) than the wire is cut. To maintain dimensional accuracy, the wire is fed continuously from the take-off spool at a rate of 2.5–150 mm/s. Cutting progresses at about 40 mm^2/min (4 in^2/h). A second cut may be taken with high-frequency ac current to eliminate the damaged surface. The quality of cut is adequate for many sheet-metalworking dies and other applications.

8-9-4 High-Energy Beam Machining

Materials can be machined—mostly cut or drilled—by melting and/or vaporizing the substance in a controlled manner. Processes are offshoots of joining processes, such as plasma-arc (Sec. 9-3-5), electron beam, or laser welding (Sec. 9-3-7) and will be discussed in Sec. 9-6. They are useful not only for metals but also for materials that are otherwise difficult to machine, for example, plastics and ceramics.

8-10 MACHINING OF POLYMERS AND CERAMICS

While our discussion has centered on metals, the principles discussed can be applied, with appropriate modifications, to plastics as well as ceramics.

8-10-1 Machining of Plastics

Even though plastics have a molecular rather than atomic structure, chip-forming processes can be applied to them if allowance is made for the differences in properties.

1 Compared to metals, plastics have a low elastic modulus and deflect easily under the cutting forces; therefore, they must be carefully supported.

2 Because of the viscoelastic behavior of thermoplastics, some of the local elastic deformation induced by the cutting edge is regained when the load is removed. Therefore, tools must be made with large relief angles and tools must be set closer than the finished size of the part.

3 In general, plastics have low thermal conductivity (Table 7-2); therefore, the heat buildup in the cutting zone is not distributed over the body and the cut surface may overheat. In a thermoplastic resin the glass-transition temperature T_g may be reached and the surface smeared or damaged, while thermal breakdown and cracking may occur in thermosetting resins. Therefore, friction must be reduced by polishing and honing the active tool faces and by applying a blast of air or a liquid coolant (preferably water-based, unless the plastic is attacked by it).

4 Since the shear zone is shortened and the cutting energy is reduced with a large rake angle (Sec. 8-1-1), cutting tools are made with a large positive rake angle (Table 8-4). This is permissible because the strength of plastics is low compared to metals. However, at excessive rake angles the cutting mechanism changes into *cleaving* in which coarse, disjointed fragments are lifted up and a very poor surface is produced.

5 Twist drills should have wide, polished flutes, a low ($> 30°$) or even zero helix angle, and a 60–90° point angle, particularly for the softer plastics.

6 Plastics can be surprisingly difficult to machine when reinforced by fillers. Glass fibers are particularly hard on the tool and it is not uncommon that only carbide or diamond tools can stand up.

In general, molding and forming methods (Sec. 7-7) produce acceptable surface finish and tolerances, and design usually aims at avoiding subsequent machining. Occasionally, however, machining is a viable alternative to molding (e.g., for PTFE, which is a sintered product and not moldable by the usual techniques).

Wood is a natural polymer of highly directional structure. Its relatively low strength allows working with highly positive rake angles, however, cleavage (splitting) ahead of the tool is a problem when the grain direction encourages cleavage at an angle that produces an increasing undeformed chip thickness.

Sheets of nonmetallic materials such as plastics, natural polymers (leather), wood, and friable ceramics and composites can be cut with a high-pressure water jet. The jet stream is pressurized to 440 MPa (64 kpsi), although higher pressures can be obtained. The nozzle is from 0.1–0.4 mm in diameter, and the high-velocity jet produces a clean, damage-free cut surface. The process is particularly advantageous when melting of the material is to be avoided.

8-10-2   Machining of Ceramics

Most ceramics are hard and act as abrasives themselves. Therefore their machining is very often limited to abrasion by a yet harder ceramic. Thus diamond can be used to dress grinding wheels or to finish tool bits or ceramic (e.g., Al_2O_3) components. All abrasive processes, including grinding, lapping, polishing, ultrasonic machining, gritblasting, and hydrohoning, are employed for both overall finishing and for localized shaping of ceramic (including glass) parts. Ceramics that are susceptible to chemical attack (as glass is to HF acid) can be chemically machined (etched).

Metal-matrix composites, such as carbide tool bits, can be conventionally ground with diamond, or the electrical conductivity of the matrix can be exploited by electrodischarge or electromechanical machining and grinding.

High-energy beam machining, and, particularly laser-beam cutting and drilling and scribing are suitable for ceramics and silicon crystals.

8-11   PROCESS LIMITATIONS AND DESIGN ASPECTS

Apart from shape limitations (to be discussed in Sec. 12-2-2), there are also dimensional limitations arising from the elastic deflection of workpieces under the forces imposed by the cutting tool and the work-holding devices. The problem is particularly serious for thin shells and slender shafts. For the same force, the deflection increases for a material of lower elastic modulus, and better means of support becomes necessary. Further limitations are mostly economic in nature; even though it may be possible to mill or turn a very hard material, the cost becomes extremely high (see Table 12-8). Similarly, machining of a very complex shape may be possible, but only by a sequence of expensive operations.

Design must, as always, take into account the manufacturing process. The sequence by which a part may be machined must be envisaged, and design features must be provided that ensure easy machining, preferably without the need of transferring the part to a different machine tool. In general, the following points need consideration:

1 The workpiece must have a *reference surface* (an external or internal cylindrical surface, flat base, or other surface suitable for holding it in the machine tool or a fixture).

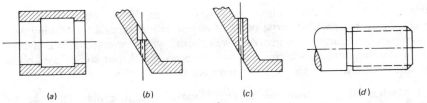

FIGURE 8-55
Some features of design for machining: (*a*) undercut chape, (*b*) hole at an angle to surface, spot-faced and (*c*) redesigned, and (*d*) run-out for thread-cutting tool.

2 If at all possible, the part shape should allow finishing in a single setup; if the part needs gripping in a second, different position, one of the already-machined surfaces should become the reference surface.

3 The shape of slender parts should permit adequate additional support against deflection.

4 Deflections of the tool in drilling, boring, or milling of internal holes and recesses limit the depth-to-diameter (or depth-to-width) ratio. Deep recesses or holes call for special (and more expensive) techniques, or a sacrifice in tolerances.

5 Undercuts (Fig. 8-55*a*) can be machined if not too deep, but they increase costs.

6 Radii (unless set by stress-concentration considerations) should accommodate the most natural cutting tool radius; the nose radius of the tool in turning and shaping, the radius of the cutter in milling a pocket, the sharp edge of the cutter in slot milling, or the rounded edge of a slightly worn grinding wheel or EDM tool.

7 Features at an angle to the main machine-movement direction call for a more complex machine, transfer to a different machine, interruption of the main machining action, and special tooling or special attachments, and should be avoided (except when a multiaxis machining center will be used).

8 Holes and pockets placed at an angle to the workpiece surface deflect the tool and necessitate a separate operation such as spot-facing (Fig. 8-55*b*) or redesign of the part (Fig. 8-55*c*).

9 One must not forget that a tool cannot be retracted instantly, and appropriate runout provisions must be made (Fig. 8-55*d*).

General features of conventional machining processes are given in Table 12-6.

8-12 SUMMARY

Material removal is, in a sense, admission of defeat: One resorts to it when other processes fail to provide the requisite shape, tolerances, or surface finish, or when the number of parts is too small to justify a more economical process. How often we fail is shown by the fact that more man-hours are spent on machining than on other unit processing methods. Machining is still indispensable for finishing the surfaces of some shafts; journal bearings; bearing balls, rollers, and races; accurately fitting machine parts; bores of engine cylinders; and for making parts of complex configuration in general. Even though valuable material is converted into low-value chips, machining can still be economical, but some basic points must be observed for both economy and quality.

1 Machinability of materials is highly variable, and a material condition ensuring low TS, n, and q should be chosen whenever possible.

2 Machining is a competition for survival between tool and workpiece; process conditions and tool materials must be chosen to prolong tool life while maximizing material removal rates. Tool life drops steeply with increasing temperatures. Temperatures in the cutting zone rise with cutting speed; therefore, material-removal rates are best increased by increasing first the depth of the cut, then feed, and only finally speed.

3 Finishing must be directed toward producing a surface that is within tolerances, has the proper roughness and roughness texture, and is free of damage (tears, cracks, smeared zones, excessive work hardening, and changes in heat-treatment conditions) and of harmful residual stresses.

4 Machining, as all other manufacturing processes, must be treated as a system in which the workpiece and tool materials, the lubricant (coolant), process geometry, and machine tool characteristics interact. This approach is essential because some of the interactions may easily go undetected, to the detriment of quality, production rates, and economy.

5 Substantial increases in productivity result when time lost in loading, unloading, checking, and tool change is minimized. NC and CNC machine tools, in combination with part and tool handling systems, can achieve very high capacity utilization. The advantages of NC machining can be fully realized only if the capabilities of advanced tool materials are exploited; for this, the machine tool must have sufficient rigidity.

PROBLEMS

8-1 Discuss, with the aid of sketches, the effect of rake angle on chip formation, cutting force, and power. (Consider rake angles ranging from highly positive to negative.)

8-2 Discuss the conditions that lead to BUE formation and the advantages and disadvantages of cutting with a BUE. Suggest ways in which the BUE can be avoided.

8-3 Make a sketch to show the effects of lead angle on undeformed chip thickness. Consider the variation of undeformed chip thickness in cutting with a ball-end mill.

8-4 An unidentified carbide insert is used for cutting AISI 1045 steel at the recommended speed. The tool wears extremely rapidly. Discuss the possible cause(s) of the problem.

8-5 Discuss the functions of cutting fluids and methods of their application.

8-6 Chips are found to clog up the machining space. Suggest remedies for cutting (a) ductile and (b) less-ductile materials.

8-7 Derive Eq. (8-3).

8-8 Make sketches to show the essential features of turning, boring, facing, and parting off. Always show the cross section of the undeformed chip, distinguishing chip thickness and depth of cut.

8-9 Discuss methods suitable for making deep holes, their relative advantages and disadvantages.

8-10 Discuss what material conditions should be specified for ease of machining (*a*) 1100 aluminum, (*b*) 7075 aluminum, (*c*) pure copper, (*d*) free-machining brass, (*e*) 1018 steel, (*f*) 1045 steel, (*g*) 1080 steel, and (*h*) Hastelloy X (refer to Tables 4-2 and 4-3 for material properties).

8-11 Find the conversion factor for K_1 (Eq. (8-13)) from SI to conventional units.

8-12 A lathe is driven by a 5-hp motor. Estimate the possible metal removal rates if the workpiece material is steel (HB 270), an aluminum alloy, and a nickel-based superalloy.

8-13 Take $n = 0.15, 0.25,$ and 0.4 for HSS, carbide, and ceramic tooling, respectively. (*a*) Find the change in tool life when increasing the cutting speed v_c to twice its initial value v_i. (*b*) From the answer to (*a*) would you recommend that experiments aimed at higher cutting speeds should be conducted with HSS in preference to carbide?

8-14 The full form of Taylor's equation, Eq. (8-18′), is valid in any measurement system. In SI units, v and v_r will be in m/s, and t and t_{ref} in s. However, a reference tool life of $t_{ref} = 1$ s would be quite unreasonable; therefore $t_{ref} = 60$ s is used. (*a*) Develop a general formula for converting C into v_r (in SI units). (*b*) If $C = 500$ ft/min, what is the value of v_r?

8-15 If $n = 0.25$ and $C = 1000$ ft/min, what should be v for a life of 1 h?

8-16 The threaded bushing shown in the illustration is to be made in large quantities (over 10 000 pieces per month). Determine the optimum screw-machine operation sequence by going through the following steps: (*a*) Clarify missing dimensions, tolerances, and surface finish specifications (in the absence of consultation with the designer, make common-sense assumptions), (*b*) choose the starting material dimensions and metallurgical condition, (*c*) determine which way the part should be oriented to allow finishing from the bar, (*d*) select the basic operations required, listing the appropriate speeds and feeds and machining times, and (*e*) determine the operation sequence, keeping in mind the possibility of simultaneous machining and of spreading lengthy operations over several positions.

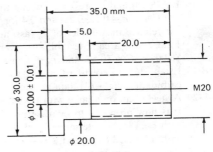

Material: AISI 1117 steel

8-17 Repeat the above procedure for the instrument screw shown.

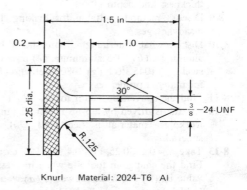

Knurl   Material: 2024-T6 Al

8-18 Rank the following materials in order of their anticipated machinability: 1008 steel, 1045 steel, 302 stainless steel, Muntz metal, 1100 Al, and nickel, all in the annealed condition. In formulating your judgements, refer to the appropriate equilibrium diagrams and to numerical data contained in this volume.

8-19 Holes of 0.25-in diameter and 1-in depth are to be drilled into 304 stainless steel. Determine the recommended (*a*) cutting speed (ft/min), (*b*) drill r/min, (*c*) feed (in/r), (*d*) feed rate (in/min), and (*e*) power requirement (hp). (*f*) Check whether the recommended values are feasible on commercially available equipment.

8-20 Holes of 12-mm diameter and 36-mm depth are to be drilled into 7075-T6 aluminum alloy. Determine the recommended (*a*) cutting speed (m/s), (*b*) drill r/min, (*c*) feed (mm/r), (*d*) feed rate (m/min or m/s), and (*e*) power requirement (kW). Check whether (*f*) the recommended values are feasible on commercially available equipment.

8-21 A rectangular slab of $4 \times 10 \times 1$ in dimensions, of 2024-T6 aluminum alloy, is to be face milled on the two larger surfaces. Determine, for a finishing cut, the (*a*) speed, (*b*) feed, and (*c*) net cutting time t_c (remember that the face-milling cutter has a diameter D_m and that it is usual, for uniformity of surface finish, to run the cutter clear off the ends of the workpiece).

8-22 The cutting speed for minimum-cost v_c is given by Eq. (8-31). (*a*) Substitute v_c into Eq. (8-18) and express the cutting time t for minimum-cost operation; denote it t_c. (*b*) Using $n = 0.12, 0.25$, and 0.4 for HSS, carbide, and ceramic, respectively, express the relative magnitudes of t_c (assuming identical tool and tool-changing costs). (*c*) Find t_c for a carbide tool with overhead cost $R_o = \$20.00/\text{h}$, machine cost $R_m = \$8.00/\text{h}$, $C_t = \$0.50$, and $t_{ch} = 1$ min.

8-23 Establish the lowest-cost cutting speed for rough turning a cast iron of HB 200 with carbide tooling. Cost data are the same as in Example 8.10.

8-24 A steel is cut on a lathe with HSS tooling. The cutting speed is taken from Fig. 8-44. It is now proposed to increase production rates by 30%. Using the constants given with Eq. (8-20), determine the change in tool life if increased production is obtained with increased (*a*) speed, (*b*) feed, or (*c*) depth of cut.

8-25 Describe open-loop and closed-loop control as applied to a milling machine. Distinguish between CNC and DNC as applied to a group of machining centers.

8-26 Discuss the mechanisms typical of abrasive grain–workpiece encounters. Identify those that contribute to metal removal.

8-27 A hardened tool steel (HRC 55) surface of 2×10 in dimensions is to be ground to a surface finish of 16–20-μin AA. Past experience shows that a suitable wheel (A-46-H-V) will produce the requisite finish with a cross feed of 0.020 in/pass and a grinding depth of 0.003 in. If the table speed is 60 ft/min and the wheel is 1.0 in wide, calculate (a) the number of passes taken (count table movement back and forth as separate passes, and take into account that the wheel starts and finishes outside the workpiece width); (b) the total time taken, remembering that the wheel must run out at each end of the workpiece (make a sketch to calculate wheel position at end); (c) the material removed, total and per minute; (d) the horsepower requirement.

8-28 The shape of the part in Example 12-6 (as defined by the solid lines) is to be sunk into a forging-die block by electrodischarge machining. Ninety-nine percent of the total material is removed at a high rate but the final one percent is removed at a low rate to produce a finish of 30 μin AA, free from surface damage. Calculate the total machining time.

FURTHER READING

A Detailed Process Descriptions

ASM: *Metals Handbook*, 8th ed., vol. 3, *Machining*, American Society for Metals, Metals Park, Ohio, 1967.

Wick, C. (ed.): *Tool and Manufacturing Engineers Handbook*, 4th ed., vol. 1, *Machining*, Society of Manufacturing Engineers, Dearborn, Mich., 1983.

B Recommendations for Machining Conditions

König, W., I. K. Essel, and I. L. Witte: *Specific Cutting Force Data for Metal Cutting*, Verlag Stahleisen, Düsseldorf, 1982.

Machining Data Handbook, 3d ed., Machinability Data Center, Metcut Research Associates, Cincinnati, Ohio, 1980.

C Texts

Armarego, E. J. A., and R. H. Brown: *The Machining of Metals*, Prentice-Hall, Englewood Cliffs, N.J., 1969.

Arshinov, V., and G. Alekseev: *Metal Cutting Theory and Cutting Tool Design*, Mir Publishers, Moscow, 1976.

Boothroyd, G.: *Fundamentals of Metal Machining and Machine Tools*, Scripta/McGraw-Hill, Washington, D.C., 1975.

Kaczmarek, J.: *Principles of Machining*, Peter Peregrinus, Stevenage, England, 1976.

Kronenberg, M.: *Machining Science and Application*, Pergamon, Oxford, 1966.

Shaw, M. C.: *Metal Cutting Principles*, Oxford University Press, Oxford, 1984.

Trent, E. M.: *Metal Cutting*, 2d ed., Butterworths, London, 1984.

D Materials and Cutting Tools

ASM: *Cutting Tool Materials*, American Society for Metals, Metals Park, Ohio, 1981.

ASM: *Influence of Metallurgy on Hole Making Operations*, American Society for Metals, Metals Park, Ohio, 1978.

ASM: *Influence of Metallurgy on Machinability*, American Society for Metals, Metals Park, Ohio, 1975.

ASM: *Machinability Testing and Utilization of Machining Data*, American Society for Metals, Metals Park, Ohio, 1979.

Brooks, K. J. A.: *World Directory and Handbook of Hard Metals*, 2d ed., Engineer's Digest, London, 1979.

Kalpakjian, S. (ed.): *New Developments in Tool Materials and Applications*, Illinois Institute of Technology, Chicago, 1977.

Komanduri, R. (ed.): *Advances in Hard Material Tool Technology*, Carnegie Press, Pittsburgh, 1976.

Komanduri, R.: *Tool Materials*, in *Kirk-Othmer Encyclopedia of Chemical Technology*, vol. 23, 3d ed., Wiley, New York, 1983, pp. 273–309.

Mills, B., and A. H. Redford: *Machinability of Engineering Materials*, Applied Science Publishers, London, 1983.

Kane, G. E. (ed.): *Modern Trends in Cutting Tools*, Society of Manufacturing Engineers, Dearborn, Mich., 1982.

Roberts, G. A., and R. A. Cary: *Tool Steels*, 4th ed., American Society for Metals, Metals Park, Ohio, 1980.

Sarin, V. K. (ed.): *High-Productivity Machining: Materials and Processes*, American Society for Metals, Metals Park, Ohio, 1986.

Thompson, R. W. (ed.): *The Machinability of Engineering Materials*, American Society for Metals, Metals Park, Ohio, 1983.

Weck, M.: *Handbook of Machine Tools*, 4 vols., Wiley, New York, 1984.

Williams, R. (ed.): *Machining Hard Materials*, Society of Manufacturing Engineers, Dearborn, Mich., 1982.

E Abrasive Machining

Andrew, C., T. D. Howes, and T. R. A. Pearce: *Creep Feed Grinding*, Industrial Press, New York, 1985.

Bhateya, C., and R. Lindsay (eds.): *Grinding: Theory, Techniques and Troubleshooting*, Society of Manufacturing Engineers, Dearborn, Mich., 1982.

Farago, F. T.: *Abrasive Methods Engineering*, American Society for Metals, Metals Park, Ohio, 1976.

McKee, R. L.: *Machining with Abrasives*, Van Nostrand Reinhold, New York, 1982.

Shaw, M. C.: *New Developments in Grinding*, Carnegie Press, Pittsburgh, 1972.

F Nontraditional Machining

Gillespie, L. K.: *Deburring Technology for Improved Manufacturing*, Society of Manufacturing Engineers, Dearborn, Mich., 1981.

Harris, W. T.: *Chemical Milling: The Technology of Cutting Materials by Etching*, Oxford University Press, Oxford, 1976.

Hoare, J. P., and M. A. LaBoda: *Electrochemical Machining*, in *Comprehensive Treatment of Electrochemistry*, vol. 2, Plenum, New York, 1981.

McGeough, J. A.: *Principles of Electrochemical Machining*, Chapman and Hall, London, 1974.

Metzbower, E. A. (ed.): *Applications of the Laser in Metalworking*, American Society for Metals, Metals Park, Ohio, 1979.

Weller, E. J. (ed.): *Nontraditional Machining Processes*, 2d ed., Society of Manufacturing Engineers, Dearborn, Mich., 1984.

G Specialized Books

Deutschman, A. D., W. J. Michels, and C. E. Wilson, Jr.: *Machine Design: Theory and Practice*, Macmillan, New York, 1975.

Kobalyashi, A.: *Machining of Plastics*, McGraw-Hill, New York, 1967.

Kokmeyer, E. (ed.): *Better Broaching Operations*, Society of Manufacturing Engineers, Dearborn, Mich., 1984.

Lambert, B. K. (ed.): *Milling Methods and Machines*, Society of Manufacturing Engineers, Dearborn, Mich., 1982.

Milling Handbook of High-Efficiency Metal Cutting, Carbaloy Systems Department, General Electric Company, Detroit, Mich., 1980.

Schlesinger, G., F. Koenigsberger, and M. Burdekin: *Testing Machine Tools*, 8th ed., Pergamon, Oxford, 1978.

Turning Handbook of High-Efficiency Cutting, Carbaloy Systems Department, General Electric Company, Detroit, Mich., 1980.

H Journals

American Machinist
Cutting Tool Engineering
Machine and Tool Blue Book
Modern Machine Shop

CHAPTER 9

JOINING PROCESSES

Joining differs from previously discussed processes in that it takes parts produced by other unit processes and unites them into a more complex part; therefore, it could also be regarded as a method of assembly. Some joints are purely mechanical, and within this category those devices that establish semipermanent joints (such as screws and bolts) are properly dealt with in a discussion of assembly. Other joints create permanent interatomic bonds, and the product becomes a continuous unit. In accordance with the emphasis of this book on unit processes, only the latter group of techniques (Fig. 9-1) will be discussed in any detail. The product may replace a part that could have been made by other techniques (e.g., a cast machine-tool frame may be replaced by a welded frame) or it may be of a kind that can be produced only by joining processes (e.g., a monocoque automobile body, an automotive radiator, or a bicycle frame). While most joining methods are practiced on metals, ceramics and polymers may be joined by similar techniques.

9-1 MECHANICAL JOINING

In addition to the semipermanent screw joint, there are several techniques for establishing a joint by *mechanical* means.

1 The most common mechanical fastener is the *rivet*. Whether it be solid or hollow (Fig. 9-2a and b), it makes a joint by clamping the two parts between heads. One head is usually formed in a prior operation; the resulting rivet is fed through predrilled or punched holes, and the second head is produced by

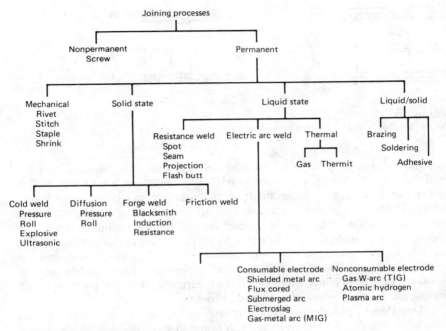

FIGURE 9-1 Classification of joining processes.

FIGURE 9-2
Permanent mechanical joints: (*a*) rivet, (*b*) tubular rivet, (*c*) staple, and (*d*) seam.

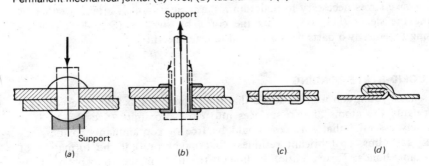

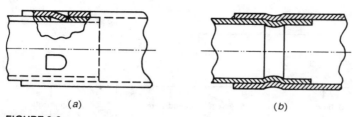

(a)   (b)

FIGURE 9-3
Permanent mechanical joints: (a) lanced tab and (b) crimped joint.

upsetting, either cold or hot (for upsetting, see Sec. 4-4-1). On a hollow rivet, the head is formed by flaring, an operation related to flanging of a tube (Sec. 5-3-4).

The hole represents a discontinuity in the structure and could cause fatigue failure. Therefore, the edges are deburred to remove stress raisers; for more critical applications, the hole is reamed or, to induce residual compressive stresses, the hole is slightly expanded by passing a larger pin through it. Riveting has lost its dominance in building construction and in the manufacture of automotive frames, but it is still of great importance. For example, tens of thousands of riveted joints are made on many airplanes. For greater consistency, riveting can be mechanized or entrusted to robots.

2 Thin sheets can be joined without preliminary drilling by *stitching or stapling* (Fig. 9-2c). Stapling is extensively used to fasten sheet to wooden backing.

3 *Seams* (Fig. 9-2d) are produced by a sequence of bends on tight ($\frac{1}{2}$ sheet thickness) radii (Sec. 5-3). Lock seams can be made impermeable with or without fillers such as adhesives, polymeric seals, or solder. Some seams are along straight lines, as the seams of radiator tubes and the side seam of three-piece beverage cans; others are along the edges of circular parts, such as can lids (Fig. 5-21), and then the process is related to flanging.

4 Joints are also produced by creating mechanical interference with the aid of plastic deformation, as in twisting or bending of *lanced protrusions* (Fig. 9-3a) and *crimping* (Fig. 9-3b).

5 *Shrinking* a sleeve onto a core is applicable mostly to round parts. The compressive stress necessary to maintain a permanent joint is attained either by heating the sleeve (and/or cooling the core), by swaging (Sec. 4-4-6), or by pressing together two parts with a conical interference fit.

9-2 SOLID-STATE BONDING

In discussing adhesion we mentioned that interatomic bonds may be established by bringing the atoms of two surfaces into close proximity (Sec. 2-2-1). It is absolutely essential that surfaces should be free of contaminants (oxides, adsorbed gas films, or lubricant residues) that might prevent the formation of interatomic bonds. This condition is difficult to satisfy in the atmosphere of the Earth (see Fig. 2-16a) and then measures must be taken to neutralize the effects of surface films:

1 *Relative movement* between surfaces helps to break up surface films. Roughening of the surface by wire brushing is helpful because, on joining, the ridges deform.

2 Plastic deformation of the contacting bodies causes a growth, *extension of the interfacial surface*, and, if surface films are unable to follow the extension, new, fresh surfaces are exposed which then form solid-state welds.

3 While theoretically no pressure would be required for bonding perfectly mating and clean surfaces, in practice a certain *normal pressure* is necessary to ensure conformance of the contacting surfaces and to break up surface films.

4 *Heat* is not an essential part of the basic bonding process, but softening of the materials promotes intimate contact and diffusion of atoms helps to achieve bonding. Diffusion is objectionable, however, when two dissimilar metals form intermetallic compounds that embrittle the joint.

In principle, any two materials can be bonded and, indeed solid-state bonding is often applied when other techniques fail. Nevertheless, the best bonds are obtained between metals when there is atomic registry (i.e., atoms of the two components are similarly spaced and crystallize in the same lattice structure). This means that metals bond best to themselves and to other metals with which they form solid solutions.

9-2-1   Cold Welding

The term *cold welding* (CW) is used loosely to describe processing at room temperature.

1 *Lap welding* relies on a 50–90% expansion of surfaces when indentors penetrate the sheets to be joined (Fig. 9-4a). Shoulders on the indentors limit distortion and promote welding.

2 *Butt welding* of wires establishes the joint by upsetting the wire ends to cause surface expansion (Fig. 9-4b). Welding is further aided when a twist is given.

3 *Roll bonding* (Fig. 9-4c) is highly effective because large extensions can be secured. Bonding can be locally prevented by depositing a parting agent such as graphite or a ceramic in a predetermined pattern. Inflation by pressurized air or fluid then yields parts such as evaporator plates (Fig. 9-4d).

4 In *explosive welding* (EXW) deformation of the interface is ensured by placing the sheets or plates at an angle to each other (Fig. 9-5a). When the explosive mat placed on top of the inclined cladding sheet (*fly plate*) is detonated, the sheet joins to the base metal by forming tight whirls or vortexes at the interface. This technique is useful also for in situ expansion of tubes into the head plates of boilers and tubular condensors.

5 Relative movement of the interface is induced in *ultrasonic welding* (USW) by tangential vibration (Fig. 9-5b). There is no massive deformation, and the process is suitable for lap welding foils and delicate instrument and electronic

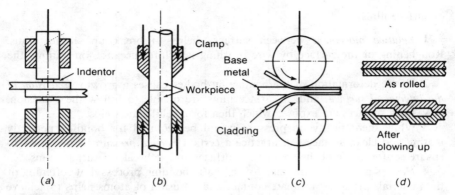

FIGURE 9-4
Parts may be joined in the solid state by: (a) cold lap welding, (b) butt welding, and (c) roll bonding. (d) Passages may be created by inflating roll-bonded assemblies.

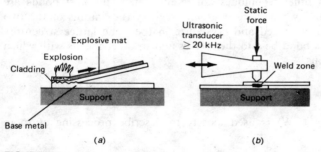

FIGURE 9-5
Solid-state joints are also established in (a) explosive joining and (b) ultrasonic welding.

components. When the welding tip is replaced by a roller, seam welds can be produced.

Example 9-1

In replacing silver-alloy quarters and dimes, nickel was unacceptable in the United States where vending machines test for magnetic properties. Therefore, a copper core is clad with cupronickel (75Cu–25Ni), giving the desired whiteness and lack of ferromagnetism. A sandwich of 7.5-mm thickness is rolled to 1.36 mm; the surface extension, together with the solubility of nickel in copper (Fig. 3-4), ensures a permanent bond. Calculate the proportion of new, atomically clean surface, assuming that preexisting surface films do not expand at all.

From constancy of volume (Eq. (2-2))

$$V = h_0 A_0 = h_1 A_1$$

$h_0/h_1 = A_1/A_0 = 7.5/1.36 = 5.515$; thus, the new surface occupies $(5.515 - 1)/5.515 = 81.9\%$ of the interface.

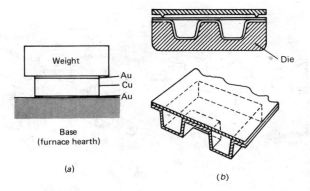

FIGURE 9-6
Elevated-temperature bonding is possible without substantial deformation in (a) diffusion bonding which may be (b) combined with superplastic forming to make complex structural parts.

9-2-2   Diffusion Bonding

Generally, better bonding is obtained when the temperature is high enough to ensure diffusion—typically, above $0.5T_m$ (Sec. 4-1-3).

Diffusion welding (DFW) is not new; for centuries the goldsmith has made filled gold by placing a weight on top of a sandwich composed of gold face sheets over a silver or copper core (Fig. 9-6a). When this sandwich is held in a furnace for a prolonged time, a permanent bond is obtained. The required pressure may also be generated in a press, or by restraining the assembly with a fixture made of a lower-expansion material (frequently, molybdenum).

In the 1970s, the technique was extended to airframe construction. Parts of complex shape are made of titanium alloys; as mentioned in Sec. 6-2-1, titanium dissolves its oxides. Simultaneous deformation further contributes to the development of a sound joint; therefore, *diffusion bonding combined with superplastic forming* (Fig. 9-6b) has proven most successful. The assembly is placed in an evacuated box (retort) which is held in a press.

In applying the techniques to aluminum alloys, deformation is vital to break up the stable, brittle oxide. Many composites are also made by diffusion welding (e.g., titanium or aluminum reinforced with boron fiber).

When bonding would not take place because of lack of solubility or would result in the formation of a brittle intermetallic compound, interlayers that are mutually soluble can be used in the form of foils or thin sheets. For example, superalloys are bonded after electroplating with a Ni–Co alloy.

9-2-3   Forge Welding

In a general sense, the term *forge welding* can be used to describe welding by deformation in the hot-working temperature range.

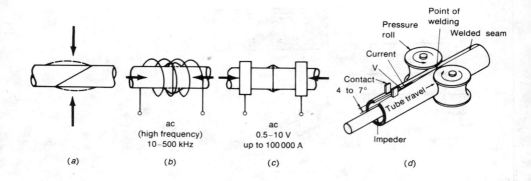

FIGURE 9-7
Elevated-temperature solid-state joining is obtained in (*a*) forge welding, (*b*) induction welding, (*c*) electric butt welding, and (*d*) high-frequency resistance (longitudinal butt seam)
welding. (*Part (d) from Metals Handbook, 9th ed., vol. 6, American Society for Metals, 1983,
p. 760. With permission.*)

1 In a more specific sense, *forge welding* (FOW) refers to the oldest industrial
welding process. The bond is created by substantial local deformation of the joint.
The hot, preshaped workpieces, usually of iron or steel, are forged together to
squeeze out oxides, slag, and contaminants, and ensure interatomic bonding (Fig.
9-7*a*). The technique was used not only for joining (e.g., forging of chain links)
but also for the welding of tubes, and for building up—layer by layer—medieval
swords and even very large objects, such as anchors.

2 Forge welding in which the ends of workpieces are pressed together axially
(*butt welding*) is possible, but the joint quality tends to be poor.

3 In more recent variants of the process, heat is provided by induction heating
to minimize oxidation (Fig. 9-7*b*). Much less deformation is sufficient; therefore,
butt welding of workpieces becomes practicable. An important application is to
the manufacture of pipe and tubing. The tube is formed by roll forming (Sec.
5-3-3) and the longitudinal seam is then made by *high-frequency induction welding*
(HFIW). The operating frequency is chosen to give optimum penetration; higher
frequencies penetrate to a lesser depth.

4 The heat may also be generated by passing a current through the compressed
faces (Fig. 9-7*c*). *Electric butt welding* has now largely been replaced by flash butt
welding (Sec. 9-3-4). However, *high-frequency resistance welding* (HFRW) of
longitudinal seams (Fig. 9-7*d*) is an important method for making large quantities
of tubing, structural members, and wheel rims. In both HFIW and HFRW there
may be a localized molten zone formed and then immediately squeezed out by
compression.

5 *Hot roll-bonding* (the high-temperature version of Fig. 9-4*c*) has been used
extensively to create composites of low cost or high performance. Thus Alclad

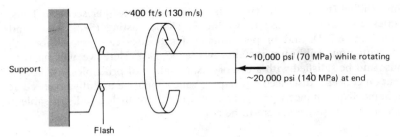

FIGURE 9-8
In friction welding heat is generated by rotating the workpieces against each other, and the bond is established by upsetting.

combines the corrosion resistance of a pure aluminum cladding with the high strength of a precipitation-hardened aluminum core. Stainless steel is clad to mild steel for corrosion protection, and alloys of different heat expansion are bonded to make thermostat strips.

9-2-4 Friction Welding (FRW)

The frictional work generated when two bodies slide on each other is transformed into heat; when the rate of sliding is high and the heat is contained in a narrow zone, welding occurs.

In one form of friction welding (*continuous-drive* FRW, Fig. 9-8), one part is firmly held while the other (usually of axial symmetry) is rotated under the simultaneous application of axial pressure. Temperature rises, partially formed welded spots are sheared, surface films are disrupted; rotation is suddenly arrested and a further upset force is applied when the entire surface is welded. Some of the softened metal is squeezed out into a flash, but it is not fully clear whether melting actually takes place. The heated zone is very thin; therefore, dissimilar metals are easily joined. For example, mild steel shanks can be fastened to high-speed-steel tools.

In *inertia-drive* FRW rotation is imparted by a flywheel, the energy of which is calculated so that the weld is completed when rotation stops.

9-3 LIQUID-STATE WELDING

In the great majority of applications, the interatomic bond is established by melting. When the workpiece materials (*base* or *parent materials*) and the *filler* (if used at all) have similar but not necessarily identical compositions and melting points, the process is referred to simply as *welding*.

Welding is closely related to casting processes. Heat is provided to melt the base metal and filler. The melt is physically contained in the melt zone where, through its contact with the surrounding base metal of high thermal conductivity, it is rapidly chilled (typically, at tens or hundreds of degrees per second). Thus, cooling rates are between those prevailing in shape casting (Sec. 3-8-2) and atomization (Sec. 6-2-1), and proper control of processes calls for a thorough familiarity with nonequilibrium metallurgy (Sec. 3-3-3). Here we will have to compromise and be satisfied with a cursory examination of principles, based to a great extent on concepts discussed in Chap. 3, which could be usefully reviewed at this time. Some acquaintance with material effects is nevertheless indispensable if limitations of various processes are to be appreciated.

9-3-1 The Welded Joint

A welded joint is far from homogeneous. The degree of inhomogeneity and complexity increases from pure metals to multiphase alloys, and is also a function of heat input per unit distance. Larger heat input gives greater penetration and thus a larger weld size; by heating deeper, it reduces the cooling rate and changes both the metallurgical structure and stress distribution in the weld. The following discussion is, therefore, generalized and does not necessarily apply for all rates of heat input.

Single-Phase Materials A section through the joint in a pure metal (Fig. 9-9), such as aluminum or copper, welded with a rod of identical composition, shows that the applied heat has melted some of the workpiece material, and the base

FIGURE 9-9
Fusion melting of a cold-worked pure metal or solid-solution alloy results in decreased strength in the heat-affected zone (HAZ).

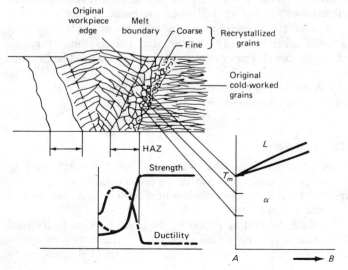

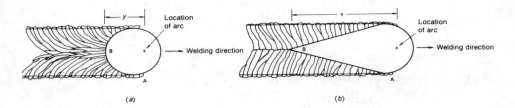

FIGURE 9-10
In making line welds, heat input per unit length determines the shape of the weld pool and the structure of the solidified weld bead: (*a*) at low travel speeds grains are interwoven, but (*b*) at high speeds a weak center plane forms if low-melting constituents are present.

material adjacent to the melt boundary bears evidence of exposure to high temperature.

If the workpiece material was originally cold worked and therefore of highly elongated grains, the *heat-affected zone* (HAZ) will show recrystallization. For a given cold work, grain size increases with increasing annealing temperature (Fig. 4-6); therefore, the very coarse grains found at the melt boundary gradually change to finer ones until, at the edge of the HAZ, only partial recrystallization is evident. If the workpiece was of annealed material, the further heat input during welding just coarsens the grains. In either case, a coarse-grained structure exists at the melt boundary. Solidification begins at this boundary by *epitaxial growth*, i.e., by the deposition of atoms in the same crystal orientation as the surface crystals (from the Greek *epi* = upon, *teinen* = arranged, and axis). This leads to the development of coarse, columnar grains in the weld material itself.

Viewed from the top, there is a weld pool at the point of maximum heat input (Fig. 9-10). When making a line weld, the heat source is moved along at a set travel speed. When the speed is low, the solidifying grains turn to follow the heat source, and the center of the weld bead has interwoven, independently nucleated grains (Fig. 9-10*a*). However, at high travel speeds the weld pool becomes elongated and solidification proceeds almost perpendicular to the melt boundary toward the center. If there are low-melting constituents, they segregate at the boundaries of cast grains and hot cracking may result.

Complicating factors enter in solid-solution alloys. Melting occurs over the $T_L - T_S$ temperature range (Fig. 3-4) and there will be a partially melted, mushy zone at the melt boundary. Minor concentrations of low-melting alloying elements or contaminants can segregate at this boundary to cause hot-shortness and cracking during cooling. Dendrites form in the weld metal, just as they do in casting (Sec. 3-1-2) but, as a result of faster cooling, the secondary dendrite arm spacing is smaller.

Two-phase Materials Most technically important alloys have a two-phase or multiphase structure, and their suitability for welding is determined by the events taking place in both the liquid and solid states.

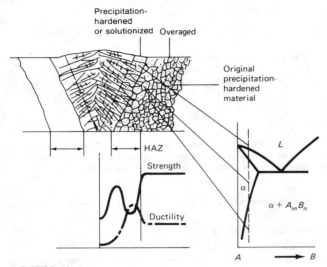

FIGURE 9-11
In welding a precipitation-hardened material, the weld bead is both weaker and less ductile.

1 Materials of eutectic composition present little problem because of their favorable solidification mode (Sec. 3-1-3).

2 Precipitation hardening alloys (Sec. 3-5-3) can be readily welded in the annealed state; if their freezing range is short, the effects of welding are mostly washed out on subsequent solution treatment and aging of the whole weldment. If the weldment is too large for this, fusion welding is possible in either the solution-treated or aged condition, but then the strength advantages of heat treatment are lost in the weld zone (Fig. 9-11). The weld itself may contain the intermetallic constituent in a coarse form, resulting in low strength and ductility. The immediately adjacent HAZ has been rapidly heated and quenched and would thus be in the solution-treated condition; however, heat conducted from the weld zone (back-heating) during cooling may cause overaging. Therefore, the HAZ is fully heat treated but of a coarse grain, with high strength and moderate ductility. Farther away, the original structure becomes overaged and soft. Apart from the poor strength of the joint, the locally varying composition can also lead to corrosion problems, e.g., in aluminum and magnesium alloys (a compromise solution is to use a different filler rod).

3 Solid-state phase transformations lead to complex changes. To take a medium-carbon steel in the annealed state as an example (Fig. 9-12), the parent metal has a microstructure consisting of pearlite colonies alternating with ferrite grains. The weld itself has the usual coarse, cast structure. Next to it the material has been heated high into the austenitic temperature range and cooled from there relatively slowly, resulting in coarse grains. With decreasing temperatures, the austenite grains also become finer; thus, finer-grain ferrite and pearlite are found

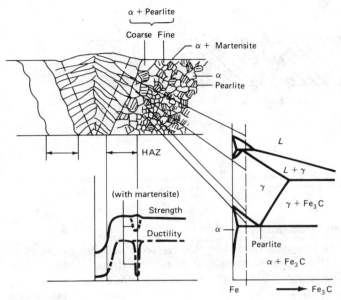

FIGURE 9-12
Heating into the austenitic temperature range can lead to martensite formation and resultant loss in ductility when welding carbon steels.

in the transformed structure. The edge of the HAZ has been heated just above the eutectoid temperature. Fast cooling rates will convert any austenite into martensite, and a large drop in ductility may result. With higher (above approximately 0.5%) carbon content, martensite will inevitably form, as it will in alloy steels.

Dissimilar Materials The situation becomes further complicated when two dissimilar materials are used, either in the workpiece or the filler. This is frequently the case because nonmatching fillers make it easier to obtain crack-free welds. The heating and cooling history is compounded by the effects of alloying, and events in the weld zone are determined by the equilibrium diagrams relating to both materials. Nonequilibrium solidification complicates the matter further.

As might be expected, alloys that form solid solutions present no problems. Eutectics tend to be more brittle, although they are favorable when both phases of the eutectic are ductile. Intermetallic compounds invariably embrittle the structure to make it useless; thus, for example, copper cannot be joined to iron. It is possible, however, to use a mutually compatible interface, in this case nickel, which forms solid solutions with both copper and iron.

In many instances the difference between melting points is very large, so that the parent metal does not melt, and then it is customary to classify the process as brazing or soldering (Sec. 9-4).

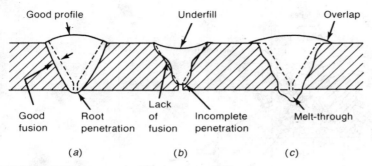

FIGURE 9-13
Fusion welds made with filler metal may show several characteristics: (*a*) a well-formed bead, (*b*) lack of fusion, lack of penetration, underfilling, or (*c*) melt-through, overlap.

9-3-2 Weldability and Weld Quality

It is obvious from the foregoing discussion that the term *weldability* denotes an extremely complex collection of technological properties. If any of the requirements for producing a sound weld are not met, *welding defects* may occur. They can exist in welds of any geometry and origin, but for illustrative purposes a fusion weld with filler metal may be used (Fig. 9-13*a*).

1 Fusion welding is a melting process and must be controlled accordingly (Sec. 3-7). Melting temperature determines—together with specific heat and latent heat of fusion—the required heat input. High thermal conductivity (or, rather, thermal diffusivity) of the base metals allows the heat to dissipate and therefore requires a higher rate of heat input and leads to more rapid cooling. The rate of heat input (usually expressed as *heat input per unit length*) must be matched to the thickness of the workpieces, the rate of weld metal deposition, and the travel speed (the speed of moving along the weld bead). Insufficient heat input causes *lack of fusion* and, in thicker cross sections, *lack of penetration* (Fig. 9-13*b*). Insufficient rate of weld metal deposition causes *underfilling*. Excessive heat input can lead to *melt-through*, and high deposition rates result in *overlap* which gives a notch effect (Fig. 9-13*c*).

2 Surface contaminants including oxides, oils, dirt, paint, metal, platings, and coatings incompatible with the workpiece material result in *lack of bonding* or lead to *gas porosity*, and must be kept out by mechanical and/or chemical preparation of the surfaces.

3 Undesirable reactions with surface contaminants and with the atmosphere are prevented by sealing off the melt zone with a vacuum, a protective (inert) atmosphere, or a slag. As in melting for casting, the slag is formed by dissolving oxides in a flux. Care must be taken to prevent *slag entrapment* which would lead to weakening of the weld by inclusions.

4 Gases released or formed during welding (e.g., CO) can lead to porosity which weakens the joint and acts as a stress raiser. High fluidity of the melt is

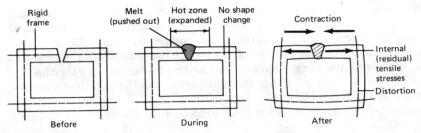

FIGURE 9-14
Expansion during welding followed by contraction on cooling can lead to distortion and residual stresses.

helpful in allowing slag and gases to rise to the surface. Particularly dangerous is hydrogen which originates from atmospheric humidity or damp flux. It causes porosity in aluminum alloys and leads to *hydrogen embrittlement* in steel. According to one explanation, hydrogen accumulates at microcracks, changes into the molecular form, and the hydrogen gas thus formed builds up high enough pressures to cause embrittlement of steels in which martensite forms upon cooling.

5 Solidification shrinkage coupled with solid shrinkage imposes internal tensile stresses on the structure (Fig. 9-14) and may lead to *distortion* and *cracking*. The problem is magnified when the structure is not free to shrink, in other words, when mechanical restraints are imposed. The properties of the base and filler materials are important: High thermal expansion results in greater distortion and residual stresses; materials that contain a low-melting phase are in greater danger of cracking; this can be alleviated with a less-alloyed, more ductile filler that reduces hot-shortness. Design is often based on the fracture-mechanics approach (Sec. 2-1-5) and the allowable flaw size is determined experimentally or theoretically. Residual stresses can also lead to stress-corrosion cracking.

6 Metallurgical transformations, discussed earlier, are of great importance especially when they lead to the formation of brittle phases such as martensite. The weld material is transformed if deposited in more than one pass (*multipass welding*).

7 The absolute and relative thicknesses of parts to be joined and the design of the joint have a powerful influence on heating and cooling, and thus on weldability.

8 The welding process itself imposes varying conditions, and thus the weldability of the same material is different in different processes.

Some of the above-mentioned difficulties can be alleviated by various measures:

Preheating the weld zone or the entire structure reduces the energy input required for completion of the weld (important for materials of high diffusivity such as aluminum or copper); reduces cooling rates in the weld and HAZ (allowing the welding of hardenable steels and other materials in which fast cooling results in the formation brittle phases); aids in diffusing H away; and reduces differential shrinkage, distortion, and residual stresses.

Peening (hammering or rolling) of the weld bead improves the strength of welds. In multipass welding it removes slag that might become entrapped and it induces recrystallization of earlier layers (however, the last pass is not peened because ductility would be lost).

Postwelding heat treatment of the entire welded structure is often essential for a variety of reasons:

1 Stress-relief anneal (Sec. 3-5-2) reduces residual stresses to acceptable levels, making the structure dimensionally stable and not susceptible to stress-corrosion cracking. It tempers martensite and eliminates the danger of cold cracking in steels and leads to overaging in precipitation-hardening materials. Thus it also tends to increase ductility and fatigue strength with only minor loss in strength.

2 Normalizing a steel (Sec. 3-5-4) wipes out most undesirable effects of welding. Recrystallization of the weld and parent material in the austenitic temperature range is followed by controlled cooling to give a structure consisting of ferrite and pearlite.

3 Full heat treatment (quenching and tempering of steels, solution treatment and aging of precipitation-hardenable alloys) gives highest properties in the entire structure, but may result in distortion.

4 An aging treatment of a precipitation-hardening material is sufficient if the weld bead ends up in the solutionized condition because of very fast cooling rates (as in electron beam welding).

Quality control is a vital part of all welding processes. Destructive techniques are useful for establishing process parameters, but production quality can often be controlled only by nondestructive inspection. Visual inspection is always mandatory and is supplemented by the whole arsenal of NDT techniques (Sec. 2-5). The most critical welds are often 100% inspected by radiography.

9-3-3 Weldable Materials

Generalizations are more dangerous for welding than for other processes, but some guidelines can be formulated. In actual production situations specialized reference volumes, computer databases, and industry standards should be consulted.

Ferrous Materials *Ferritic steels* are readily welded, but martensite formation is a danger in pearlitic steels. In general, increasing hardenability means that martensite is formed at lower critical cooling rates; thus, it indicates an increasing danger of martensite formation and a decreasing weldability. Martensite is not only hard and brittle, but its formation proceeds with a volume increase which imposes further stresses on the structure, reducing the strength of the weld.

Preheating and, if possible, postheating are necessary when martensite or bainite formation is unavoidable.

Alternatively, the structure may be heated into the austenitic range, cooled to a temperature above M_s (Fig. 3-25), and welded before transformation begins. The completed structure is then cooled. Such *step welding* makes even tool steels amenable to welding. It should be noted that fully heat-treated alloy steels may be welded but the weld must be rapidly quenched to obtain a tempered martensite.

We already mentioned the danger of hydrogen embrittlement. Sulfur creates porosity and brittleness and, while the welding of resulfurized free-cutting steels is possible, they are often brazed in preference.

Stainless steels always contain chromium, which forms an extremely dense Cr_2O_3 film. Welding conditions must be chosen to prevent its formation. Apart from this, austenitic steels (containing both Cr and Ni) are weldable but chromium carbides formed at high temperatures reduce the dissolved chromium content below the level required for corrosion protection, and subsequent corrosion (*weld decay failure*) is a danger. To avoid this, the carbon content should be very low, or the steel must be stabilized (Ti, Mo, or Nb added to form stable carbides), or the structure must be heated above 1000 °C (1830 °F) after welding and then quenched to retain the redissolved carbon and chromium in solution.

Stainless steels containing only chromium are either ferritic or martensitic in structure. Ferritic steels (over 16% Cr) can be welded but the coarse grain will weaken the joint. Martensitic steels form a martensite with a hardness depending on carbon content; careful preheating is followed by postheating to above 700 °C (1300 °F) to change the martensite into ductile ferrite with embedded chromium carbide precipitates.

Cast iron. The weldability of cast irons (Sec. 3-6-1) varies greatly, but many cast irons are welded, especially by arc welding. A high-nickel filler is frequently used to stabilize the graphitic form. Preheating and slow cooling are also helpful. In welding gray iron the welding rod is enriched in silicon, and—to ensure spheroidal graphite formation—Mg is incorporated in the rod for welding nodular cast iron. Malleable iron reverts to brittle white iron, reducing the toughness of the weld. When toughness is important, the weldment is heat-treated or the joint is made by brazing.

Nonferrous Materials *Low-melting materials.* Tin and lead are easily welded, provided that the heat input is kept low enough to prevent overheating. Zinc, on the other hand, is one of the most difficult materials to weld, because it oxidizes readily and also vaporizes at a low temperature (at 906 °C (1663 °F)). It can be resistance welded and stud welded, although soft soldering is more usual.

Aluminum and *magnesium* share a number of characteristics. Most alloys are readily weldable, particularly with an inert gas envelope. Otherwise the oxide film must be removed with a powerful flux, which in turn must be washed off after welding to prevent corrosion. Moisture (H_2O) must be kept out, because it reacts to give an oxide and hydrogen (which embrittles the joint by causing porosity).

The high thermal conductivity and specific heat yet low melting point of these alloys require high rates of heat input and adequate precautions against overheating. The high thermal expansion coefficient necessitates preheating of hot-short materials. Because of the difficulties encountered with precipitation-hardening materials, such alloys are often heat-treated after welding or, if this is not possible, a different filler is used (very often, Al–Si for aluminum alloys).

Copper-based alloys. Deoxidized copper is readily welded, especially if the filler contains phosphorus to provide instant deoxidation. Tough-pitch copper cannot be welded because its oxygen content (typically, 0.15%) reacts with hydrogen and CO to form water and CO_2, respectively, both of which embrittle the joint by generating porosity.

Brasses can be welded but zinc losses are inevitable; therefore, either the filler is enriched in zinc, or Al or Si is added to form an oxide that reduces evaporation.

Tin bronze has a very wide solidification range and is thus exceedingly hot-short. Phosphorus in the welding rod prevents oxidation, while post-heating is necessary to dissolve the brittle nonequilibrium intermetallic phase. Aluminum bronzes present no problem but the oxide formed must be fluxed, just as with pure aluminum.

Nickel and its solid-solution alloys are readily welded. The precipitation-hardening superalloys contain Cr, Al, and Ti, and the oxide must be fluxed or its formation prevented. All nickel alloys are very sensitive to even the smallest amount of sulfur which forms a low-melting eutectic and results in hot cracking. Some precipitation-hardening alloys have a low-ductility temperature range and may crack too.

Titanium and *zirconium* alloys are also weldable but an inert atmosphere is essential to prevent oxidation.

Refractory metal alloys (W, Mo, Nb) can be welded but the volatility of the oxides makes special techniques (e.g., electron-beam welding) mandatory.

9-3-4 Resistance Welding

Electric resistance welding represents, in some ways, a transition from solid-state welding to liquid-state welding. After pressing together the two parts to be joined, an alternating current is passed through the contact zone. Since this zone represents the highest resistance in the electric circuit, power losses are concentrated there. The energy is converted into heat, and the current is left on until melting occurs at the interface between the two parts. Pressure is then kept on until the weld solidifies.

According to Joule's law, the heat generated (in joules) is

$$J = I^2Rt \qquad (9\text{-}1)$$

where I is the current (A), R is the resistance (Ω), and t is the duration of current application in seconds. The voltage can be low, typically 0.5–10 V, but currents are very high (for examples, see Fig. 9-15).

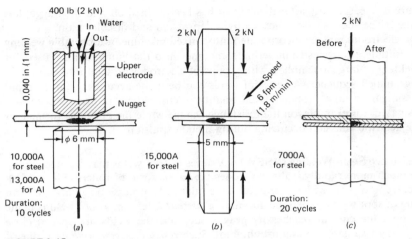

FIGURE 9-15
Only partial melting is aimed for in (*a*) resistance spot welding, (*b*) resistance seam welding, and (*c*) projection welding. (For illustrative purposes, process conditions are given for 1-mm-thick low-carbon steel sheet.)

Since heat must be concentrated in the weld zone, resistance away from this zone should be low, especially at the points where the current is supplied to the workpieces by the electrodes. Materials of high heat conductivity and specific heat (such as aluminum or copper) call for very high currents to prevent dissipation of heat.

Surface cleanliness is important but not quite as vital as in solid-state joining because some of the contaminants are expelled from the melt. Nevertheless, scale, thick oil films, and paint must be removed, but relatively simple surface preparation is adequate and zinc-coated steel can be welded too. Quality control is most important. Welds are destructively tested to establish optimum process parameters; thereafter, in-process inspection includes surface temperature measurement (from which weld zone temperatures can be extrapolated), ultrasonics, and acoustic emission techniques. Closed-loop and adaptive control are possible.

Resistance Spot Welding (RSW)   Because of the widespread application of sheet metal parts, *resistance spot welding* has acquired a prominent position, from attaching handles of cookware to assembling whole automobile bodies (there are some 8000 to 10 000 spot welds per car). Two, usually water-cooled electrodes press the two sheets together (Fig. 9-15a). Electrodes are made of materials of high conductivity and hot strength, such as copper with some Cd, Cr, or Be additions, or copper–tungsten or molybdenum alloys. The current is then applied for a predetermined number of cycles (in the automotive industry, 20–30 cycles), whereupon the interface heats up and a molten pool (*weld nugget*) is formed. The pressure is released only after the current has been turned off and the nugget has solidified. The sheet surface shows a light depression and discoloration.

The electrodes may be incorporated into a fixed machine or a portable welding gun. Multiple electrodes (numbering sometimes into the hundreds) are used for the welding of large assemblies, with groups of electrodes brought into contact in a programmed sequence. A series of welds can be made accurately and in rapid succession by welding robots. Weld quality is ensured by in-process measurements based, for example, on the resistance change while the nugget is formed, or on the acoustic emission occurring during metal expulsion.

Resistance Seam Welding (RSEW) A series of spot welds may be made along a line much more rapidly if the electrodes are in the form of rollers (Fig. 9-15*b*). The current is switched on and off in a planned succession, giving uniform spacing of spot welds. When the alternating current is left on, a spot weld is made every time the current reaches its peak value, and the welds are spaced close enough to give a gas- and liquid-tight joint. Such *resistance seam welding* is one of the methods of producing the body of a can and is used for the manufacture of beams and box sections.

Projection Welding (RPW) The extent of the weld zone is better controlled, and several welds can be made simultaneously with a single electrode, when small dimples or *projections* are embossed (Sec. 5-5-1) or coined (Sec. 4-4-5) on one of the sheets. When the current is applied, the projections soften and are pushed back in place by the electrode pressure as the weld nuggets form (Fig. 9-15*c*). Projections forged or machined onto solid bodies allow welding to a sheet or other solid body. A form of projection welding is practiced when grids formed by crossed wires are resistance welded.

Flash Butt Welding Butt welding in the general sense means joining the end faces of two bodies. In the narrower sense, it is applied to resistance welding in which, just as in spot welding, the pressure is applied prior to switching on the current (Sec. 9-2-3).

Much more widespread today is *flash butt welding*, in which the current is applied during the approach of the two parts; thus, extremely rapid heating takes place when surface irregularities first make contact. Molten metal is violently expelled and burns in air, and some arcing occurs; hence the name flashing. A substantial length may be burnt off to ensure a good weld, but all liquid metal is expelled and the weld is formed by upsetting the hot, solid metal surfaces (Fig. 9-16). Thus, the strength of the joint is not impaired by the presence of residual as-cast weld metal. Therefore, the process can be regarded as a transition between solid- and liquid-phase processes. A good weld will be produced only if the rate of approach, current and voltage, total travel, and upsetting pressure are closely controlled. Manual control is now often replaced with automatic, adaptive control.

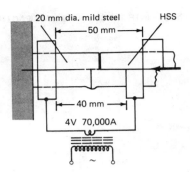

FIGURE 9-16
Flash-butt welding, illustrating
critical process parameters.

The end faces to be joined are often chamfered so that melting moves from the center outward to squeeze out contaminants into the flash. Preheating is possible by switching on the current after the faces have been pressed together, but the faces are then again slightly separated to induce flashing. For uniform heating the two parts should have equal cross-sectional areas, but their composition can be different, for example, in joining a low-carbon steel shank to a HSS tool bit. Bars and sections bent into the shape of rings (Fig. 5-17b), tubes, and sheet structures are often welded edge-to-edge, and the process is extensively used for joining the ends of wire and sheet coils to allow the continuous operation of processing lines.

9-3-5 Electric Arc Welding

Electric arc welding differs from electric resistance welding in that a sustained arc generates the heat for melting the workpiece (and, if used, the filler rod) material. The workpiece is connected to one terminal of a power source, the electrode to the other. Upon briefly touching the workpiece with the electrode, an arc is drawn; thereafter, current is conducted mostly by electrons that are stripped off one electrode and off the atoms of the gas occupying the space between electrode and workpiece. The positively charged gas ions flow to the negative pole whereas electrons flow to the positive pole.

High temperatures are maintained for some time; therefore, materials of low melting or boiling temperature—or materials coated with a low-melting metal—cannot be welded. Furthermore, complete protection from the atmosphere is essential. In some processes and with some materials, there is also need for a flux that dissolves oxides and removes them from the melt zone. Very broadly, arc welding processes comprise both consumable and nonconsumable electrode methods.

Stud Welding (SW) A special form of arc welding is *stud welding* (Fig. 9-17a) in which the arc is maintained between a projection of one workpiece and the surface of the other workpiece (typically, a plate). When the projection melts,

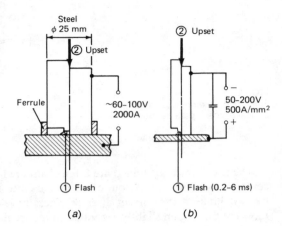

FIGURE 9-17
An arc is drawn in (*a*) stud and
(*b*) percussion welding.

(*a*)

(*b*)

pressure is applied to join the stud to the plate. An expendable ceramic *shielding ferrule* concentrates the heat of the arc, protects against oxidation, and confines the melt. Applications are found in building construction, attachment of handles and feet to appliances, and electric panel construction.

In a variant of the process, *capacitor-discharge stud welding* (Fig. 9-17*b*), the energy stored in a condenser is used for heating. Discharge takes place just before or during approach to the surface. The intense, localized heat allows the joining of widely differing cross sections and also of dissimilar materials. Timing and motion control are critical. Studs can be welded to thin sheets, even to those coated with paint or PTFE on the other side, allowing the fastening of instrument panels, nameplates, and auto trim.

The term *percussion welding* (PEW) is used to describe capacitor-discharge welding applied to joining wires to terminals and other flat surfaces. Since the two terminals must be separate prior to impact, rings cannot be welded.

Consumable-Electrode Welding This group of processes is characterized by the use of a metal electrode which melts to become part of the weld seam. The weld zone is protected either by a flux (Fig. 9-18) or by gas (Fig. 9-19).

Shielded Metal-Arc Welding (SMAW) The arc, is struck between the filler wire or rod (consumable electrode) and the workpieces to be joined (Fig. 9-18*a*). The current may be ac or dc. In the latter case, the electrode may be negative (direct current, electrode negative, DCEN or *straight polarity*) or positive (DCEP or *reverse polarity*). The weld dimensions depend on polarity, current intensity, voltage, flux, electrode size and orientation, and the rate of travel along the weld seam.

An essential element of the process is the coating applied to the outside of the filler wire (*coated electrode*). The coating fulfills several functions: combustion

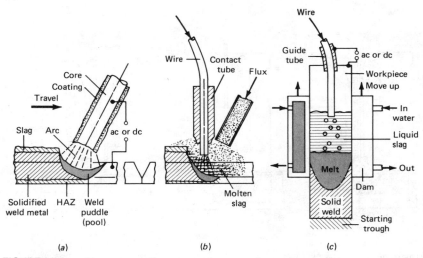

FIGURE 9-18
A sustained arc, shielded by molten slag, is maintained in consumable-electrode welding by the (*a*) shielded metal-arc, (*b*) submerged arc, and (*c*) electroslag methods.

and decomposition under the heat of arc creates a protective atmosphere; melting of the coating provides a molten slag cover on the weld; the sodium or potassium content of the coating readily ionizes to stabilize the arc. Also, alloying elements may be introduced from the coating. Choice of the electrode is critical to the success of the process. During welding, the electrode melts at a rate of approximately 250 mm/min (10 in/min) while the coating melts into a slag which must be removed if more than one pass is required to build up the full weld thickness.

Since the coating is brittle, straight sticks of typically 450-mm (18-in) length are generally used, making this process suitable only for hand operation, at relatively slow rates, but still at a low cost. The process is versatile and suitable for field application, but requires considerable skill. Welding in all positions, including overhead welding is possible if the metal and slag solidify fast enough.

Flux-Cored Arc Welding (FCAW) Basically the same result is obtained with the flux inside a tube. However, the welding wire can now be coiled, and automatic, continuous welding becomes possible. Sometimes additional shielding is provided with a gas, and then the process resembles gas metal-arc welding, to be discussed later.

Submerged Arc Welding (SAW) The consumable electrode is now the bare filler wire and the weld zone is protected by a granular, fusible flux supplied quite independently from a hopper (Fig. 9-18*b*) in a thick layer that covers the arc. The flux shields the arc, allows high currents and great penetration depth, acts as a deoxidizer and scavenger, and may contain powder-metal alloying elements. SAW

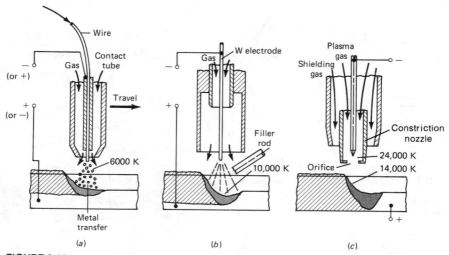

FIGURE 9-19
The arc is shielded by gas in the (a) gas metal-arc, (b) gas tungsten-arc, and (c) plasma-arc welding processes. Note that the depth of penetration increases with increasing arc temperature.

is primarily an automatic welding process with high travel speeds. Tandem electrodes can be used to deposit large amounts of filler material.

The weld position must be horizontal; thus, it is suitable for steel line pipes, cylinders, and also for circular welds if the workpiece is rotated. Double submerged-arc welding (with one weld from the inside, the other from the outside) is used in making spiral-welded pipelines. It can also be used with welding robots, with the workpiece manipulated into appropriate positions.

Electroslag Welding (ESW) In this process, used extensively for welding thick (25 mm (1 in) or over) plates and structures, electrode wire is fed into a molten slag pool (Fig. 9-18c). An arc is drawn initially but is then snuffed out by the slag, and the heat of fusion is provided by resistance heating in the slag. Water-cooled copper *shoes* (*dams*) close off the space between the parts to be welded to prevent the melt and slag from running off. The welding head must be raised as the weld deposit builds up. In a variant of the process, a consumable guide tube of steel is used which melts into the weld pool; thus, the welding head need not be moved. If the part is rotated, circumferential welds can also be made.

Gas Metal-Arc Welding (GMAW) The consumable metal electrode, fed through the welding gun, is shielded by an inert gas, thus, the acronym MIG (metal inert gas) welding (Fig. 9-19a). It is suitable for most metals. No slag is formed and several layers can be built up with little or no intermediate cleaning. Argon is a suitable gas for all materials; helium is sometimes preferred—because of its higher ionization potential and, therefore, higher rate of heat generation—for the welding of aluminum and copper; Ar with 20–50% CO_2 or pure CO_2 is generally used for carbon steels; speciality gases are also being introduced, tailored to specific tasks.

The electrode is usually connected to the positive terminal (DCEP or reverse polarity). At high current densities metal is transferred from the electrode to the weld zone in a fine spray (*spray transfer*, Fig. 9-19*a*); at low currents and voltages transfer is in blobs that drop by gravity (*globular transfer*). Particularly with CO_2 as the shielding gas, a *short-circuiting mode* of operation occurs with drops of liquid metal transferring by gravity and surface tension. The latter operation is preferable for thin sections or sheets because of the lower heat. The advantage of CO_2 (with small amounts of oxygen) for steel is the low cost and high transfer rate.

The wire electrode can be supplied in long, coiled lengths which allow uninterrupted welds in any welding position. In semiautomatic welding the welder guides the gun and adjusts process parameters; in automatic welding all functions are taken over by the welding machine or robot. On-site welding can be difficult because drafts blow the shielding gas away from the weld zone.

Electro-Gas Welding (EGW)　This is an outgrowth of electroslag welding. The electrode wire is solid or flux-covered, and protection is provided by a gas (typically 80% Ar, 20% CO_2). The molten pool is again retained with copper dams.

Nonconsumable-Electrode Welding　In these processes the electrode does not melt and the weld metal is supplied by the flow of the parent metal or from a separate filler rod.

Gas Tungsten-Arc Welding (GTAW)　The arc is maintained between the workpiece and a tungsten electrode protected by an inert gas (hence the name tungsten inert gas or TIG welding). A filler may or may not be used (Fig. 9-19*b*). The protective atmosphere may be given by argon, which maintains a stable arc or—for deeper penetration and a hotter arc—by helium, or by a mixture of the two.

To strike an arc, electron emission and ionization of the gas are initiated by withdrawing the electrode from the work surface in a controlled manner, or with the aid of an initiating arc. High-frequency current superimposed on the alternating or direct welding current helps to start the arc and also stabilizes it.

Both hand and automatic operations are possible. The process demands considerable skill but produces very high-quality welds on almost any material, in any welding position, and also on thinner gages (below 6 mm (0.25 in)). The weld zone is visible, and there is no weld spatter or slag formation, but electrode particles may enter the weld.

Twin-Arc Welding　This was the forerunner of present-day nonconsumable-electrode processes. The arc is drawn between two tungsten or carbon electrodes. It is now used with hydrogen, which dissociates into the atomic form and gives high-quality welds for many materials.

Plasma-Arc Welding (PAW)　In the space between the tungsten electrode tip and the workpiece, the high temperature strips off electrons from gas atoms; thus, some of the gas becomes ionized. The mixture of ions and electrons is known as a *plasma* (for a more detailed discussion, see Sec. 10-3-2). The plasma gets hotter by resistance heating from the current passing through it. If the arc is constrained by an orifice, the heat intensity and, thus, the proportion of ionized gas increase and a *plasma arc* is created (Fig. 9-19c). This provides an intense source of heat and ensures greater arc stability. The shielding gas is argon or argon–helium. The arrangement shown in Fig. 9-19c is typical of the *transferred plasma arc* method of operation; electrons flow to the workpiece which is connected to the positive terminal. In the *nontransferred arc* technique the constriction nozzle is connected to the positive terminal; the arc is drawn between electrode and nozzle, and the arc heats the workpiece by radiation. This technique is used also for *plasma spraying* (coating of surfaces, Sec. 9-5). PAW is particularly useful for the welding of thin sheets. Both manual and mechanical forms are practiced, and filler metal may be used if an extra material supply is needed.

Automation of Welding Processes   We already remarked on the possibilities of automation for various arc welding processes. Developments have taken two directions.

First, the process itself is increasingly controlled automatically. The shape of current pulses is controlled by electronic devices. The optimum current, travel rate, electrode feed rate, etc., are set and controlled, and adaptive control—utilizing noncontacting temperature sensors, electrode-to-workpiece distance sensors, etc.—is introduced.

Secondly, manipulation of the workpiece and/or the welding gun relative to each other is increasingly automated. The task is relatively simple when welding along a straight line, helical, or other readily defined path, and mechanization is often possible. It was, however, only with the appearance of multiaxis, teachable robots that welding along complex spatial lines became possible. Here too adaptive control has been gaining an increasing role, for example, in using a laser to find the location of the seam. Welding robots achieve an arc time of 80%, whereas a manual welder seldom maintains 30%. Repeatability is much higher too, and it is not surprising that welding is the single largest field of robot application.

9-3-6   Thermal Welding

The heat required for fusion may be provided by a *chemical heat source*.

Gas Welding   In the most widespread form of gas welding, *oxyfuel gas welding* (OFW), heat is produced by the combustion of acetylene (C_2H_2) with oxygen. Both are stored at high pressure in gas tanks and are united in the welding torch. After ignition, a temperature of approximately 3700 K (3400 °C or 6200 °F) is generated in the flame. Three zones may be distinguished (Fig. 9-20a). Primary combustion takes place in the *inner zone* and generates two-thirds of the heat by

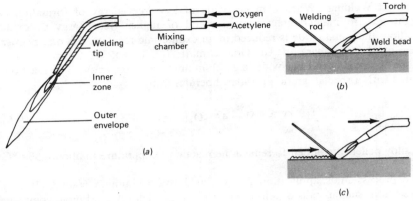

FIGURE 9-20
A thermal heat source is employed in (*a*) oxyfuel gas welding. A wider bead is produced in (*b*) forehand than in (*c*) backhand welding.

the reaction

$$2C_2H_2 + 2O_2 \rightarrow 4CO \rightarrow 2H_2 \tag{9-2}$$

These reaction products predominate in the *second zone*, and, thus, provide a reducing atmosphere favorable for the welding of steel. Complete combustion takes place in the *outer envelope* by the reactions

$$4CO + 2O_2 \rightarrow 4CO_2 \tag{9-3}$$

$$2H_2 + O_2 \rightarrow 2H_2O \tag{9-4}$$

The flame protects low-carbon steels, lead, and zinc sufficiently, but a flux is needed for most other materials. The relatively low flame temperature, the ability to change the flame from oxidizing to neutral and even reducing, and the flexibility of manual control make the process suitable for all but the refractory metals and reactive metals such as titanium and zirconium. The process has the advantage of portability and is suitable for all welding positions.

The welder (assumed to be right-handed) holds the torch at an angle to the surface, and may pull the filler rod (with its flux coating) away from the *weld puddle* in *leftward* or *forehand welding* (Fig. 9-20*b*), thus, preheating the joint area and obtaining a relatively wide joint. When the filler is moved above the weld bead (*rightward* or *backward welding*, Fig. 9-20*c*), the weld puddle is kept hot for a longer time and a narrower weld, often of better quality, results.

Other gases, including propane, natural gas, or hydrogen are also used as the heat source, particularly for aluminum and lower-melting metals.

Thermit Welding When a metal oxide of low free energy of formation is brought into intimate contact with a metal of higher free energy of oxide formation, the metal oxide is reduced in an exothermic reaction. *Thermit powder* (a registered trade mark of Th. Goldschmidt AG, Essen, W. Germany) is a mixture of a metal and an oxide, e.g., aluminum and iron oxide. The reaction, initiated with a special ignition powder, liberates iron

$$3Fe_3O_4 + 8Al \rightarrow 4Al_2O_3(\text{slag}) + 9Fe \qquad (9\text{-}5)$$

Iron alloy pellets are added to reduce the reaction temperature to about 2500 °C (4500 °F).

The process finds application for joining heavy (minimum 60-cm^2 (10-in^2) area) sections such as rails or reinforcing rods in the field. A sand mold complete with sprue, gate, and riser is built around the joint area, the thermit powder is ignited in a crucible placed on top of the mold, and the resulting iron is tapped—through a bottom hole of the crucible—directly into the mold. After solidification, the mold is destroyed and the still-hot excess steel is chiselled off. Large electrical bus bars are similarly welded, using aluminum and Cu_2O.

9-3-7 High-Energy Beam Welding

The heat of fusion may be provided by converting the energy of impinging electron or light beams into heat.

Electron Beam Welding (EBW) The energy source for EBW is an electron gun (Fig. 9-21a) similar to a vacuum tube. The cathode emits masses of electrons that are accelerated and focused to a 0.25–1-mm- (0.01–0.04-in-) diameter beam of high energy density (up to 10 kW/mm^2 (6 MW/in^2)). The kinetic energy of electrons is transformed into thermal energy, sufficient to melt and vaporize the workpiece material; molten metal ahead of the vapor hole flows around to fill the gap; thus, narrow gaps can be welded without a filler (although filler rods may be used). The heat-affected zone is very narrow and energy-conversion efficiency is high, about 65%.

The electron gun is always in high vacuum. Deepest penetration and best weld quality are obtained when the workpiece too is enclosed in *high vacuum* (10^{-2} to 10^{-3} Pa, EBW-HV), but pumping down the welding chamber takes several minutes. *Medium vacuum* (1–10 Pa, EBW-MV) still permits welding many metals with a pumping time of less than 1 min. With specially constructed vacuum traps, the electron beam can emerge from the gun into a shielding gas, and such out-of-chamber or *nonvacuum welding* (EBW-NV) can produce high-quality welds in many materials.

The process is extremely adaptable and excels in welding both thin gages and thick sections, parts of dissimilar thicknesses, hardened or high-temperature materials, and dissimilar materials. It lends itself to automatic control.

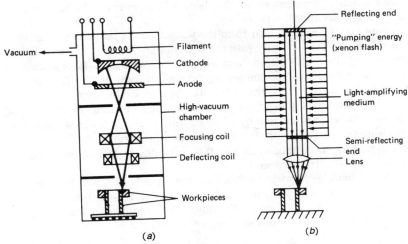

FIGURE 9-21
Very high energy densities are obtained in high-energy beam welding: (a) electron beam welding and (b) laser beam welding.

Laser Beam Welding (LBW) The word laser stands for light amplification by stimulated emission of radiation. Some materials (*lasing media*) emit a highly collimated, coherent, monochromatic light beam when excited (*pumped*) by some appropriate energy source.

The first lasers utilized ruby (an Al_2O_3 crystal with Cr ions) as the lasing medium and such lasers are still useful for tasks such as alignment and measurement. For manufacturing purposes, two kinds of lasers have found widest application:

1 *Solid-state lasers.* Small concentrations of neodymium ions incorporated into yttrium aluminum garnet (YAG) or glass cylinders emit radiation of 1.06-μm wavelength when pumped with high-intensity white light from a xenon or krypton lamp (Fig. 9-21b).

2 *Gas lasers.* These contain a mixture of gases in which CO_2 is the lasing medium, excited by an electric discharge between electrodes placed in the discharge tube. The emitted light is of 10.6-μm wavelength, in the far-infrared range.

YAG lasers develop up to 500 J/pulse of 0.1–20-ms duration, with an overall energy conversion efficiency of 2%, whereas CO_2 lasers can develop up to 20 kW in the continuous mode at 15% efficiency.

Because the beam is highly collimated, peak energy densities of 80 kW/mm^2 (50 MW/in^2) are reached. The beam can be focused with lenses made of materials transparent to the particular wavelength (conventional glass optics for

YAG, zinc selenide or germanium for CO_2 lasers). With mirrors, the beam can be directed to various locations.

The energy may be used to heat from the surface of the material (*conduction-limited mode welding*), or—mostly with the high-power CO_2 lasers—to penetrate the full depth of the joint (*deep-penetration mode* or *keyhole welding*). Because heating is a function of surface emissivity, the shorter-wave Nd:YAG lasers are more suitable for highly reflective materials but cannot be used on glass or polymers. The laser has the advantage that vacuum is not necessary. The workpiece usually needs protection by argon or helium, except for spot welding in which exposure time is very short. Oxygen blown on the surface of metals reduces light reflection and, in cutting, increases removal rates by oxidation; inert gas increases heat transfer for nonmetals.

The laser is finding growing application, particularly for thin-gage metals. Welding speeds of about 2.5 m/min (100 in/min) are achieved on steel sheet 1.5 mm (0.060 in) thick. It is suitable for automation, and either the workpiece, the laser, or the beam may be moved along prescribed paths.

9-4 LIQUID–SOLID-STATE BONDING

When the joint is established without melting the base metal, the main source of strength is adhesion between filler and base metal, developed in the absence of contaminant surface films. The strength of the joint is higher than that of the filler, provided that the filler is applied in a thin layer so that it is restrained by the base metal. Dissimilar materials can be joined, as can parts with greatly differing wall thicknesses. A great advantage is that there is no need for access to all parts of the joint, because surface tension draws in the filler metal. Hence, complicated assemblies, including those consisting of many parts, can be simultaneously joined. The techniques are suitable for the manufacture of automotive radiators, plate-and-tube heat exchangers, impellers, fans, appliance parts, and for the joining of wires. The distinction between brazing and soldering is somewhat arbitrary: When the filler metal melts below 425 °C (800 °F), one speaks of soldering.

FIGURE 9-22
Several components are simultaneously joined in brazing; the filler metal is distributed through capillary action. (*From Metals Handbook, 8th ed., vol. 6, American Society for Metals, 1971, p. 607. With permission.*)

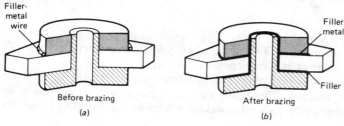

9-4-1 Brazing

In *brazing* the filler metal is placed into position prior to heating the base metal (the workpieces to be joined). It may be placed outside the future joint (*local*, Fig. 9-22a) or inside the joint (*preplaced*).

Brazing Conditions Several criteria must be satisfied to achieve a strong bond:

1 Surfaces must be free of contaminants that would prevent adhesion. Thus, scale is removed by mechanical or chemical means, and heavy oily residues are removed by degreasing.

2 Oxides formed during heating would prevent adhesion. To prevent this, several measures may be taken:

a Brazing is performed in vacuum or an appropriate (neutral or reducing) atmosphere, or heating times are kept very short.

b Wetting of the surface is often ensured by the application of fluxes. A good flux melts at a low enough temperature to prevent oxidation of the base and filler materials; it has a low viscosity so that it is replaced by the molten filler metal; it may react with surfaces to facilitate wetting; it shields the joint while the filler is still liquid; and it is relatively easy to remove after solidification of the filler. Fluxes are in the form of powder, paste, or slurry, and are composed of borates, fluorides, chlorides, and similar materials in various proportions, tailor-made for specific applications.

In general, strongly reducing atmospheres or more active fluxes are needed when stable oxides form, such as on aluminum, magnesium, heat-resistant alloys, and stainless steels. Surface carbon would prevent wetting of cast irons and must be removed by electrolytic or molten salt-bath treatment of the parts.

3 The parts to be joined are most often made of sheet metal but may also be forged, extruded, or cast components. They are assembled and temporarily held together in the correct position by fixtures or mechanical fastening such as expanding, staking, swaging, etc. The design of some assemblies ensures correct positioning and makes them self-fixturing.

4 The *filler metal* (*braze metal*) must wet the base metal, must have high fluidity to penetrate crevices (but not so high that it would run out of the joint), should preferably have a narrow melting range, and must not lead to galvanic corrosion during service. It is applied in the form of wire, strip, preforms, powder, or paste to the joint area which is then heated, either locally or by heating the entire assembly (Fig. 9-22). Alternatively, the filler metal is preapplied to the surface of one of the contacting parts as a coating (cladding), often by rolling (Fig. 9-4c), electrolytic deposition, or hot dipping. Typical filler metals are listed in Table 9-1 for various classes of base metals. In general, filler metals of higher melting point give higher strength, but the high brazing temperature may affect the strength of the base metal. Some materials, such as graphite are not readily wetted and must first be coated with a wettable layer (such as copper) or a filler must be used that contains carbide-forming elements.

TABLE 9-1 FILLER METALS USED FOR BRAZING*

AWS designation	Brazing filler metal Composition, wt%	Brazing temperature, °C	Base metal
BCu-1	99.9Cu	1100–1150	Steel (< 0.3% C); low-alloy steel
BAg-1	45Ag, 15Cu, 16Zn, 24Cd	620–760 ⎫	Carbon steel; low-alloy steel; cast iron;
BAg-5	45Ag, 30Cu, 25Zn,	740–840 ⎭	stainless steel; copper alloys
RBCuZn-A	59Cu, 40Zn, 0.6Sn	910–950	Low-carbon steel; low-alloy steel
BNi-1	74Ni, 14Cr, 3B, 4Si, 4Fe, 0.7C	1060–1200	Stainless steel
BAu-1	37Au, 63Cu	1015–1090	Stainless steel
BCuP-2	92.75Cu, 7.25P	820–870	Copper alloys
BAlSi-2	91Al, 7.5Si, 0.2Zn, 0.25Cu	600–620	Aluminum alloys

*Compiled from *Metals Handbook Desk Edition*, American Society for Metals, 1985.

5 The *clearance* between mating surfaces is critical. It must be large enough for the filler to penetrate by capillary action, but not so large that the strength of the joint would be lost. The clearance depends on brazing alloy and is typically 20–120 μm (0.001–0.005 in). This gap must exist at the brazing temperature; therefore, differential expansion must be taken into account when dissimilar metals are joined.

6 *Controlled cooling* is necessary to ensure rapid solidification without causing distortion of the assembly or cracking of the joint.

Brazing Methods A great variety of heating techniques can be employed, each with some special advantages. Most techniques are readily mechanized or automated.

1 *Furnace brazing* is a mass-production process, with assemblies placed into a box furnace or, for highest production rates, conveyed through a continuous furnace on belts, roller gangs, or suspension hooks. In the brazing of steel with copper, the clearance is zero or even negative; with other materials, brazing proceeds with the usual small, positive clearance.

2 *Dip brazing* derives heat from immersion of the assembly into molten salt. Heating rates are high and the salt may perform the fluxing function.

3 *Torch brazing* allows selective heating of the joint area by an acetylene, propane, or natural gas flame. It is suitable for manual operation but is easily mechanized.

4 *Resistance brazing* utilizes the same equipment as resistance welding (Fig. 9-7c) but a filler is placed in the joint. Rapid heating minimizes oxidation and the heat-affected zone is small.

5 *Induction brazing* with a typically 10–460-kHz power supply has the same advantages as resistance brazing. It cannot be applied to aluminum or magnesium alloys because the melting point of the filler metal is too close to that of the base metal.

6 *Laser brazing* and *electron beam brazing* are justified mostly for precision assemblies of high-value, relatively high-temperature materials.

7 *Braze welding* differs from brazing in that a much wider gap is filled with the brazing metal (mostly brass) with the aid of a torch, and thus capillary action plays no part. Besides assembly, it is also used for repair of steel and iron castings.

9-4-2 Soldering

Soldered joints are of lesser strength than brazed ones, but parts can be joined without exposing them to excessive heat. The lower temperatures make good wetting more critical than in brazing. Therefore, surface preparation by mechanical and chemical means and the use of fluxes are essential.

The filler metal (*solder*) may be chosen by reference to the relevant phase diagrams. Solid solubility and, in particular, the formation of intermetallics is a sign of good wettability, although the presence of surface films greatly modifies the behavior. A surface that appears uniformly coated by the solder may still de-wet on reheating. Thus, *solderability* is a technological property that can only be determined experimentally. In general, precious metals and copper are readily soldered; iron and nickel require more aggressive fluxes; the tenacious oxides of aluminum and chromium make soldering of aluminum alloys and high-chromium steels more difficult. Cast iron, titanium, magnesium, and ceramics (including graphite) require preplating.

The most widely used solders are tin–lead alloys. A low ($< 5\%$) tin content gives higher strength and is suitable for automotive radiators and lock-seam cans (Fig. 5-21) and tubes made of tinplate. The wide freezing range of the 35% Sn alloy makes it ideal as a wiping solder for the joining of copper tubes. The eutectic composition (63% Sn) has, by definition, the lowest melting point and solidifies at a constant temperature, making it most suitable for electric connections.

For general work, the *flux* is a water solution of zinc (or $Zn + Na + $ ammonium) chloride and must be washed off to prevent corrosion. Noncorrosive organic rosins are essential for electrical connections where corrosion could create high local resistance and even loss of conduction. Coated metals, especially tinplate, facilitate soldering. A large variety of other solders are used in specific applications. In particular, tin–zinc alloys with 9–100% Zn have been developed for the soldering of aluminum in conjunction with special fluxes.

Soldering may be carried out by all the techniques previously described for brazing (flame, hot dip, resistance, induction). Because of the low temperatures involved, a heated copper or iron-plated copper bit (*soldering iron*) is used extensively. *Radiant heating* (infrared heating) is possible. Agitation of the molten solder bath by ultrasonic waves improves wetting in *dip soldering*, a technique particularly useful in soldering aluminum, as are techniques in which the surface to be bonded is rubbed (e.g., by an ultrasonic transducer) while the solder is flowed onto it. A special method, *wave soldering* is of great value in assembling

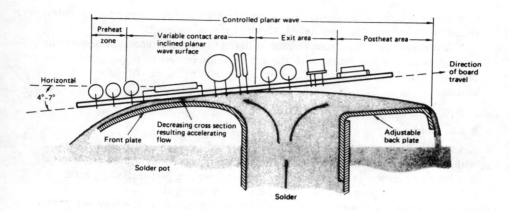

FIGURE 9-23
Planar wave soldering showing adjustable backplate to control wave configuration and flow pattern. (*Courtesy Electrovert Consulting Services, Elmsford, N.Y.*)

printed-circuit boards. Molten solder is pumped through a narrow slot (nozzle) so that fresh solder moves in a gently flowing wave. A typical arrangement is shown in Fig. 9-23.

9-5 SURFACE COATINGS

We have seen many instances where it is desirable to produce a hard (and sometimes brittle) layer on a lower-strength but tough body. Applications include not only repair but also the initial manufacture of cutting tools, rock drills, cutting blades, parts for earth-moving equipment, valves and valve seats for diesel engines, saw guides, forging dies, screws for extrusion of plastic and food products, and, in general, applications requiring wear resistance. Several techniques are suitable:

1 *Hard facing* is related to welding. The alloys used resemble cutting-tool materials (Sec. 8-3-1), with very hard phases (mostly carbides) embedded in a more ductile matrix (iron, nickel, or cobalt alloy). They often have such a high alloying-element concentration that they cannot be manufactured into welding rods; the ingredients are then incorporated in the flux coating or packed inside tubular rods, and the alloy is formed in the welding process itself. Hard particulates such as WC can also be deposited in a metal matrix.

Most welding processes, including various forms of arc welding, plasma-arc welding (often with combined transferred and nontransferred arcs) as well as oxyfuel gas welding, may be used. Deposition of WC by an electric arc is called *spark hardening*, useful for cutting tools. When thick layers are deposited, one speaks of *weld overlays*. The rate of metal deposition is increased two- to fivefold by supplying metal powder to the heat zone or by substituting strip for the welding wire (*strip overlay welding*).

2 *Spraying* differs from hard facing by welding in that a powder is applied to the surface and melted, usually by an oxyfuel torch (OFSP) or nontransferred plasma arc (PSP). A special process forms a thin (5-μm) film by shooting powder by detonating an acetylene-oxygen mixture (*detonation gun*). The techniques are used for both metallic and ceramic (such as oxide, carbide, nitride, etc., Sec. 6-3-6) coatings on wear-resistant parts, rocket-motor nozzles, and dies for the hot extrusion of steel.

3 *Chemical vapor deposition* (CVD) and *physical vapor deposition* (PVD) are gaining in popularity because a wide variety of metals and ceramics may be deposited in films of well-controlled thicknesses (with thickness measured in micrometers). By appropriate control of the processes (Sec. 10-3-2), good adhesion to the substrate is assured. Typical applications are to steel and cemented-carbide cutting and forming tools and to wear components.

4 *Diffusion coatings* are produced by diffusing an alloying element into the surface where it often forms a hard intermetallic compound. We already saw examples of diffusion of carbon and/or nitrogen into steel (Sec. 3-5-5). Other high-temperature processes are used for boronizing and chromizing.

Much lower temperatures are generated in *ion nitriding*, a process involving bombardment of the steel surface by low-energy (less than 1 keV) nitrogen ions produced in a plasma (see Sec. 10-3-2 for a discussion of plasmas). Pressures of 260–700 Pa are maintained in a nitrogen–hydrogen gas mixture; the presence of hydrogen ensures an oxide-free surface and thus facilitates the diffusion of nitrogen into the metal. *Ion implantation* (Sec. 10-3-3) relies on the penetration of high-energy (10–500-keV) ions into the surface. Nitrogen is the most frequently used species on tools but other ions may also be implanted. There is virtually no dimensional change with either process. Ion implantation is often found to give better wear resistance.

9-6 CUTTING

A very important application of fusion welding processes accomplishes an exactly opposite function: headers, risers, and flash are removed from castings, forgings, or moldings, or workpieces of varying shape are *cut* out from sheet, plate, and even heavy sections. The heat required for melting may be provided by a chemical heat source, electric arc, or high-energy beam.

1 Steels, preheated to 850 °C (1600 °F), burn if oxygen is blown on them. *Oxygen cutting* is widespread in steel mills, for cleaning up surfaces (*scarfing*) and cutting up billets while the steel is still hot.

2 For general cutting applications, the steel is first preheated with an oxyfuel gas flame, and a stream of high-pressure oxygen is then directed onto the heated spot. Oxidation generates more heat, melting the steel. The melt is flushed away

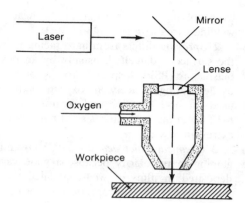

FIGURE 9-24
Laser cutting is greatly accelerated when oxygen is supplied to the cutting zone.

by the flame and oxygen (*flame cutting* or *oxyfuel gas cutting*, OFC). Plates of 5–1500-mm (0.2–60-in) thickness are cut. When iron powder is added to the gas stream (*powder oxyfuel cutting* or *powder metal cutting*, POC), oxidation of the powder provides the heat to melt oxidation-resistant materials.

3 In general, cuts of better surface quality are produced by variants of welding processes.

a *Plasma-arc cutting* (PAC) with a transferred arc has acquired great importance in cutting all metals in thicknesses up to and even over 25 mm (1 in). The quality of the cut edge is better than in flame cutting.

b *Electron-beam machining* utilizes the energy of electrons. Since an electron beam can be deflected with an electromagnetic coil (Fig. 9-21*a*), cuts of high quality can be made in complex patterns, on practically any material.

c High-quality edges are also produced by *laser cutting*. Depending on energy density, some material is melted (ablated), some is evaporated. Machining to a controlled depth is also possible. Frequently, oxygen is supplied to the surface (Fig. 9-24) to increase energy adsorption. The exothermic oxidation reaction also supplies heat and accelerates melting; furthermore, the oxide melts at a lower temperature and is blown away. Steel plates of up to 10-mm (0.4-in) thickness are cut. Heat-treated steels may be cut: The edge quality is excellent, suitable even for punches and dies used in sheet metalworking.

In the cutting of plastics, polymer chains break down, thermoplastic polymers melt, and thermosets decompose (char). Clean edges are produced, therefore, lasers are extensively used for trimming and cutting out parts. A great advantage in cutting fiber-reinforced polymers is that there is no fraying and the edges are sealed. Light in the far-ultraviolet range (200 nm) causes ejection of small molecules without melting (*heatless laser etching*). High-quality cuts are also made on wood if the energy density is high enough to cause vaporization. Lasers are also used for cutting and scribing of ceramics and marking of all materials.

All cutting processes are eminently suitable for making contour cuts, using tracer mechanisms, NC, or CNC. Plasma or laser cutting is often incorporated into CNC punching centers (Sec. 5-2-4), greatly increasing the versatility of these machines. Productivity can be very high in most applications, making the laser competitive despite the high capital costs.

9-7 ADHESIVE BONDING

All processes discussed hitherto used a metal to establish a joint between two metallic or ceramic parts. *Adhesive bonding* differs in that the material of the joint is a polymer or, less frequently, a ceramic. Alloying is not possible, and bond strength relies entirely on the adhesion of the adhesive to the metal or polymer substrate and on the strength of the adhesive itself. Adhesion between metal and polymer surfaces is not fully understood but it is clear that only secondary bonding forces can come into play. Bond strength is, in general, lower than when primary (metallic, covalent, or ionic) bonds are established, and the design of joints must take this into account.

Adhesive bonding has many advantages: Only low temperatures are involved, thus no undesirable changes occur in most substrate materials; the exterior surface remains smooth; dissimilar materials and thin gages can be joined; the adhesive contributes to energy adsorption in shock loading and in the presence of vibrations; and, not least, complex assemblies can be made at low cost.

Adhesive bonding technology has advanced to the point where load-bearing structures can be built reliably. Applications include such critical structures as control surfaces in aircraft, entire aircraft bodies, and countless applications in the automotive, appliance, and consumer goods fields. In addition, adhesives are also used for sealing, vibration damping, insulating, and other nonstructural applications. Our concern here is with load-bearing applications, for which a class of adhesives termed *structural adhesives* are used.

9-7-1 Characteristics of Structural Adhesives

The terminology is somewhat imprecise but a structural adhesive is one used to establish a *permanent joint* between two higher-strength parts (*adherends*). The joint is expected to retain its load-bearing ability over a long period of time, under a wide variety of—and often hostile—environmental conditions. Thus, a structural adhesive must possess a number of desirable attributes; many of these are linked to properties of polymers, and a review of Chap. 7 may be useful at this point.

1 The adhesive must have sufficient *cohesive strength* at service temperatures. Shear strength is governed by composition (Fig. 9-25) but, for long-term applications, creep strength—which is generally lower than short-term strength—is of great concern. Many adhesives are thermosets in which the degree of cross-linking

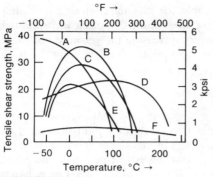
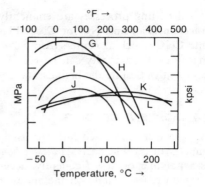

FIGURE 9-25

The shear strength of adhesives declines rapidly at some temperature which depends on composition, limiting their range of application. (*a*) Paste and liquid adhesives, (*b*) tape, film, and solvent-based adhesives. A: two-part urethane + amine; B: one-part rubber-modified epoxy; C: one-part epoxy, general-purpose type; D: one-part epoxy, heat-resistant type; E: two-component room-temperature-curing epoxy-polyamide; F: silicone sealant; G: nylon-epoxy; H: nitrile-epoxy; I: nitrile-phenolic; J: vinyl-phenolic; K: epoxy-phenolic; L: polyimide. (*Reprinted from J. C. Bolger, in Adhesives in Manufacturing, Dekker, 1983, p. 142, by courtesy of Marcel Dekker Inc.*)

can be increased to increase creep resistance, although usually at the expense of bond flexibility. For example, in epoxies an average cross-link separation of < 2 nm gives a strong but brittle bond; some toughness is obtained by increasing the separation to approximately 3 nm, and at > 3 nm the bond becomes soft and flexible. Creep resistance can be increased also by the addition of fillers. These are particularly important in thermoplastics but are also used to improve the impact properties of thermosets.

2 The adhesive must be able to *distribute stresses* imposed in service. Flexibility helps to reduce stress concentrations arising from mechanical loading, makes the adhesive joint more resistant to fatigue, and prevents sudden catastrophic separation in the event of joint failure. Even in the absence of mechanical loading, stresses are generated during thermal cycling because polymers have, in general, higher thermal-expansion coefficients than metals (Table 7-2). Flexibility is again useful, and fillers can be used to reduce thermal expansion by as much as 75%.

3 The adhesive must not suffer *degradation* such as splitting of chains by water (hydrolysis) or ultraviolet radiation, burning, cracking, loss of coherence at operating temperature (thermal degradation), or stress-corrosion cracking or dissolution in certain media.

4 Many adhesives are applied as liquids and must then have a low enough *viscosity* to flow into joint areas but not so low as to be lost from the joint. With thermoplastics, this requires heating above T_g; with thermosets, a prepolymer is employed which cross-links and polymerizes after application. A temporary decrease of viscosity on shearing (*thixotropy*) is useful.

5 Development of the adhesive bonds requires *wetting* of the adherend surfaces. Wetting is a surface phenomenon, depending greatly on the nature of the adherend surface, the presence of adsorbed surface films, and the migration of certain components of the adhesive to the surface. In some instances, sudden, catastrophic debonding may take place after some time.

9-7-2 Adhesive Types and Their Applications

High-quality structural joints can be established only if all phases of the process are fully controlled.

1 *Joint design* must recognize the limitations of adhesive bonding. Adhesives are strongest in shear (Fig. 9-26a) and under normal tensile stress (Fig. 9-26d), and weakest in cleavage (Fig. 9-26e) or peeling (Fig. 9-26f). For example, a rigid epoxy may have a lap shear strength of 21 MPa (3 kpsi), but a peel strength of only 30-N/cm (17-lb/in) lap width. Thus, lap joints are most frequently used, just as they are favored for soldered or brazed joints. Joint strength is directly proportional to lap width but increases less steeply with overlap; hence, laps are made as wide as possible. Localized shear stresses are higher than the calculated average stress (Fig. 9-26a) and joints must be designed with this in mind. When lap joints between flexible adherends are loaded in tension, deformation of the joint (Fig. 9-26b) results in cleavage stresses. These stresses can be reduced by tapering the ends of the sheets, allowing their deformation (Fig. 9-26c).

2 *Surface preparation* is crucial. Organic surface films are harmful with most (but not all) adhesives. Oxide films are harmful if loosely bonded, whereas a strongly bonded, porous oxide film or conversion coating is helpful. Special

FIGURE 9-26
Some adhesive joints are (a) loaded in shear, (b) resulting in deformation, (c) which can be alleviated by tapering. Other joints are loaded in (d) tension, (e) cleavage, or (f) peeling.

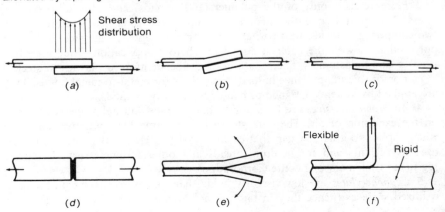

surface preparation techniques have been developed to promote adhesion and also to prevent undesirable reactions that can lead to bond separation in the long term. Primers and adhesion promoters are also used.

3 *Surface roughness* can be desirable because it gives a larger contact area and provides some mechanical interlocking. However, surface films are difficult to remove from valleys; air may become entrapped too, and bonding is then limited to asperities.

Application Methods   A good adhesive bond is, in general, stronger than the strength (cohesive strength) of the weaker body, which may be one of the adherends or the adhesive itself. Therefore, maximum strength is usually obtained with very thin—but not starved—adhesive films. This is ensured by the proper application method.

1 *Hot-melt adhesives* are thermoplastic polymers. The polymer is heated above T_g and applied in beads or webs, by appropriate equipment, to the surfaces to be joined. Rapid cooling quickly establishes a bond. The flexibility of the bond is controlled by choosing an appropriate polymer. Water resistance is high but heat and creep resistance tend to be low; polyimides are suitable for high-temperature applications. There are many applications in the construction, packaging, furniture, and footwear industries.

2 *Tapes and films* of thermosets are applied at room temperature and cured at elevated temperature. They are composed of a high-molecular-weight backbone polymer such as a flexible vinyl, neoprene, or nitrile rubber, and of a low-molecular-weight cross-linking resin such as an epoxy or phenolic cured by a catalyst. To prevent starvation, a support mesh of reinforcing material such as fiberglass may be used. Reliable joints of good toughness are produced. Several thousand kilograms of such adhesives are used in the construction of a large aircraft. The technique is used also for the bonding of brake linings, lamination of windshield glass, and a variety of construction applications.

3 *Pastes* consist only of the polymer (100% solids) and often also contain fillers. The most frequently encountered classes are room-temperature curing, two-component epoxies, heat-curing one-component epoxies, acrylics, or flexible polyurethanes, with the catalyst added just before application. They are often used to bond elastomers, fibers, fabrics, and fiber-reinforced plastics to steel in the automotive and appliance industry, as well as for metal-to-metal bonds. They are applied by systems designed to handle high-viscosity fluids.

4 *Anaerobic adhesives* are 100% solids that have the unusual property of curing on the exclusion of air. They are stored in air-permeable containers and cure when applied to a narrow gap that excludes oxygen. Hence, they are extensively used as sealants and for holding threads and other fasteners (e.g., Loctite, a registered trademark of Loctite Corporation).

5 *Cyanoacrylates* are low-viscosity fluids that cure on contact with moist or oxidized surfaces (hence they instantly bond human skin). They have low peel strength and temperature resistance but are useful in the electronics industry and for joining rubber or plastic to metal.

6 *Liquids* usually contain a solvent. Thus, PVC plastisols, sometimes fortified with an epoxy, are used for assembling car hoods, trunk lids, roofs, furniture, and appliance cabinets. They can be applied by brush, spray, printing, etc.

7 *Water emulsions* of polymers, such as polyvinyl acetate (white glue) need a porous substrate to allow the removal of water.

It is evident from the foregoing discussion that the adhesive joint is a system composed of adherends, metal oxide, primer, adhesive, and the environment. Therefore, in all instances, total control of application is crucial.

In critical applications, such as aircraft construction, the presence and continuity of bonds is verified by nondestructive testing techniques.

9-8 JOINING OF PLASTICS

Parts made of polymers may be joined to each other or to metals by variants of the techniques discussed hitherto.

1 *Mechanical fastening* employs metal or plastic screws, driven into bosses or holes molded (sometimes cut) in the part, or into metallic inserts molded into the plastic part.

2 *Thermal sealing and bonding* of thermoplastic polymers relies on heat and pressure. As in the pressure welding of metals, surfaces must be clean, and localized deformation that breaks up adsorbed surface films is helpful. However, in contrast to metals, melting of the polymer usually occurs. Several techniques are used:

a *Heated tooling or rollers* are suitable mostly for thinner parts since polymers are poor heat conductors. For example, LDPE is heat sealed with PTFE-coated, electrically heated tooling.

b *Friction joining* relies on heat generation at the interface, either by rotation (*spin welding*, as in Fig. 9-8) or by vibration. Vibration may be low-frequency oscillation imposed by mechanical means, or ultrasonic (as in Fig. 9-5*b*).

c *Hot wire welding* provides localized heating by incorporating a resistance wire into the joint area. The wire ends must stick out so that current may be passed through. A closed wire loop may be heated inductively (as in joining knife blades to handles). Once the polymer melts, the parts are pressed together.

d *Dielectric heating* results when polymers, which are insulators, are placed into an electromagnetic field. Frequencies of 1–200 MHz (most often 27.12 MHz, the frequency allocated by the Federal Communications Commission) are used; therefore, one speaks also of radio-frequency (rf) or high-frequency (hf) heating. Polymers containing polar molecules, such as PVC, nylon, polyurethane, and rubber can be through-heated and are most suitable.

e *Electromagnetic bonding* and *magnetic heat sealing* are made possible by embedding very small (around 1-μm-diameter) magnetic particles into the polymer. On applying a high-frequency field, the polymer is heated and melted by heating of the particles. Particles are often limited to a bead of thermoplastic, placed at the future joint.

f *Hot-gas welding* is the equivalent of oxyfuel gas welding (Fig. 9-20), except that hot air (or inert gas) is the heat source. The joint is prepared, beveled as on metals, and a filler rod, of the same thermoplastic as the parts, is used.

3 *Adhesive bonding* is the most versatile of all joining methods, suitable for thermoplastics, thermosets, dissimilar materials, polymer-metal and ceramic-metal combinations, and even for joining man-made and natural structures (e.g., metal or ceramic crowns on teeth or implants into bone). The cement may be a monomer (as for PMMA), an elastomer, or a thermoset. The optimum adhesive is chosen with regard to service requirements, and the method of application (brush, roller, screen printing, spraying, tape, etc.) is chosen according to manufacturing considerations.

4 *Solvent welding* is possible when the polymer (especially, an amorphous one) has a specific solvent as, for example, ABS, PVC, PMMA, polycarbonate, and polystyrene.

9-9 DESIGN ASPECTS

As in all design, the limitations and advantages of joining processes must be taken into account together with service requirements. Limitations imposed by the chosen process will be sensed from an understanding of the processes themselves. For example, the cross sections to be joined must be equal when heat generation is a function of cross section, as in butt welding and flash butt welding, whereas they may be quite dissimilar in projection welding, electric arc welding, brazing, soldering, and adhesive bonding. The joint configuration is governed by several considerations:

1 *Butt joints* present the smallest joint area between two parts and are thus the weakest joints in tension. Nevertheless, they are completely adequate when the joint itself is strong, as in liquid- and solid-state welding. In liquid-state welding the soundness of weld is ensured by appropriate preparation (Fig. 9-27) which vitally affects the fatigue performance of the joint.

a Square grooves are adequate for thinner stock (Fig. 9-27a) but grooves shaped for greater weld penetration and controlled bead formation are essential for gas and arc welding of thicker gages (Fig. 9-27b and c). No groove is needed (square preparation) for electron beam welding.

b Corner joints with square grooves (essentially, a butt joint) are suitable only for thinner material; for thicker gages, groove preparation is essential (Fig. 9-27d).

2 Butt welding may not give sufficient strength in thin sheet. Several solutions are available:

a *Flared joints* give larger surface area (Fig. 9-27e). In fusion welding the weld bead provides the strength; in other processes, including brazing and adhesive

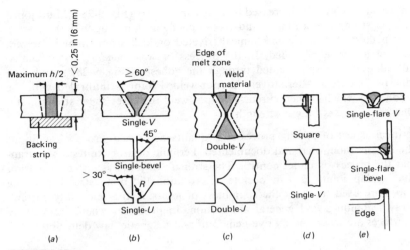

FIGURE 9-27
Joints must be prepared for arc and gas welding by providing: (*a*) square, (*b*) single-V, (*c*) double-V, (*d*) corner, or (*e*) edge grooves.

joining, the joint area is increased by bending the sheet. However, such joints are susceptible to failure by peeling separation.

b A better joint is usually obtained in nonfusion processes by creating a large contact area loaded in shear (*lap joint*, Fig. 9-28*a* and *b*). Better stress distribution in the joint area is achieved by tapering the sheets (Fig. 9-26*c*). If a smooth surface is required, the edges must be prepared (Fig. 9-28*c*). In corner or *T* joints,

FIGURE 9-28
Adhesive joints are designed to give maximum adhesive joint area: (*a*) simple lap joint, (*b*) offset lap joint, (*c*) smooth-surface lap joint, (*d*) corner joint with inserts, (*e*) tongue-and-groove butt joint, and (*f*) weld-bond joint.

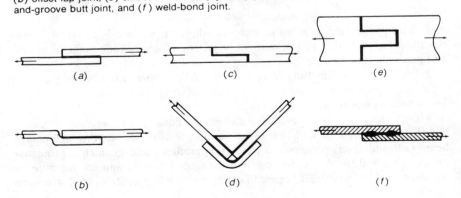

the contact surface may be increased by fillers and covers (Fig. 9-28d). Butt joints have greater strength in a tongue-and-groove configuration (Fig. 9-28e).

c For highest strength, a mechanical (riveted or lock seam) joint is brazed, soldered, or adhesively bonded. In the process of *weld bonding* sheets coated with an adhesive are spot-welded through the adhesive (Fig. 9-28f), and then the adhesive is cured. Alternatively, a spot-welded joint is infiltrated with the liquid adhesive which is then cured. The spot weld increases shear strength and the adhesive increases fatigue strength.

A critical aspect of joining processes is quality assurance. Process conditions must be strictly monitored and documented. Periodic destructive testing guarantees that the process is under control. The strength of joints is tested in tension and that of lap joints also in peel tests. Additionally, nondestructive testing techniques are used to check the soundness of joints. Ultrasonic, eddy current, and x-ray techniques find general application. In processes where cracks may form, surfaces are inspected by dye-penetrant and magnetic flaw detection.

9-10 SUMMARY

Joining expands the scope of all manufacturing processes; castings, forgings, extrusions, plates, sheet metal, and machined parts can all be joined to make more complex shapes or larger structures. Welded constructional girders, machine frames, automobile bodies, tubing and piping of all sizes, containers, and cans are all around us. Welding is often the most economical and practical repair for broken machinery. Brazed bicycle frames, heat exchangers, and soldered radiators and plumbing joints abound. Adhesive joining is increasingly used in the aircraft, automotive, appliance, and other industries.

Joints, by their very definition, provide a transition between two not necessarily similar materials. Quality control is even more important than in other processes because oxidation, surface films, slag inclusions, porosity, gaps, undercuts, hot cracks, cold cracks (embrittlement), and residual stresses could cause dangerous delayed failures. Nevertheless, good-quality joints, sometimes equal to the parent (workpiece) material in strength, can be obtained through a variety of means:

1 Solid-phase welding relies entirely on adhesion; while it is extremely sensitive to surface contaminants, it does allow joining of a very wide range of similar and dissimilar materials.

2 Highly localized melting in resistance welding represents a transition from solid- to liquid-phase processes; it is still sensitive to contaminants but the heat-affected zone is small.

3 Deep or through-the-thickness melting in liquid-phase welding processes broadens the heat-affected zone in the parent metal in all but the high-energy beam (EB and laser) processes. Surface preparation, weld geometry, protective atmospheres and/or slag and fluxes, and the rate of heating and cooling must be simultaneously controlled. Localized heating and cooling make welded structures

susceptible to distortion and cracking under the influence of internal stresses. Welding processes generate very high temperatures, and eye protection is absolutely essential.

4 In brazing and soldering, solid workpieces are joined—without melting the parent metal—with a lower-melting metal. The joint again relies on adhesion, making surface cleanliness and fit the critical factors.

5 Polymers can be formulated to fit the particular task in adhesive joining. Control of surface preparation and adhesive application are crucial to success.

6 Techniques related to welding can also be employed to deposit surface layers (particularly for wear resistance), to cut and otherwise shape metals, ceramics, and polymers, and to heat treat (anneal or harden) metals. High rates of heat input are most useful in limiting heating to a surface layer of controlled thickness (Sec. 3-5-5).

7 Many joining techniques present considerable hazards. Light emitted by high-temperature arcs and laser beams calls for eye protection or total enclosure; heat, molten metal, and sparks flying from a weld zone require face shields, goggles, gloves, and other protective clothing; irritating or toxic fumes emanating from welding, brazing, soldering, and adhesive bonding operations and toxic solvents and monomers necessitate exhausts and other safety and health measures.

General characteristics of welding processes are given in Table 12-7.

PROBLEMS

9-1 A small, complex aircraft part is to be manufactured of 6061 aluminum alloy. The highest strength obtainable with this alloy must be maintained in the entire part. (*a*) Check in the literature (e.g., *Metals Handbook*) what metallurgical condition will give the highest strength. (*b*) Determine whether this material would permit manufacturing the part by welding two less-complex parts together. (*c*) If a way can be found, specify the best welding process (in justifying your choice, work by the process of elimination). (*d*) Describe what postwelding treatment, if any, is needed.

9-2 It is proposed to join aluminum to low-carbon steel. After reviewing the equilibrium diagram, consider the feasibility of establishing a reasonably ductile joint by: (*a*) cold-welding, (*b*) diffusion bonding, (*c*) spot welding, and (*d*) flash butt welding. Justify your judgements.

9-3 A single-phase resistance-welding machine is used to join two steel sheet parts with 20 projection welds made simultaneously. One assembly is joined every 10 s. If the steel sheets are 1 mm thick and individual projection welds are similar to the one shown in Fig. 9-15*c*, calculate: (*a*) the welding current required; (*b*) the duty cycle, expressed as the fraction (or percentage) of time the current is on; (*c*) the required kVA for each welding cycle, if the desired current is attained at a voltage of 4 V; (*d*) the kVA rating of the transformer, taking into account the duty cycle; (*e*) the kW rating, allowing for a power factor of 0.5 (for phase shift due to inductive loading in the transformer); (*f*) the electrical energy consumption for each weld in kWh; and (*g*) the force requirement.

9-4 Two 5052 Al-alloy sheets of 1-mm thickness are to be joined by spot welding. (*a*) Review the appropriate phase diagram and deduce if the operation is feasible. (*b*) If the answer to (a) is yes, calculate the transformer rating and the power consumption as in Prob. 9-3, but with one spot weld made every 2 s.

9-5 Two 0.25-in-thick carbon steel plates are to be joined by oxyacetylene welding in a single-V groove (Fig. 9-27*b*). Using data from Table 12-7, determine (a) whether this is feasible, and, if the answer is yes, (b) what welding rate (in/min) one might expect at the top metal deposition rate.

9-6 Stainless steel sheets of 0.4-mm thickness are to be joined (butted) on their edges. (a) What processes could be considered, and (b) which of these is likely to be most frequently available.

9-7 Explain why an off-eutectic solder is often preferred to a eutectic one.

9-8 Define brazing, braze welding, and soldering.

9-9 A steel tube of 25-mm OD is to be adhesively joined to another steel tube of 25-mm ID, using a general-purpose epoxy. The assembly will operate at room temperature (15–30 °C) and an axial tensile load of 12 kN will have to be supported with a factor of safety of 2. Calculate the minimum length l of the lap joint (i.e., the depth of interpenetration of the tubes).

FURTHER READING

A Joining, General

ASM: *Metals Handbook*, 9th ed., vol. 6, *Welding, Brazing, and Soldering*, American Society for Metals, Metals Park, Ohio, 1983.

Lindberg, R. A., and N. B. Braton: *Welding and Other Joining Processes*, Allyn and Bacon, Boston, 1976.

Parmley, R. O.: *Standard Handbook of Fastening and Joining*, McGraw-Hill, New York, 1977.

Schwartz, M. M.: *Metals Joining Manual*, McGraw-Hill, New York, 1979.

Welding Handbook, 5 vols., 7th ed., American Welding Society, New York, 1976 on.

B Welding

Arata, Y. (ed.): *Plasma, Electron, and Laser Beam Technology*, American Society for Metals, Metals Park, Ohio, 1986.

Blazynski, T. Z. (ed.): *Explosive Welding, Forming and Compaction*, Applied Science Publishers, London, 1983.

Burgess, N. T. (ed.): *Quality Assurance of Welded Construction*, Applied Science Publishers, London, 1983.

Carey, H. B.: *Modern Welding Technology*, Prentice-Hall, Englewood Cliffs, N.J., 1979.

Davies, A. C.: *The Science and Practice of Welding*, 7th ed., Cambridge University Press, Cambridge, 1977.

Esterling, K. E.: *Introduction to the Physical Metallurgy of Welding*, Butterworths, London, 1983.

Gray, T. G. F., J. Spence, and T. H. North: *Rational Welding Design*, Butterworths, London, 1975.

Houldcroft, P. T.: *Welding Process Technology*, Cambridge University Press, Cambridge, 1977.

Lancaster, J. F.: *Metallurgy of Welding*, 3d ed., Allen and Unwin, London, 1980.

Masubushi, K.: *Analysis of Welded Structures—Residual Stresses and Distortion and Their Consequences*, Pergamon, Oxford, 1980.

Metzbower, E. (ed.): *Applications of the Laser in Metalworking*, American Society for Metals, Metals Park, Ohio, 1981.

Mohler, R.: *Practical Welding Technology*, Industrial Press, New York, 1983.

Romans, D., and E. N. Simons: *Welding Processes and Technology*, Pitman, London, 1974.

Saperstein, Z. P. (ed.): *Control of Distortion and Residual Stress in Weldments*, American Society for Metals, Metals Park, Ohio, 1977.

Schwartz, M. M. (ed.): *Innovative Welding Processes*, American Society for Metals, Metals Park, Ohio, 1981.

Schwartz, M. M. (ed.): *Source Book on Electron Beam and Laser Welding*, American Society for Metals, Metals Park, Ohio, 1981.

C Brazing and Soldering

Schwartz, M. M. (ed.): *Source Book on Brazing and Brazing Technology*, American Society for Metals, Metals Park, Ohio, 1980.

Manko, H. H.: *Solders and Soldering*, 2d ed., McGraw-Hill, New York, 1979.

Roberts, P. M.: *Brazing*, Oxford University Press, Oxford, 1975.

Thwaites, C. J.: *Capillary Joining—Brazing and Soft Soldering*, Wiley, New York, 1982.

D Adhesive Bonding

Brewis, D. M. and J. Comyn (eds.): *Advances in Adhesives*, Warwick, Birmingham, 1983.

Cagle, C. V. (ed.): *Handbook of Adhesive Bonding*, McGraw-Hill, New York, 1973.

DeLollis, N. J.: *Adhesives, Adherends, and Adhesion*, Krieger, Melbourne, Fla., 1980.

Landrock, A. H.: *Adhesives Technology Handbook*, Noyes, Park Ridge, N.J., 1985.

Patrick, R. L. (ed.): *Treatise on Adhesion and Adhesives*, Dekker, New York, 1981.

Schneberger, G. L. (ed.): *Adhesives in Manufacturing*, Dekker, New York, 1983.

Shields, J.: *Adhesives Handbook*, 3d ed., Butterworths, London, 1984.

Skeist, I. (ed.): *Handbook of Adhesives*, 2d ed., Van Nostrand Reinhold, New York, 1977.

Wake, W. C.: *Adhesion and the Formulation of Adhesives*, 2d ed., Applied Science Publishers, London, 1982.

Wu, S.: *Polymer Interface and Adhesion*, Dekker, New York, 1982.

E Journals

Welding Design and Fabrication
Welding in the World
Welding Journal

10

MANUFACTURE OF SEMICONDUCTOR DEVICES

In Sec. 1-1-3 we discussed the immense changes brought about by the development of microelectronic devices. Since the invention of the *transistor* by Bell Telephone Laboratories scientists John Bardeen, Walter Brattain, and William Shockley in 1947–1948, the field has grown by leaps and bounds and is still in a phase of rapid development. This rapid growth is attributable to concentrated efforts in developing both the physical, chemical, and materials science background *and* the manufacturing technologies needed for making solid-state products.

Transistors were first made as individual devices, essentially as straightforward replacements for electron tubes. They were then wired, with other components, into complete circuits. A revolutionary change came after 1958–1959, when the monolithic, *integrated circuits* (ICs) were invented by J. St. C. (Jack) Kilby of Texas Instruments and R. N. (Bob) Noyce of Fairchild Camera. All components of a circuit are now placed on a single chip cut from a thin (perhaps 200-μm-thick) wafer of doped silicon or other semiconductor material. With advances in manufacturing technology, more and more components—making up circuits of increasing complexity—have been accommodated on a single chip. This allowed the development of that most powerful of devices, the microprocessor, in the early 1970s. The minimum feature length gradually decreased and the size of chips increased (although most still measure only a few millimeters on their edges), and the industry moved from small-scale integration (SSI) to medium-scale (MSI), large-scale (LSI), and very-large-scale (VLSI) integration (Fig. 10-1). The number of components per chip increased hundredfold between 1972 and 1981, and millions of components can now be fitted on a single chip.

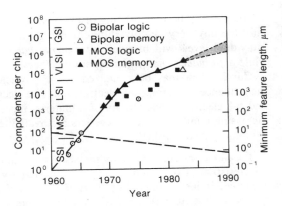

FIGURE 10-1
Over the years, the minimum feature length steadily decreased while the number of components per chip increased —very steeply until the early 1970s and more slowly thereafter. (*Adapted from G. Moore, Electron Aust. 42:(2)18 (1980). With permission. Magazine Promotions Australia, Melbourne.*)

Almost unparallelled in the history of industrial development, these advances came at a lower price; the cost of one bit of memory was halved every two years. Hundreds of millions of integrated circuits are now manufactured every year, to be used not just in calculators and computers, radios and television sets, but also in all kinds of machinery and toys, articles of production and consumption. All this could not have been possible without the development of mass-manufacturing techniques.

A study of manufacturing processes would be incomplete without at least a general outline of the most important techniques used in microelectronics. Some processes are modifications or further developments of processes previously used in other fields, but others have been developed specifically for the purpose and have, over the years, found applications in other fields too. Compared to most other manufacturing processes, the production of microelectronic devices places extreme demands on the absence of unintentional elements (impurities) in the starting materials, and on cleanliness and strictest control of process parameters during all phases of manufacturing. To understand the necessity for these measures, it is important to have at least a nodding acquaintance with the physical principles underlying the technology.

10-1 ELEMENTS OF SEMICONDUCTOR DEVICES

Within the scope of this book, it is impossible to cover all semiconductor devices serving various purposes. The emphasis will be on devices used in logic (digital) circuits, with only passing reference to analog devices.

10-1-1 The Semiconductor

It will be recalled from physics that electrons in an isolated atom can occupy only discrete energy levels and that only two electrons can share the same energy level.

Therefore, in a body consisting of many atoms, the outermost (valence) electrons must occupy very slightly different energy levels, and the very closely spaced individual levels will broaden into *energy bands*. Some of these bands overlap, while a *forbidden energy gap* separates others.

Current flow may be visualized as the physical movement of electrons under the influence of an imposed electric field. The electron is a negative charge carrier, and the electrical conductivity of a material increases (resistivity decreases) if an increasing number of charge carriers are available. Three classes of materials are commonly distinguished:

1 *Conductors* are typified by metals. As mentioned in Sec. 2-2-2, they have very low resistivities (e.g., aluminum has a resistivity of $\rho = 3 \times 10^{-8}$ $\Omega \cdot$ m; in the electronic industry, the old cgs unit of $\Omega \cdot$ cm is frequently used, and resistivity is quoted as 3×10^{-6} $\Omega \cdot$ cm). We saw that metals may be visualized as positively charged ions glued together by valence electrons. These electrons are free to move and can be set in motion on imposing the slightest voltage, because the conduction bands are only partially filled (Fig. 10-2a). All disturbances of the lattice—including dislocations, solute atoms, vacancies, and thermal excitation of atoms—present obstacles to electron motion; therefore, the resistivity of metals *increases* with temperature.

2 *Insulators* are typified by ceramics such as SiO_2 which have a resistivity of 10^{14} $\Omega \cdot$ cm. We already saw (Fig. 6-10) that all valence electrons of Si are used up in completing the outer shell of oxygen atoms, and are thus firmly held in the covalent bond. The valence band is completely filled (Fig. 10-2c) and an electron can be brought into the conduction band only if an energy of 8 eV is supplied so that an electron can jump the band gap. For an electron to acquire this 8 eV, a very large energy must be applied to the solid.

3 *Semiconductors*, as the name implies, stand between conductors and insulators. The most frequently used semiconductor is still Si. Silicon is a tetravalent

FIGURE 10-2
Energy bands: (*a*) unfilled valance bands in metals; (*b*) filled valance bands separated by a small energy gap from the conduction band in semiconductors, and (*c*) large energy gap in insulators.

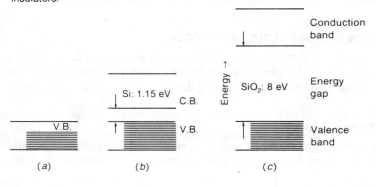

element and all electrons are again taken up in completing the outer electron shell of neighboring atoms, creating covalent bonds. The valence band is filled as in insulators, but the bond energy is lower, and the energy gap is only 1.15 eV (Fig. 10-2b). When external energy is supplied, a number of electrons can acquire enough energy to break away (jump into the conduction band); therefore, the resistivity of semiconductors *decreases* with increasing temperature. Very pure Si is a poor conductor because at room temperature only one electron in 10^{13} has the probability of acquiring enough energy to jump the gap; hence, the conductivity of silicon (an *intrinsic semiconductor*) is low (resistivity is approximately $2.5 \times 10^5 \; \Omega \cdot cm$).

However, the properties of Si can be dramatically altered by introducing very small quantities, on the order of 1 part per million (1 ppm) of foreign elements, impurities, also called *dopants*. Two possibilities exist:

a Pentavalent elements (from group V of the Periodic Table) such as N, P, As, or Sb can be introduced to form a substitutional solid solution in Si. One electron per atom will now be surplus, available to conduct electricity; hence, these are called *donor* elements. Since the electron is a negative charge carrier, one speaks of an *n-type extrinsic semiconductor*.

b Trivalent elements (from group III of the Periodic Table), such as B, Al, Ga, or In also form solid solutions but there is now one electron missing in the valence band. Such a *hole* can be regarded as a positive charge carrier; hence, one speaks of *p-type extrinsic semiconductors*. The hole can accept an electron; therefore, the trivalent elements are also called *acceptors*.

When an electron (*n* carrier) finds a hole (meets a *p* carrier), the carriers recombine and disappear. However, it takes time for carriers to recombine, and one speaks of carrier lifetime.

Even a very small amount of donor or acceptor element provides many charge carriers. For example, 1.0-ppm P amounts to 4.5×10^{16} atoms/cm^3 of Si; by injecting the same number of electrons, it reduces resistivity to about 0.15 $\Omega \cdot cm$ (Fig. 10-3). Hence, the silicon must be extremely pure if unintended conducting paths (and leakage currents) are to be avoided.

Crystallographic defects (vacancies and dislocations) also interfere with the intended operation of devices because they reduce charge mobility; furthermore, dopants migrate to them and provide low-resistivity leakage paths. Hence, the silicon must be not only very pure but must also be free of defects (unless defects are intentionally introduced to capture, *getter*, contaminants). Therefore, the great majority of semiconductor devices are made on silicon single crystals, although polycrystalline silicon (often called *polysilicon*) is used for specific purposes.

Germanium, the substrate for early transistors, has an energy gap of only 0.7 eV, and is now of limited industrial significance. Semiconducting devices are, to a still limited extent, made also on group III–V substrates, particularly GaAs (energy gap: 1.435 eV). Compared to silicon devices, GaAs devices are more expensive; however, they operate at higher speeds, have higher resistance to

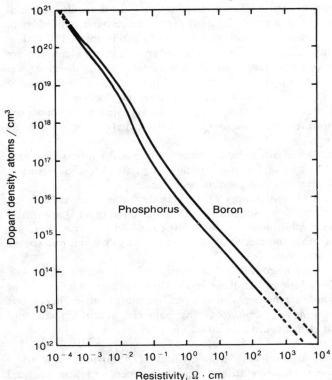

FIGURE 10-3
The resistivity of pure silicon drops rapidly even with small concentrations of dopant (impurity) atoms. (*From W. R. Turber, R. C. Mattis, and Y. M. Liu, in Semiconductor Characterization Techniques, Electrochemical Society, Pennington, N.J. 1978, p. 89. This figure was originally presented at the Spring 1978 Meeting of The Electrochemical Society, Inc. held in Seattle, Washington. With permission.*)

radiation, lower power consumption, and function at temperatures from -200 to $+200$ °C (as compared to the -55 to $+125$ °C of silicon devices). Our discussion will concentrate on silicon semiconductors.

10-1-2 Semiconductor Devices

The variety of devices based on semiconduction is very large and is still growing; here we will mention only a few.

Diodes When adjacent regions of a silicon single crystal are doped with *p*- and *n*-type dopants, a *p-n* junction is formed (Fig. 10-4*a*). There will be excess electrons in the *n* region and excess holes in the *p* region; in the *n* region,

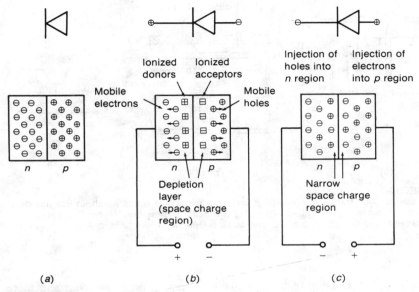

FIGURE 10-4
The simplest semiconducting device is the diode which (*a*) consists of *n* and *p* zones. On connecting a current source, (*b*) virtually no current flows when reverse bias is applied but (*c*) the diode conducts when forward bias is applied.

electrons are *majority carriers* and holes are *minority carriers*. The reverse is true of the *p* region. When a *diode* is connected to an electric current source, one of two situations may exist:

1 *Reverse bias* (Fig. 10-4*b*) causes the charge carriers to move away from the interface, an insulating zone forms, and practically no current can flow.

2 *Forward bias* (Fig. 10-4*c*) results in the movement of majority charge carriers into the opposing halves. Current flows as in a conductor, although some electrons and holes are annihilated in the recombination zone (around the interface).

In the described form the diode can serve as a rectifier. In other diodes, made of III–V-type semiconductors, recombination may result in the emission of light (*luminescence*). Among the *light-emitting diodes* (LEDs) GaAs emits red light and GaP emits green light.

Transistors A *transistor* is a device formed by coupled pairs of *p-n* junctions. They are used in two basic forms:

1 *Bipolar transistors* are two junctions in series. The *p-n-p* transistor was originally more widespread because of easier fabrication, but the *n-p-n* transistor (Fig. 10-5) is now preferred because the mobility of electrons is some three times

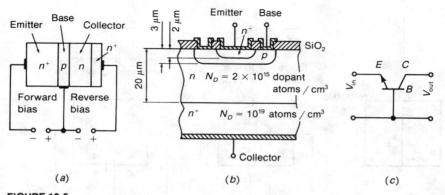

FIGURE 10-5
A bipolar junction transistor consists of (*a*) *n-p-n* (or *p-n-p*) zones which, in integrated circuits, are made (*b*) in a planar form. (*c*) Symbolic presentation. (*From E. S. Yang, Fundamentals of Semiconductor Devices, McGraw-Hill, New York 1978, p. 229. With permission.*)

higher than the mobility of holes, making the *n-p-n* transistor faster and more suitable for operation at high frequencies and fast switching rates.

The three regions are called *emitter, base*, and *collector* (Fig. 10-5*a*). In the *n-p-n* transistor a very thin, lightly doped *p* region is sandwiched between two moderately doped *n* regions. The emitter junction is forward biased to drive electrons into the narrow *p* region, where some electrons recombine with holes but the majority pass through and are injected into the collector. The collector current increases exponentially with increasing input emitter voltage; hence, this transistor acts as an *amplifier* or it can be used as a *switch*. Since both *n* and *p* charge carriers are involved, one refers to *bipolar junction transistors*. Metal connections to *n*-doped silicon would make other transistors. To avoid this, the crystal is heavily doped (indicated by the plus sign) where connections (*ohmic contacts*) will be made.

In the early phases of development the transistor physically looked somewhat similar to Fig. 10-5*a*; however, the fabrication method soon changed to *planar devices*, in which connections are brought to the wafer surface (Fig. 10-5*b*). Irrespective of its physical appearance, the transistor is symbolically shown as in Fig. 10-5*c*.

2 *The insulated-gate field-effect transistor* (IGFET) or *metal-oxide-semiconductor field-effect transistor* (MOSFET) is a *unipolar* device because only one charge carrier plays a role. In the *n*-channel MOSFET shown in Fig. 10-6*a*, a channel of *L* length is formed, in a lightly doped *p* substrate, between the heavily doped (and hence marked n^+) *source* and *drain*. The magnitude of current is controlled by the *gate*, which is insulated by a thin (< 100-nm-thick) SiO_2 layer. On applying a positive voltage to the gate, the electric field repels holes from the substrate and, when the applied voltage exceeds a threshold value, a mobile

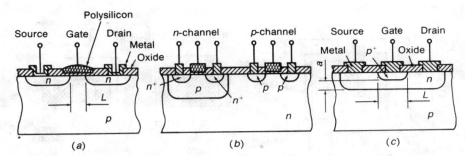

FIGURE 10-6
Only one kind of charge carrier is involved in the (*a*) metal-oxide-semiconductor field-effect transistor (MOSFET), (*b*) complementary MOS (CMOS), and (*c*) junction field-effect transistor (JFET).

electron layer is set up: An *n*-type conductive channel now exists in the zone between source and drain. If the drain is made positive, an electron current will flow from source to drain, the magnitude of which is controlled by the gate voltage. Since the current is carried by electrons, the structure consists of *m*etal, *o*xide, and *s*emiconductor, and the current is controlled by the field set up under the gate, one speaks of an *n-channel* MOSFET. The metal gate is most often replaced by a highly doped, *n*-type polysilicon stripe.

MOSFETs are slower than bipolar transistors, but they dissipate much less power; they are self-insulating, thus occupy less space; they are easier to manufacture; they can be used as memory elements; and the channels acts as resistors, eliminating the need for separate resistors in MOSFET circuits. Because of their many advantages, MOSFET circuits were used in the first hand-held calculators and in microprocessors, and are used in many VLSI devices. For this purpose, a further advantage is that MOSFET circuits lend themselves to CAD.

In many applications, especially for computers, wrist watches, and the like, complementary MOS (CMOS) are used in which an *n* channel and a *p* channel are paired in adjacent positions (Fig. 10-6*b*).

Less frequently used is the *junction field-effect transistor* (JFET, also simply called FET). In the *n*-channel FET shown in Fig. 10-6*c*, an *n*-channel of *a* width and *L* length is formed in a lightly doped *p* substrate. The magnitude of current is controlled by the heavily doped p^+ gate. Again, current is controlled by the space charge set up by the gate.

Integrated Circuits   We mentioned that transistors were first made as discrete devices which were then wired into complete circuits. Printed circuits produced by ECM (Sec. 8-9-2) and the development of wave soldering (Fig. 9-23) increased production rates and improved reliability. With the introduction of the integrated circuit, all circuit elements—including transistors, resistors, and capacitances—are fitted on a single silicon chip, at an ever increasing density (inductances cannot be made).

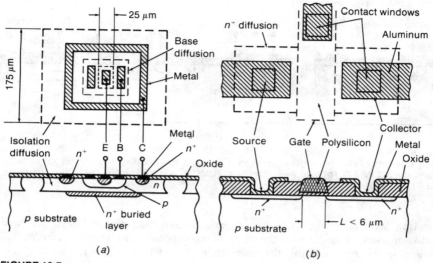

FIGURE 10-7
The physical appearance of an (a) insulated bipolar junction transistor and (b) MOSFET. (*Adapted from E. S. Yang, Fundamentals of Semiconductor Devices, McGraw-Hill, 1978, pp. 220, 267, and 271. With permission.*)

1 We saw in Figs. 10-5b and 10-6 some typical cross sections of transistors. In plan view, they appear as small, electrically isolated areas on the wafer surface, interconnected by metallic patterns, referred to as *metallization*.

a Bipolar transistors need isolation by dielectric insulating regions or by reverse-biased *p-n* junctions. An example is shown in Fig. 10-7a; the *n-p* junction is reverse biased and provides the isolation. Because of the planar arrangement, the active part of the transistor is far from the collector, and a highly doped n^+ *buried layer* is introduced to provide a low-resistance path. Other technologies are available that allow higher density of devices.

b We mentioned that MOSFETs are self-isolated; in plan view they appear as in Fig. 10-7b.

2 Resistors can be formed as thin sheets (stripes) of doping elements diffused into a silicon substrate. The resistance of one square (Fig. 10-8a) is

$$R = \rho \frac{a}{ha} = \frac{\rho}{h} \tag{10-1}$$

and is independent of a, the side of the square. Hence, it is quoted in units of $\Omega/\square$ (ohm per square). Resistivity ρ depends on doping element concentration (Fig. 10-3) and, for a given doping depth h, it can be increased by diffusing over a greater length l, i.e., by creating the equivalent of several squares in series. For a given resistance, the length of the resistor can be reduced by reducing a. It should

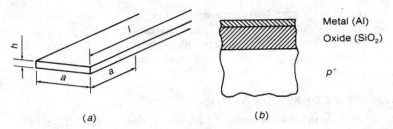

FIGURE 10-8
Passive devices: (*a*) typical dimensions of diffused resistors and (*b*) capacitor.

be noted that a *p-n* junction is also formed. As mentioned, a MOS transistor can also be used as a high-value resistor (the channels have a sheet resistance on the order of 10 kΩ/$\square$).

3 Capacitors are formed by depositing a dielectric (insulating) layer such as SiO$_2$ on top of a heavily doped p^+-type Si substrate. The other plate of the capacitor is formed by a deposited metal (aluminum) film (Fig. 10-8*b*). Obviously, a MOS structure can also serve as a capacitor.

Miniaturization   The first integrated circuits contained only a few devices and were quite large. There are powerful incentives to reduce the size of devices so that more of them can be fitted on a single chip. Signals propagate at the rate of about 5 ns/m; hence, speed of operation increases with decreasing device size. Power consumption drops too and thus the problem of cooling is also alleviated. With larger-scale integration, the number of external interconnects and the failures associated with them decrease and the cost of complete circuits drops.

Miniaturization hinges on reducing the feature size. The minimum feature size is defined as the average of the minimum line width and spacing (in other words, one-half of the pitch).

It is anticipated that certain limitations imposed by physical phenomena will be reached when feature sizes are reduced below 0.25 μm. Before these limits are reached, various difficulties have to be overcome. First of all, the design of very complex circuits becomes a formidable task; fortunately, the concurrent development of CAD has made this task possible. Computer programs are used not only for the design of circuit layout but also for the modeling of circuit performance. Manufacturing challenges are also of great magnitude, and in the remaining part of this chapter we will explore some solutions. Problems multiply with increasing integration because the likelihood of a defect falling within a given circuit becomes higher. Nevertheless, it is believed that gigascale integration—with several hundred million components on a single chip—will be reached with three-dimensional (multilevel) chips or with wafer-scale structures.

There are innumerable and ever newer devices designed, and there are corresponding developments in manufacturing technology. Because the technology is still fluid, our discussion will be limited to some of the basic technologies employed for the manufacture of integrated circuits, in particular, *monolithic integrated circuits* (ICs). As shown by the Greek derivation (monolithic = single stone), these contain all circuit components on a single *chip* (also called *die*; plural, *dice*). In most instances the chip is cut from a wafer of a silicon crystal.

10-2 MANUFACTURE OF SILICON WAFERS

We saw that a semiconductor must be free of accidental doping elements and that the tolerable concentration of impurities is very low. In general engineering applications, a 99.99% pure metal (e.g., superpurity aluminum or electrolytic zinc) is regarded as very pure, yet it contains 0.01% or 100-ppm impurities. In electronic-grade silicon (EGS) the concentration of doping elements is on the order of *parts per billion* (ppb), so low that their presence can only be inferred from resistivity measurements.

10-2-1 Production of EGS

Production of EGS involves the following steps:

1 Quartzite (pure, natural SiO_2) is reduced with carbon in a submerged-arc electric furnace. The resulting 98%-purity silicon is suitable for metallurgical alloying but must be further refined for electronic purposes.

2 Silicon is hard and brittle. It can be broken up, pulverized, and reacted at 300 °C with anhydrous HCl

$$Si \text{ (solid)} + 3HCl \text{ (gas)} = SiHCl_3 \text{ (gas)} + H_2 \text{ (gas)} \qquad (10\text{-}2)$$

Trichlorsilane ($SiHCl_3$) and chlorides of impurities condense to a liquid at room temperature. The boiling point of $SiHCl_3$ is 32 °C; it can be separated from the chlorides of impurities by fractional distillation.

3 EGS is obtained by chemical vapor deposition (CVD, see Sec. 10-3-2), in this case by reaction with H_2

$$2SiHCl_3 \text{ (gas)} + 2H_2 \text{ (gas)} = 2Si \text{ (solid)} + 6HCl \text{ (gas)} \qquad (10\text{-}3)$$

Deposition takes place on thin (4-mm-diameter) resistance-heated silicon rods (*slim rods*, Fig. 10-9a), over several hours, until polycrystalline silicon rods of up to 200-mm diameter grow.

The processes are power intensive but the entire world demand for EGS amounted to only 3000 ton in 1982; hence, the microelectronics industry consumes very little power per unit value produced.

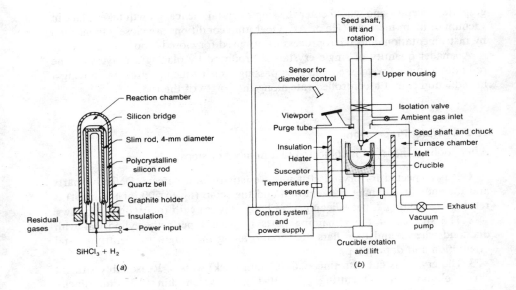

FIGURE 10-9
The growth of (a) polycrystalline silicon rods by CVD and of (b) single-crystal ingots by the
Czochralski technique. (*Part (a) after L. D. Crossmann and J. A. Baker, in Semiconductor
Silicon 1977, Electrochemical Society, 1977, Pennington, N.J. p. 18. This figure was originally
presented at the Spring 1977 Meeting of The Electrochemical Society, Inc. held in Phila-
delphia, Pennsylvania. With permission; part (b) after C. W. Pearce, in VLSI Technology, S. M.
Sze (ed.), McGraw-Hill, 1983, p. 25. With permission of Bell Telephone Laboratories.*)

10-2-2 Crystal Growing

Silicon crystallizes in the diamond lattice (Fig. 6-9*a*) which can be visualized as
two interpenetrating fcc lattices. Microelectronic devices generally require that
lattice defects be minimum; therefore, the aim is to grow near-perfect single
crystals, usually in the $\langle 111 \rangle$ or $\langle 100 \rangle$ orientation.

Crystals are grown in a scaled-up version of the laboratory technique known as
the *Czochralski method* (Fig. 10-9*b*). Broken pieces of EGS are loaded into a
crucible (usually made of pure SiO_2) supported by a graphite susceptor. Heat is
provided by induction or resistance heating. Silicon melts at 1414 °C. To grow a
crystal of fixed orientation, a single crystal of Si (the *seed crystal*) is partly
immersed in the melt, and then pulled slowly upward with simultaneous rotation,
while the crucible also rotates. Solid silicon deposits on the seed in the same
orientation to form a cylindrical single crystal. A melt of 60 kg yields a
3-meter-long, 100-mm-diameter *crystal* (also called *boule* or *ingot*). Crystals up to
150- and even 200-mm diameter can be grown. Oxygen would burn up the

graphite susceptor and would enter the single crystal; hence, growth takes place in vacuum or in an inert gas (He or Ar). Operating conditions are closely monitored by instrumentation, and microprocessors are used for closed-loop control.

A smaller quantity of single crystals is produced by placing a polycrystalline rod onto a single-crystal seed and then passing a melt zone upwards, by moving a hf induction coil at a controlled rate along the length of the rod.

10-2-3 Wafer Preparation

After cooling, the ingot is subjected to a number of steps:

1 It is inspected for crystal structure, resistivity, and defects. Defective parts and the ends of the ingot are removed and the scrap (up to 50% of the ingot) is returned to melting. Since silicon is very hard, it is cut with rotary diamond saws.

2 The surface of the remaining parts is ground perfectly cylindrical with diamond wheels, and then flats are ground along the length to identify crystal orientation and doping type.

3 The crystal is cut into thin (0.6–0.7-mm-thick) wafers. Reasonably straight cut is obtained in ID cutting (Fig. 10-10), in which thin (0.325-mm-thick), stainless steel, annular blades are used, coated with diamond powder on their ID. Here some one-third of the silicon is again lost. Blades of typically 200-mm ID are rotated at 2000 r/min, and are fed into the crystal at 0.5 mm/s.

4 Flatness and parallelism are improved by simultaneous, two-sided lapping of the surfaces, using Al_2O_3 in glycerine as the lapping medium. Some 20 μm per side is removed.

5 The edges are rounded with diamond wheels to prevent chipping.

6 The wafers are chemically etched to remove the mechanically damaged surface layers.

FIGURE 10-10
Single crystal ingots are cut into wafers by the ID cutting process.

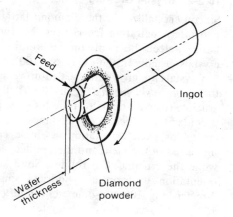

7 The wafers are polished in a slurry of fine (10-nm) SiO_2 particles in an aqueous solution of NaOH. Some 25 μm is removed per side by a combination of mechanical action and the oxidation of silicon by NaOH.

8 The wafers are cleaned in a special solution (*slice clean*) to remove all residues. After a rinse with deionized water, they are rigorously inspected.

10-3 DEVICE FABRICATION

The silicon wafers serve as the substrate on which planar devices are made by a sequence of process steps. Since an integrated circuit contains tens or hundreds of thousands of devices, it is necessary to build test circuits or conduct computer simulations. Once the design is fixed, the manufacturing sequence is determined.

10-3-1 Outline of Process Sequence

As shown in Fig. 10-7, circuits consist of intricate patterns of semiconducting, insulating, and conducting features which must be created in several superimposed layers. This can be achieved either by deposition of these layers in the requisite pattern, or by deposition over the entire surface and removal of the unwanted portions.

The major steps of processing are shown in Fig. 10-11. The substrate of all devices is either an *n*-type or a *p*-type wafer (Fig. 10-7). The impurity element concentration is set either during crystal growing or, after wafer preparation, by overall diffusion. The substrate is now ready for developing circuit features. For each layer of the device, a pattern is made which is optically reduced in size to make an optical mask. To transfer the pattern onto the surface of the wafer, the surface of the wafer is oxidized and a photosensitive resist is deposited and exposed through the optical mask. Unwanted portions of the photoresist are dissolved, and the oxide film thus exposed is etched away. Thus, an oxide mask is generated which then controls the diffusion of doping elements into the substrate. This sequence is repeated several times until the device is constructed. Metallic conducting paths are deposited on the surface and the circuit thus created is then provided with connections to the outside world. The IC is then packaged to protect it against damage and the effects of the environment.

In the following, each processing step will be discussed from the manufacturing point of view, and then the integration of processes will be indicated.

10-3-2 Basic Fabrication Techniques

It will be useful to review some of the equipment and techniques that find a variety of applications in semiconductor manufacture. It should be noted that basically all processing is done in batches. However, by the simultaneous processing of tens and even hundreds of wafers (each containing up to several hundred ICs), and by a high degree of automation, the cost of individual devices can be kept very low.

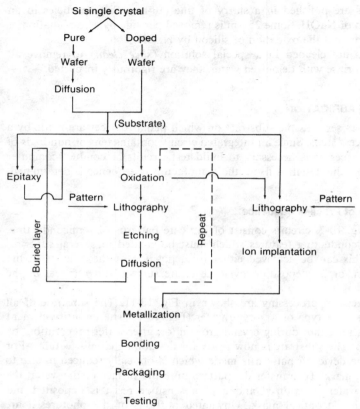

FIGURE 10-11
General outline of fabrication sequence for integrated circuits.

Heating   Much of wafer processing takes place at elevated temperatures. Two approaches are usual:

1 In *cold-wall processing* the wafer rests on a susceptor which is heated by resistance, induction, or radiant heat. Gases (or vacuum) necessary for processing are contained in a chamber with cooled walls.

2 In *hot-wall processing* the wafers are placed into a furnace which is heated to the processing temperature.

When thermal or chemical reactions take place in the cold-wall enclosure or furnace, it is usual to speak of a *reactor*.

Pyrolysis   The atomic species required is obtained by the thermal decomposition of a compound (often a halide) of the element. The compound is introduced into the furnace where decomposition and deposition on the substrate (wafer) take place.

Chemical Vapor Deposition (CVD) In the broader sense, this term includes pyrolysis; in the narrower sense, it refers to the deposition of an element (or compound) produced by a vapor-phase reaction between a compound of the element and a reactive gas, with the formation of by-products that must be removed from the reactor. We already saw an example of CVD in Fig. 10-9a and Eq. (10-3). Provisions are often made for the introduction of various reacting compounds in succession.

Physical Vapor Deposition (PVD) A substance such as a metal or oxide may be evaporated by the application of sufficient heat. The atoms or molecules liberated move away from the source in all directions; when they come into the range of atomic or molecular attraction of the workpiece (substrate), they condense on it. To facilitate bonding, the workpiece is often heated, below its melting point. Several methods of operation are practiced:

1 In the basic form of PVD (Fig. 10-12a), the substance such as a metal is evaporated by one of several techniques; placing small wire hangers onto a refractory-metal (typically, W) filament (Fig. 10-12b); heating in a crucible by induction (Fig. 10-12c); or by the impingement of an electron beam on the surface of the metal (Fig. 10-12d). In the latter method, heating is confined to a small zone of the metal and no contaminants are introduced from crucibles.

FIGURE 10-12
Physical vapor deposition (PVD): (a) basic arrangement, in which the source may be a (b) wire evaporated on a tungsten filament, (c) melt heated by rf induction, or (d) metal impinged by an electron beam. (After D. B. Fraser, in VLSI Technology, S. M. Sze (ed.), McGraw-Hill, 1983, p. 357. With permission of Bell Telephone Laboratories.)

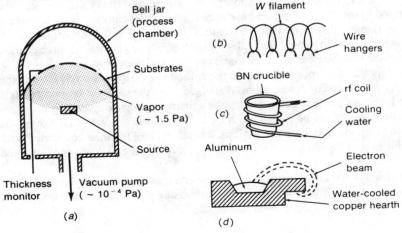

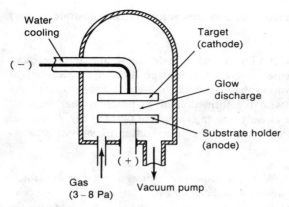

FIGURE 10-13
Arrangement for sputtering. (*After D. Roddy, Introduction to Microelectronics, 2d ed., Per-gamon Press, Oxford, 1978, p. 137. Reprinted with permission.*)

Collision with atoms of air or other gas would reduce the efficiency of the system; therefore, PVD is conducted in a bell jar, evacuated to typically 10^{-4} Pa (5×10^{-6} torr). Thickness is usually monitored by placing a quartz crystal into the bell; as the deposit builds up, the natural frequency of the crystal and, thus, the frequency of a quartz-controlled oscillator changes.

2 The rate of deposition is accelerated by the application of a dc electric field, making the metal the cathode and the substrate the anode. Material is removed from the electrode in the form of negatively charged ions which are accelerated toward the substrate by the substrate's positive charge. For better bonding, the substrate is often heated.

3 The method of operation changes when the bell jar is partially backfilled (to 3–8 Pa (0.02–0.05 torr)) with a heavy inert gas such as argon (Fig. 10-13). A high voltage (2–6 kV) is imposed which ionizes the argon. The heavy positive ions are accelerated in the electric field and are hurled against the cathode at such velocity that the impact dislodges (sputters) atoms from the cathode surface. Such *sputtering* is useful in several ways. If the specimens are given a negative charge, their surface can be cleaned of all contaminants and adsorbed films. For PVD, the substance to be deposited is made the cathode and atoms are released. If the gas is oxygen instead of argon, atoms dislodged from the cathode are immediately oxidized, and an oxide is deposited on the substrate (*reactive sputtering*).

4 Insulators cannot be treated with dc PVD because a positive charge builds up on the cathode, repelling the Ar ions. Therefore, PVD is then conducted with ac in the radio-frequency range. In such *rf sputtering* the cathode is discharged during each reverse-polarity half cycle.

5 In *magnetron sputter deposition* the source (cathode) is surrounded by a magnetic field which captures electrons, increasing the ionizing efficiency, and thus increasing the rate of sputtering.

Deposition rate is constant for points lying on the inside surface of a sphere; therefore, substrates are attached to rotating, spherical sections. Typical deposition rates are 1 μm/min for aluminum.

Plasmas Because of their importance to IC processing, plasmas deserve our special attention.

Gases are not conductors at atmospheric pressure. However, electrons may be removed from an atom by supplying energy (in the form of electron impact, x-ray, UV light, heat). In Sec. 9-3-5 we already mentioned that in the intense heat of a welding arc some atoms lose their electrons and thus a *plasma*—a neutral, gaseous mixture of electrons and positive ions—is formed. Such high-temperature plasma would be useless for the production of semiconductors.

A plasma can also be obtained by imposing an electric field of some critical magnitude on a gas. Many of the electrons released by the energy input are captured by positive ions, but some survive and are elastically scattered. The impact of these electrons with the surrounding atoms excites these atoms into higher quantum states. When the atoms drop back to the stable state, the energy released is emitted as photons, leading to the characteristic *glow discharge*. If the energy of electrons is high enough, they ionize other atoms; therefore, above a critical voltage (the *breakdown potential*), the gas becomes ionized. Typically, of the 10^{16} atoms/cm^3 that are present at atmospheric pressure, only 10^9–10^{12} atoms/cm^3 are ionized. Hence, even though the mean electron temperature is between 10^4 and 10^5 K, the gas itself remains cool, a great advantage in semiconductor fabrication.

For processing purposes, a plasma can be formed by imposing a dc potential of several hundred volts on two electrodes placed in a partially evacuated (approximately 5-Pa (0.03-torr)) envelope. Positive ions are accelerated by the electric field and, on impact with the cathode, eject secondary electrons and even atoms; thus, a plasma can be used for sputtering.

Alternatively, and preferably, an ac field in the radio-frequency range is applied (the frequency of 13.56 MHz is internationally assigned for industrial and scientific use). A great advantage is that, because of the periodic reversal of the electric field, the electrode can be covered even with an insulator, allowing the processing of SiO$_2$ and other ceramics.

We may now proceed with the application of the above basic techniques to microelectronic circuit fabrication.

10-3-3 Changing the Composition of the Surface

The composition of the wafer may have to be changed either over its entire surface or only in locations determined by the resist pattern. Basically, two kinds of changes may be made: impurity atoms are introduced into the surface (diffusion or ion implantation) or reaction with oxygen is induced (thermal oxidation).

Diffusion We already discussed in Sec. 3-1 the role of diffusion in metals. Diffusion of doping elements into silicon has been one of the most important methods of changing the composition of the wafer in a controlled manner, and it is unavoidable whenever the wafer is heated for any length of time. Diffusion can serve one of two purposes (Fig. 10-11):

1 Overall diffusion to change the characteristics of the wafer and, thus, of the entire substrate. Silicon crystals can and are being grown with controlled amounts of doping elements, particularly As, P, Sb, or B, to give resistivities from 0.0005–50 $\Omega \cdot$ cm. However, pure EGS crystals are also grown, and dopants are then introduced into the wafers at the beginning of processing.

2 Localized diffusion may serve several purposes: develop p and n regions in active devices; create stripes for resistors; and develop highly doped regions for interconnects and for pads to which metallic contacts will be applied (ohmic contacts).

The most common technique is thermal diffusion, conducted at 800–1200 °C, using a gaseous, liquid, or solid source of the dopant. For example, B can be diffused from boron tetrabromide. The first step is oxidation

$$4BBr_4 \text{ (liquid)} + 3O_2 = 2B_2O_3 + 16Br \tag{10-4}$$

Boron is incorporated into the silicon surface with the formation of a thin SiO_2 film

$$2B_2O_3 \text{ (gas)} + 3Si \text{ (solid)} = 4B + 3SiO_2 \tag{10-5}$$

Alternatively, BN slices can be placed between the silicon wafers; at elevated temperatures, the concentration gradient drives B into the silicon. Good control is obtained by depositing a doped oxide and then driving the dopant into the silicon at elevated temperatures.

We saw in Sec. 3-1-2 (Eq. (3-2)) that diffusion is greatly accelerated with increasing temperature; hence, diffusion is conducted at high temperatures. In general, smaller atoms diffuse more rapidly, and differences in diffusion rates can be used to control the depth of impurity penetration in various steps of processing. All atoms diffuse more readily and preferentially along grain boundaries and at other disturbances of the crystal structure, and this is one of the reasons for using single crystal wafers, free of defects.

Diffusion is omnidirectional; hence, diffusion away from the edges of masks cannot be avoided, and this puts a lower limit to the length of features (width of lines) that can be produced.

The concentration of dopant is highest at the surface and may reach the limit of solid solubility (Sec. 3-1-2). Concentration falls off rapidly and, if a more uniform distribution is desired, deposition (now called *predeposition*) is followed by subsequent heating (*drive-in diffusion*). To prevent the escape of dopants through the surface during drive-in diffusion, a thin oxide film is formed after predeposition. Much of the diffusion takes place interstitially, and holding at

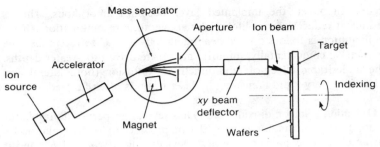

FIGURE 10-14
Ion implantation is the dominant technique in VLSI technology.

temperature allows dopants to enter into substitutional positions and, thus, become electronically active.

As mentioned, diffusion—wanted or unwanted—takes place every time the temperature is raised in the course of processing, and this must be taken into account in designing the process sequence.

Ion Implantation A highly controlled method of introducing dopants is *ion implantation*. The essential features of the equipment are shown in Fig. 10-14.

The *ion source* usually consists of a gas feed and hot oven. Atoms of gases emerging from the oven enter a noble-gas discharge plasma where they become ionized. Thus, ions of the doping element are generated. After *acceleration*, ions are subjected to a strong magnet (*analyzing* or *mass-separating magnet*) which deflects foreign ions so that only the desired species passes through a slit (*resolving aperture*). An energy of 10–500 keV accelerates the ions sufficiently to penetrate some 10–1000 nm below the surface of silicon, even if a thin oxide film is present (indeed, such a film is often intentionally grown to protect the surface from unwanted reactions). Voltages applied to *deflection plates* cause the beam to sweep over the *target* (wafer) area (or the beam is fixed and the target is mechanically moved).

The ions lose their energy in collisions with target nuclei and electrons, and come to rest some distance below the surface. If the ion enters in perfect alignment with a major crystallographic direction, it penetrates deeply. To avoid such *channelling* effect, the ion beam is directed slightly obliquely to the crystal axis, and a fairly uniform depth distribution from the surface is then obtained.

Ion implantation has several advantages: The dose can be readily controlled, usually in the range of 10^{14}–10^{18} atoms/cm³ (but even to concentrations exceeding the solubility limit); a very steep concentration gradient is obtained at the edge of masks and, thus, line width can be reduced; and the temperature of the wafer remains low.

The high-energy impacts greatly disturb the wafer surface. Silicon atoms are knocked out of their lattice locations, vacancies form, and substantial structural damage occurs. In effect, the implanted layer becomes amorphous. This is undesirable since it reduces carrier lifetime by accelerating recombination; therefore, ion implantation is followed by annealing at 200–800 °C to reestablish the crystalline structure; coincidentally, impurity atoms are driven to greater depths. This has the undesirable effect of omnidirectional diffusion (both lateral and in-depth); therefore, very rapid annealing by laser beam is often favored.

Thermal Oxidation Silicon dioxide (SiO_2) has several properties desirable for IC fabrication. It is an insulator, hence, it can provide dielectric isolation of devices; it is an essential component in MOS devices (Fig. 10-6a and b); and it isolates conductors in multilevel structures. It adheres well to silicon and it has a similar thermal expansion, thus avoiding thermal stresses. Diffusion of P, Sb, As, and B is very slow in it; thus, even a thin film (on the order of 0.1–1 μm, more usually, around 0.5 μm) can serve as a diffusion barrier. It is used as a mask for doping by diffusion, and also as a barrier against the loss of previously deposited dopants, as in the drive-in stage of two-stage diffusion. However, Ga, Al, Zn, Na, and O diffuse fast in the oxide and, for these elements, a silicon nitride layer must be used as a barrier, often with a thin oxide interface on the silicon surface.

The thermal oxidation of Si may be conducted in dry oxygen

$$Si + O_2 = SiO_2 \tag{10-6}$$

but steam oxidation is more suitable for thicker (over 0.5-μm-thick) films

$$Si + 2H_2O = SiO_2 + 2H_2. \tag{10-7}$$

Growth rates are greatly accelerated by moderately high pressures; increasing the pressure from atmospheric to 20 atm (2 MPa) causes a tenfold increase in oxidation rate.

Plasma oxidation in a pure oxygen discharge has the advantage of keeping temperatures below 600 °C.

10-3-4 Deposition of Surface Films

These techniques differ from the previous ones in that layers, films, are formed entirely by material deposited on the surface, with no intentional reaction with the surface.

Epitaxy We saw in Sec. 9-3-1 that epitaxial growth refers to the growth of a crystal in the same orientation as the substrate: Essentially, the new layer becomes a continuation of the substrate. This was also the case in growing a silicon crystal from the liquid on a seed crystal (Sec. 10-2-2), and also in annealing after ion implantation. *Epitaxy* as a term used for silicon microcircuit fabrication refers to the process in which a thin layer of doped silicon is grown on the substrate, at temperatures below T_m. The layer may be grown over the entire surface prior to forming local features (as the n layer in Fig. 10-5b), or after a local feature has been formed (Fig. 10-11); this local feature then becomes a *buried layer* (as the n^+ layer in Fig. 10-7a).

Cleanliness of the substrate surface is critical; therefore, the previously cleaned wafers are first etched, in situ, with anhydrous HCl at 1200 °C. Deposition then follows in the same equipment (reactor).

The most frequently used technique is CVD by hydrogen reduction of the halides of Si. Thus, silicon is deposited by passing, for example, silicon tetrachloride ($SiCl_4$) over the surface of wafers at 1150–1250 °C

$$SiCl_4 \text{ (gas)} + 2H_2 \text{ (gas)} = Si \text{ (solid)} + 4HCl \text{ (gas)}. \qquad (10\text{-}8)$$

The layer grows at a rate of 0.2–0.3 μm/min. Doping elements are codeposited from halides, such as arsine (AsH_3) which decomposes on the hot surface and becomes entrapped in the epitaxial silicon layer. Dopant atoms will also diffuse from the substrate into the epitaxial layer (*autodoping*) and this sets a lower limit on the thickness of epitaxial layers that can be grown with a controlled impurity concentration.

Some of the compounds (such as arsine) are highly toxic; therefore, reactors are constructed and operated with the utmost care.

Another method of deposition is *molecular beam epitaxy* (MBE). This is conducted in ultra-high vacuum (10^{-6}–10^{-8} Pa (10^{-8}–10^{-10} torr)).

Deposition of Dielectrics and Polysilicon Dielectrics (insulators) and polycrystalline silicon (or, simply, *polysilicon*) do not take an active part in the semiconductor action but are essential for the functioning of devices.

Deposition normally takes place by CVD with reactant gas flowing over the surface of wafers in cold-wall reactors. Deposition is more uniform, at reduced pressures (30–250 Pa), in hot-wall reactors (low-pressure CVD or LPCVD). Lowest temperatures (100–400 °C) are typical of plasma-assisted deposition (or, simply, plasma deposition), because the energy for the chemical reaction is supplied primarily by the glow discharge.

Dielectrics are deposited for various purposes: to provide electrical insulation between successive layers; to present barriers to diffusion; to serve as masks for diffusion and ion implantation; to passivate the surface (tie down the free surface bonds), and protect devices from environmental and mechanical damage. The most frequently used dielectrics are as follows:

1 Silicon dioxide (SiO_2) is usually deposited in the pure form, although phosphor-doped silica (P-glass) is also frequently grown. With 4–7% P, it has the advantage of flowing in a viscous manner at 1100 °C, giving good step coverage with a smooth surface.

2 Silicon nitride. The stoichiometric form Si_3N_4 is deposited by CVD, from a reaction of silane (SiH_4) with ammonia (NH_3), at 700–900 °C. Plasma-assisted deposition produces a nonstoichiometric nitride (SiN). Both kinds of nitride are excellent barriers to water and sodium diffusion. In plasma deposition the temperature is low enough to allow deposition over the completed device (i.e., encapsulation for protection against the environment and mechanical damage).

3 Polysilicon is deposited by LPCVD, at 600–650 °C, by the pyrolytic decomposition of silane

$$SiH_4 = Si + 2H_2 \qquad (10\text{-}9)$$

Deposition proceeds at the rate of some 10 nm/min. The layer may be doped by adding dopant gases during deposition, or after deposition by ion implantation or diffusion. Doped to high concentrations, polysilicon can serve as: gate elements in MOS devices; conductive contacts (ohmic contact) to crystalline silicon; high-value resistors; and as an intermediate product, namely as a diffusion source for the formation of shallow junctions. In MOSFET devices it is used to self-align source and drain diffused regions (Sec. 10-3-7) and enables a high level of integration.

Film deposition processes require stringent controls. Most reactants, including silane, are toxic and flammable, and the properties of films—including composition, resistivity, and internal stresses—are greatly affected by process conditions.

Metallization Metallic films replace the wiring used in conventional electronics, and as such have made integrated circuits practical. Their purpose is to provide highly conductive current paths between devices. The most frequently used metal is aluminum and its alloys; it adheres well to SiO_2 and silica glasses, and it makes low-resistance contacts with the highly doped p^+ and n^+ regions of transistors and with polysilicon. Gold is used to a more limited extent for back-side chip contacts.

Most metallization is carried out by physical vapor deposition (PVD, Fig. 10-12). Deposition is followed by annealing to form an alloy interface with Si.

CVD also finds use primarily for tungsten, which is not only a reasonably good conductor but also resists high temperatures. It may be formed by the hydrogen reduction of WF_6

$$WF_6 + 3H_2 = W + 6HF \qquad (10\text{-}10)$$

or by pyrolytic decomposition

$$WF_6 + \text{energy (thermal, plasma, or optical)} = W + 3F_2. \qquad (10\text{-}11)$$

A disadvantage of W (and Mo) is rapid oxidation. Therefore, silicides such as WSi_2, $MoSi_2$, $TiSi_2$, and $TaSi_2$ are used as MOSFET gate electrodes, alone or with heavily doped polysilicon. Deposition is by simultaneous evaporation (or sputtering) of the refractory metal and silicon, or by CVD. Relative to polysilicon, silicides have the advantage of lower resistivity, important for VLSI devices.

10-3-5 Lithography

Having reviewed the basic techniques of film formation and deposition, we are ready to see how the spatial distribution of features is controlled. It is evident from Fig. 10-7 that an IC containing thousands of devices on a small chip must have a very intricate pattern indeed. To control deposition and removal in such fine detail, a modern version of an old technique is used.

At the end of the 18th century, *lithography* (from the Greek *lithos* = stone) was developed for printing. The design is put on a flat stone (or metal) surface with grease. First water and then an oil-base ink is applied; the ink is repelled by the wet parts but is adsorbed by the greasy surfaces. Thus an image of the pattern can be transferred to paper. For microcircuits, the pattern is transferred photographically; hence, the term *photolithography*.

Generation of Masks The first step is the preparation of a *photomask* through which a photosensitive film, deposited on the wafer, will then be exposed. For this, patterns are made for each layer of the IC.

For circuits of moderate complexity, patterns can be large-scale drawings (also called *artwork*), magnified some 100–2000 times. The drawing is photographically reduced to $10 \times$ magnification onto a glass plate, which in turn is reduced again to $1 \times$ size, reproduced repeatedly (step-and-repeat) in exact locations, until a plate corresponding to the wafer surface area is completely covered with tens or hundreds of identical patterns.

For VLSI circuits, the complexity of design demands that the pattern be generated and verified by CAD, and the digital data can be used to drive directly a $1 \times$ or $10 \times$ pattern generator. Best definition is obtained with an electron beam (such as is found in a TV tube, but highly collimated to give a spot size below 1 μm).

A mask prepared by photographic techniques consists of a glass plate covered by a photographic emulsion, which is soft and susceptible to damage. For this reason, masks are also made with a thin (100–200-nm), patterned film of a hard material such as chromium metal or iron oxide.

Pattern Transfer The pattern is then transferred to the surface of the wafer that has been coated with a resist. A *resist*, in the most general sense, is a substance in which wave energy produces chemical or physical changes, making it resistant to acids. A portion of the resist is removed with a solvent, either in the exposed areas, creating a positive of the pattern (*positive resist*) or in the unexposed areas, creating a negative image (*negative resist*).

A *photoresist* is a light-sensitive substance, typically a polymer. It is applied in a thin layer: A few drops of the resist are placed on the wafer which is then spun (at up to 12 000 r/min) to spread out the resist. Residual solvent is driven off by baking; the resulting film is typically 0.3–1.0 μm thick. Negative resists are polymers that cross-link upon irradiation; hence, the exposed area becomes insoluble. Positive resists suffer chain scission (e.g., PMMA), and the exposed area becomes soluble. Negative photoresists are faster and wafers can be produced in seconds, but swelling of the polymer during development limits resolution. Positive resists are slower, but do not swell and give a finer line width.

If there already are step-like features on the wafer from previous processing stages, the steps may have to be filled with a thicker polymer film until a level surface is produced on which a thin photoresist can then be deposited.

Several techniques of exposure are used:

1 UV light lithography. Exposure is made with a mercury lamp, emitting light of 310–450-nm wavelength.

a In some ways the simplest is the *contact method* in which the mask is laid upon the wafer. However, a single trapped silicon dust particle can damage the mask and all subsequently exposed wafers; therefore, extremely high standards of cleanliness must be maintained. Resolution is around 1 μm and is limited by diffraction of light.

b The chance of damage is less when the mask is held at a distance of 10–25 μm from the surface (*proximity method*), but resolution drops to 2–4 μm.

c No damage occurs at all when the mask is held away from the wafer and the image is projected on the mask (*projection method*). The mask is either full-size or enlarged (typically, 10 ×). By exposing only a small part of the wafer at a time, resolutions of 1 μm can be achieved.

2 Exposure with "deep UV" light of 200–300-nm wavelength gives resolutions of 0.5 μm, which represents the lower limit attainable by light lithography. To minimize light absorbtion, the mask is made on a quartz plate.

3 Proximity printing with x-rays of 0.4–5-nm wavelength reduces diffraction effects and allows higher resolutions, down to 0.3 μm. The mask must be made on material transparent to x-rays (such as polyimide, Si, SiC, Si_3N_4, or Al_2O_3). Only a few square millimeters can be exposed at a time and the step-and-repeat process must be used. It takes typically 1 min to process a 125-mm wafer.

4 Patterns can be directly written with an electron beam, focused to 0.01–0.5-μm diameter. The resists are the same as in x-ray lithography. Resolutions of 0.1 μm (100 nm, 1000 Å) are possible with resists such as PMMA. Further resolution is limited by backscatter. The main drawback is the relatively low output rate; typically, it takes minutes to hours to write a 125-mm wafer.

Techniques are continually developing, allowing a reduction of feature size with all techniques, and some more exotic techniques, not discussed here, are also being developed.

Since all ICs are multilayer devices, several lithography steps are involved in succession. Alignment features are incorporated into each pattern to facilitate registry relative to previously developed layers.

The photoresist is destroyed by high heat; therefore, for high-temperature processing steps such as diffusion, the pattern must be developed in a heat-resistant film such as SiO_2. An example of photolithography is given in Fig. 10-15 for the preparation of a SiO_2 layer that will serve as the mask for a subsequent processing step.

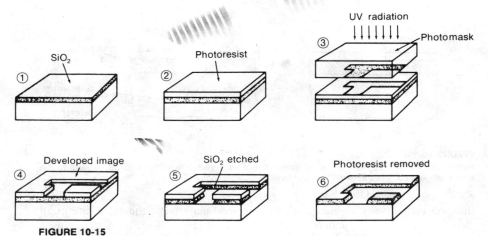

FIGURE 10-15
Typical sequence of preparing SiO₂ masks by photolithography. (*After W. C. Till and J. T. Luxon, Integrated Circuits, Materials, Devices, and Fabrication,* © *1982, p. 258. Reprinted by permission of Prentice-Hall, Englewood Cliffs, N.J.*)

The example given in Fig. 10-15 is that of the *subtractive method* of pattern transfer: Unmasked portions of a film are removed by etching. In the *lift-off method* the lithographic mask (the photoresist pattern) is made first; the surface layer (most often metallization) is deposited on the wafer, and then the unwanted portions of the film are lifted off by dissolving the mask.

10-3-6 Etching

Etching is the most frequently performed process step. It may serve one of several purposes:

1 Remove (strip) a film such as an oxide from the entire wafer surface.

2 Remove material over selected areas, as defined by a photoresist. Thus, a silicon oxide or silicon nitride film formed on the wafer is locally etched away in preparation for diffusion, ion implantation, or surface layer deposition. Alternatively, a film such as aluminum deposited over the entire surface is locally etched away to leave only the metal needed for interconnections.

Basically, two approaches are used:

1 *Wet etching.* The wafers are submerged in aqueous solutions of chemicals that selectively dissolve one or the other material. Thus, HF dissolves SiO₂ but not aluminum, whereas phosphoric acid dissolves Al without attacking SiO₂ or Si. Neither of them attacks photoresists which, because they are organic polymers, must be removed (stripped) with a solvent such as acetone.

Wet etching has lost favor for several reasons. Reaction products build up in the etchant and this creates the problem of periodic disposal. A further problem is

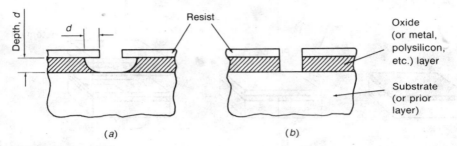

FIGURE 10-16
Patterns of etching: (*a*) fully isotropic material removal in wet (chemical) etching creates much less-well-defined features than (*b*) anisotropic plasma etching.

that wet etchants are *isotropic*, i.e., they remove material at the same rate in all directions, causing an undercut under the resist (Fig. 10-16*a*). In designing a photomask this can be taken into account, but features still must have relatively large dimensions: dry etching has become the dominant technique for VLSI devices.

2 *Dry etching* relies on the energy created by a plasma. A great advantage of plasma etching is that no large quantities of effluents are produced. Furthermore, if the bombardment is directional, a high degree of *anisotropy* of etching can be achieved: The rate of etch becomes much greater in depth than in lateral directions; hence, undercuts are minimized and fine features, with walls perpendicular to the wafer surface (Fig. 10-16*b*) can be produced. Such directional bombardment is obtained in planar reactors (Fig. 10-17). Basically, three methods of operation are possible:

a *Sputter etching.* Material is removed by the impact of energetic ions of a nonreactive gas such as argon.

b *Plasma etching.* Ionization of some molecular gases may produce highly reactive fragments. Thus, chemical etching is obtained when gases containing halogen atoms are ionized, e.g., carbontetrafluoride produces very aggressive free F

$$CF_4 + e = CF_3^+ + F^- + e \qquad (10\text{-}12)$$

Carbontetrafluoride does not react with Si but F forms the volatile SiF_4.

c *Reactive ion etching* occupies a place between sputter etching and plasma etching. The system is arranged so that both occur simultaneously. In sputter etching gas pressure is low and rf energy high; in plasma etching gas pressure is higher and rf energy lower; in reactive ion etching both parameters are intermediate.

A great advantage of these processes is that gases can be chosen to accomplish specific purposes. Thus, Si, SiO_2, and silicon nitride can be plasma etched with F as the active ion. Etch rates range between 10–1000 nm/min. Pure Si etches

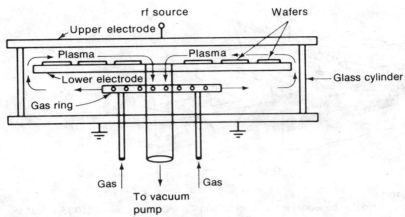

FIGURE 10-17
Highly directional, anisotropic etching is obtained in planar plasma reactors in which wafers lay flat on the lower electrode, with the plasma directly above them. (*From Till and Luxon, as Fig. 10-15, p. 264.*)

rather slowly and etching rates can be greatly accelerated by adding some 10% O_2 to CF_4. It is these techniques that made VLSI possible; grooves of only 1-μm width and several μm in depth can be etched into oxide films.

Because gas temperatures are low, conventional photoresist can be used as masks. After plasma etching is completed, the reactive gas is replaced with oxygen and the plasma is used to strip the resist.

10-3-7 Process Integration

As indicated in Sec. 10-3-1, the techniques described in Secs. 10-3-3–10-3-6 are employed in succession to generate several superimposed layers. Not surprisingly, microprocessor and computer control of individual steps is widely practiced, and great advances have been made in the complete closed-loop automation of processing sequences. This has the benefit of increased productivity and, since contamination of wafers through human contact is avoided, the yield is also greatly improved. In the following, typical process sequences will be illustrated in some detail for bipolar and MOS devices.*

Fabrication of Bipolar Devices The buried-layer transistor shown in Fig. 10-18 is produced in the following steps; the paragraph numbers correspond to the device features indicated in Fig. 10-18.

---

*After W. C. Till and J. T. Luxon, *Integrated Circuits: Materials, Devices, and Fabrication*, Prentice-Hall, Englewood Cliffs, N.J., 1982.

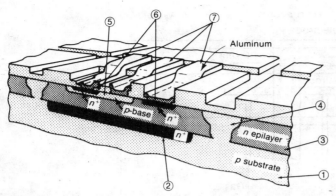

FIGURE 10-18
Features of a bipolar, buried-layer transistor. (*From Till and Luxon, as Fig. 10-15, p. 281.*)

1 A *p*-type substrate is used.

2 After slice clean, a thermal oxide layer is grown; photoresist is applied for etching windows into the oxide (as in Fig. 10-15) at positions of the buried layers; a liquid containing As or Sb is sprayed or spun on the surface; the wafer is heated to diffuse the impurity elements to a depth to 2–5 μm.

3 The oxide mask is dissolved, together with the oxide grown during diffusion, in dilute HCl. After slice clean, the *n*-type epitaxial layer is grown to a thickness of 3–25 μm.

4 A thermal oxide film is grown to make the mask for the isolation zones. After slice clean, boron is deposited and diffused into the wafer. The oxide now formed remains to prevent out-diffusion; hence, subsequent films show the steps visible in the oxide of Fig. 10-18.

5 Photoresist is applied to etch windows into the oxide for the deposition and diffusion of boron into the *p*-type base zone (and any resistors that may be needed).

6 Photoresist is now applied to open windows for diffusion into the n^+ emitter region and also for the n^+ ohmic contact in the *n*-type collector for the future metallization. To minimize diffusion of already-diffused impurity elements, a fast-diffusing species such as P or As is used.

7 A photoresist is used to open windows in the oxide layer over the emitter, base, and collector, and aluminum is deposited over the entire wafer. A new photoresist allows selective removal of the aluminum, leaving only the interconnections.

The IC is now complete but vulnerable. To protect it, a low-temperature glass is deposited (not shown on Fig. 10-18), e.g., by the pyrolitic decomposition of silane (Eq. (10-9)). A photoresist is again needed to etch holes where connections to the Al metallization will be made. A plasma-deposited silicon-nitride film may be used to provide further protection.

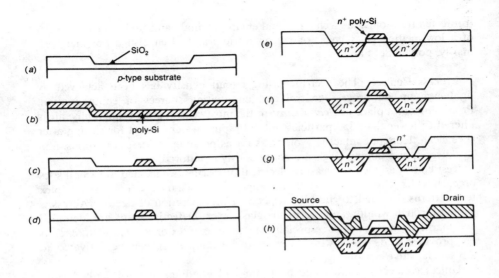

FIGURE 10-19
Simplified sequence of silicon-gate MOSFET device fabrication. See text for process steps.
(*After Till and Luxon, as Fig. 10-15, p. 295.*)

Fabrication of MOS Devices   The example given here (Fig. 10-19) is for the fabrication of a silicon-gate n-channel MOSFET device. The technique, called the *self-aligned silicon gate process*, avoids the difficulty of aligning masks on devices with very small feature sizes, in particular, the alignment of the gate (Fig. 10-7b) with the underlying channel. Preliminary steps are not shown in Fig. 10-19.

A p-type substrate is oxidized (Fig. 10-19a); the thickness of the film is reduced in the channel region. A polysilicon film is vapor deposited over the entire surface (Fig. 10-19b). After photolithography, the polysilicon is etched away everywhere except the gate stripes (Fig. 10-19c). Further photolithography allows removal of the oxide in the source and drain regions (Fig. 10-19d). Heavy n^+ diffusion makes the polysilicon gate stripe conducting, and at the same time the n^+ source and drain regions are developed in the substrate (Fig. 10-19e). The gate oxide prevents diffusion of dopant into the underlying channel; hence, channel and gate are automatically aligned without the need for masking. The device is protected by vapor deposition of P-glass (Fig. 10-19f). After reflowing the glass, photolithography is used to open contact windows (Fig. 10-19g) and the metallization pattern is developed (Fig. 19-19h). In plan view, the device appears as in Fig. 10-7b.

In detail, many more operating steps are involved in LSI and VLSI chip manufacture, but the basic steps are the same. They are just repeated several times to form the many layers of these structures. The wafer is passivated (protected from outside mechanical and environmental damage) by the deposition of a low-melting glass layer, or silicon nitride, and often also a polymer (primarily, polyimide) film.

Process Control The ever-increasing circuit density has been achieved by tightening up on process control and eliminating sources of contamination. Processing takes place in *clean rooms*; in the most critical area, lithography, air is filtered to fewer than 350 particles/m^3. Ultrapure water is used for rinsing after slice clean. Human contact is avoided as far as possible. Process conditions, film thicknesses, and doping levels are continuously monitored.

Test chips designed to measure electrical properties are located at several sites on each wafer. Circuits are tested on completed wafers, using often very complex test patterns designed to check all aspects of the operation of devices. Access is gained through windows etched in the glass layer. Defective chips are identified and marked. On some chips, especially the more complex ones, redundant circuits are provided and, by blowing fuses in the metallization, can be activated to replace defective circuits.

Unless connections are made by beam-lead bonding, to be described later, individual chips are now separated. Because silicon is hard and brittle, it can be scribed with diamond-tipped scribers, diced with diamond-impregnated disks, or scribed with a pulsed laser beam, and then snapped apart. Complete separation with diamond-coated wheels is less frequent. Sorting may take place at this point. Defective chips are discarded: good chips may be attached to a backing to facilitate automatic processing. The chips are now ready for packaging.

10-3-8 Packaging

Here we enter more familiar territory. The delicate chip must be electrically connected to the outside world and must be protected from mechanical damage and environmental influences. Therefore, packaging involves interconnections and encapsulation. The techniques used are adaptations of processes described in earlier chapters. This does not mean that packaging is free of problems or that it is inexpensive; indeed, often the cost of packaging exceeds that of chip fabrication.

Physically, the package may appear in various forms. One of the most popular ones is the dual in-line package (DIP, Fig. 10-20). The IC, measuring only a few millimeters on its sides, sits on or in a recess of the substrate to which it is attached (bonded) by a metallic or polymer layer. Contact with the outside world is made with pins or terminals, typically, on 2.54-mm (0.1-in) centers, or, for higher pin densities, at smaller centers (all measured in decimal inches). These terminals must be connected to the pads provided on the metallization of the chip (die). The whole package is hermetically sealed against the outside environment. Thus, there are three areas of major interest to us: attachment of the chip to the substrate (*die bonding*), attachment of leads (*interconnecting*), and protection of the assembly (*encapsulation*).

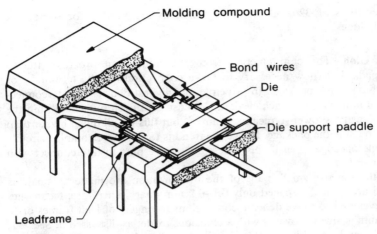

FIGURE 10-20
Sectional view of a DIP package shows the chip (die) bonded to a lead frame and encapsulated in plastic. (*J. R. Howell, in Proc. Int. Rel. Phys. Symp. p. 105, © 1981, IEEE. With permission.*)

Die Bonding The chip must be attached to a sturdier substrate which also helps to remove heat. The substrate may be alumina, which is an electrical insulator but has low heat conductivity (Table 7-2). Beryllia (BeO) has a high heat conductivity but is more expensive and its dust is highly toxic. Alternatively, the substrate is a metal, such as a Cu alloy or Kovar (a Fe–Ni–Cr alloy with a thermal expansion matching that of glass). Metal substrates are often made up in the form of lead frames that provide conducting paths to pins. Attachment is made by one of three techniques:

1 When the thermal expansion of the substrate is much different from that of the die, a ductile, high-lead solder (of a liquidus between 300 and 315 °C) is used.

2 Bonding to Kovar lead frames or ceramics is usually done with pure Au or Au–2Si foil, of less than 50-μm thickness, cut to size (*preform*). To ensure good wetting, the die and the substrate are plated with Au or, for metal lead frames, also with Ag. In the Au–Si system there is a eutectic at 3.6% Si. On heating above the eutectic temperature of 370 °C, the eutectic forms by alloying between die and preform. As in all joining processes, surface films must be broken through and mechanical scrubbing is helpful.

3 For most less-demanding applications, attachment is by polymers, primarily epoxy or polyimide loaded with Ag powder. The expansion coefficient is high (Table 7-2), but thermal stresses are low because the elastic modulus of the polymer is low. Polymers are easily dispensed pneumatically or by printing, processes which lend themselves to automation. The low (125–175 °C) curing temperature prevents damage to delicate active devices. The cured epoxy can stand temperatures of 320 °C for short times, as are needed for some wire bonding techniques.

Interconnection The chip emerges from the fabrication process with Al connecting pads. It is now necessary to make connections to the terminals (or connecting pins) of the package. In principle, the highest density of interconnection ($50–1000$ mm^{-2}) can be achieved, at low cost, on the chip itself. Within the package the density is lower (typically, 1 mm^{-2}) but this is still much higher than can be made outside the package. Hence the trend is to put more complex circuits within a single package, even if this necessitates more pins to interconnect with the outside world.

To keep the die size small, pads of the metallization are made as small as possible, and are typically spaced only 0.1–0.2 mm apart. Therefore, techniques had to be developed to make delicate connections to fingers of lead frames or to pins. Aluminum is always covered with a tenacious oxide; as discussed in Sec. 9-2, reliable bonds can be established only if surface films are broken up, and methods of achieving this goal are central to all interconnecting techniques.

1 *Wire bonding.* Individual, thin (25-μm- (0.01-in-) diameter) Al or Au wire is bonded to the pads by one of two techniques:

a *Thermocompression bonding* is an example of solid-state welding (Fig. 9-4) by hot flattening a wire. Obviously, no lubricant can be used and, as in all unlubricated upsetting, sticking friction prevails over the contact surface (Fig. 4-18d). The oxide remains undisturbed in the dead-metal zone, and heavy deformation is needed to increase the d/h or L/h ratio; this ensures sliding and thus breaks up the oxide, at least away from the central dead-metal zone. To reduce the required pressure and facilitate diffusion, bonding is performed with a heated tool (*thermode*) and the pressure is kept on for 0.5–2 s.

b *Thermosonic bonding* is an example of ultrasonic joining (Fig. 9-5b). The substrate and die are heated to 150 °C, and the wire is compressed while the bonding tool is ultrasonically vibrated. The ultrasonic energy input heats and softens the wire, rubbing is effective in breaking up oxides, and good quality joints result at a relatively low temperature.

The application of these techniques is shown in Fig. 10-21.

2 *Tape automated bonding* (TAB) is a technique of making all joints simultaneously. The pattern of connectors is etched, by photolithography, into 33- or 66-μm-thick Cu foil, which is usually Au plated to improve bonding. It may be backed with a polymer such as polyimide. Alternatively, the pattern is plated on polymer tape. Either way, it is supplied in a long tape, allowing automation. The bond is established against *bumps*; these are tiny columns of approximately 25-μm height, of Au-plated Cu or of pure gold. They are deposited either on the Al pads of the die or at the corresponding locations on the tape. Bond is then established by one of two techniques:

a Thermocompression bonding, with a thermode in the shape of the connector pattern, heated to 450–550 °C, applying a pressure of 275–480 MPa (the joint does not reach this temperature).

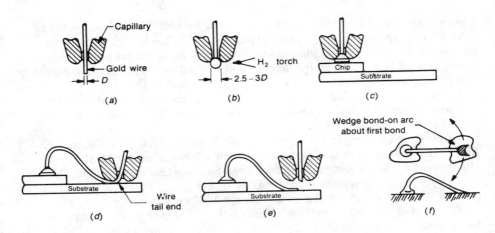

FIGURE 10-21
Thermosonic ball-wedge bonding involves: (a) feeding gold wire, (b) forming ball with hydrogen torch, (c) bonding the ball to the pad on the chip, and (d) bonding the wire to the lead frame. (*Adapted from J. W. Stafford, Semiconductor International, May, 1982, p. 82. With permission Cahners Publishing Co., Chicago.*)

b Eutectic bonding to a tin-plated copper tape. The Au–Sn eutectic forms at 280 °C. Under the applied pressure, rapid alloying takes place.

3 *Flip chip technique.* This also involves the formation of Au-plated bumps on the die, but a layer of Pb–Sn is now evaporated on top of the bumps. Heating makes the solder flow into a ball. The chip thus prepared is then flipped over the tape, and the bond is established by heating.

4 *Beam-lead bonding* differs from all the above techniques in that the gold leads to the outside world are deposited on the wafer, and the chips are then separated by etching away the silicon. The leads overhang the chip, appearing like tiny beams (hence their name); connections to the Al pads can be made by wire bonding.

Packaging Once the connections are established, the IC is complete and needs only protection. The degree of protection depends on the application; for example, packages must be mechanically stronger and more temperature-resistant in an industrial or military application than in a home computer. Apart from the required electrical characteristics, the package must also aid in removing heat. The following are the major packaging techniques used:

1 *Plastic molding.* The completed assembly is placed into the cavity of a transfer-molding die (Fig. 7-16b) and is enveloped with a thermosetting (e.g., epoxy, epoxy-silicone, or silicone) or thermoplastic (e.g., polyphenylene sulfide) polymer. For reduced shrinkage and thermal expansion and for increased strength, the polymer is filled with SiO_2 or Al_2O_3 (Table 7-2). As indicated in Sec. 7-7-3, shear heating reduces the viscosity of the polymer so that it does not damage the delicate connecting wires. Molding pressures can be kept low (around 6 MPa (1 kpsi)) and exposure to temperature ((typically, 175 °C) is limited to 1–5 min. Such *postmolding* is relatively inexpensive, fast, and is the dominant technique for mass-produced ICs. However, the difference in thermal expansion between polymer, silicon, and gold sets up stresses during molding and in service.

2 The chip is spared the exposure to molding stresses and temperatures when the mold is premade with an appropriate cavity (*premolds*). Pin connections are usually made with a lead frame.

3 *Ceramic packaging.* In Secs. 6-1-2 and 6-3-4 we saw that ceramics, mostly Al_2O_3, are tape-cast into thin strips from which substrates can be blanked out. Holes through which electrical connections will be made (*via holes*) are punched out, wiring paths are printed on the surface in the form of a refractory metal (usually W) powder slurry, and the via holes are filled with metal. Several ceramic blanks are assembled into a sandwich which is then fired. The W is nickel plated to facilitate brazing of Kovar lead wires or lead frames, using a eutectic Cu–Ag brazing alloy. The process requires strictest control of shrinkage. It is expensive but essential for highest performance requirements. Beryllia substrates offer higher heat conductivity but at a higher price.

4 *Glass-sealed refractory packages* again use ceramic substrates but the lead frame is sealed into a glass (Table 6-1) above 400 °C. To prevent alloying of Au and Al (which would lead to the formation of a brittle intermetallic compound), an all-aluminum construction is used.

10-4 THICK-FILM AND THIN-FILM TECHNOLOGIES

The completed IC package or other semiconductor device may be installed directly into some industrial product and then it only needs a socket or other means of connection. Frequently, however, it is part of a yet larger circuit, and then it is combined with thick-film or thin-film circuitry to make up a *hybrid circuit*. The distinction between thick and thin films is rather arbitrary and is based more on the method of manufacture than on actual thickness, although *thin films* are typically from 10 nm–1 μm thick whereas *thick films* are 10–25 μm thick. These films are used primarily to provide resistors and capacitors which would use too much chip surface, as well as for the interconnection of several discrete components (including inductors, diodes, transistors, etc.) and ICs, usually in mass production where discrete components assembled on a printed circuit board would be too expensive.

10-4-1 Thin-Film Fabrication Methods

The technology is essentially the same as for ICs, but a variety of film materials are used. The circuit is designed as for ICs and, if features are fine, photolithography is used to develop an SiO_2 mask (Fig. 10-15). Metals are then deposited by PVD or sputtering. Alternatively, metals are deposited over the entire surface and etched away with the aid of a photomask. Various devices can be formed:

1 *Resistors.* For low-value resistors, a Nichrome (Ni–Cr) alloy, tin oxide, or tantalum nitride is deposited for resistivities of 10–1000 $\Omega/\square$. For high-value resistors, cermets (usually Ta–Cr silicides) of 100–20 000 $\Omega/\square$ are used. Stability is achieved by heating (baking).

Exact values are obtained by *trimming*, i.e., subjecting the resistor to a treatment while its resistance (or the intended operation of the circuit) is monitored. The method of trimming depends on the resistor material. Oxides can be further oxidized until the desired resistance value is reached. Nichrome is mechanically trimmed; the line width is locally reduced by a diamond scribe, spark erosion, or high-energy beam (EB or laser).

2 *Capacitors* are constructed by the successive deposition of metal (usually Al or Au), insulator (SiO_2, Al_2O_3, or, for higher dielectric constant and capacitance, HfO_2), and again metal films. For an insulating film of 100-nm thickness, the capacitance is 800 pF/mm^2 with Al_2O_3 and 3500–7000 pF/mm^2 with HfO_2. Adjustment is possible by providing tabs (spurs) which can be trimmed off.

10-4-2 Thick-Film Circuits

The basic difference between thick-film and thin-film techniques is that, when the fine definition given by thin-film fabrication is not needed, thick-film circuits can be produced at a lower cost. This comes from the replacement of photolithography with *silk-screen printing*, a technique adopted from the textile and graphic arts industries. In the original form, a dye or ink is squeezed through the holes of a silk screen to form a predetermined pattern on fabric or paper. In thick-film technology the silk is replaced with a metal screen. The following steps are followed:

1 Circuits are designed as in IC fabrication, except that circuits are often less complex and, because they contain much coarser features (typically, 50–500-μm line width), the artwork is only 5–20 × magnified. It is drawn on paper or cut out of a plastic film by hand or computer. A photomask is then prepared.

2 The screen is usually stainless steel of typically 8 to 325 mesh (woven of wires of 0.94–0.028-mm diameter, giving openings of 0.118–0.051 mm). In one of the processes, the so-called *direct emulsion steel screen process*, a photosensitive emulsion is deposited on the screen, exposed through the photomask, and the unexposed areas are dissolved. The circuit pattern is defined by the nonblocked holes of the screen.

3 The screen is held, with controlled tension, some 0.5–1.0 mm off the ceramic substrate, and an ink is applied with a squeegee (Fig. 10-22). The *ink* is formulated to have appropriate rheological properties: it must not flow under gravity (hence it must have a well defined yield point as a Bingham fluid does, Fig. 3-20b, line D), yet it must flow at a low shear stress when it is spread (hence it must be pseudoplastic, Fig. 3-20b, line C). The ink contains several powders dispersed in a polymer-solvent system. The ingredients are: for a conductor, metal

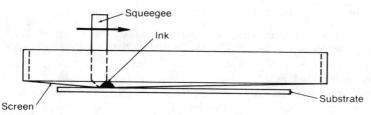

FIGURE 10-22
Thick-film circuits are deposited by silk-screen printing.

powder (usually, Pt/Au, Pd/Ag, or similar alloy) with some oxides; for resistors, also a metal powder-oxide mix but with a higher proportion of oxide; for dielectrics, only ceramic (such as $BaTiO_3$) powder. For the depostion of solders, the ink contains Sn–Pb (or, to prevent loss of Ag from conductors by diffusion, a Sn–Pb–Ag) solder powder, together with wetting and fluxing compounds. Inks of glass dispersions are used for encapsulation. To establish bonding between the metal and oxide powders and the substrate, a low-melting glass powder (glass frit) is always incorporated in the ink.

4 The printed substrates are dried at 125 °C to drive off the solvent, then fired, usually in a continuous furnace, where the polymer is first burnt off; the temperature is then raised to about 850 °C to sinter the metal and oxide particles and to fuse them, with the aid of the glass, to the substrate. The circuits are then allowed to cool in a programmed manner to prevent cracking and oxidation.

5 Circuit performance can be optimized by trimming resistors as in thin-film technology. Air-abrasive trimming is also possible: the width of the resistor is adjusted by abrasive blasting the edge of the resistor stripe, using fine (approximately 50-μm) powder (typically Al_2O_3) in high-pressure air.

10-5 SUMMARY

Microelectronic devices are the agents of the Second Industrial Revolution. They control machines from robots to microwave ovens, from automotive engines to the landing of aircrafts; they aid in computation, from hand-held calculators to supercomputers; they allow communication, from telephones to satellites to fiber optic devices; they help to entertain and teach, from radio to television to computer-aided instruction; they are at the heart of CAD/CAM and CIM.

Microelectronic devices are based primarily on electrical phenomena taking place in semiconductor materials such as doped Si and GaAs. Analog and digital circuits are formed in a planar arrangement on the surface of wafers. By shrinking the size of individual features, integrated circuits containing hundreds of thousands of components can be fitted on a single chip, the sides of which measure only a few millimeters. Advances in design and manufacturing have allowed increased performance at steeply reducing costs, at a rate unparallelled in other manufacturing fields.

The enormous complexity of circuits characteristic of VLSI can be mastered with the aid of CAD. The density of components demands that features be fabricated with minimum sizes below 1 μm, and a host of manufacturing techniques had to be developed—some from laboratory techniques, others from more generally used manufacturing processes.

The operation of devices depends on the incorporation of impurity (doping) atoms in a highly controlled manner. Unintended impurities, crystal defects, localized damage to the circuit or crystal all defeat operation. Therefore, manufacturing is conducted in an exceptionally clean environment, with a high degree of automation, using techniques that minimize chance events.

A sequence of manufacturing steps is needed to develop more complicated devices. Miniaturization hinges on the development and strictest control of lithographic and film deposition and implantation techniques. By further reducing minimum feature sizes, increasing die sizes, and the formation of multilevel (three-dimensional) structures, giga-scale integration (GSI), with hundreds of millions of devices on a single chip, will become possible. The impact of such complex, powerful devices on our lives cannot be perceived.

FURTHER READING

A General Coverage

Brodie, I., and J. J. Muray: *The Physics of Microfabrication*, Plenum, New York, 1982.

Doane, D. A., D. B. Fraser, and D. W. Hess (eds.): *Semiconductor Technology*, Electrochemical Society, Pennington, N.J., 1982.

Ghandhi, S. K.: *VLSI Fabrication Principles*, Wiley-Interscience, New York, 1983.

Gise, P. E., and R. Blanchard: *Semiconductor and Integrated Circuit Fabrication Techniques*, Reston Publishing Co., Reston, Va., 1979.

Gupta, D. C. (ed.): *Semiconductor Processing*, STP 850, American Society for Testing and Materials, Philadelphia, 1984.

Kerridge, C. C.: *Microchip Technology*, Wiley, New York, 1983.

Labuda, E. F., and J. T. Clemens: *Integrated Circuits*, in *Kirk-Othmer Encyclopedia of Chemical Technology*, 3d ed., vol. 13, 1981, pp. 621–648.

Roddy, D.: *Introduction to Microelectronics*, 2d ed., Pergamon, Oxford, 1978.

Symposium on VLSI Technology, The IEEE Electron Device Society and the Japan Society of Applied Physics. Annual since 1981.

Sze, S. M. (ed.): *VLSI Technology*, McGraw-Hill, New York, 1983.

Till, W. C., and J. T. Luxon: *Integrated Circuits: Materials, Devices, and Fabrication*, Prentice-Hall, Englewood Cliffs, N.J., 1982.

Yang, E. S.: *Fundamentals of Semiconductor Devices*, McGraw-Hill, New York, 1978.

Veronis, A.: *Integrated Circuit Fabrication Technology*, Reston Publishing Co., Reston, Va., 1979.

B Individual Processes

Auciello, O., and R. Kelly (eds.): *Ion Bombardment Modification of Surfaces*. Elsevier, Amsterdam, 1984.

DeForest, W. S.: *Photoresist*, McGraw-Hill, New York, 1975.

Dembovsky, V.: *Plasma Metallurgy*, Elsevier, Amsterdam, 1984.

Harper, C. A. (ed.): *Handbook of Thick Film Hybrid Microelectronics*, McGraw-Hill, 1975.

Horne, D. F.: *Photomasks, Scales, and Gratings*, Hilger, Bristol, 1983.

Lyman, J. (ed.): *Microelectronics Interconnection and Packaging*, McGraw-Hill, New York, 1983.

Meiksin, Z. H.: *Thin and Thick Films for Hybrid Microelectronics*, Lexington Books, Lexington, Mass., 1976.

Newman, R. (ed.): *Fine Line Lithography*, North-Holland, Amsterdam, 1980.

Oskam, H. J. (ed.): *Plasma Processing of Materials*, Noyes, Park Ridge, N.J., 1984.

Morgan, R. A.: *Plasma Etching in Semiconductor Fabrication*, Elsevier, Amsterdam, 1985.

Thompson, L. F., C. G. Willson, and M. J. S. Bowden (eds.): *Introduction to Microlithography*, American Chemical Society, Washington, 1983.

Thompson, L. F., C. G. Willson, and J. M. J. Frechet (eds.): *Materials for Microlithography*, American Chemical Society, Washington, 1984.

Vossen, J. L., and W. Kern (ed.): *Thin Film Processes*, Academic Press, New York, 1978.

CHAPTER 11

MANUFACTURING
SYSTEMS

In Sec. 1-4 we already outlined the major activities involved in the totality of manufacturing, and in Chaps. 3–10 we investigated the principles related to individual processes. Armed with this knowledge, we may now proceed to explore elements of some vital technological and organizational features of manufacturing systems, including the movement of material within a plant, organization of production facilities for mass and batch production, quality assurance with emphasis on statistical process control, and manufacturing management.

11-1 MATERIAL MOVEMENT

The movement of materials, parts, and tools is an essential element of all manufacturing operations, including the production of parts and the assembly of parts into subassemblies or finished products. Studies of batch-type shop operations have shown that, for some 95% of the total production time, parts are being transported from one place to another or are just waiting for something to happen. Even of the 5% of the time that they spend on a machine tool, they are actually worked upon only some 30% of the time, while the rest of the time is absorbed in loading and unloading, positioning, gaging, or idling for some extraneous cause (Fig. 11-1). If productivity is to be increased, first the methods of material movement, loading, positioning, clamping, and unloading must be improved, and only then will it make sense to worry about speeding up the process itself. Conversely, if in-process time is already high, there is little incentive to improve upon material movement. There are several ways of moving material.

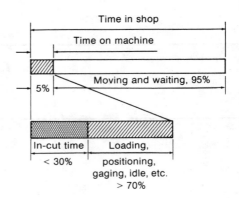

FIGURE 11-1
In batch production, the average workpiece is actually worked upon only a small fraction of the total time it spends in the shop. *(After C. F. Carter, in Proc. 2d Int. Conf. on Product Development and Manufacturing Technology, pp. 125–141, Macdonald, London, 1972. With permission.)*

11-1-1 Attended Material Movement

Operators can move objects with the least capital expenditure but this is usually also the least efficient and most costly technique. Efficiency can be increased by loading smaller parts into baskets, but this has the disadvantage that parts have to be picked out for the next operation. Parts can be arranged on *pallets* (or platforms or trays) and, if desired, oriented, so that they become more accessible for the next operation.

Forklift trucks facilitate material movement while retaining flexibility, but need unobstructed passageways.

Cranes also ensure flexibility and need no floor space but may interfere with each other.

11-1-2 Mechanized Material Movement

The term material movement usually refers to transportation between production units.

1 Roller gangs, endless belts, carousels, overhead conveyors, towline carts (moved by below-floor chains), and similar devices can be highly efficient for moving parts as well as pallets but can be reorganized only by changing the physical layout of the system.

2 Flexible transportation is ensured with *automated guided vehicle systems* (AGVS). The vehicles move, like forklift trucks, on the factory floor, but follow any one of several paths, for example, under inductive guidance provided by wire (cable) guides embedded in the floor. The path of individual vehicles can be readily programmed and reprogrammed for changing production requirements, avoiding collisions while optimizing the path of each vehicle. Sensors stop the vehicles when they encounter an obstruction.

Material movement to and from machine tools often requires that the part be turned, oriented, gripped, and placed into a predetermined position. Manual operation is the most flexible but it is also most prone to error from operator fatigue, especially if the task is repetitious or involves the movement of very small or very large (and heavy) parts. Heat, smoke, fumes, gases, or particulates may make the environment unpleasant or unhealthy. Therefore, there are powerful incentives to mechanize and automate the loading and unloading of parts. Simple and relatively low-cost automation has long been employed.

1 Purely *mechanical devices* of varying degrees of complexity are, in general, highly efficient but inflexible. They may combine material movement with machine loading and unloading. We already saw examples of mechanized transportation between successive stages of progressive machine tools such as cold headers, transfer presses, and automatic screw machines.

2 Parts can be individually *palletized*. Alignment in the machine tool is automatically obtained if: the pallet is made with the precision of a fixture; the part is held (clamped) in exact position (usually with the aid of locating holes); and the pallet is accurately located (with the aid of locating pins) on the machine-tool bed. The cost of pallets is reduced by the use of *modular fixtures* constructed on precision base plates.

3 Small parts are often handled effectively with simple mechanical devices. Vibratory belts and bowls, reciprocating forks, rotary disks, or magnetic devices are combined with simple but ingenious work-orientation devices from which the parts progress through feed tracks to the machine tool, where a metering device (such as a mechanically actuated escapement) releases the part at the proper time.

4 Loading and unloading can often be done with mechanical arms, generally referred to as *manipulators*. They can be divided into several subgroups. Classification is possible by the method of control:

a *Manipulators*, in the more restricted sense of the term, are mechanical arms under manual control, with the aid of push buttons, joy sticks, or devices that take the motions of the operator's hand and transform it into equivalent motions of the mechanical arm. Their lifting capacity ranges from a few grams to hundreds of tons. Examples are *remote manipulators* used in dangerous environments (e.g., in the atomic industry) and *forge manipulators* used in the open-die forging of large ingots. In a sense, programming is totally flexible in response to the operator's commands.

b *Fixed-sequence manipulators* advance to preset positions in a preset sequence. Position and sequence are set by limit switches, limit stops, and relays. Limit switches are also used to sense whether an action—by the manipulator or the machine it serves—has indeed been taken, thus providing a primitive version of feedback. Typical examples are manipulators that move sheet metal parts between presses, and unloading shuttles used with die-casting and injection-molding machines.

c *Robots* are defined by the Robot Institute of America as "programmable multifunctional manipulators designed to move material, parts, tools, or specialized devices through variable programmed motions for the performance of a variety of tasks." Thus they differ from fixed-sequence manipulators only in their variable programming. In the simplest form, this can be achieved by the use of plug boards, but the complexity of programming is then limited and such variable-sequence robots are often lumped together with fixed-sequence manipulators. The so-called *pick-and-place robots* are reprogrammable, for example, with a programmable logic controller, but often lack a feedback system, thus could be classified as programmable fixed-sequence manipulators.

11-1-3 Robots

Robotic devices consist of two elements:

1 A *mechanical structure* which includes:

a A *base* with movable parts, articulated in such a way that one or more (usually up to six) degrees of freedom are attained. Many devices comprise simple shuttles moving along guide rods; each shuttle has one degree of freedom. In its most familiar form, the robot has an arm which may be articulated in various ways (Fig. 11-2). A rigid arm moving up and down and swivelling around a column has two degrees of freedom. An arm that moves (or tilts) up and down, rotates (swivels), moves radially in and out, and has a wrist with twist (swivel), bend (pitch), and yaw movements, possesses six degrees of freedom (Fig. 11-3).

b The *gripper* (*hand, jaw* or, more generally, the *work-holding device* or *end effector*) which holds and moves the part or tool.

c *Drive elements* that provide the motive power for the various motions. Drives are usually pneumatic, hydraulic, or electric, and sometimes combinations of these. Mechanical devices such as cams, levers, and linkages are less frequently encountered because of their relative inflexibility of programming. Air or, more frequently, hydraulic cylinders or ball screws generate linear motion. Air, hydraulic, or stepping motors, or ac or dc servomotors provide rotation.

d Most robots are *fixed* to the floor, but there are some that move on ground or overhead rails or pneumatic tires (*mobile robots*).

It should be noted that a robot may either move a part relative to the tooling or move the tooling relative to the part (both versions are used, for example, in painting).

2 A *control system*. True robots are driven by servomechanisms which incorporate closed-loop control (as in Fig. 1-8c). Sensors measure displacements and feed a signal back to the controller, so that the gripper is positioned accurately, usually within 1.0 mm or better. In their simpler forms, they move from point to point along their axes, without following a defined path. Continuous-path robots follow a defined path, thus can be used for such operations as spray painting and arc welding. Robots can be used in the greatest variety of applications, including the loading of machine tools and presses, inspection, and assembly.

Robots are programmed in various ways:

Play-back robots can be programmed or "taught" using the "walk-through" method in which a robot arm (or a substitute "training arm") is manually moved through the required path, with control commands inserted whenever some

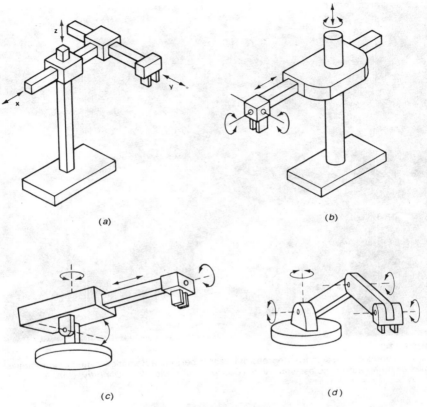

(a)

(b)

(c)

(d)

FIGURE 11-2
The mobility of robots depends on prismatic (P) and revolute (R) joints. Their combination
gives (a) Cartesian (PPP), (b) cylindrical (RPP), (c) polar (RRP), and (d) revolute (RRR), also
called articulated, configurations. (*After L. L. Toepperwein et al., Technical Report
AFWAL-TR-80-4042, Air Force Wright Aeronautical Laboratories, Dayton, Ohio.*)

particular action (switching a tool on or off, or waiting for a machine tool to
perform a given action) is needed. In the "lead-through" method a control panel
("teach pendant") is used to position the arm. Either way, the controller stores
the instructions thus received and plays them back, but without the interruptions,
delays, and hesitations typical of manual control.

Other robots are programmed off-line, the same way as NC machine tools (Sec.
1-4-4). In the most advanced form, the database established in CAD/CAM is
used to preprogram all robot motions and actions. Several programs may be
stored and called upon when the appropriate part is presented. Part identification
may be given by a previous work station or by reading bar codes (similar to those
used in supermarkets) applied to the part or pallet.

FIGURE 11-3
Industrial robots possessing five degrees of freedom perform automatic welding of automotive bodies along an automatic assembly line. (*Courtesy Unimation Inc., Danbury, Conn.*)

An *intelligent robot* or *sensory robot* is an NC robot equipped with some form of artificial intelligence which allows it to cope with nonfixed situations (randomly oriented parts, parts not presented in exact positions) and to perform adaptive control of operations. Some forms of sensing are already fairly widely available:

1 *Visual sensing* requires cameras that usually contain a grid (matrix) of light-sensitive elements (picture elements or pixels) or a linear array of elements. These are rapidly scanned to acquire information on the distribution of light intensities. This information, when converted into the required digital form, can be processed in a computer for image recognition (*image processing*). More complex tasks call for the processing of images obtained from several cameras simultaneously.

2 *Tactile sensing*, in the simplest form, requires force-sensing elements built into the end effector. There are many possibilities for further feedback. For example, infrared light may be ducted through fiber optic bundles into the jaws of the end effector; when the jaws come close enough to the part for light to be reflected, jaw movement is slowed and the part is gripped with a preset force. Systems more closely approaching human senses are being continually developed.

3 Adaptive control links the actions of the robot to the information obtained by sensors. Among others, force or torque sensing elements are involved in adaptive control. For example, a deburring robot may move along the edge of a part at a high rate while searching for a burr. Increased deflection of the tool (increased force on the toolholder) indicates the presence of burr, whereupon feed rate is reduced until the burr is removed. Similarly, in a polishing operation the correct polishing pressure can be maintained, irrespective of part shape, by feedback from a force sensor.

Although robots can often be introduced into an existing plant, some changes are most likely needed. The robot is less tolerant of variations in shape and dimension of parts than a human operator, and it often pays to redesign parts to suit the limitations of the robot. However, robots can perform tasks tirelessly and, if adequately protected and maintained, reliably, even in hostile, dangerous, or unpleasant environments.

11-2 PRODUCTION ORGANIZATION

We already saw in Sec. 1-4-1 that problems of manufacturing are best treated as a system. Within this system, parts production is organized for maximum efficiency and minimum cost, consistent with the required quality standards. There is no single form of organization that satisfies all requirements, and the choice depends on the characteristics of production.

11-2-1 Production Characteristics

Two important factors in the choice of processes and their organization are the *total number* of parts to be produced and the *rate of production* (i.e., the number of units produced in a time period such as an hour, day, month, or year). Total production quantity and production rate together define the justifiable expenditure on special machinery and tooling.

The total production quantity is often insufficient to keep a production unit continuously occupied, and production proceeds in lots representing a fraction of the total number of parts. The *batch size* or *lot size* is the number of units produced in an uninterrupted run. There are no strict definitions, but it is customary to speak of *small-batch* (1–100 units), *batch* (over 100), and *mass* (over 100 000 or even million units) *production*. In general, a larger batch size justifies the choice of processes with inherently higher production rates and thus more favorable economies.

Lot size is not determined by purely technical considerations. The cost of setting up (changing over) must be weighed against the cost of stocking (warehousing) parts between production runs; the move to just-in-time delivery has had the effect of reducing lot sizes.

In evaluating the number of parts produced and the rates of production, it is best to consider all parts that show any similarities in features and operating sequences. Close similarities may allow grouping of parts for processing by more

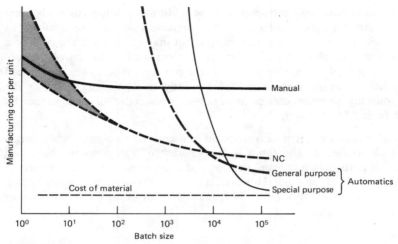

FIGURE 11-4
The most economical approach to production depends on batch size.

productive techniques; absence of similarities will require that great flexibility of operation be retained.

11-2-2 Optimum Manufacturing Method

The *optimum manufacturing process* is selected from a knowledge of process capabilities and limitations, tempered by restraints imposed by the requisite production rates and batch sizes. The choice of machine tool depends on cost factors, and break-even charts (similar to Fig. 11-4) can be constructed to show where one machine tool becomes more profitable than another.

1 Stand-alone machines with manual control require the smallest capital outlay, but their operation is labor-intensive. Labor costs do not drop significantly with increasing batch size (Fig. 11-4); thus, such machines are best suited to one-off and small-batch production. The operator may be a highly skilled artisan or, in repetitive production, may be semiskilled, with a setup person providing the necessary skills.

2 Properly chosen stand-alone NC or CNC machines are most suitable for small-batch production, although with the trend toward increasingly user-friendly programming devices and with the application of group technology, they become competitive with manually operated machinery.

Once the workpiece is clamped in place on the NC machine tool table and a reference point is established, machining, bending, welding, cutting, etc., proceed with great accuracy and repeatability. Nonproductive setup time is practically nil; therefore, NC can become economical even for very small lots (Fig. 11-4) widely separated in time. The operator may again be highly skilled, this time with some programming knowledge, or the programs may be provided to the machine by a programmer who may be working from the database of a CAD/CAM system; in this case the operator performs machine supervision and service functions such as pallet loading.

3 In large-batch production programmable automatics are most economical, while special-purpose (and often hard-programmed) automatics are limited to mass-production facilities.

Automobiles, appliances, and consumer goods generally fall into the latter two categories and have been relatively efficiently produced with traditional methods. Machine tools, off-the-road and railroad equipment, heavy machinery, and aircraft usually fall into the batch-production categories. Products of the latter industries are characterized by large expenditures for the main components of complex shape, and a relatively small expenditure on the much more numerous, mass-produced, often purchased components (Fig. 11-5). Obviously, greatest economy will be ensured by organizing effective production of the complex main parts. Thus, the organization of manufacturing has traditionally followed different philosophies for mass production and batch production.

FIGURE 11-5
A large proportion of the total value added originates in the batch manufacture of complex main parts, whereas standard parts, usually of smaller size, are mass produced. (*H. Opitz and H. P. Windahl, International J. Prod. Res.* **9**:181–203 (1971). *With permission.*)

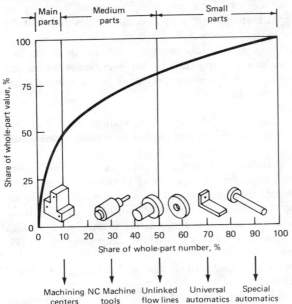

11-2-3 Organization for Mass Production

The large lot sizes typical of mass production make the installation of special-purpose machines economical. When parts are of identical, simple shapes, they can be easily held in the correct position relative to the tool. Parts of irregular shape can be worked upon after establishing a reference surface, hole, or boss (*qualifying the part*) for example, by high-speed machining. The part is clamped to a base, pallet, or fixture, or moved on its own from machine tool to machine tool.

Part movement is by fixed means (such as conveyors, carousels, mechanical arms) between machine tools organized in the sequence of operations. In such *transfer lines* each machine performs only one (or a related group) of operations, and is controlled by fixed (hard) automation (cams, levers, or relays). Setup is manual, time-consuming, and calls for a highly skilled setup person. Product changes cannot be accommodated without substantially rebuilding the production line.

Lines must be carefully balanced to equalize the output of various stages; otherwise, the most time-consuming station would slow down the entire line. Tooling and process conditions are selected so that all tools can be changed at the same time, avoiding costly, random shut-downs. Sensors, gaging heads, and probes are built into the line at appropriate places to confirm that an operation has indeed taken place and that the next operation can proceed. In the event of difficulty, one of several actions may be provided: signal lights or alarms alert the line attendants; defective parts are automatically marked with paint or ink; the line is slowed down; the line is shut down entirely.

Fixed production lines are of very high productivity but of virtually no flexibility. Because of this, input material and in-process inventories must also be large so as to provide *buffers* against unexpected shut-downs.

Increasing international competition, rapidly changing customer demands, and the high cost of money have made it imperative that even mass-production facilities become more flexible. Several approaches, alone or in combination, allow a move to the *flexible transfer line*:

1 The production line is grouped into *sections* of 5–12 stations, with a smaller buffer storage in between, so that breakdown, tool change, or setting in one group does not stop the whole line.

2 Operations that would upset the line balance are performed on *branch lines*.

3 Fixed machine tools are replaced by *powerhead production units* which consist of a base with feed mechanism, a drive unit (power spindle), and several interchangeable attachments, so that various operations (drilling, tapping, turning, milling, etc.) can be carried out according to need. These units are examples of *modular* (sometimes also called *metamorphic*) hardware. Sometimes even CNC machining centers are incorporated.

4 *Quick-change tool holders* allow rapid tool change or the change of the tool holder complete with a preset tool.

5 Parts belonging to the same family and differing only in the presence or absence of some feature (such as a hole) can be processed on the same line if the parts are identified on entering the line (e.g., by a boss provided for sensing, or a bar code on the part or pallet); the appropriate station is then activated or deactivated.

Such flexible transfer lines can be operated under logic control, increasing their flexibility by the ease of reprogramming.

11-2-4 Organization for Batch Production

Batch production differs from mass production not only in batch size but also in the speed of response to changing demands. The epitome of batch production is the job shop which makes its living by providing service to a large number of customers.

Functional Layout Batch production has traditionally taken place in shops organized around individual machine tools. Parts are moved by some flexible means (manually, by overhead conveyors, cranes, forklift trucks) from machine to machine. This results in complex and often disorganized, time-consuming material movement.

Separating machine tools of one kind into groups (*functional layout*) hardly improves the situation, since different parts are made in different production sequences (Fig. 11-6a) and material movement remains chaotic.

Management of such a plant is also highly demanding. Each machine is manned by one operator; production plans must be drawn up that ensure full utilization of machine and operator time while at the same time assuring that parts are produced in the correct number for scheduled delivery. This usually turns out to be impossible. Frequent setups would take up most of the production time; therefore, batch sizes are increased whenever possible, even at the expense of increasing the in-process inventory. This, however, increases processing time and reduces the plant's ability to respond to changing customer needs.

Group Technology Many problems can be resolved if parts to be produced can be classified according to the principles of *group technology* (GT, Sec. 2-4-2). The potential of GT can be fully exploited only if the plant is reorganized. All the equipment necessary to produce a family of parts is grouped into a *cell*. In a more modern plant a cell may comprise, for example, a complex (and expensive) machine such as a CNC machining center, supported by several special-purpose and, therefore, lower-cost machines.

The parts are transferred with minimum movement and wasted time from one unit to the other. For larger batch sizes, machines are laid out along a line (or U or L shape) in the sequence of operations (Fig. 11-6b), creating a transition between cell and modular-construction transfer line.

Typically, such cells are still attended, but one operator may take care of several machines. Thus productivity increases while also making the task of the operator much more varied and interesting.

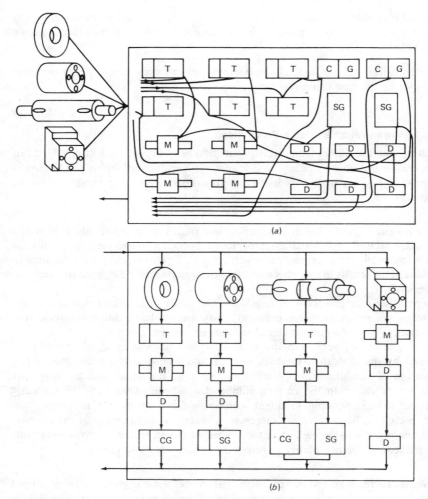

FIGURE 11-6
Comparison of material flow in plants with (*a*) functional and (*b*) group layout. T, turning; M, milling; D, drilling; SG, surface grinding; CG, cylindrical grinding. (*C. C. Gallagher and W. A. Knight, Group Technology, Butterworths, London, 1973. p. 2. With permission.*)

There are multiple benefits, many of them flowing directly from the application of GT principles:

1 The variety and quantity of starting material as well as in-process inventory are reduced.

2 Production planning is simplified, and better information can be collected for production control and planning.

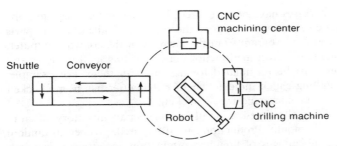

FIGURE 11-7 Typical layout of a flexible manufacturing cell.

3 Tooling costs can be reduced by standardization, and setup times are minimized.

4 Total processing time is reduced, lead times are shorter, response to customer needs is faster, and competitiveness increases.

Flexible Manufacturing Cells (FMC) A further increase in flexibility is obtained if several operations are combined into one (or more) highly flexible CNC machine which is served by some flexible means of material movement, such as a robot or pallet changer (Fig. 11-7). In such a *flexible manufacturing cell*, FMC, the task of the operator is reduced to: loading the racks from which the robot will pick up the parts; clamping parts on pallets; removing finished parts; and changing the tools and other supplies in magazines. Most FMCs are for machining and comprise CNC lathes, milling machines, machining centers, grinders, etc. Sheet-metalworking cells comprise sheet stacking, CNC punching, laser or EB cutting, bending, and parts stacking units. Many FMCs incorporate automatic inspection. (The principle of FMC is not limited to the manufacture of hardware; one of the early applications was for the making of cakes.)

Compared to attended cells, FMCs are more demanding: machine tools must be more rigid, foundations more substantial for greater stability and better alignment, and preventive maintenance must be strictly enforced. Still, FMCs offer several advantages:

1 Machine tools of greater flexibility are more expensive but may replace several conventional tools. Therefore, investment may be 70–130% of attended cells of similar capacity.

2 In-process time increases from the less than 5% typical of stand-alone attended tools to 75 or 80%; thus, productivity, expressed as output per machine, is much higher, and deliveries are much faster.

3 Productivity is higher also in terms of output per person per hour. Still, the requisite investment can often be justified only if the FMC is operated 24 hours a day, seven days a week, with operators in attendance only in one shift. Five to tenfold productivity increases have been realized. For unattended operation, the operator loads or palletizes numerous parts; the robot then identifies the parts and calls up the appropriate program from the memory of the control computer. Provisions must be made to detect malfunctions or impending malfunctions, and appropriate strategies must be formulated for dealing with them. For example, sensors built into a milling chuck indicate whether the cutter has been chucked correctly and, if not, give a command for rechucking.

4 The greater flexibility allows reduction of in-process part inventory, often to one-quarter of the usual amount. Production can, if required, proceed in random order; in a sense, the advantages of flow-line production are attained in batch production and small runs become profitable. In principle, manufacturing of at least the sɯaller parts can be geographically distributed to small centers, creating jobs in many locations.

5 Quality improves because human error is eliminated as a source of problems. Unattended operation requires though that quality be routinely (and often 100%) checked by automatic inspection. Cleanliness becomes of paramount importance: fluids, chips, and dust create problems in fixturing and also interfere with the proper functioning of sensors.

Flexible Manufacturing Systems (FMS)   When all the FMCs (and automatic inspection) of a plant are interlinked, a *flexible manufacturing system*, FMS is created. This is a huge undertaking which requires that many elements of CIM be already in place. The complexity of computer control becomes substantial and, to ensure real-time control and response to situations, several (4–6) levels of hierarchical control are usually needed. An essential feature of FMS is the *automatic storage and retrieval* (ASR) warehouse.

FMS is often implemented by installing several FMCs first. New plants can be designed as FMS but experience has shown that complete cooperation between user and supplier of the FMS is essential and that the software must be developed as a joint effort. Once an FMS is up and running, it outperforms attended operations by a substantial margin.

A feel for the ranges of application for different manufacturing systems may be gained from Fig. 11-8. It is important to remember that flexibility is relative and that it costs money. Therefore, many FMSs are designed to deal with only 10 or so products of the same family; they are flexible only relative to the fixed automation they replace. One of the difficulties is communication between machines and controllers of different manufacture, although great progress has been made by the worldwide introduction of a standard set of communication specifications (*manufacturing automation protocol*, MAP), using General Motor's MAP document as a starting point.

In principle, it is possible to build a fully automated factory in which all unit processes, tool changing, material movement, and inspection are accomplished without operator assistance. This is true also of assembly, although some assembly operations remain difficult to automate.

Examples:

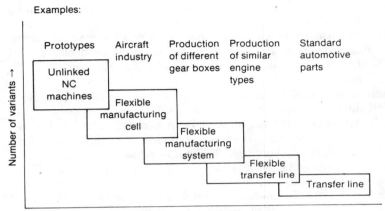

FIGURE 11-8
The suitability of various manufacturing systems for specific production tasks depends on the variety of products and the number of workpieces. (*After W. Eversheim and P. Herrmann, J. Manufacturing Systems 1:139–148 (1982). With permission of the Society of Manufacturing Engineers, Dearborn, Mich.*)

11-2-5 Organization of Assembly

In the final phase of manufacturing, individual components are assembled into the end product. This presents a wide range of problems, depending on production quantities.

Manual assembly is still the only method practicable for small-batch production. In larger production quantities, the repetitive nature of work, the danger of making errors when hundreds of parts are involved, and the low overall efficiency have led to early attempts at organizing and mechanizing assembly operations.

Assembly Lines In assembling a complex machine, great progress can be made by breaking down the operation into smaller units; this also facilitates material handling by ensuring that all parts can be supplied in their proper place and sequence.

This concept led to the *synchronous assembly line*, pioneered by Henry Ford in 1913. The units to be assembled move on a conveyor at a preset rate while operators stationed along the conveyor perform their assigned tasks. This assembly method, more than any individual advance, has made possible the mass production of consumer goods that previously were regarded as a luxury. However, the monotony of work has led to some dissatisfaction with the system, resulting in attempts at replacing it with alternative yet similarly productive methods.

Nonsynchronous assembly lines permit operator judgement; units are passed on when finished. Another potentially attractive solution entrusts an entire assembly (e.g., an automobile engine) to a group of operators who are given considerable freedom in organizing themselves and who also perform the quality-control function. The alternative is, of course, mechanization and automation.

Mechanization of Assembly   Some types of assembly operations lend themselves to fairly simple mechanical methods of assembly. Thus, screws or bolts can be driven and parts placed, crimped, or riveted with mechanical devices. The cost is reduced while productivity and consistency of product increase, but only if the reliability of mechanization is very high. The cost of off-line repairs can quickly cancel all savings. A crucial factor of success is *in-line inspection* to pinpoint and remove imperfect assemblies, either during or at the end of assembly operation. Many of the elements used in automatic assembly are the same as in mechanized production and workpiece handling, and may be purely mechanical, electromechanical, or numerically or computer controlled.

During assembly the unit may move continuously; *indexing workheads* move with it and retract after completing their task, backtrack, and repeat the operation on the next unit to come along. Alternatively, the line itself indexes and *stationary workheads* perform the operation while the line is at rest.

In all instances, assembly may be performed in-line, that is, along a conveyor on which parts move (if necessary, on pallets that ensure accurate positioning). The line may be laid out in a straight line or U or L shape. Pallets are returned below the line. The assembly line may also be oval or circular, so that pallets return to their starting position. When the total number of stations is not too large, a rigid carousel may be used to carry the unit from station to station.

Operation of a line may be *synchronous*, in which case each unit is moved at the same time and any local holdup affects the entire line. Breaking down the line into smaller modules, with storage buffers between modules, makes operation of the line less critical.

Greater freedom is secured by *nonsynchronous* movement of units. Each unit is moved on command at the completion of operation; the total number of units is larger than the number of assembly stations so that there is always a buffer between each station.

Branch lines feeding into the main line help in maintaining output when a subassembly operation is more time consuming.

Flexible Assembly Systems (FAS)   The same pressures that have forced the evolution toward FMC and FMS have contributed also to the development of *flexible assembly systems*, FAS. Such systems bring the economy of mechanized assembly to batch production. Many of the techniques used for attended and mechanized assembly are incorporated. The difference is that many—and sometimes all—operators are replaced by flexible assembly machines, usually robots. These robots range from pick-and-place devices to complex, fully articulated robots. By definition, they are under computer control, and sometimes they have artificial intelligence, particularly pattern recognition, to pick randomly oriented parts from a bin or conveyor belt and to locate parts in the assembly in the correct position. The robot itself may perform the assembly operation, or it may present the part to an assembly machine such as a press or nut driver.

Automatic assembly is successful only if existing product designs are modified to take into account the limitations and capabilities of automatic assembly.

11-2-6 Scheduling of Assembly

In the traditional mode of operating an assembly line, large stocks of parts are *warehoused* for several reasons. First, the supplier of parts achieves economy of production by shipping larger batches. Second, these stocks provide a buffer in the event of a disruption of supply for whatever reason (technical problems, labor disputes, transportation breakdown). Third, if there is no assurance that all parts received actually meet requirements, parts can be rejected as assembly proceeds.

Many industries find the cost and quality implications of this approach unacceptable and have turned to the *just-in-time* (JIT) *delivery* system developed in Japan. Deliveries are made to the assembly line frequently (daily or several times a day) so that only the parts immediately needed are stocked directly at the line. The system pulls supplies as required by consumption on the assembly line, instead of pushing on the basis of forecasts. This presents several advantages as well as challenges:

1 Investment in factory space and inventory drops drastically, increasing the economy of production.

2 Assembly becomes more flexible because production schedules can be changed in the absence of large inventories. However, assembly also becomes more sensitive to any malfunction on the assembly line.

3 Reliability of supply becomes of particular concern because, in the event of an interruption in supplies, the parts required for a temperary change in production profile may simply not be available.

4 Suppliers of parts must adopt flexible manufacturing techniques so that they can respond to demand. Quality problems must be corrected immediately: equipment maintenance and good labor relations become critical; failure to supply on time may mean a permanent loss of business.

5 There is often neither time nor inclination to inspect incoming material, and there is no buffer that would help to tide over a bad batch. Responsibility for quality assurance is shifted, to a large extent, to the supplier.

6 Frequent deliveries necessitate that suppliers be within reasonably close geographic locations (with distance depending on the transportation mode). Parts are shipped in carefully designed, often reusable packaging to ensure that parts arrive in perfect condition. In the ultimate development, parts are produced on branch lines feeding the assembly line on demand; this represents a true zero-inventory system.

JIT delivery can function only if there is the closest cooperation between producer and user of parts. This often leads to joint efforts in statistical process control, technology transfer, and even in the design of parts. Competitive bidding based purely on price is not used as the basis of awarding contracts.

11-3 QUALITY ASSURANCE

The aim of manufacturing is the creation of reliable products, i.e., products that will perform their intended function, under stated conditions, for a specified period of time. *Reliability* is the probability that a product will do so, and is usually expressed as a percentage. Thus, when we say that a product has 98% reliability for a 1000-hr service, we mean that 98 out of 100 units will perform without breakdown.

Assuming that the product had been correctly designed and that the proper manufacturing operations were chosen, steps must be taken so that the planned reliability will indeed be achieved. For this, the quality of the product must be controlled. *Quality* is usually defined as conformance to written specifications, but it should be recognized that there are aspects of quality that are difficult to define exactly yet can be readily judged subjectively.

Quality control is concerned primarily with inspection and the analysis of defects; *quality assurance* is a broader activity, beginning with interaction with design, and spreading over most aspects of manufacturing, including the formulation and auditing of quality-control programs.

Quality assurance starts from the recognition that all properties of manufactured products are subject to random variations. Allowance for this is made when the designer specifies dimensional tolerances (Sec. 2-4-3) or requests some minimum strength, knowing well that the strength of individual parts will show some spread above that minimum. In dealing with random variations, quality assurance draws heavily on concepts of statistics; hence, one often speaks of *statistical quality control* (SQC). It will, therefore, be necessary to review some elements of statistics.

11-3-1 Statistical Aspects of Manufacturing

In Sec. 2-4-3 we already stated that no part can ever be made to exact dimensions. This is true also of other measurable *variables* (such as surface finish or mechanical or electrical properties) and of *attributes* (qualitative data such as surface blemishes, dents, defects in welded seams, etc.).

Variations are changes in the value of measured characteristics. They are of two kinds: *Assignable* (*attributable, special*) *causes* are intermittent, unpredictable, and can be followed up to remedy the situation. Even then, there will remain small, random variations. Such *common-cause variations* are inherent in any

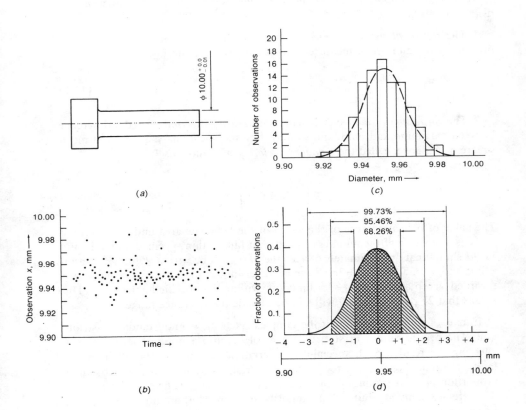

FIGURE 11-9
Machine capability in turning (*a*) a simple shaft is established by (*b*) taking 100 consecutive samples, (*c*) plotting the observations in the form of a bar graph and evaluating it in terms of the (*d*) normal distribution.

process and as long as only these random variations exist, the process is stable and is said to be *in statistical control*. Analysis of variations is best illustrated on the example of turning a cylindrical part (a shaft) on a lathe (Fig. 11-9*a*).

A sample of, say, 100 consecutive parts is taken in the course of a smooth production run. Eccentricity in the lathe spindle, variations of cutting force and, hence, tool deflection, etc., will cause random variations in the diameter of the shaft (Fig. 11-9*b*). The data points can be organized by grouping them into narrow dimensional ranges and plotting a bar graph (Fig. 11-9*c*). For a process in statistical control, the distribution is bell-shaped and approximates the so-called *normal distribution* (the curve of which can be mathematically derived). Such curve has two important characteristics: one is the mean, the other is the dispersion of data.

1 The *statistical average* or *mean* $\bar{x}$ is simply the sum of all measured values x divided by the number of measurements n (this is the centerline of the bell)

$$\bar{x} = \frac{\sum x}{n} \tag{11-1}$$

2 The *dispersion of data* can be characterized by the *range R* (the difference between maximum and minimum values). For statistical analysis, the root mean square of variance, called the *standard deviation* σ, is most useful

$$\sigma = \left[\frac{\sum (x - \bar{x})^2}{n - 1} \right]^{1/2} \tag{11-2}$$

The value of σ is a measure of the width of the bell. The area under the normal curve is a measure of the number of parts that falls within specified limits. Figure 11-9d shows that approximately 68% of the parts will be within $\pm 1\sigma$, 95% within $\pm 2\sigma$; and 99.73% within $\pm 3\sigma$ of the average. In general industrial practice it is regarded acceptable if the $\pm 3\sigma$ limit fits within the specified tolerance. (This means that 27 parts in 10 000 will still be outside the tolerance range.)

It may, of course, happen that the samples taken show a nonstandard distribution. The process is not in statistical control because a nonrandom variable is affecting the results. Such special-cause variation may usually be attributed to faulty machines, operator error, wrong material, worn tool, etc. A substantial proportion of parts produced may then fall outside the specification limits.

There are many attributes that cannot be measured but are no less important. For example, machining may reveal slag inclusions in some of the shafts. We can then count the number of defects in one part (unit), or the number of defective parts (units) in a production lot.

11-3-2 Acceptance Control

The approach to quality control has changed over time. In preindustrial days quality was assured by the pride of the artisan and the control exerted by guilds. In earlier days of the industrial revolution, responsibility for quality was divided between operator and management. Following the work of F. W. Taylor,* tasks were organized with clearly identifiable responsibilities, and the task of quality control shifted to quality-control departments. Beginning with the 1920s, statistics was applied for process control. The huge increase of production during the Second World War made a systematic, unified approach necessary, and a number

---

*F. W. Taylor, *Scientific Management*, Harper and Row, New York, 1919.

of military specifications were drafted that are still extensively used. Ideally, specifications such as these would assure that no defective parts are delivered for assembly and that no defective assemblies are delivered to the final customer. *Acceptance control* is based on tests applied to parts and completed products by the producer and/or customer. The tests may involve:

1 *Hundred-percent inspection.* When conducted by inspectors, there is no guarantee that 100% of the defective parts will be found; because of fatigue and boredom, human inspection is only 80–85% effective. With the wider availability of automatic inspection techniques (Secs. 2-4-6 and 2-5), 100% inspection can be fully effective and, in many instances, also economical.

2 *Acceptance sampling.* Hundred-percent inspection may be uneconomical and, if inspection involves destructive testing, impossible. A limited number of samples will then be tested. The number of samples and the methods of obtaining them are prescribed on the basis of probability theory. The sampling plan is drawn up so that the risk of rejecting good lots (the producer's risk) is minimized without an undue increase in the risk of accepting bad lots (the consumer's risk).

This approach is essentially *reactive*. Inspection is made after the fact; if an unacceptably high proportion of nonconforming parts or assemblies is detected, the sampling rate is increased and, if necessary, out-of-specification parts are separated by sorting. Correction involves rework, reinspection, or scrapping. The origin of rejects may be difficult to trace and, by the time the information gets back to the producer, further lots—of good or bad quality—may already have been produced. In a sense, *quality is inspected into the product*. Even though quality control may be performed by highly trained people employing the best equipment and inspection plan, nonconforming parts will be allowed to go through; the purpose of inspection is only to ensure that the proportion of nonconforming parts does not exceed a predetermined value. Assemblies consisting of many parts may turn out to be defective in a significant proportion of cases, destroying the competitiveness of the product. Tighter quality standards can usually be satisfied only at a higher cost.

11-3-3 Statistical Process Control

A much greater probability of success is attained if quality control is applied in the course of production itself. As mentioned, the beginnings of *statistical process control* (SPC) go back to the 1920s, but its execution was usually assigned to quality-control departments, with inspectors drawing samples during production runs. This often led to an adversarial relationship with the operators and reduced the effectiveness of control.

Beginning with the 1950s, American experts, notably W. E. Deming, introduced the technique in Japan in a modified form, with the execution of quality control entrusted to the operator. This approach recognizes that without appropriate guidance, the operator is not able to control quality; he or she may take actions when they are not needed and, conversely, take no action when action is needed. Therefore, management (usually through the quality assurance department) must provide the appropriate tools: not just measuring instruments, but also the plan of control based on principles of statistics. Thus, the operator can perform the measurements, evaluate the significance of results, and take immediate corrective action. The key role played by the operator is often recognized by the name *operator process control* (OPC). The phenomenal success of this approach has also prompted its acceptance in North America. The principle of OPC is straightforward:

1 Quality is not regarded as a separate issue, to be controlled in isolation from the process. Instead, the entire production system is reviewed, and factors affecting quality at various work stations are evaluated.

2 From this review, critical variables or attributes are identified and control limits are set. Measurements and inspections are prescribed at time intervals (or number of production units) that ensure statistical significance.

3 The operator performs the prescribed measurements and immediately plots the data on appropriate charts. If trends suggest that deviations may reach values approaching the control limits, remedial action is taken before any nonconforming parts would be produced.

Thus, even though a chance event may lead to the production of a defective part, this will be extremely rare; the vast majority of parts will be acceptable without further inspection: *quality is built into the product*. The producer benefits because all parts that are manufactured can also be shipped. The purchaser benefits too; the need for reinspection and for the rework or repair of assemblies disappears. Competitiveness increases through higher quality, and the costs drop because of higher productivity. The technique even allows improvements to the process. By identifying variations that are attributable to a definite source (assignable cause), means of removing these sources can be found, control limits tightened, and parts of higher quality produced.

Application of SPC The basic elements of the approach can be illustrated by returning to the example of producing a shaft on a lathe (Fig. 11-9a). The operator controls the diameter of the shaft and aims to keep it well within the specified limits of 9.90 and 10.00 mm. A micrometer is provided for measurement. Several approaches are possible. In one approach, the following steps are taken:

1 *Evaluate the measuring device.* This task is usually performed by the quality assurance department, for example, by measuring two calibration-accuracy plug gages, of 9.900- and 10.000-mm diameter. The measuring device must have graduations corresponding to one-tenth of the tolerance or better (in this case, graduations of 0.01 mm).

2 *Evaluate reproducibility.* Once the micrometer is found to be accurate, the combined reproducibility and repeatability of measurement is evaluated in the plant (separation of the two requires more extensive testing). Two operators (A and B) measure, only once, five shafts selected at random. Estimates of readings are rounded to the nearest half graduation. Gage error is then calculated as shown

TABLE 11-1
CHECKING GAGE ERROR

Part no.	Operator A	Operator B	Range, mm
1	9.945	9.960	0.015
2	9.925	9.930	0.005
3	9.940	9.940	0.000
4	9.935	9.930	0.005
5	9.955	9.945	0.010

Sum of ranges $\Sigma R = 0.035$
Average range $\bar{R} = \Sigma R / n = 0.035 / 5 = 0.007$ mm
Tolerance range $= 10.00 - 9.90 = 0.10$ mm
Gage error (gage repeatability and reproducibility, GRR) =
$\quad 4.33 \bar{R} = 0.0303$
GRR as percent of tolerance $= (0.0303 / 0.100)100 = 30.3\%$.

in Table 11-1 (the multiplier 4.33 comes from statistical theory and is a function of the number of operators and samples). If the error is greater than 10% of the tolerance range, the micrometer is adjusted, changed, or—if the gage error cannot be reduced—other means of measurement must be found. In less critical applications, an error of 20% is tolerable.

3 *Determine process capability* (i.e., whether the process is capable of meeting specifications at all.) To obtain a snapshot in time, *short-term capability* is established by sampling. To ensure that the results will be statistically significant, the sample size should be large; for economy, it must be kept small. A good compromise is $n = 5$ pieces (units) per sample (subgroup). At least 10, but preferably 20 samples (100 consecutive pieces) are taken; no adjustments must be made during this period. The measured diameters (Table 11-2) are plotted in a histogram (Fig. 11-9c). If the distribution is close to normal, proceed with the analysis. If the normal distribution is centered off 9.95 mm, reset the tooling and repeat sampling. If distribution is bimodal, find the source of the effect (e.g., if the shaft is turned down in two stages, analyze each stage). If the distribution is skewed, find the source (e.g., a drift due to tool wear). If skew cannot be eliminated, proceed with the analysis but note the cause of drift.

From the data, calculate the average diameter $\bar{x}$ (Eq. (11-1)), and range R for each five-piece sample (Table 11-2; note that the measurements are shown only to the nearest hundredths and that only 10 of the 20 samples are included in the table; hence, the averages and standard deviation are not quite the same as indicated in Fig. 11-9). Calculate the average of averages (grand average)

$$\bar{\bar{x}} = \frac{\Sigma \bar{x}}{(n)} \tag{11-3}$$

and the average $\bar{R}$ of ranges R for the $(n) = 20$ samples (again, only 10 samples

TABLE 11-2
EVALUATION OF AVERAGES AND RANGES*

| | | Sample number | | | | | | | | | |
		1	2	3	4	5	6	7	8	9	10
Data	1	9.92	9.95	9.98	9.99	9.93	9.96	9.97	9.94	9.95	9.96
from	2	9.94	9.96	9.94	9.92	9.95	9.97	9.94	9.96	9.93	9.92
sample	3	9.93	9.96	9.95	9.93	9.99	9.94	9.92	9.95	9.97	9.92
x,	4	9.95	9.97	9.97	9.94	9.93	9.98	9.92	9.96	9.96	9.94
mm	5	9.93	9.94	9.93	9.92	9.94	9.99	9.93	9.92	9.98	9.93
Total	Σx	49.67	49.78	49.77	49.70	49.74	49.84	49.68	49.73	49.79	49.67
Average	$\bar{x}$	9.934	9.956	9.954	9.940	9.948	9.968	9.936	9.946	9.958	9.934
Range	R	0.03	0.03	0.05	0.07	0.06	0.05	0.07	0.04	0.05	0.04

*Only 10 samples out of 20 shown. $\bar{\bar{x}} = 9.947 = 9.95$ mm; $\bar{R} = 0.049$ mm.

are shown in Table 11-2). An estimate of the standard deviation may now be obtained from

$$\sigma = \frac{\bar{R}}{d_2} \qquad (11\text{-}4)$$

where d_2 is taken from Table 11-3. Since a dispersion of $\pm 3\sigma$ is expected in manufacturing, it is usual to compare 6σ to the tolerance to express *machine capability* (MC)

$$MC = \frac{6\sigma}{\text{tolerance}} 100 \quad (\%) \qquad (11\text{-}5)$$

If MC = 133%, capability is excellent. At MC = 100%, the process is just acceptable, and at MC < 100% it is not acceptable (or the parts produced would have to

TABLE 11-3
FACTORS FOR CONTROL LIMITS

Number of observations in sample, n	d_2	A_2	D_3	D_4
2	1.128	1.880	0	3.267
3	1.693	1.023	0	2.575
4	2.059	0.729	0	2.282
5	2.326	0.577	0	2.115
6	2.534	0.483	0	2.004
8	2.847	0.373	0.136	1.864
10	3.078	0.308	0.223	1.777
15	3.472	0.223	0.348	1.652
20	3.735	0.180	0.414	1.586
25	3.931	0.153	0.459	1.541

be sorted). Process capability is often expressed by the reciprocal of MC; an acceptable process then has the capability of 1 (100%) or less.

A shift of $\bar{\bar{x}}$ from the center of the tolerance band may cause some parts to be outside the tolerance limits, but resetting the tool will correct the matter.

4 *Calculate control limits.* If the process is capable, control limits can be calculated for the averages:

$$\text{UCL}_{\bar{x}} = \bar{x} + 3\sigma = \bar{x} + A_2 \bar{R} \tag{11-6}$$

$$\text{LCL}_{\bar{x}} = \bar{x} - 3\sigma = \bar{x} - A_2 \bar{R} \tag{11-7}$$

and for the ranges:

$$\text{UCL}_R = D_4 \bar{R} \tag{11-8}$$

$$\text{LCL}_R = D_3 \bar{R} \tag{11-9}$$

where A_2, D_3, and D_4 are taken from Table 11-3. In our example

$$\text{UCL}_{\bar{x}} = 9.95 + (0.577)(0.049) = 9.976 \text{ mm}$$

$$\text{LCL}_{\bar{x}} = 9.95 - (0.577)(0.049) = 9.919 \text{ mm}$$

$$\text{UCL}_R = (2.115)(0.049) = 0.1036 \text{ mm}$$

$$\text{LCL}_R = (0)(0.049) = 0 \text{ mm}$$

These limits are then plotted on so-called $\bar{x}$-R (x bar-R) charts, Fig. 11-10; it should be noted that these charts *do not* show the specification limits. For a process in control, the points are randomly distributed within the limits.

5 *Operator quality control.* At this point, the operator takes over and plots, on the control charts provided, $\bar{x}$ and R for samples taken at predetermined intervals (Fig. 11-10). As long as variations are random and stable, the process remains in a state of statistical control; the sampling interval can be lengthened with the agreement of the quality control department. The $\bar{x}$-R charts alert the operator to conditions that lead to greater than normal variations, so that remedial action can be taken *before* out-of-tolerance parts are produced; some examples are shown in Fig. 11-11. The purchaser of parts is assured that no out-of-specification part will be delivered. Increasingly, purchasers will buy parts only from producers who maintain statistical process control and are willing to cooperate with suppliers in installing a meaningful process-control procedure.

In some processes, with appropriate equipment, it is possible to do 100% inspection. Analysis of the data then provides important clues regarding the effects of process variables, allowing the tightening of control limits. When processes are run under statistical control, a very small spread of R indicates that either something is wrong with the measurement or that the process is better

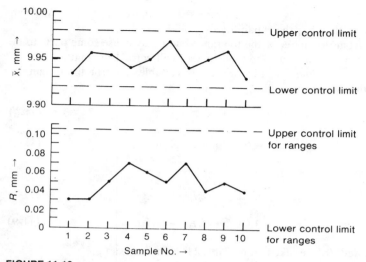

FIGURE 11-10
The process is in statistical control because none of the sample averages or ranges fall outside the established control limits.

FIGURE 11-11
Statistical process control gives valuable clues: (a) process is in statistical control, (b) improperly set tooling, (c) rapid tool wear, (d) mixed material lot of two hardnesses, (e) process out of control, and (f) something is going very well and is worth investigating.

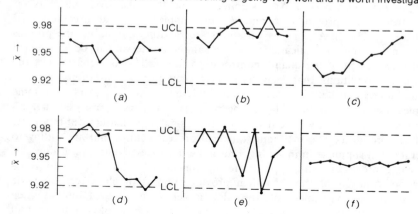

controlled than normally; by following up clues, improvements can again be made.

In some instances tolerances are so tight that no existing process is capable of satisfying them. Then parts must be separated by 100% inspection into subgroups for *selective assembly*, and parts in the assembly are no longer interchangeable.

In many applications direct measurement of a variable is not possible, and process control must be based on the number or ratio of defectives in a sample or on the number of defects in a unit. Quality control charts can be set up for such attributes too.

11-4   MANUFACTURING MANAGEMENT

A plant equipped with efficient machine tools and served by the best means of material movement and assembly can still lose money. For profitable production, technology and the physical means of production must be effectively managed. *Management* is not an end in itself; without constantly advancing technology, the best-managed company cannot survive.

11-4-1   Company Organization

Neither company organizations nor the terminology used to describe organizational elements are standardized. Nevertheless, the *organization chart* in Fig. 11-12 is fairly typical of North American practice.

FIGURE 11-12
Example of a possible organizational structure.

All manufacturing concerns have, at the company level, finance, personnel, purchasing, and sales departments. *Research and development, marketing,* and *market development* departments are essential for growth. Depending on the nature of operations there may also be *product engineering* which takes care of the development, design, testing, and evaluation of new products, with the full involvement of the manufacturing group of the company or outside vendors.

At the plant level, all facets of production are under the direction of the plant manager. Actual production is headed by the *manufacturing* superintendent, with the assistance of superintendents, general supervisors, and supervisors. Their job is to keep production going at peak efficiency, a task so complex as to need support from several other departments.

1 *Production planning and control* determines economical lot sizes; establishes schedules for manufacturing and assembly; controls the inventory of raw materials and in-process parts; dispatches materials, tooling, and equipment to the plant locations specified by the schedule; keeps track of progress; and often provides program evaluation review. It is here that MRP (Sec. 1-4-3) resides.

2 The *methods and work standards* group provides not only time and motion studies and sets norms, but very often analyzes the whole process, sometimes attached to or working with the *process engineering department*.

3 *Plant engineering* is responsible for preventive maintenance of equipment, replacement of machinery, and provision of services (power, heating, lighting, etc.). Maintenance is particularly critical with computerized equipment and robots, because downtime results in large economic losses.

4 *Inspection and quality control* are a vital activity, reporting to a high level of management while also increasingly integrated into production.

5 *Manufacturing engineering* is central to our considerations and will be discussed separately.

It should be noted that the organization shown in Fig. 11-12 is by no means universally adopted and is not necessarily the best. It tends to encourage separation of departmental efforts and, for this reason, there is a growing trend to a *horizontal* organization scheme in which all departments report to the plant manager, thus ensuring greater interdepartmental cooperation.

11-4-2 Manufacturing Engineering

The *manufacturing engineering* group, also called *process engineering*, is usually headed by the *chief engineer*. It is in this group that technological awareness resides, and the competence of its people determines whether the company will be competitive and profitable. In the present environment, people in this group must understand processes and their control, and must be versed in materials, mechanics, electronics, computers, and systems analysis.

Typical tasks include evaluation of manufacturing feasibility and cost; selection of optimum processes and process sequences; production equipment; tooling, jigs and fixtures (their design and manufacture, and control of the tool room);

material movement methods and equipment; and plant layout. In addition to unit processes discussed in this book, the group also issues specifications for auxiliary and finishing processes such as heat treatment, cleaning, painting, plating, coating, and, in general, treating of individual parts and finished assemblies. Cost analysis (Sec. 12-1) is often performed here, and the group also has a central function in value analysis (Sec. 12-4). The group cooperates closely with research and development specialists.

With increasing mechanization and automation, and particularly with the special demands set by numerical and computer control, the activities of the manufacturing engineering department account for a substantial share of the total production cost, and there are convincing reasons why at least some of these activities should be regarded as direct rather than indirect labor (Sec. 12-1-3). The true cost-effectiveness of new manufacturing technologies can then be judged and the need for a thorough reorganization of production facilities revealed. This is particularly true of numerical (computer) control, which often results in no savings unless the entire manufacturing concept is changed.

11-5 SUMMARY

The unit processes discussed in Chaps. 3–10 can lead to competitive production only if they are integrated into a manufacturing system. Of special importance are the following points not previously discussed:

1 Material movement can be most time-consuming and may limit the productivity of machine tools. Therefore, parts and tools must be transported so that they are immediately available to machine tools as the need arises for them. Mechanical devices such as roller gangs, endless belts, overhead conveyors, and towline carts are relatively inflexible but of high productivity; manual material movement is flexible but often of low productivity. Programmable devices such as robots (especially those fitted with elements of artificial intelligence) and automated guided vehicles can be both flexible and productive. Fixturing or palletization of parts allows quick setup.

2 Production quantities, production rates, and similarities in features and operational sequences determine the most appropriate method of production organization. Small-batch production of highly varying units requires stand-alone machine tools; as the similarity of features and operational sequences increases, FMC or even FMS becomes feasible and economical. Mass production is still best performed in dedicated automatic machines and production lines, although some flexibility may be introduced even in these. Group technology is a most important aid in the organization of production.

3 Assembly too may be manual or automatic; automation may be flexible or fixed, depending on the number and variety of products. Some flexibility is often incorporated even into mass production, aided by the introduction of just-in-time delivery of smaller batches of components.

4 Lowest-cost production is attained when each and every part produced satisfies quality requirements. Therefore, statistical process control—which ensures that quality is built, and not inspected, into the product—can be most profitable and also helps to secure the market in a competitive environment.

5 The best process will fail unless supported by strong management; conversely, the best management will do nothing for a fundamentally noncompetitive process.

FURTHER READING (SEE ALSO CHAP. 1)

A Automation

Groover, M. P.: *Automation, Production Systems, and Computer-Aided Manufacturing*, Prentice-Hall, Englewood Cliffs, N.J., 1980.

Wang, P. C. C. (ed.): *Automation Technology for Management and Productivity: Advancements through CAD/CAM and Engineering Data Handling*, Prentice-Hall, Englewood Cliffs, N.J., 1983.

B Material Movement and Robots

Ayres, R. U., and S. M. Miller: *Robotics Applications and Social Implications*, IFS (Publications) Ltd., Bedford, England, 1983.

Ballard, D. H., and C. M. Brown: *Computer Vision*, Prentice-Hall, Englewood Cliffs, N.J., 1982.

Bonney, M., and Y. F. Young (eds.): *Robot Safety*, IFS (Publications) Ltd., Bedford, England, 1985.

Craig, J. J.: *Introduction to Robotics: Mechanics and Control*, Addison-Wesley, Reading, Mass., 1986.

Critchlow, A. J.: *Introduction to Robotics*, Macmillan, New York, 1985.

Dorf, R. C.: *Robotics and Automated Manufacturing*, Reston Publishing Co., Reston, Va., 1984.

Faugeras, O. D. (ed.): *Fundamentals in Computer Vision*, IFS (Publications) Ltd., Bedford, England, 1983.

Gevarter, W. B.: *Intelligent Machines: An Introductory Perspective of Artificial Intelligence and Robotics*, Prentice-Hall, Englewood Cliffs, N.J., 1985.

Hunt, D.: *Industrial Robotics Handbook*, Industrial Press, New York, 1983.

Husband, T. M.: *Education and Training in Robotics*, IFS (Publications) Ltd., Bedford, England, 1985.

Kafrissen, E., and M. Stephans: *Industrial Robots and Robotics*, Reston Publishing Co., Reston, Va., 1984.

Morgan, C.: *Robots—Planning and Implementation*, IFS (Publications) Ltd., Bedford, England, 1984.

Muller, T.: *Automated Guided Vehicles*, IFS (Publications) Ltd., Bedford, England/Springer, Berlin, 1983.

Parent, M., and C. Laurgeau: *Robot Technology*, Vol. 5: *Logic and Programming*, Prentice-Hall, Englewood Cliffs, N.J., 1985.

Pham, D. T., and W. B. Heginbotham (eds.): *Robot Grippers*, IFS (Publications) Ltd., Bedford, England, 1986.

Pugh, A. (ed.): *Robot Vision*, IFS (Publications) Ltd., Bedford, England, 1983.
Pugh, A. (ed.): *Robot Sensors*: vol. 1; *Vision*; vol. 2: *Tactile and Non-Vision*, IFS (Publications) Ltd., Bedford, England, 1985.
Ranky, P., and C. Y. Ho: *Robot Modelling—Control and Applications with Software*, IFS (Publications) Ltd., Bedford, England, 1985.
Recent Advances in Robotics, Wiley, New York, annual.
Rehg, J. A.: *Introduction to Robotics: A Systems Approach*, Prentice-Hall, Englewood Cliffs, N.J., 1985.
RIA Robotics Glossary, Robot Institute of America, Dearborn, Mich., 1984.
Snyder, W. E.: *Industrial Robots: Computer Interfacing and Control*, Prentice-Hall, Englewood Cliffs, N.J., 1985.
Tanner, W. R. (ed.): *Industrial Robots*, 2 vols., Society of Manufacturing Engineers, Dearborn, Mich., 1979.
Warnecke, H. J., and R. D. Schraft (eds.): *Industrial Robots: Application Experience*, IFS (Publications) Ltd., Bedford, England, 1982.
Zeldman, M.: *What Every Engineer Should Know About Robots*, Dekker, New York, 1984.

C Fixtures

Boyes, W. (ed.): *Jigs and Fixtures*, 2d ed., Society of Manufacturing Engineers, Dearborn, Mich., 1982.
Sedlik, H.: *Jigs and Fixtures for Limited Production*, Society of Manufacturing Engineers, Dearborn, Mich., 1970.
Wilson, F. W. (ed.): *Handbook of Fixture Design*, McGraw-Hill, New York, 1984.

D Group Technology

Burbridge, J. L.: *The Introduction of Group Technology*, Heinemann, London, 1975.
Gallagher, C. C., and W. A. Knight: *Group Technology*, Butterworths, London, 1973.
Ham, I., K. Hitomi, and T. Yoshida: *Group Technology: Applications to Production Management*, Kluwer-Nijhoff, Boston, 1985.
Hyer, N. L. (ed.): *Group Technology at Work*, Society of Manufacturing Engineers, Dearborn, Mich., 1984.
Proceedings of Group Technology Conference, Institution of Production Engineers, London, annual from 1971 on.

E Flexible Manufacturing

Hartley, J.: *FMS At Work*, IFS (Publications) Ltd., Bedford, England/North-Holland, Amsterdam, 1984.
Holland, J. R. (ed.): *Flexible Manufacturing Systems*, Society of Manufacturing Engineers, Dearborn, Mich., 1984.
Proceedings of International Conference on Flexible Manufacturing Systems, IFS (Conferences) Ltd., Bedford, England, from 1982.
Ranky, P. G.: *The Design and Operation of FMS*, IFS (Publications), Ltd./North-Holland, Amsterdam, 1983.
Warnecke, H. J., and R. Steinhilper (eds.): *Flexible Manufacturing Systems*, IFS (Publications) Ltd./Springer, Berlin, 1985.

F Quality Assurance

Batchelor, B., D. Hill, and D. Hodgson (ed.): *Automated Visual Inspection*, IFS (Publications) Ltd., Bedford, England, 1984.

Charbonneau, H. C., and G. L. Webster: *Industrial Quality Control*, Prentice-Hall, Englewood Cliffs, N.J., 1978.

Deming, W. E.: *Quality, Productivity, and Competitive Position*, Massachusetts Institute of Technology Center for Advanced Engineering Study, Cambridge, Mass., 1982.

Enrick, N. L.: *Quality Control and Reliability*, 7th ed., Industrial Press, New York, 1977.

Feigenbaum, A. V.: *Total Quality Control*, 3d ed., McGraw-Hill, New York, 1983.

Halpern, S.: *The Assurance Sciences: An Introduction to Quality Control and Reliability*, Prentice-Hall, Englewood Cliffs, N.J., 1978.

Juran J. (ed.): *Quality Control Handbook*, 3d ed., McGraw-Hill, New York, 1974.

_____ and F. M. Gryna, Jr.: *Quality Planning and Analysis*, 2d ed., McGraw-Hill, New York, 1980.

Klippel, W. H. (ed.): *Statistical Quality Control*, Society of Manufacturing Engineers, Dearborn, Mich., 1984.

Murdoch, J.: *Control Charts*, Macmillan, New York, 1979.

Pryor, T., and W. North (eds.): *Applying Automated Inspection*, Society of Manufacturing Engineers, Dearborn, Mich., 1985.

Sharpe, R. S., J. West, D. S. Dean, D. A. Taylor, and H. A. Cole: *Quality Technology Handbook*, 4th ed., Butterworths, London, 1984.

Simmons, D. A.: *Practical Quality Control*, Addison-Wesley, Reading, Mass., 1979.

Stout, K. J.: *Quality Control in Automation*, Prentice-Hall, Englewood Cliffs, N.J., 1985.

The Tool and Manufacturing Engineers Handbook, Vol. 4, *Assembly, Testing, and Quality Control*, Society of Manufacturing Engineers, Dearborn, Mich., 1986.

G Production Planning and Management

Aft, L. S.: *Productivity Measurement and Improvement*, Reston Publishing Co., Reston, Va., 1983.

Bertrand, J. W. M., and J. C. Wortmann: *Production Control and Information Systems for Component Manufacturing Shops*, Elsevier, Amsterdam, 1981.

Chang, T. C., and R. A. Wysk: *An Introduction to Automated Process Planning Systems and TIPPS*, Prentice-Hall, Englewood Cliffs, N.J., 1985.

Chase, R., and N. Aquilano: *Production and Operations Management*, 3d ed., Irwin, Homewood, Ill., 1981.

Enrick, N. L., and H. E. Mottley, Jr.: *Manufacturing Analysis for Productivity and Quality Cost Enhancement*, 2d ed., Industrial Press, New York, 1983.

Gardiner, K. (ed.): *Systems and Technology for Advanced Manufacturing*, Society of Manufacturing Engineers, Dearborn, Mich., 1983.

Hales, L. E.: *Computer-Aided Facilities Planning*, Dekker, New York, 1984.

Hendrick, T. E., and F. G. Moore: *Production / Operations Management*, 9th ed., Irwin, Homewood, Ill., 1985.

King, J. R.: *Production Planning and Control*, Pergamon, Oxford, 1975.

Menipaz, E.: *Essentials of Production and Operations Management*, Prentice-Hall, Englewood Cliffs, N.J., 1986.

Niebel, B. W.: *Motion and Time Study*, 7th ed., Irwin, Homewood, Ill., 1982.

Oliver, S.: *The Management of Production Technology*, Mechanical Engineering Publ., London, 1978.

Peterson, J.: *Industrial Health*, Prentice-Hall, Englewood Cliffs, N.J., 1977.

Peterson, R., and E. A. Silver: *Decision Systems for Inventory Management and Production Planning*, Wiley, New York, 1979.

Radford, J. D., and Richardson, D. B.: *The Management of Manufacturing Systems*, Macmillan, New York, 1977.

Smith, G. L.: *Work Measurement: A Systems Approach*, Macmillan, New York, 1978.

Tersine, R. J.: *Productions / Operations Management*, 2d ed., North-Holland, Amsterdam, 1985.

Tool and Manufacturing Engineers Handbook, vol. 5: *Manufacturing Engineering Management*, Society of Manufacturing Engineers, Dearborn, Mich., 1987.

H Assembly

Boothroyd, G., C. Poli., and L. E. Murch: *Automatic Assembly*, Dekker, New York, 1982.

Den Hamer, H. E.: *Interordering: A New Method of Component Orientation*, Elsevier, Amsterdam, 1980.

Heginbotham, W. (ed.): *Programmable Assembly*, IFS (Publications) Ltd., Springer, Berlin, 1984.

Owen, T.: *Assembly with Robots*, Prentice-Hall, Englewood Cliffs, N.J., 1985.

Prenting, T. O., and N. T. Thomopoulos: *Humanism and Technology in Assembly Line Systems*, Spartan Books, Hayden Book Co., Rochelle Park, N.J., 1974.

Rathmill, K. (ed.): *Robotic Assembly*, IFS (Publications) Ltd., Bedford, England, 1985.

Riley, F. J.: *Assembly Automation*, IFS (Publications) Ltd., Bedford, England, 1983.

Tool and Manufacturing Engineers Handbook, vol. 4: *Assembly, Testing, and Quality Control*, 4th ed., Society of Manufacturing Engineers, Dearborn, Mich., 1986.

Treer, K. R.: *Automated Assembly*, Society of Manufacturing Engineers, Dearborn, Mich., 1979.

I Journals

Engineering Management International

The Journal of Product Innovation Management

CHAPTER 12

COMPETITIVE ASPECTS OF MANUFACTURING PROCESSES

We argued in Sec. 1-2 that the economic well-being of a nation is critically dependent on a competitive manufacturing industry, and that productivity is a key issue in securing a competitive position in the world.

In this book the emphasis has been on unit processes that are needed to make discrete parts. We showed in Chaps. 1 and 11 that these processes must be organized into a well-coordinated manufacturing system in which the functions of design and manufacturing are interwoven; whenever appropriate, these points were also emphasized in Chaps. 3–10. There is a danger that the inevitable mass of details obscures the larger picture; it is now necessary to take a bird's eye view and examine the competitive aspects of manufacturing. Competition is often taken to exist between corporations or nations. The discussion that follows will show that there is also competition between manufacturing processes. There is no absolute and eternal solution to any manufacturing problem. Only the function of the product is defined, and even this function may be continually redefined. For products of well-defined function, there is not only a never-ending competition between various processes and process sequences, but there is also a competition between various materials and the processes implied in the material choice. Competition is based on serving the function at minimum cost; hence, cost considerations will always enter into the choice of manufacturing processes.

12-1 MANUFACTURING COSTS

In choosing a manufacturing process, no decision—not even a preliminary one—can be made without considering at least the major cost elements. A complete discussion of *manufacturing cost analysis* is outside the scope of this book and only an outline of the approach can be given here.

12-1-1 Cost and Productivity

In manufacturing, *cost* is usually expressed in a monetary unit (such as the dollar) per unit output. *Productivity* is a more difficult term to define. In the simplest form, it is taken as the value of production per employee. This *labor productivity* is useful as a very general measure, but it does not take into account the capital employed in generating the output. This aspect has become particularly significant with the increase of automation. The *productivity of capital* (holding the quality and quantity of other inputs fixed) increases only if the efficiency of new capital goods grows faster than the price of these goods. Capital productivity is much more difficult to measure and, therefore, more controversial than labor productivity. What is more easily seen is that there must be an *economic limit* on automation where further investment yields *diminishing returns* in terms of cost reduction.

To illustrate these points, it will be instructive to look at an example taken from the automotive industry. It should be noted that the dollar values attached to various units do not necessarily bear any relation to present reality.

Example 12-1

More than half the total weight of a production-type automobile is still steel, much of it in the form of pressworked sheet-metal parts. To take one part as an example, the outer door panel is a fairly sophisticated product with a shallow curvature, complex outer profile, and several cutouts.

Several car models of a manufacturer share the same door, and the production quantity over a typical seven-year model period may account to, say, 6 million units for a mass-produced car. This justifies a special-purpose production line (a transfer line) in which several presses perform the required sequence of operations, with mechanical means of passing the part from one press to another. With the increasing need for fast response to customer demands, the line may be equipped with quick-change die sets, and robots may be employed in material movement. If we assume that the line is tended by three operators in each of three shifts, their productivity can be expressed as the number of units produced per year (Table 12-1). A more generally applicable measure of labor productivity is value produced, and if we assume (without further justification) that the door panel is worth $3.00 for the manufacturer, productivity can be expressed in value produced per operator (Table 12-1).

At the other end of the scale, some luxury cars are made by specialized companies in small numbers, say, 10 000 units over 10 years. Equipment and tooling costs have to be reduced by stretch-forming and low-cost blanking (such as rubber forming) techniques, for example. The plant operates perhaps on only one shift and we can assume that five operators produce the requisite 1000 door panels per year. This may, of course, be split up into several lots spread over the year, and we assume that equivalent parts are produced for the rest of the time for a total of 1.2×10^4 units per year. The door panel, in the luxury market, is now worth $26.00 to the manufacturer, and productivities can again be calculated (Table 12-1).

When productivities are compared on either a unit or dollar basis, there would seem to be no room for the batch-produced automobile even in the luxury market. However, our calculation ignored several important factors. First of all, the nine operators of the transfer line are backed up

TABLE 12-1
COMPARISON OF PRODUCTIVITIES AND COSTS
IN A SHEET-METALWORKING OPERATION

	Mass production	Batch production
Production, units	6×10^6	1.2×10^4
Value generated, $	1.8×10^7	3.1×10^5
Number of operators	9	5
Number of employees	27	7
Productivity, units/operator	6.7×10^5	2.4×10^3
$/operator	2×10^6	6.25×10^4
Productivity, units/employee	2.2×10^5	1.7×10^3
$/employee	6.4×10^5	4.4×10^4
Capital outlay, total $	10×10^6	4×10^5
Amortization, $/annum	4×10^6	8×10^4
Tooling cost, total $	5×10^5	1×10^5
Amortization, $/annum	3×10^5	4×10^4
Cost of tool changes, $/annum	1×10^4	1×10^4
Labor cost, $/employee	2×10^4	2×10^4
Total $/annum	5.4×10^5	1.4×10^5
Production cost, $/annum	4×10^6	8×10^4
	$+\ 3 \times 10^5$	$+\ 4 \times 10^4$
	$+\ 1 \times 10^4$	$+\ 1 \times 10^4$
	$+5.4 \times 10^5$	$+1.4 \times 10^5$
	$\overline{4.85 \times 10^6}$	$\overline{2.7 \times 10^5}$
Production cost, $/unit	0.86	22.50
Material cost, $/unit	1.50	2.00
Total cost, $/unit	2.36	24.50

Note: All values are fictitious and bear no relation to any existing manufacturer's operation.

by a team of, say, 18 people in maintenance, supervision, quality control, programming, and various levels of management. The five operators of the batch-production plant perform some of these functions themselves and are supported by only two people. Productivities now change (Table 12-1).

These numbers reveal nothing about *costs* or *profits*. For these, the means of production must also be considered. Capital is invested in the purchase of presses and auxiliary equipment, their installation, and the requisite buildings and services. This capital must be regained through some time period that formed the basis of investment decision. The actual annual repayment (amortization) depends on interest rates, tax treatment, and accounting procedures. The same argument applies to the cost of tooling, except that it has to be repaid over a shorter time period.

The *production cost* per unit is still vastly different (Table 12-1) between mass and batch production and the difference remains when the cost of material (somewhat higher in batch production because of the greater losses in rubber blanking and stretching) is added. After deducting other costs not accounted for in Table 12-1 (e.g., carrying an inventory of parts), the profit for the two operations is obtained. Economies would change again if the press line(s) were replaced with huge transfer presses, further increasing productivity.

This example, while fairly crude, illustrates that several cost elements must always be considered. It also shows that before any meaningful cost calculation can be made, the starting material and the major production sequence must be identified. The cost elements associated with them can then be taken into account.

12-1-2 Operating Costs (Direct Costs)

Direct costs can be clearly allocated to the product, and they are proportional to the number of units produced.

Material Costs The *net material cost* is the cost of purchasing the *raw material* (whether it be a casting, forging, rolled section or plate, metal or ceramic powder, polymer, or any other starting material), less the value of the *scrap* produced. Thus the weight of the purchased starting material, W_0 (including the weight loss in cutting off, etc.), and of the finished part, W_f is determined. If the unit price of the starting material is C_0 and that of the scrap is C_S (separated into heavy and light scrap if their resale value is widely different), the material cost per piece C_P is

$$C_P = W_0 C_0 - (W_0 - W_f)C_S \qquad (12\text{-}1)$$

In general, a process that generates less scrap, or generates it in a more valuable (separated by composition and usually heavier) form, is more economical.

Labor Costs With a possible process sequence and the appropriate equipment settled, the *number and skill of operating personnel* can also be predetermined. From experience, time studies, or a breakdown of operator functions into identifiable physical and mental action elements, the time required for completion of one piece can be calculated.

Strictly speaking, the *net production time* is the time period during which the material is actually shaped or processed; thus, in machining (Sec. 8-7-4) the net cutting time t_c is the time during which material is actually removed. However, in other processes where the truly productive time is only a fraction of a longer but essential cycle, the total cycle time is usually taken as the productive time. For example, in press working the time during which the press actually moves (and not just forms) is taken as the productive time.

Whether the time required for moving the material from the floor to the machine and off again is regarded as productive or nonproductive is a matter of philosophy; a truer picture of productivity is obtained if it is classified as nonproductive. In the cost analysis, it usually appears in the total or floor-to-floor time, and is charged at direct labor rates, especially if the movement is performed by the machine-tool operator personally.

When *energy costs* are a significant fraction of the total cost, they too are allocated directly to production units, as are tools and dies if used only for that particular purpose.

12-1-3 Indirect Costs

As seen in the example of loading and unloading, the distinction between direct and indirect cost may be a fuzzy one.

Indirect costs arise from the functions and services that contribute to the efficient performance of the actual production process. Traditionally, they include: indirect labor (including material movement, cleaning services, etc.); repair and maintenance; supervision (from supervisor to plant superintendent); engineering (manufacturing and industrial engineering, quality control, laboratory, etc.), research and development; sales; the entire management hierarchy of the company; lighting and heating (and sometimes all energy supplies and materials not directly used in production); office and sales expenses, etc. Some of these are somewhat flexible and can vary with the volume of production, but the relationship is never as direct as with operating costs.

If properly controlled, indirect activities represent an indispensable part of the total production effort. In the ultimate development of a fully automated manufacturing process, there would be no direct labor cost in the classical sense. Yet, many of the indirect costs—including programming and other activities required for the operation of a computer-integrated manufacturing system—obviously have to be regared as directly related to production. Similarly, development—and even long-range research if properly managed—is a vital part of the production process, and it could be argued that even these should be regarded as direct costs.

Indirect costs are also called *overhead* or, sometimes, *burden*. Unless properly controlled, indirect cost can indeed become a burden that finally nullifies the productivity gains in direct production. Therefore, breakdown of indirect costs into their elements is essential, and this is one of the purposes of CIM, discussed in Sec. 1-4-3. All too often, the actual production process is carefully analyzed, improved, and made more productive, while the indirect cost sector is allowed to grow out of proportion. This finally impairs the competitive position of the company; for that matter, unchecked growth of indirect costs can destroy a national economy.

In manufacturing cost estimates the indirect costs may appear as a multiplying factor applied to direct labor costs, as a fixed cost per unit product, or a fixed cost per hour worked on a machine.

12-1-4 Fixed Costs

Fixed costs include the costs of equipment, buildings, and of total facilities in general, taking into account depreciation, interest, taxes, and insurance. For a given capital outlay, the fixed costs depend on the interest rates prevailing at the time of purchase and during the life of the equipment, the tax treatment accorded to investments, and the useful life of the production facility. In general, the life of equipment can be extended, but only at the expense of rapidly rising maintenance costs and at the danger of obsolescence, and a replacement decision must sooner or later be made.

In manufacturing cost estimates, fixed costs are allocated on the basis of anticipated equipment utilization. Thus, if a press was purchased on the premise that it will be used in a two-shift operation, the fixed cost per unit production doubles if only a one-shift operation can be sustained. Conversely, the capital outlay for flexible manufacturing systems can usually be justified by the three-shift, seven-day operation of such systems. In the simplest estimating procedure, the fixed cost appears as a machine-hour rate or burden.

12-2 COMPETITION BETWEEN MANUFACTURING PROCESSES

It must be obvious from the discussion of various processes in Chaps. 3–10 that all processes are subject to limitations in terms of shape complexity, minimum and maximum dimensions, tolerances, and surface finish. It must be also clear that these limitations are highly dependent on workpiece material.

FIGURE 12-1
The choice of possible manufacturing methods is aided by classifying shapes according to their geometric features.

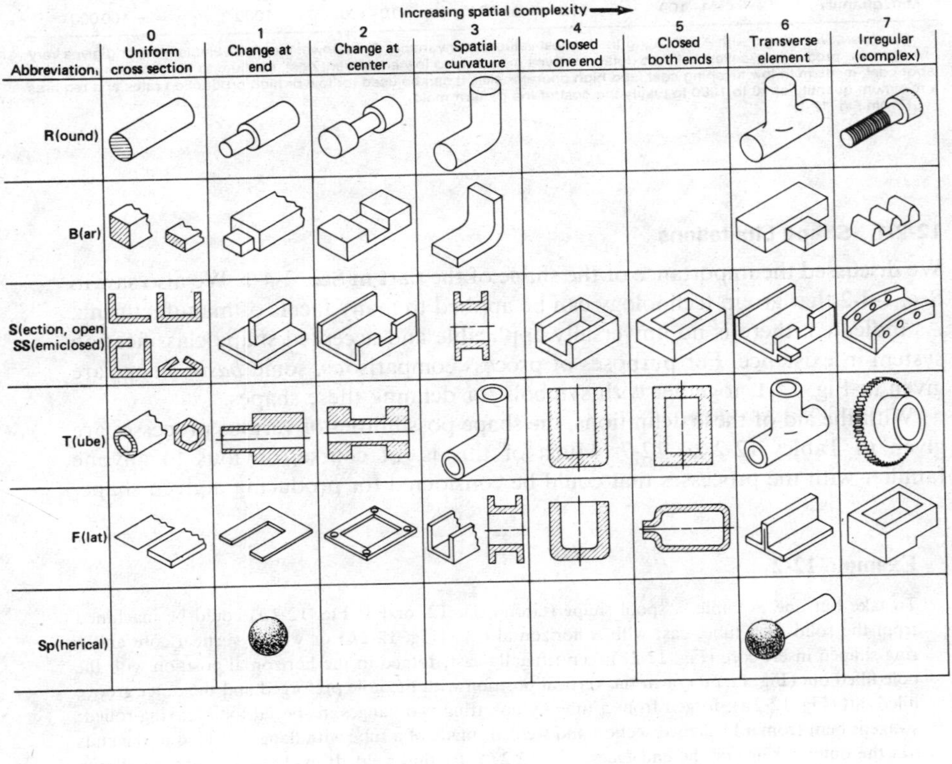

Increasing spatial complexity
Abbreviation
0 Uniform cross section
1 Change at end
2 Change at center
3 Spatial curvature
4 Closed one end
5 Closed both ends
6 Transverse element
7 Irregular (complex)
R(ound)
B(ar)
S(ection, open
SS(emiclosed)
T(ube)
F(lat)
Sp(herical)

TABLE 12-2   GENERAL CHARACTERISTICS OF CASTING PROCESSES

Characteristics	Casting process					
	Sand	Shell	Plaster	Investment	Permanent mold	Die
Part:						
Material (casting)	All	All	Zn to Cu	All	Zn to cast iron	Zn to Cu
Porosity*	C–E	D–E	D–E	E	B–C	A–C
Shape†	All	All	All	All	Not T3, 5, F5 with solid core	Not T3, 5, F5
Size, kg	0.01–300 000	0.01–100	0.01–1000	0.01–10(100)	0.1–100	< 0.01 to 50
Min. section, mm	3–6	2–4	1	1	2–4	0.5–1
Min. core diam., mm	4–6	3–6	10	0.5–1	4–6	3 (Zn: 0.8)
Surface detail*	C	B	A	A	B–C	A–B
Cost:						
Equipment*	C–E	C	C–E	C–E	B	A
Die (or pattern)*	C–E	B–C	C–E	B–C	B	A
Labor*	A–C	C	A–B	A–B	C	E
Finishing*	A–C	B–D	C–D	C–D	B–D	C–E
Production:						
Operator skill*	A–C	C	A–B	A–B	C	C–D
Lead time	Days	Weeks	Days	Hours–weeks	Weeks	Weeks–months
Rates (piece/h · mold)	1–20	5–50	1–10	1–1000	5–50	20–200
Min. quantity	~ 1–100	~ 100	~ 10	~ 10–1000	~ 1000	~ 100 000

*Comparative ratings, with A indicating the highest value of the variable, E the lowest (e.g., investment casting gives very low porosity, produces excellent surface detail, involves moderate to low equipment cost, medium to high pattern cost, high labor cost, medium to low finishing cost, and high operator skill. It can be used for low or high production rates and requires a minimum quantity of 10 to 1000 to justify the cost of the pattern mold.
†From Fig. 12-1.

12-2-1   Shape Limitations

We discussed the importance of the shape of the part in Sec. 2-4-1. We also saw in Sec. 2-4-2 that group technology can be applied to many facets of manufacturing. As indicated, there is no universally applicable and accepted shape classification system in existence. For purposes of process comparisons, some *basic shapes* are given in Fig. 12-1, together with symbols for defining these shapes.

With the aid of these definitions, the shape possibilities of various processes are given in Tables 12-2 to 12-7. Much of this is, of course, obvious to anyone familiar with the processes that could be considered for producing a given shape.

Example 12-2

To take but one example, a spool shape (Group S3, T2, or F3; Fig. 12-2*a*) could be machined from the solid or a tube; cast with a horizontal core (Fig. 12-2*b*) or with a vertical core and a ring-shaped insert core (Fig. 12-2*c*); centrifugally cast; forged in the horizontal position with the hole filled out (Fig. 12-2*d*) or in the vertical position with the hole preforged and the outer groove filled out (Fig. 12-2*e*); forged from a tube by upsetting two flanges in special tooling; ring-rolled; swaged; bent from a U-shaped section and welded; made of a tube with flanges welded to the ends (on the outer surface or the end faces, Fig. 12-2*f*); friction-welded; made by powder metallurgy

TABLE 12-3 GENERAL CHARACTERISTICS OF BULK DEFORMATION PROCESSES

Characteristics	Hot forging		Hot extrusion	Cold forging, extrusion	Shape drawing	Shape rolling	Transverse rolling
	Open die	Impression					
Part:							
Material (wrought)	All	All	All	All	All	All	All
Shape†	RO-3; B, T1,2; FO; Sp6	R; B; S; T1,2,4; (T6,7); Sp	R; B; S; SS, T1,4; Sp	As hot	RO; BO; SO; TO	RO; BO; SO	R1-2,7; T1-2; Sp
Size, kg	0.1–200000	0.01–100	1–500	0.001–50	10–1000	10–1000	0.001–10
Min. section, mm	5	3	1	(0.005)1	0.1	0.5	1
Min. hole diam., mm	(10)20	10	20	(1)5	0.1		
Surface detail*	E	C	B–C	A–B	A	A–B	A–C
Cost:							
Equipment*	A–D	A–B	A–B	A–C	B–D	A–C	A–C
Die*	F	B–C	C–D	A–B	C–D	A–C	A–C
Labor*	A	B–D	B–C	C–E	C–E	C–E	C–E
Finishing*	A	B–C	C–D	D–E	E	E	D–E
Production:							
Operator skill*	A	B–C	C–E	C–E	D–E	B	B–C
Lead time	Hours	Weeks	Days–weeks	Weeks	Days	Weeks	Weeks–months
Rates, pieces/machine	1–50 per h	10–300 per h	10–100 per h	100–10000 per h	10–2000 m/min	20–500 m/min	
Min. quantity	1	100–1000	1–10	1000–100000	1000 m	50000 m	1000–10000

*Comparative ratings with A indicating the highest value, E the lowest (see example in Table 12-2).
†From Fig. 12-1.

TABLE 12-4 GENERAL CHARACTERISTICS OF SHEET METALWORKING PROCESSES INCLUDING BAR AND TUBE BENDING

Characteristics	Blanking	Forming process				
		Bending	Spinning	Stretching	Deep drawing	Rubber forming
Part:						
Material (wrought)	All	All	All	All	All	All
Shape†	F0-2; T7	R3; B3; S0,3,7; SS; T3; F3,6	T1,2,4,5; F4,5	F4; S7	T4; F4,7	As blanking, bending, deep drawing
Max. thickness, mm	>10	>100	>25	>2	>10	2
Min. hole diam.	$\frac{1}{2}$–1 thickness				<3	50 (for $h =$ 1 mm Al)
Cost:						
Equipment*	B–D	C–E	B–D	B–C	A–C	A–C
Die*	B–D	B–E	B–D	A–C	A–B	C–D
Labor*	C–E	B–E	B–C	B–E	C–E	A–D
Finishing*	D–E	D–E	D–E	C–E	D–E	C–E
Production:						
Operator skill*	D–E	B–E	B–C	B–E	D–E	C–E
Lead time	Days	Hours–days	Days	Days–months	Weeks–months	Days
Rates, pieces/h	10^2–10^5	10–10^4	10–10^2	10–10^4	10–10^4	10–10^2
Min. quantity	10^2–10^4	1–10^4	1–10^2	10^5	10^3–10^5	10–10^2

*Comparative ratings, with A indicating the highest value, E the lowest (see example in Table 12-2).
†From Fig. 12-1.

TABLE 12-5 GENERAL CHARACTERISTICS OF POLYMER PROCESSES

	Manufacturing process						
	Molding					Thermoforming	
Characteristics	Compression	Transfer	Injection	Extrusion	Casting	Vacuum	Pressure
Part:							
Material	Plastics, glass	Plastics	Plastics	Thermoplastics	Plastics, glass	Thermoplastics	Thermoplastics, glass
Preferred Shape†	Thermosets All but T3,5, 6 and F5	Thermosets All but T3,5, and F5	As transfer	As transfer	Thermosets All	T4; F4,7	T4,5; F4,5,7
Min. section, mm	(0.8)1.5	(0.8)1.5	0.4 thermoplastic, 1 thermoset	0.4	4	<1	<1
Cost:							
Equipment*	B–C	B–C	A–C	A–B	D–E	B–D	B–D
Tooling*	A–C	A–C	A–C	A–C	B–E	B–C	B–C
Labor*	C–E	C–E	D–E	D–E	A–C	B–E	B–E
Production:							
Operator's skill*	D–E	D–E	D–E	D–E	B–E	B–E	A–E
Lead time	Weeks	Weeks	Weeks	Weeks	Days	Days–weeks	Days–weeks
Cycle time(s)	20–600	10–300	10–60	10–60		10–60	(1)10–60
Min. quantity	100–1000	100–1000	1000	10000	1	10–1000	10–1000

*Comparative ratings with A indicating the highest value, E the lowest (see example in Table 12-2).
†From Fig. 12-1.

TABLE 12-6 GENERAL CHARACTERISTICS OF MACHINING PROCESSES

Characteristics	Machining process						
	Lathe turning	Automatic screw machine	Shaping, planing	Drilling	Milling	Grinding	Honing, lapping
Part:							
Material	All*	All*	All*	All*	All*	All	All
Preferred		Free-machining		Free-machining	Free-machining	Hard	Hard
Shape†	R0-2,7; T0-2,4,5; Sp	As turning	B; S0-2; F0	T0	B; S; SS; F0-4,7	As turning, shaping, milling	R0-2; T0-2; 4-7; F0-2; Sp
Min. section, mm	< 1 diam.	< 1 diam.	< 2	0.1 (hole diam.)	< 1	< 0.5	< 0.5
Surface configuration	Axially symmetrical	As turning	Straight generatrix	Cylindrical	Three-dimensional	All (mostly flat, axially symmetrical)	Flat, cylindrical, three-dimensional
Cost:							
Equipment‡	B–D	A–C	B–D	D–E	A–C	A–C	B–D
Tooling‡	D–E	A–D	D–E	D–E	A–D	B–D	A–E
Labor‡	A–C	D–E	B–D	B–D	A–B	A–E	B–D
Production:							
Operator skill‡	A–C	D–E	B–D	B–E	A–B	A–D	C–E
Setup time‡	C–D	A–C	C–D	C–E	A–C	B–D	C–E
Rates (pieces/h)	1–50	10–500	1–50	10–500	1–50	1–1000	10–1000
Min. quantity	1	500	1	1	1	1	1

*Except most ceramics.
†From Fig. 12-1.
‡Comparative ratings with A indicating the highest value, E the lowest (see example in Table 12-2).

TABLE 12-7 GENERAL CHARACTERISTICS OF WELDING PROCESSES

Characteristics	Welding process						
	Arc welding					Electron beam	Oxyacetylene
	Shielded metal	Flux cored	Submerged	Gas-metal	Gas-W		
Part (assembly)							
Material	All but Zn	All steels	All steels	All but Zn	All but Zn	All but Zn	All but refractory metals
Preferred	Steels	Low-C steels	Low-C steels	Steels; non HT Al; Cu	All but Zn	All but Zn	Cast iron, steels
Thickness, min., mm	(1.5)3	1.5	5	0.5	0.2	0.05	0.6
Single pass max.	8–10	3–6	40	5	5	75	10
Multiple pass max.	> 25	> 15	> 200	> 25	> 6		> 20
Unequal thickness	Difficult	Difficult	Very difficult	Difficult	Difficult	Easy	Difficult
Distortion*	A–B	A–C	A–B	B–C	B–C	C–E	B–D
Jigging needed	Minimum	Minimum	Full	Variable	Variable	Full	Minimum
Deslagging for multipass	Yes	Yes	Yes	No	No		No
Current: type	Alternating or direct	Direct (reverse polarity)	Alternating or direct	Direct (reverse polarity)	Alternating or direct (straight polarity)		
Volts	40 or 70↓	40 to 70↓	25–55	20–40 or 70↓	60–150	30–175 kV	
Amperes	30–800	30–800	300–2500	70–700	100–500	0.05–1	
Costs:							
Equipment*	D	B–D	B–C	B–C	B–C	A	D–E
Labor*	A	A–D	B–D	A–C	A–C	A–D	A
Finishing*	A–B	A–C	A–C	B–D	B–E	C–E	A
Production:							
Operator skill*	A	A–D	C–D	A–D	A–D	A–D	A
Welding rate, m/min	(1–6 kg/h)	0.02–1.5	0.1–5	0.2–15	0.2–1.5	0.2–2.5	(0.3–0.6 kg/h)
Operation	Manual	All	Automatic	All	All	All	Manual

*Comparative ratings, with A indicating the highest value, E the lowest (see example in Table 12-2).

with isostatic compaction in a flexible die, or compaction in a multipiece permanent die); or by a combination of several of these methods. The optimum method depends on a great many factors, among them the material, size, wall thickness, wall-thickness ratio, and slenderness (length-to-diameter ratio) of the part.

Example 12-3

The economies attainable by changing the process sequence are shown in Fig. 12-3. Sparkplug shells are made of low-carbon steel and are needed by the tens of millio..s. Traditionally they were machined of hexagonal bar stock, on automatic screw machines, with one shell finished every 4–6 s. Much of the material was turned into low-value chip (typically, worth only one-twentieth of the price of the bar). The process became totally noncompetitive when the introduction of the phosphate-soap lubrication system allowed the cold extrusion of steel. The starting material is round wire (bar), sheared and cold upset (preformed) in a cold header at the rate of 250–400 per minute to form a slug which, after application of the phosphate-soap system, is cold extruded at the rate of 160 per min in a single stroke. (Many shells are also progressively cold formed at similar

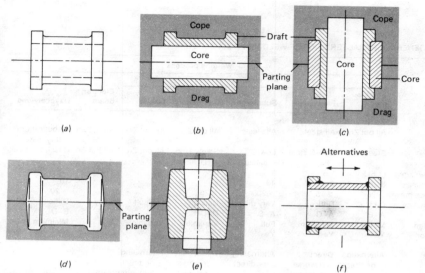

FIGURE 12-2
Some possible methods of making a spool-shaped part: (*a*) machining from bar or tube, (*b*) casting in a horizontal position, (*c*) casting in a vertical position, (*d*) forging in a horizontal position, (*e*) forging in a vertical position and, (*f*) welding.

rates.) The shell is finished, at the rate of 60 per min, with minimum machining, saving 70% of the material once used. Similarly, cost reduction through elimination or reduction of machining is the driving force in the aerospace industry for a shift to net-shape and near-net shape production by forging, casting, powder metallurgy, and diffusion bonding and superplastic forming.

12-2-2 Size Limitations

The *maximum size that can be produced* by any one technique is most often limited simply by the availability of large-size equipment. In some processes, however, there are limitations due to process conditions themselves. Thus, a casting mold may not stand up to the excessive solidification times imposed by very heavy walls, a welding process may be limited to a maximum metal thickness if only single-pass welding is permissible, or the thickness of plastic parts may be limited by the low heat conductivity of the plastic.

More frequently, however, the limitation is on the *minimum size that can be produced* or on *wall thickness*. Thus, the wall thickness of a casting is limited by the fluidity of the metal, and that of a forging by the die pressures developed with increasing d/h ratios. The limits given in Tables 12-2 to 12-7 reflect both practical and fundamental limitations and as such are not inflexible. Thinner, smaller, and larger parts may well be made, but usually under special circumstances and at extra expense. With the development of technology, these limits are continually pushed outward.

FIGURE 12-3
The savings achieved by changing the manufacturing method are illustrated by the spark plug body. (a) Originally it was produced by machining from hexagonal bar stock; now (b) a round slug is upset and (c) cold extruded into a shell of (d) thin wall, for a material saving of 70%. (*Courtesy Braun Engineering Co., Detroit.*)

Since the difficulty of filling a mold or deforming a part increases with the distance over which material must be moved, minimum attainable wall thicknesses are also a function of the extent (width) of the thinnest section; more detailed guidance is given in Fig. 12-4.

12-2-3 Surface Finish and Tolerances

We already stated in Secs. 2-4-3, 2-4-4, and 2-4-5 that unnecessarily tight tolerances and surface finish specifications are a major cause of excessive manufacturing costs. Each manufacturing process is capable of producing a part to a certain surface finish and tolerance range without extra expenditure. Some general guidance is given in Fig. 12-5.

It will be noted that surface finish and tolerances are interrelated. In the diagram, the surface finish and tolerances *usually* attainable in the process are indicated by the heavy lines adjacent to the name of the process. The capabilities

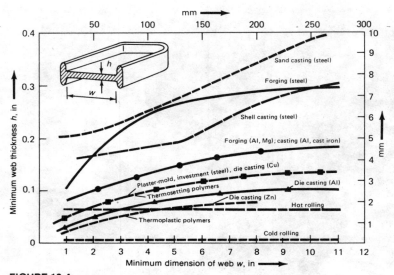

FIGURE 12-4

For a given process, the minimum web thickness increases with the distance over which material must move.

FIGURE 12-5

Under typical conditions, each manufacturing process is capable of producing parts to some characteristic tolerance and surface finish. See text for interpretation of graph.

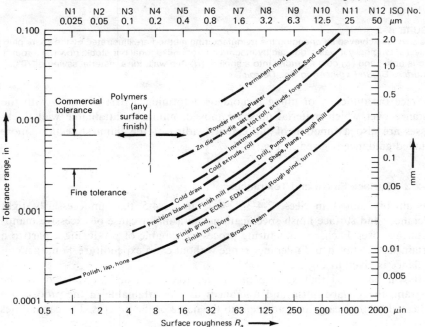

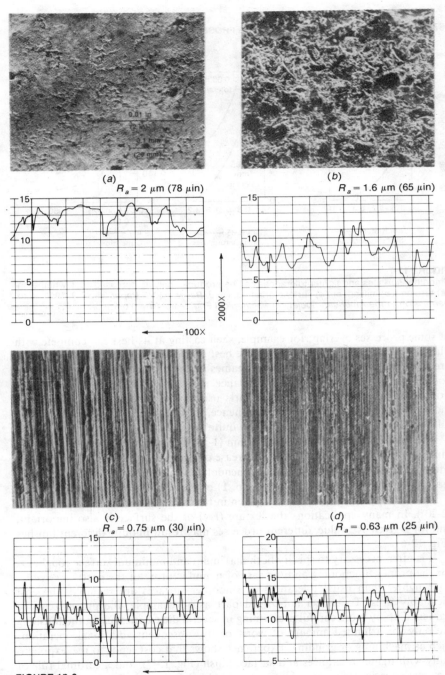

FIGURE 12-6
Scanning electron microscope images and surface profile traces (all 100 ×) reveal that quite different detail features may exist on surfaces of similar roughness averages. Random surfaces: (a) permanent-mold cast and (b) shotblasted. Directional surfaces: (c) cold-rolled and (d) ground.

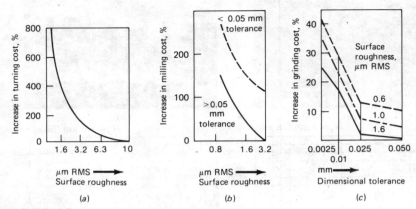

FIGURE 12-7
Smoother surfaces and tighter tolerances can be produced only at increased cost, whether in
(a) turning, (b) milling, or (c) surface grinding, (*L. J. Bayer, ASME Paper 56-SA.9, 1956. With
permission of the American Society of Mechanical Engineers, New York.*)

of some processes overlap; for example, shell casting at its best can compete with
plaster casting, but can never match the best plaster casting results. When ranges
are common to several processes, the names of these processes are separated by
commas; for example, the same tolerance and surface finish are obtained in
drilling or punching a hole. Plastic parts usually bear the surface finish of the
mold or die in which they were made; hence, they can be produced to any finish
(although fiber-reinforced parts may be quite rough).

The tolerances given apply to a 25-mm (1-in) dimension. For larger or smaller
dimensions, they do not necessarily increase or decrease linearly. In a production
situation it is best to take the recommendations published by various industry
associations (see Further Reading, Chap. 1) or individual companies.

Surface roughness in Fig. 12-5 is given in terms of R_a (arithmetic average, Eq.
(2-20)). In many applications the texture (lay) of the surface is also important,
and for a given R_a value, different processes may result in quite different finishes
(Fig. 12-6).

It is often necessary to specify both maximum and minimum surface roughness
for the proper functioning and/or ease of manufacturing of the part.

If tighter than usual tolerances or smoother surfaces are required, the cost
inevitably rises. To take a hot-forged part as an example, the process can be
tightened up by more careful preforming in several die cavities, using the finishing
die only for a limited amount of work, or by subsequent additional operations
such as machining or grinding. Experience shows that cost tends to rise exponentially with tighter tolerances and surface finish (Fig. 12-7), and a cardinal rule of
the cost-conscious designer is to specify the loosest possible tolerances and
coarsest surfaces that still fulfill the intended function. The specified tolerances
should, if possible, be within the range obtainable by the intended manufacturing
process (Fig. 12-5) so as to avoid separate finishing operations.

The limits indicated above are not inflexible. Indeed, various manufacturing industries have responded to competitive pressures by tightening tolerances, reducing minimum wall thicknesses, and generally improving quality without necessarily raising the cost of their product.

Example 12-4

The bodies of most automobiles are made of low-carbon or HSLA steel sheet. The surface roughness is usually defined by specifying minimum and maximum R_a (or R_q) together with the minimum number of peaks per mm (or in). A very smooth surface would not entrap enough lubricant; sheet-to-die contact would result in cold welding and tool pickup, and pressed parts would have to be rejected because of scoring of the surface. However, on a very rough surface there would be too few asperities on which the press-working pressure could be distributed, and controlled draw-in into the die could not be achieved. Furthermore, asperities would show through the paint film of the finished automobile, giving an unacceptable appearance.

Example 12-5

An infant respirator relies on a small-volume, fast-cycle air pump. There is a PTFE seal on the piston, and it has been found that the walls of the pump cylinder must be finished to between 0.1–0.2-μm R_a. A rougher finish results in loss of compression because of rapid wear of the seal. A very smooth finish results in higher friction between seal and cylinder; heat builds up, the PTFE melts locally, and the respirator fails.*

12-3 COMPETITION BETWEEN MATERIALS

The designer always has the opportunity to consider alternative materials. Substantial economies are sometimes possible. For example, machining costs can vary over several orders of magnitude (Table 12-8) and, if there are no other offsetting

*M. F. DeVries, M. Field, and J. F. Kahles, *Annals of CIRP*, **25**: 569–573 (1976).

TABLE 12-8
RELATIVE MACHINING COSTS FOR A GIVEN PART
Lathe turning, 60-min tool life

Material	Hardness	Machining cost
7075-T6 Al		10
1020 steel	111 BHN	25
410 stainless steel	163 BHN	40
310 stainless steel	168 BHN	55
Ti–6Al–4V		75
4340 steel	52R_C	100
Inconel X		170
Inconel 700 (aged)	400 BHN	340

*From *Profile Milling Requirements for Hard Metals 1965–1970*, Report of the Ad Hoc Machine Tool Advisory Committee to the Department of the U.S. Air Force, May 1965.

costs or constraints imposed by service requirements, great savings can result from switching from one to another metallic material.

A new dimension has entered into the competition between materials with the appearance of engineering (high-performance) plastics. Plastics first presented challenges to zinc alloys, although the zinc die-casting industry succeeded in holding onto much of the market by developing thin-wall casting methods. More recently, plastics have become attractive alternatives for sheet-metal parts, zinc- and aluminum-alloy castings, and cast iron. The competition is particularly fierce in the automotive field where weight reduction is one of the means of reducing fuel consumption. Within only a decade (from 1975 to 1985), the average weight of American cars was reduced from almost 1720 to 1230 kg; the weight of aluminum rose from 10 to 60 kg per car and that of plastic parts from 40 to 100 kg per car. The first applications were in obvious directions: fascia panels, interior fittings, and other non-loadbearing components. Now structural parts such as bumpers, brake-fluid tanks, and some body parts are also made of plastics. Even within the share of steel, shifts occurred in the direction of more extensive HSLA and galvanized sheet usage. Great weight savings are possible, for example, by replacing cast iron exhaust manifolds with stainless-steel sheet assemblies.

12-4 IDENTIFYING THE OPTIMUM APPROACH

In the ideal manufacturing situation the designer cooperates with the manufacturing engineer in ensuring that the part should have features that make it eminently producible by a selected, optimum process (an approach often referred to as *design for manufacturability*). Indeed, this increasingly happens in the largest mass-producing industries and also aerospace industries that do both their design and manufacturing in house. In addition to assembly plants, they operate their own manufacturing plants that make parts, often in competition or cooperation with specialized, independent suppliers. Process selection then follows a formalized procedure.

In one approach, representatives of groups having knowledge of and interest in alternative processes get together in a *brainstorming session*. The basic rule is that no idea is regarded as too silly; no criticism is allowed until all ideas emerge, and only then does the process of whittling down alternatives begin. Those processes that stand concentrated criticism are then evaluated in detail and the optimum is finally chosen, often after detailed scrutiny by estimators and designers working with CAD.

The same function is performed, but in a much more organized form, by *value engineering groups*, headed by a value engineer who reports directly to higher management. The task of such groups is to review the functional requirements and ease of manufacturing of the part, and attach a value to functions and processes. The group is empowered to initiate design changes that make manufacturing (including assembly) easier. In addition, economic aspects are also fully weighed. It is not unusual that, as a result of value analysis, a part should be entirely redesigned, or several parts forming an assembly be made as a single unit.

With the growth of CAD/CAM, some of these functions have shifted to the design group.

In many smaller companies the situation is quite different. The manufacturing engineer or technologist may have to rely on his or her own resources, without the benefit of interactions with specialists in other fields. There is then real danger of settling on the first obvious solution, conditioned by experience with a given process. As a minimum, the manufacturing engineer must consider alternatives in a "brainstorming session" with himself or herself. The availability of low-cost CAD packages and increased access to computer data bases greatly improves the situation.

In many instances, a company performs only assembly operations, or possesses only machining facilities of its own. The raw material is purchased in the form of forgings, castings, sheet-metal stampings, plastic moldings, and semifabricated products such as bars, tubes, sections, plate, and sheet. A vital function is fulfilled by the purchasing department that conveys the information to suppliers and often also acts as an intermediary between designer (or manufacturing specialist) and supplier. Identification of alternative production methods and location of suppliers of appropriate capacity is essential. Without proper coordination between purchasing and manufacturing, the part is often "hacked out of the solid." While there are instances in which this approach is indeed the most economical, alternatives must always be explored.

First, one would assume that the designer has taken full advantage of the innumerable standardized and semistandardized components obtainable from specialized producers who employ mass-production techniques and can guarantee highest quality at lowest cost. Screws, rivets, clips, springs, packings, seals, bearings, gears, etc., are available in almost any justifiable size and material, and should seldom if ever have to be custom made.

Second, geometrically similar parts can often be made identical without sacrificing functional performance, or they may be made into a family suitable for group-technology treatment (Sec. 2-4-2).

Third, in specifying materials, function as well as ease and cost of manufacturing must be considered. A sound decision is possible only if the true function of the part is known and no unnecessary constraints are imposed.

Example 12-6

In this example, the steps usually followed in the course of process selection are illustrated. The part is a flange, and the dimensions and material were specified by the designer.

Once the dimensions, surface finish, and material of a workpiece are specified, process selection is limited in its scope. Assuming that no communication (at least no meaningful communication) with the designer is possible, the choices can be readily identified.

If production alternatives are to be considered by one person, the danger of discarding profitable alternatives is best avoided by following the *process of elimination*; that is, one first considers all alternatives, throws out those that are obviously impossible, and then concentrates on the feasible ones.

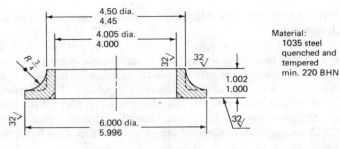

4.50 dia.
4.45

4.005 dia.
4.000

$R\frac{3}{4}$

32

32

Material:
1035 steel
quenched and
tempered
min. 220 BHN

1.002
1.000

32

6.000 dia.
5.996

32

All dimensions in inches
(Broken lines show alternative design)

Casting In the case of the flange, only casting processes capable of withstanding the casting temperatures of steel need be considered at all. This immediately excludes die casting (in anything but graphite molds) and plaster-mold casting (Table 12-2).

None of the casting processes can satisfy the tolerances and surface finish specified for the end faces and the internal surface (Fig. 12-5); therefore, machining will be inevitable, and the lowest-cost process capable of producing the shape is acceptable. Investment casting can be rejected because shape complexity does not justify the extra expense, but sand casting, shell molding, or permanent-mold casting should all be investigated.

Further narrowing of the choices requires some economic analysis and, in all likelihood, shell molding will emerge as the most suitable process for the given dimensions (Fig. 12-4). With the aid of data from appropriate books or industry publications (see Further Reading, Chap. 3), the casting can be designed.

Deformation Processes Of the deformation processes, tube rolling or extrusion could produce thick-walled tubes for further machining; these are, however, essentially semifabrication processes providing a rather high-cost input to machining. More directly, one should consider those deformation processes that yield a semifinished part.

The surface finish requirement is within attainable range for cold deformation processes (Fig. 12-5) such as sheet-metal forming and cold forging or extrusion. Subsequent coining of a hot-worked part is also possible and a part finished exclusively by deformation techniques could be economical.

Hot forging is a very reasonable proposition, particularly on a hot upsetter which would upset a flange and then pierce out the bar, without creating scrap. The feasibility of cold forging can be determined only after making pressure and force calculations (Secs. 4-4-1 and 4-4-4). Piercing (or back-extrusion) is almost certainly uneconomical because of the inevitable scrap loss in the bore. Piercing of a smaller ring followed by ring rolling, however, may be economical in larger quantities.

Machining The basic choice is between machining from a solid bar, a tube, or a cast or deformed part. The decisive factor is cost. The solid bar is by far the cheapest and can be machined on a screw-type multispindle automatic machine, but much scrap of low value is produced. The higher price of tubing may or may not be offset by reduced input weight and scrap losses. Cast or formed parts must be machined on a chucking automatic and could be turned around, although a much better solution would be machining of one face and the bore, followed by finishing the other face on a vertical-spindle, rotary table grinder.

Redesign The problem would change entirely if discussions with the designer would reveal, say, that the functions of the part call simply for a minimum YS of 550 MPa, and a minimum wall thickness of 6 mm without need for the sharp edge (broken lines in the figure).

A review of materials satisfying the minimum yield strength requirements brings to light as possible candidates: 1050 steel (cold worked), high-strength low-alloy steel (hot rolled in a controlled manner), nodular cast iron (quenched and tempered), malleable iron, and cast steel grades (quenched and tempered), as well as powder-metallurgy products.

Of the casting grades, nodular cast iron is likely to be the most economical, whereas 1050 and HSLA steels open the possibility of proceeding with cold or warm forward-extrusion and, since the sharp edge is no longer required, with sheet-metalworking techniques. Whether blanking of rings followed by flanging, or blanking of circles followed by drawing and punching out the base is more practical will be determined by the ductility (elongation capacity) of the material (Sec. 5-3). Sheared edges will not satisfy surface finish and tolerance requirements and machining is still necessary, but it could now be grinding as well as turning.

The very simple shape of this part actually complicates the selection of an optimum process because the final choice will hinge on a rather detailed economic analysis. Unless the quantities are very large, one would concentrate on hot upsetting, machining from the solid or tube, powder metallurgy, and perhaps also cold forging or drawing. Parts of more complex shape often limit the number of feasible processes (Tables 12-2 to 12-7) and simplify the search for the optimum.

12-5 SUMMARY

In this book we concentrated on the principles that underlie unit processes of manufacturing. We also indicated, although in a much more general sense, how these processes can be brought together into a viable system. In the final analysis, the success of all this hinges on the product designer; while striving to satisfy functional requirements, the designer must be aware of manufacturing implications. This demands at least some familiarity with processes and process limitations as affected by materials, and a willingness to discuss the design with manufacturing specialists. Increasingly, many of the alternatives can be weighed with the aid of specialized computer software. Some basic rules are always worth considering:

1 Specify the broadest tolerance and coarsest surface finish that still satisfy functional requirements.

2 Subject the design to value analysis, formally or informally. Start with the real functions, strip off all unnecessary constraints, and seek alternatives with the early involvement of people with knowledge of various manufacturing techniques (as opposed to narrow specialists).

3 Remember that a seemingly minor change in shape, wall thickness, or radius can make the part suitable for manufacturing by a different, more economical technique.

4 Look at the relation of parts to each other; it may well happen that several parts are much easier to make as a single unit.

Manufacturing is a peculiar blend of science, art, and economics, with very broad social implications. It would be unreasonable to expect the designer or manufacturing engineer to consider formally all these elements whenever a detail decision is made. They must, nevertheless, guide the general approach to the profession; without sound scientific foundations, manufacturing remains but a collection of myriads of isolated rules; without practical and economic sense, the best theory and research will fail to make an impact; and without conscious attention to quality and cost, the best-designed product and process will lose out against competition.

PROBLEMS

12-1 Trace the coordinate system of Fig. 12-5 on transparent paper; then construct a line corresponding to a tolerance of $20R_a$. From Fig. 12-5 judge whether a 16-μin R_a surface finish specified for a journal of 1.0000 ± 0.0004-in diameter can be produced (a) by grinding; (b) by cold drawing. (c) If the answer to (a) and/or (b) is yes, is the chosen finish reasonable? (Refer to the line constructed above.)

12-2 Collect at least five types of metal cans, selected from among containers for fizzy drinks, fruit juices, baby food, canned meat, and sardines. Make sure you have samples of two- and three-piece containers. Carefully section them and investigate their structure. Write an essay describing their method of manufacture as deduced from the evidence available. (This is an extension of Prob. 5-21, and now includes the methods of joining.)

12-3 What processes could you envisage for making (a) a high-pressure gas cylinder; (b) a CO_2 cartridge? (c) What other processes could one consider if the part would be of shape F5 (Fig. 12-1) without the need to sustain internal pressures?

12-4 A component is found to fail in service by fatigue. Laboratory examination reveals the presence of residual surface tensile stresses. Describe the methods that could be considered for changing the residual stresses to compressive, assuming that the original cause of surface tensile stresses cannot be eliminated.

12-5 Circles of 300-mm diameter are to be cut from 10-mm-thick low-carbon steel plate. (a) Make a list of all potential processes you can identify from this text. (b) Establish an order of merit based on the quality of the cut surface. (c) Choose the process that appears the most economical for production rates of 10 000 pieces per month, with no special quality requirements regarding the cut surface. (d) Choose processes that appear to be economical for production rates of 500 pieces per month, and a requirement of a perpendicular edge surface; rank them in order of anticipated quality.

12-6 Obtain two samples each of socket-head cap screws, and recessed- and slotted-head screws. With the aid of a magnifying glass or, preferably, a stereomicroscope, investigate the heads for evidence of the method of manufacture. Report your findings, including any defects discovered, in a professional manner.

12-7 Carry out the tasks described in Prob. 12-6, but collect six different self-tapping screws and include the threaded portion in your investigation. If facilities are available, mount screws in plastic and make longitudinal sections. Etch to reveal flow lines. Note any defects that may be present and draw conclusions regarding their possible effects on the performance of the screws.

12-8 Carry out the tasks of Prob. 11-7, but on six wood screws.

12-9 Inspect the bumpers of various automobiles. Make sketches of the structure; describe—as best as you can—the materials of construction and the probable methods of manufacture.

12-10 Survey the products used for merchandizing liquids in approximately 1-L units. Describe at least five products, identifying the materials of construction and the probable methods of manufacture.

FURTHER READING

DeGarmo, E. P., W. G. Sullivan, and J. R. Canada: *Engineering Economy*, 7th ed., Macmillan, New York, 1984.

Kurtz, M.: *Handbook of Engineering Economics*, McGraw-Hill, New York, 1984.

Malstrom, E. M. (ed.): *Manufacturing Cost Engineering Handbook*, Dekker, New York, 1984.

Ostwald, P. F. (ed.): *Manufacturing Cost Estimating*, Society of Manufacturing Engineers, Dearborn, Mich., 1980.

Steven, G. T., Jr.,: *Economic Analysis of Capital Investments for Managers and Engineers*, Reston Publishing Co., Reston, Va., 1983.

TERMINOLOGY OF COMPUTING APPLIED TO MANUFACTURING

1 *Computers* are electronic devices capable of accepting information, applying prescribed processes to this information, and supplying the results of these processes. Physically they consist of a central processing unit (CPU) incorporating an arithmetic-logic unit and a control unit; memory; and input and output (I/O) devices. The plan for the solution of the problem is called a *program* which, together with the supporting documentation, constitutes the *software*.

2 Because computers function in response to the presence or absence of a signal, computer machine language is based on the binary digit (*bit*) which can take only the values of 0 or 1. A *word* is a bit string considered as an entity and may be made up of 4, 8, 16, 32, or 64 bits. Various coding schemes have been devised to represent numbers and alphabetic and special characters, the most common of which is the ASCII code.

a Most programs in manufacturing are written in higher-level languages (e.g., FORTRAN, BASIC) which require no knowledge of machine language; the method of solving the problem (the *algorithm*) is written in English-like statements.

b The relatively few people who are involved in analyzing the system and writing original programs for manufacturing must be familiar not only with computer programming but, most importantly, must understand the process for which the program is to be written. The users of truly "user-friendly" programs need little or no knowledge of computers but must be fully familiar with the physical basis of processes.

c In some instances the practical problem is so complex that only experts with long experience can solve it. The knowledge, logic, and judgement of the expert can be captured in *expert programs*, developed in cooperation between the expert and system specialists (also called knowledge engineers). The program contains facts generally available to experts in the particular field; rules of thumb (heuristics), which allow the expert to make educated guesses even when data are incomplete; and inferences, i.e., rules of good judgement. Special programs are available that reduce the programming effort needed for building expert systems; nevertheless, expert programs tend to be lengthy and are expensive to produce. Once completed, they allow a less-experienced person to find the solution to the problem by interacting with the program through if-then sequences.

3 Operation of the computer requires the manipulation of many instructions and data. Both program and data are stored in *memory*. Parts of the program and other fixed elements (e.g., look-up tables) are usually permanently stored in *read-only memory* (ROM), others are loaded into *read/write memory*; both kinds can be accessed in any order (*random-access memory*, RAM).

Physically the memory elements may be semiconductor circuits which lose the information when power is interrupted (*volatile memory*). Nonvolatile storage is possible in microscopic magnetic domains (bubbles) formed in a thin crystalline magnetic film, or in special forms of semiconductor memory in which 0 and 1 states are fixed in the locations required by the program. Some nonvolatile ROMs may be customized by the user to carry out repetitive tasks: a *programmable ROM* (PROM) can be changed or initialized only once, an *erasable PROM* (EPROM) and *electrically alterable ROM* (EAROM) can be reprogrammed repeatedly. Memory capacity is given in thousands (K) of bits, thus 8 chips of 256K capacity are needed to make up the 256K memory of an 8-bit micro-computer.

Programs and data files are normally stored in large-capacity, nonvolatile *secondary storage devices* such as punched paper or plastic tape; punch card; or disks (floppy disks or hard disks) or tapes coated with a ferromagnetic material. All of these can be "written" on with devices that are within the financial reach of an individual or small business, and have storage capacities up to several megabytes. Magnetic storage devices can be easily overwritten and can have random access, whereas punched tapes or cards cannot be altered and can only be read sequentially; they do have, however, an indefinite life against the 15–20-year life of magnetic storage media. Videodisk technology, with a laser beam as the reading element, offers long-term storage of data at a very high information density.

4 Frequently, manufacturing computation must be performed and commands for action issued in *real time* (while the event takes place). It is usually more economical and technically more feasible to accomplish this with *hierarchical systems*. Mainframe computers (large, fast computers capable of executing millions of instructions per second) are used for overall planning and supervision and for many of the computations related to the MIS and to some of the more complex CAD analyses. Smaller computers (minicomputers) suffice for running selected groups of machines and provide the computing power for most CAD and CAM. Microcomputers (personal computers) can be dedicated to single production machines for control and distributed processing of information. Thus the task of programming the system becomes easier too. Networking allows all computers to communicate with each other.

5 A vital need is the *collection and entry of data*. Some of this is accomplished by the measurement of process variables with appropriate transducers. A transducer is a device that converts one form of energy into another; in the sense used here, a transducer converts physical quantities (such as distance, velocity, pressure, torque, fluid flow, temperature, etc.) into electrical signals (see Sec. 2-4-6). The signal may be continuous (analog), such as a voltage, or digital (from a numerical encoding or pulse-summation device). Analog signals are often converted into digital form with analog-to-digital (A/D) converters. Other information is entered by operators on terminals or with the aid of micros serving as "intelligent" terminals capable of performing some calculations. Communication with the computer is usually by means of a keyboard, by voice, or by a device that converts motion into digital information (mouse, ball, joystick, etc.).

6 Computers and controllers are generally based on *integrated circuits* (IC) which contain all elements of an electronic circuit on a single chip of semiconductor material, generally silicon. Miniaturization allows large-scale integration (LSI); thousands of circuit elements are placed on a single chip, thus increasing operational speed while reducing cost (the manufacture of these circuits is discussed in Chap. 10). A microprocessor is an LSI chip (or chips) containing most elements of a computer, without the I/O devices.

7 An important application of microprocessors is in *programmable controllers* (PC); to avoid confusion with personal computers, they are now called *programmable logic controllers* (PLC). The control of many processes requires sequencing, timing, counting, logic, and arithmetic functions which used to be satisfied by relay logic circuits. These had to be rewired if the logic was to be changed. Their place has now largely been taken by the PLC (Fig. A-1), the memory of which can be readily reprogrammed with a programming panel or a computer. The difference between microcomputer and PLC is narrowing all the time. A PLC is more rugged to stand up to plant conditions, and most PLCs can be programmed in the "ladder logic" familiar to people versed in relay circuits.

FIGURE A-1
Schematic of programmable logic controller.

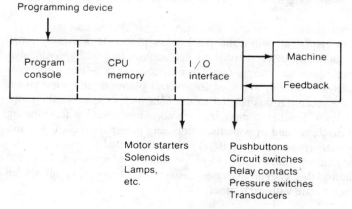

APPROXIMATE
CONVERSION
OF HARDNESS VALUES

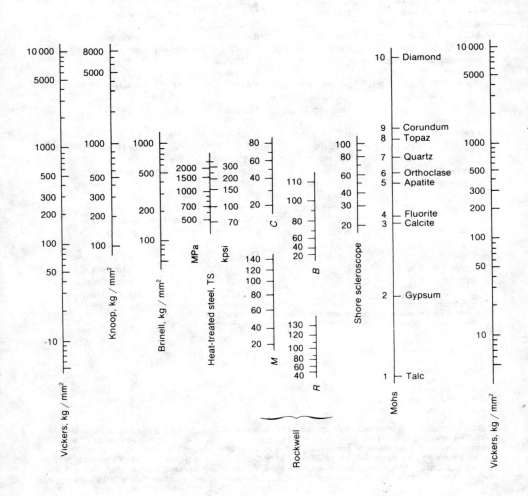

CONVERSION FACTORS FROM U.S. CONVENTIONAL UNITS TO METRIC (SI) UNITS

Quantity	To convert from USCS units		Multiply by	To obtain SI unit	
	Symbol	Name		Symbol	Name
Length	in	inch	*25.4	mm	millimeter
	ft	foot	*0.3048	m	meter
Area	in^2	square inch	*6.4516×10^{-4}	m^2	
Volume	in^3	cubic inch	1.639×10^{-5}	m^3	
	gallon	US gallon	3.785×10^{-3}	m^3	
Time	min	minute	*60	s	second
Velocity	ft/min		*5.08×10^{-3}	m/s	
Mass	lb	pound	0.4536	kg	kilogram
	ton	short ton	0.9072	tonne	tonne
Acceleration	ft/sec^2		*0.3048	m/s^2	
(gravitational)	$(32\ ft/sec^2)$		*(9.80665)	m/s^2	
Force	lbf (or lb)	pound force	4.448	N	newton
	tonf	ton force (2000 lb)	8.9	kN	
	kgf	kilogram force	*9.80665	N	
Stress (pressure)	lbf/in^2	psi	6.895×10^3	Pa	pascal ($= N/m^2$)
	ksi (or kpsi)	1000 psi	6.895	MPa	(or N/mm^2)
	torr	mm Hg	133	Pa	
Torque (work)	lbf · ft	foot-pound	1.356	N · m	newton-meter
Energy (work)	Btu	British thermal unit	1055	J	joule ($= N \cdot m$)
	cal	gram calorie	*4.1868	J	
Power	hp	550 ft · lb/sec	*746	W	watt ($= J/s$)
Viscosity	P	poise $(dyn \cdot s/cm^2)$	0.1	Pa · s	(or $N \cdot s/m^2$)
Temperature interval	F	Fahrenheit	0.5555	C or K	Celsius degree or kelvin
Temperature	t_F		$(t_F - 32)5/9$	t_C	degree Celsius
	t_C		$t_C + 273.15$	t_K or T	absolute degrees

Notes: Exact conversion factors are recorded with an asterisk.
The Celsius degree is often written °C to avoid confusion with C (coulomb).
Most frequently used multipliers:

	Prefix	Symbol
10^9	giga	G
10^6	mega	M
10^3	kilo	k
10^{-3}	milli	m
10^{-6}	micro	μ
10^{-9}	nano	n

The International Committee of Weights and Measures (CIPM) modernized the metric system in 1960. The resulting SI units are now used worldwide in the literature; all industrialized nations have already committed themselves to conversion to the International System (SI).

For a detailed discussion see, for example, *ASME Orientation and Guide for Use of Metric Units*, American Society of Mechanical Engineers, New York, or *The International System of Units*, National Bureau of Standards SP330 (SD cat. no. C13.10:330/2), Government Printing Office, Washington, D.C.

SOLUTIONS TO SELECTED PROBLEMS

CHAPTER 2

2-1 (a) $e_t = 0.213\%$ for Ti–6Al–4V, 0.123% for 4340 steel
(b) $P_{0.2} = 259$ kN for Ti, 429 kN for steel, 156 kN for Al
(c) $P_{max} = 282$ kN for Ti, 462 kN for steel, 175 kN for Al

2-2 The Ti-alloy bar is the lightest (344 g)

2-3 Energy stored: 116 J for Ti alloy, 121 J for steel, 113 J for Al alloy bar

2-4 Contribution of specimen 19.7%; contribution of machine 80.3%

2-5 Smaller machine has lower spring constant, hence initial slope is lower

2-6 1.0%

2-7 HRC 40

2-15 Ring gage: GO 24.980 mm, NOT GO 24.959 mm
Plug gage: GO 25.000 mm, NOT GO 25.033 mm

2-19 An increase of 3.1 μm

2-20 Error is 0.00155 in

2-21 Error is 270.8 μm (0.271 mm)

2-25 There will be significant creep with Zn

CHAPTER 3

3-4 Severe coring in Cu–5Sn, less in Al–5Mg, least in Cu–30Zn. $(T_L - T_S)/T_S = 0.17$, 0.07, and 0.025, respectively

3-9 HB = 3TS with deviations of less than 10%

3-15 (a), (b), (c), (f): 2% Si; (d), (e), (g): 12% Si

3-17 (a) 92 cm^3/100 g steel; (b) 87.9%; (c) 62.7 mm/min

3-18 Velocity 1.98 m/s; flow rate 198 cm^3/s

3-22 (a) $e = 0.0046$; (b) $\dot{\varepsilon} = 7.6 \times 10^{-6}$ s^{-1}; (c) $\sigma_f = 37$ MPa; (d) $\sigma = 580$ MPa, thus plastic deformation will occur

3-24 Approximately 50 tonf

CHAPTER 4

4-4 H38 temper
4-7 0.375 in
4-11 Ti–6Al–4V alloy
4-13 (*b*) 42.7% proeutectoid α, 57.3% pearlite; (*c*) spheroidized
4-14 62.5% saved
4-15 $P_c = 700$ kN
4-16 (*a*) $p_a = 233$ kpsi; $P_a = 38$ tonf
  (*b*) $p_a = 485$ kpsi; $P_a = 80$ tonf
  (*c*) D2 tool steel, HRC 58–60, or carbide
4-18 (*a*) $d_0 < 40$ (say, 38 mm); $l_0 = 84.5$ mm
  (*b*) $p_l = 1220$ MPa
  (*c*) $p_l = 440$ MPa; $p_i = 714$ MPa (or up to 1190 MPa)
4-19 (*a*) $p_a = 8$ kpsi; $P_a = 31.4$ klbf
  (*b*) $p_a = 27.3$ kpsi; $P_a = 107$ klbf
4-21 Unlubricated: $p_p = 840$ kpsi
  Lubricated: $p_p = 462$–500 kpsi
4-22 Single blow: 2000-kg hammer
  Three blows: 1000-kg hammer
4-24 (*a*) $p_i = 1488$ MPa (possibly up to 2480 MPa)
  (*b*) D2 steel, HRC 58–60; YS = 1920 MPa
  (*c*) Permissible punch pressure: for bending 1664 MPa, for compression 1920 MPa
4-25 (*a*) $\sigma_{fm} = 305$ MPa; (*b*) $P_{dr} = 3.36$ kN; (*c*) no; (*d*) draw in two passes
4-26 Drawing on a bar
4.28 $d_0 = 0.375$ in
4.29 (*a*) $P_r = 25.7$ kN; (*b*) 1.85 kW
4-30 (*a*) $\sigma_f = 76$ kpsi; (*b*) $p_r = 183$ kpsi; (*c*) yes

CHAPTER 5

5-3 Without shear: 220(TS) where TS is in N/mm^2
  With shear: half of above value
5-4 (*a*) $P_s = 14.84$ kN; (*b*) 33; (*c*) strip width 199 mm
5-7 (*a*) OS015; (*b*) OS100; (*c*) H08
5-11 (*a*) Necking: $R_b = 17.2$ mm; fracture: $R_b = 2.17$ mm
  (*b*) $\alpha_b = 91.7°$ (die angle = 88.3°)
5-12 $P_b = (\text{TS})lh^2/3L$ where $L = h + 2(\text{die radius})$
5-13 (*a*) Al alloy: lower E, greater springback
  (*b*) HSLA steel: higher YS, greater springback
5-14 (*a*) Low r; (*b*) high r
5-17 (*a*) 13.8 tonf; (*b*) 48 tonf; (*c*) 3.912-in total (outside) height
5-19 (*a*) 240 mm; (*b*) 117 mm; (*c*) $h/d = 1.17$ for LDR = 2.4; $h/d = 0.74$ for LDR = 2.0
5-20 (*a*) 301 mm (to allow for trimming the edge of the pot, 310 mm); $P_s = 886$ kN
  (*b*) $d_0/D_p = 1.58$
  (*c*) $P_d = 148$ kN
  (*d*) Ironing; power spinning

CHAPTER 6

6-3 (*a*) 8.2 g/cm^3; (*b*) 92.13%; (*c*) 7.87%

6-4 $d = 29.16$ mm; $h = 48.6$ mm

6-5 (*b*) 942 tonf; (*c*) no

6-7 (*a*) $d_1 = 1.67d_0$

6-9 (*a*) 27.1%; (*b*) 13.3%; (*c*) 8.68 × 17.34 × 130 mm; 19.54 cm^3

6-14 Fused silica safe, borosilicate marginal, others fracture

CHAPTER 7

7-1 (*a*) 1; (*b*) infinity; (*c*) $1/n = m$

7-5 Wall thickness: 0.417 mm at waist, 0.265 mm at bulge

7-8 Yes; 11.8% saving in material cost

7-9 Press size: 300 kN; clamp force: 870 kN

CHAPTER 8

8-11 Multiply in^3/hp · min by 2.73 to obtain mm^3/W · s

8-12 10 in^3/min of steel, 42 in^3/min of Al alloy, 4.5 in^3/min of superalloy

8-13 Tool life decreases to 0.01, 0.0625, and 0.177 of the original tool life with HSS, carbide, and ceramic tooling, respectively

8-14 (*a*) Multiply ft/min by 0.305 to obtain m/min
(*b*) 2.54 m/s (or 152.4 m/min)

8-15 360 ft/min

8-19 (*a*) 90 ft/min; (*b*) 963 rev/min; (*c*) 0.0025 in/rev; (*d*) 2.4 in/min; (*e*) 0.34 hp

8-20 (*a*) 1.5 m/s; (*b*) 2400; (*c*) 0.24 mm/rev; (*d*) 9.6 mm/s; (*e*) 1.27 kW

8-21 (*a*) 850 ft/min; (*b*) 0.045 in/rev; (*c*) 0.625 min

8-22 (*b*) Relative cutting times 7 : 3 : 1.5 for HSS, carbide, and ceramic, respectively; (*c*) 9.43 min

8-24 Tool life decreases by a factor of 14 with speed, 4.7 with feed, 2.9 with depth of cut

8-27 (*a*) 125 passes; (*b*) 2.43 min; (*c*) 0.060 in^3 total, 0.0247 in^3/min; (*d*) 0.48 hp

8-28 Roughing: 2.77 hr; finishing: 5.6 hr

CHAPTER 9

9-3 (*a*) 140 kA; (*b*) 3.3%; (*c*) 560 kVA; (*d*) 18.5 kVA; (*e*) 37 kW; (*f*) 0.103 kWh/20 welds; (*g*) 40 kN for 20 projections

9-5 (*b*) 130 in/hr

9-9 11 mm

INDEX

No separate glossary is provided, however, definitions of terms will generally be found on the first referenced page. Information contained in illustrations is referenced by the page number. Page numbers in *italic* indicate tables.